Manufa[illegible] Management: Principles and Systems

Richard Burman

McGRAW-HILL BOOK COMPANY

London · New York · St Louis · San Francisco · Auckland
Bogotá · Caracas · Lisbon · Madrid · Mexico
Milan · Montreal · New Delhi · Panama · Paris · San Juan
São Paulo · Singapore · Sydney · Tokyo · Toronto

Published by
McGRAW-HILL Book Company Europe
Shoppenhangers Road, Maidenhead, Berkshire, SL6 2QL, England
Telephone 01628 23432
Fax 01628 770224

British Library Cataloguing in Publication Data
Burman, Richard
Manufacturing Management:Principles and
Systems
I. Title
658.5

ISBN 0–07–709044–6

Library of Congress Cataloging-in-Publication Data
Burman, Richard
Manufacturing management : principles and systems / Richard
Burman.
p. cm.
Includes bibliographical references and index.
ISBN 0–07–709044–6
1. Manufactures–Management. 2. Industrial management.
3. Production management. I. Title
HD9720.B8 1995
670′.68–dc20

94–44158
CIP

12345 CL 98765

Typeset by TecSet Ltd, Wallington, Surrey
Printed and bound in Great Britain at Clays Ltd, St Ives plc
Printed on permanent paper in compliance with the ISO Standard 9706.

CONTENTS

PREFACE

In the manufacturing world of today, success goes to those who can manage their business more effectively than their competitors and the competition is now world-wide. Such world-class management is fairly easy to define, but very hard to achieve.

In the first chapter of this book I have tried to define effective management in terms of the characteristics involved rather than in a direct definition. For instance, it is much easier to define great painters or composers in terms of their work than to describe their physical appearance (which can often be totally misleading).

As the title states, this book is about manufacturing management principles and systems. The sequence is deliberate, for without an understanding of the principles, the systems cannot be developed. These principles are the foundations that must be in place before any attempt is made to develop the business into the highest league (world-class). They include the disciplines and procedures that—once in place—can make the implementation of computerized systems relatively easy.

Although most manufacturing companies have some form of computer system, this book is not specifically designed to be used by those who wish to develop their computer skills, and in fact many of the areas covered can be applied without a computer, or with a basic PC and standard software such as a spreadsheet or database.

Many of the computer applications that fail do so because the preparatory work has not been carried out, or has been carried out incorrectly. A computer system can never make up for the inadequacies of a badly managed business.

My objective in writing this book was to try to pass on what I have learned in manufacturing industry over the past 40 years so that the mistakes I have made and those made by others need not be repeated. The world is a much smaller place than it was, and competition is much tougher due to better communications and improved technology. For this reason, mistakes that in the past would have been of minor importance can now be serious or even fatal.

This book has been written for all those who wish to make their career in manufacturing, whether by the 'high road' of graduation via university or college or by the 'low road' of apprenticeship and day release. In either case I hope that this book will help you to achieve success for both yourself and the company you work for.

On the lighter side, it is worth noting that there are a number of widely held fallacies about manufacturing industry and the use of computers in manufacturing which I hope this book will dispel. The following are examples.

- *This business is unique, with problems that do not occur anywhere else* This is a view that is often stated (and is never true!) in order to repel a newcomer with new ideas.
- *There are no standard solutions to our company's problems* This is unlikely to be true, and is probably a form of self-preservation by the speaker.
- *When you have worked here for 40 years you might understand* This suggests that no one is qualified to tackle the problems except the speaker—obviously not true!
- *Only a computer specialist can use a computer to its best advantage* This is just not true! In fact, some of the best computer users cannot tell a bit from a byte or RAM from ROM!
- *You cannot learn about manufacturing management from books* No, but equally you cannot learn about it solely from practical experience as the speaker seems to suggest. In fact both theory and practice are necessary.
- *Yes, but we tried that 10 years ago and it didn't work* Why didn't it work? Was it badly implemented or badly managed, or was the technology not up to it? Were the users properly trained and was management fully committed? If it is a good idea, try it again!
- *Well, it might work in Japan but it wouldn't work here* Not true. Many Western countries have adapted and implemented Japanese ideas with great success and equally the Japanese have adapted and used Western ideas. If they can do it why can't we?

I hope that you will find this book 'user friendly' so that you can use it not just as a student textbook, but as a book that you will still be referring to when you are the chief executive of a large manufacturing company.

I have given considerable thought to the sequence of chapters, my objective being to introduce topics in a way that 'mirrors' the flow through a factory, so that each chapter leads naturally into the next one. In order to help with this concept I have added a linking paragraph between each pair of chapters. This is designed to show how the principles and systems covered in the previous chapter link to those in the next one, and is meant to illustrate a vital point—that a company should be not a patchwork quilt but a single seamless piece without visible joins.

I also hope that you will be able to apply the ideas in this book in your own way and in your own type of business. I have tried, where possible, to avoid the use of jargon which is often used by 'experts' to confound the rest of us. In most cases such jargon is unnecessary, but where I have had to use such words, phrases or acronyms (such as MRP or FMEA) I have always tried to include a definition and to include them in the glossary in appendix F. I have also included a list of further reading at the end of each chapter, which may be helpful to those who wish to study a specific area in more detail.

Richard Burman
Oswestry
August 1994

INTRODUCTION

SUMMARY

In order to define the areas covered by this book, it is necessary to explain what is meant by the term 'manufacturing company' and also to examine some typical examples. This can best be done by looking at the basic functions that are an essential part of such a company and the factors that influence those functions.

MANUFACTURING—DEFINITION AND CLASSIFICATION

The conventional definition of manufacture can be found in most dictionaries, for example:

> *To process or make (a product) from a raw material, especially as a large scale operation, using machinery.*
>
> Collins English Dictionary

This definition is rather general, since it can cover a wide range of manufacturing types that have different functional requirements. It is therefore necessary to define a series of manufacturing classifications, and this can be done in two ways: by industry/product type and by market type.

Industry/product type

Manufacturing industry can be classified as engineering (where raw materials such as steel or plastics are processed and assembled, with some purchased components, into a product) or processing (where raw materials are processed into a different form which is then sold on to another manufacturer or to the general public); these can then be subclassified as follows.

- *Heavy engineering* These companies manufacture products that are sold, on an infrequent basis, to large organizations such as the transportation industry (shipping, airlines, railways and commercial road vehicles), the power generation industry, the mining industry and large processing companies (chemical, steel making, etc.).
- *Medium engineering* In this category, the products are less costly but are still not sold on a regular basis. Typical such products are machine tools, passenger vehicles, contractors' plant and agricultural machinery.

- *Light engineering* These companies manufacture items that are used by both industrial and domestic consumers, some of which are purchased regularly (small components for other engineering companies) and occasionally by the general public (TVs, cookers, washing machines, etc.).
- *Large processing* This category includes products that are sold in high volumes, probably on a regular basis, such as petrochemical production, steel making and paper making where there is a high investment in capital plant.
- *Other processing* All other processors would fall into this category, which includes food and drink, textiles, paint, building materials, tobacco products and many others.

Market type

Market types can be more easily classified, in terms of answers to the following questions.

1. Who are your major customers? Are they manufacturers, processors, or the general public?
2. How do they buy? Are they regular customers who require supplies against a schedule, who buy fairly regularly on an as-required basis, or irregular customers who buy against large contracts?
3. What do they expect? Do they expect you to hold stocks or to make to order? If you are temporarily out of stock of some items will they go elsewhere or accept a reasonable wait?

OTHER DEFINITIONS IN INDUSTRIAL CLASSIFICATION

A number of terms are used to describe various aspects of industrial types, and definitions of these help to define the overall classifications. The following are examples.

- *Capital goods* Items of plant or machinery that are used to process the products. Such items are usually costly and are generally purchased infrequently, the cost being recovered over a period of years by including a small cost element in each product sold.
- *Revenue items* Items that become a part of the product such as raw materials, components and consumable tools. Such items are also known as direct materials because the quantities required are in direct proportion to the volume of production.
- *Consumer durables* Products that are purchased by the general public but are likely to last for a number of years. Items such as domestic appliances, garden tools, home entertainment and leisure products, and ceramics/glassware are classed as consumer durables.
- *White goods* A particular type of consumer durable falls into this category, such as cookers, fridges, freezers, washing machines, dishwashers and dryers.

There are a number of standard industrial classifications that cover this in much greater detail and are used to prepare statistical information for the various sectors (for example by the Board of Trade).

THE TYPICAL FACTORY

A typical factory buys in materials and components and applies a series of operations or processes to convert these into a saleable product. These operations can be considered as a series of 'process groups' as shown in Table I.1.

Table I.1 Operation of process groups

Group	Description	Examples
Forming	Changing the form or shape of an item	Metal removing, pressing, casting, extrusion, forging, cutting, drilling, rolling
Joining	Assembling items	Screwing, bolting, riveting, clipping, gluing, welding (metal or plastic)
Blending	Mixing items together to make a new item	Chemicals, dyes, paints, food and drink components, clays, fuels, metal alloys
Finishing	Changing the nature of the surface of an item	Painting, embossing, dyeing, metal deposition (plating), chemical treatment, plastic coating, grinding, polishing
Heating	Melting a material or altering its characteristics	Melting metal for casting or food for blending, heating for forging, heat treating for hardening or stress-relieving
Firing	Applying high temperature to change the nature of a product	Pottery firing, cement making, bakery ovens, steel or glass making
Drying	Reducing the moisture content of an item	By heating to dry paint, paper or clayware, or by freezing (instant coffee)
Packaging	Obtaining a measured quantity and packaging it	Filling bottles, cans or jars (food, drink and drugs), filling bags (cement or fertilizer), counting and packing (small DIY items), pallet stacking, carton filling.

A factory can therefore be visualized in terms of a sequence of events which starts with the receipt of materials and ends with the despatch of goods. Such a sequence is shown in diagrammatic form in Fig. I.1.

THE FUNCTIONS INVOLVED

Figure I.1 also shows that a number of functions are involved at various stages throughout this chain.

- *Basic materials* Four functions are directly involved in the acquisition of basic materials:
  - — *Design*, which, with Sales, determines what is to be made and designs it
  - — *Sales*, which determines the sales levels for the various products
  - — *Planning*, which determines the materials needed to make those products
  - — *Purchasing*, which buys the materials needed.
- *First operation* Four functions are again involved in the start of production:
  - — *Planning*, which determines the way work is issued to production
  - — *Purchasing*, which ensures that the required materials are available.
  - — *Production*, which provides the people and machines to perform the work
  - — *Quality*, which ensures that quality standards are met for all production.

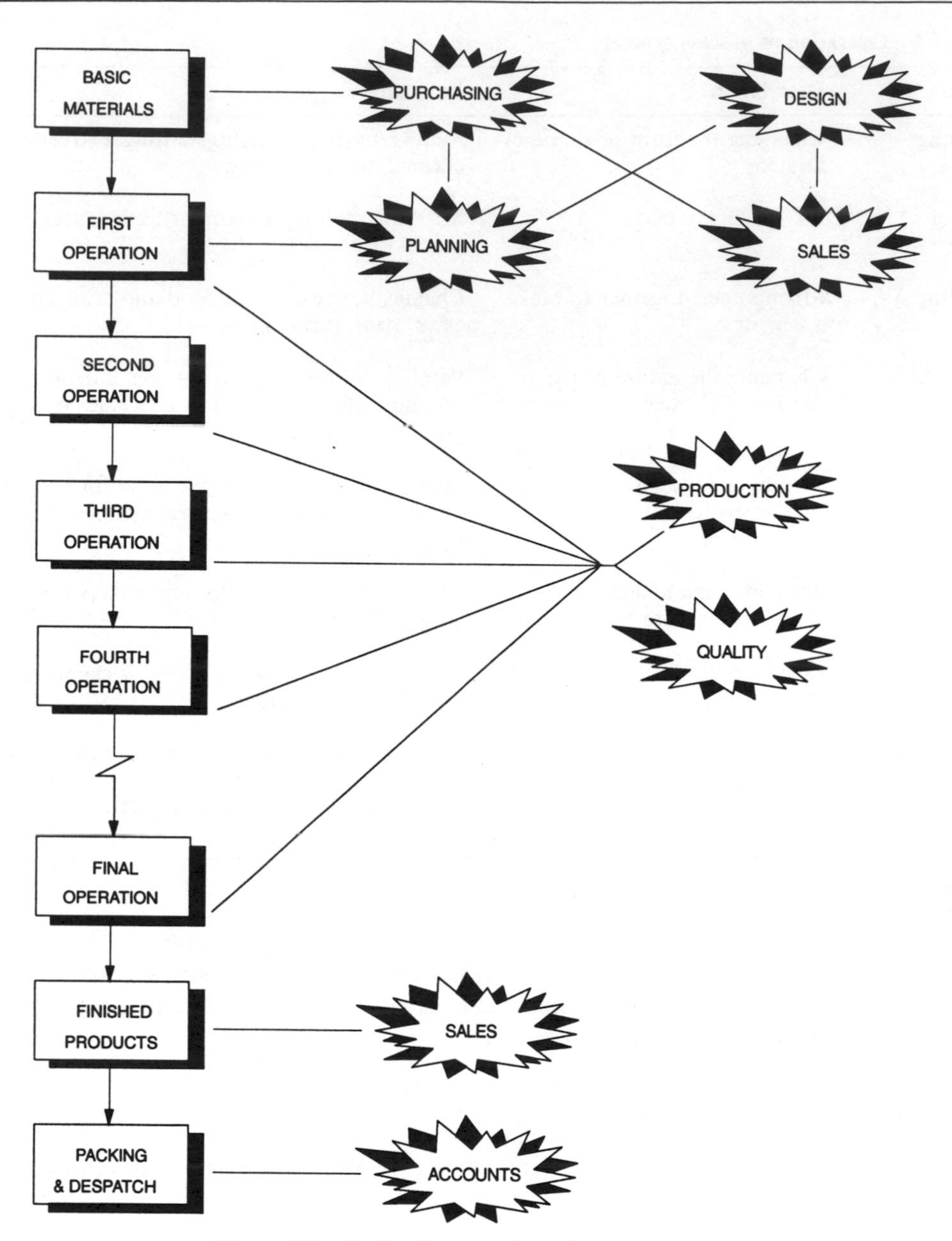

Figure I.1 Basis of a manufacturing company.

- *Other operations* Both Production and Quality are involved with all subsequent operations in the same way as they were for the first operation.
- *Finished products* Sales is involved again to ensure that finished products are allocated to the right customers and at the right time. It is also responsible for the documentation used to despatch the products and for creating invoices.
- *Packing/despatch* The invoices created by Sales are used by Accounts to maintain financial records in the sales ledger and to ensure that payments are received when due. (Accounts also

works jointly with Purchasing to control purchase expenditure through the purchase ledger, and deals with wages and salaries.)

However, the typical factory includes various other functions not included in the diagram, such as:

- *Costing* The Costing section calculates the costs of all products in terms of labour, materials, and overheads in order to determine selling prices or profit margins and to control such costs.
- *Engineering* The Engineering section maintains all plant and machinery to minimize breakdowns and other production losses (e.g. poor quality), and may also be responsible for installing new, or moving existing, machines.
- *R&D* The Research and Development unit works closely with the Design and Quality functions to eliminate operational faults from new designs and to suggest changes based on its findings.
- *Personnel* The Personnel section deals with all human problems and is usually responsible for recruitment, training, health and safety, and disciplinary matters.

The size of the company will determine whether other functions are carried out internally or on an 'as-required' basis by an outside agency (advertising, press relations, computer and systems development, external transport, etc.).

INFLUENCING FACTORS

The nature of a manufacturing organization is set by a number of factors that influence the way in which it operates, and these factors can be defined by examining the answers to a series of questions, for example:

1. What are the nature and derivation of the basic materials used?
2. What is the market?
3. What is the added value as a percentage of material costs?
4. What type of orders are received from customers?
5. How are products delivered to customers?

Basic materials

There are three types of basic material, each of which can influence the type of company. These are as follows.

- *Natural* Such materials occur naturally and include agricultural products (cereals, latex, cotton, etc.), animal derivatives (hides, wool, eggs, milk, meat/fish, etc.) and minerals and ores (metals, sand, clays, oil, etc.). A company that depends on such items must be aware of market fluctuations and be prepared to buy when prices are low, regardless of its immediate needs. It must also be aware that such materials are variable in quality and therefore it needs to employ technicians who can analyse an incoming batch and modify the processes accordingly.
- *Processed* Such materials are purchased from a processor (who probably buys natural materials) and typical examples include flour, leather, textile yarns, steel sheets, rolled sections

and strip; chemicals; refined sugar; paints and dyes; etc. A company that buys such items is less likely to be affected by market fluctuations and quality variations, but must be willing to change suppliers if quality, service or price becomes uncompetitive.
- *Components* Where the manufacturer is mainly an assembler, components will be purchased from another manufacturer. These can be standard items that are widely available, or specials that are designed to suit the purchaser. A company that buys standard components must be prepared to modify or design other items to suit or alternatively work closely with the supplier on the design of special components. Considerable thought must therefore be given to the advantages and disadvantages of the two options, taking account of such factors as price, availability, and the effect on other components.

Market

There are two basic types of market (customer): domestic sales and industrial sales. There are also subclassifications, for example, home/export and institutional/general public. Very few manufacturers sell directly to the general public; they prefer to arrange this through agencies such as wholesalers or distributors.

- *Domestic* These market types include products that are used in the home, the garden, or for leisure activities and personal use (food, clothing, furniture, sports goods, etc.). A company in this type of market must be aware of the fickle nature of public loyalty and be prepared to advertise extensively, to offer a wide range of consumer choice (based on market research), to develop new products quickly and frequently, and to operate an effective system for dealing with customer complaints and problems.
- *Industrial* These market types include products that are incorporated into other manufacturers' products (components and materials) or are used indirectly to manufacture those products (plant and machinery, tools, office equipment). A company in this type of market must be willing to work closely with its customers to supply the right products (backed by suitable research and development facilities) and be able to meet their needs in terms of quality and flexible delivery schedules.

In addition, if the supplier sells in export markets, it must be aware of the implications of exchange rate fluctuations and various types of local legislation (e.g. on emissions from car exhausts). Thus special designs/products may be required for specific markets. Also, if the market is institutional it will be necessary to negotiate special contract conditions and to meet special administrative and other requirements.

Added value

The amount of work put into a product can be expressed in terms of added value, i.e. the difference between the cost of the materials purchased and the selling price of the product before profit is added. This value can show how efficient a company is at converting purchases into sales, as shown in Table I.2.

Note Non-value added costs are costs incurred due to inefficiency, for example, non-productive labour, machine down-time, energy losses or excess rectification and rejects. They do not add anything to the value of the product!

Table I.2 The concept of added value

Costs, etc. (£k)	Company A	Conpany B
Materials	100	100
Labour	150	150
Overheads	50	50
Value added total	200	200
Non-value added costs	50	100
Total costs	350	400
Profit	50	Nil
SELLING PRICE	400	400

The added value figures can also show whether the company is 'labour-intensive' and to some extent the type of business, if the percentages are calculated for labour, materials and overheads against total costs. For instance:

- if labour costs are over 40 per cent the business is probably labour-intensive and may need to invest in more mechanization or even automation
- if material costs are over 40 per cent the business is probably material-intensive and extra effort should be employed to reduce these costs by minimizing stocks and wastage and negotiating better prices
- if overhead costs are over 40 per cent the business is carrying heavy overheads, and if these are fixed (i.e. not related to output volume) the company may have a high break-even factor (i.e. the sales level at which profit turns to loss). This can cause problems if demand fluctuates, and efforts should be made to reduce these fixed costs.

Naturally, every effort should be made to eliminate non-value added costs.

Order types

The way in which customers' orders are placed can be a further indication of the type of business: some typical examples are shown in Table I.3. Each of these types requires a different sort of organization, as in the following examples.

- *Tele-sales* A highly organized sales office, well equipped with computer systems for instant information on stocks and prices. Production closely linked to stock requirements; stocks well managed and controlled; deliveries on a planned basis.
- *Schedules* Sales works closely with customers, possibly with electronic links between computers; Production Planning ensures that flexibility allows for changes in schedules at the latest possible time (just-in-time operation); some buffer stocks may be required but these should be minimized; Purchasing builds close partnerships with its suppliers; Quality Control to eliminate rejects (zero defects).
- *Repeat* Production Planning works on the basis of MRP (materials requirements planning) to minimize stocks, and every effort is made to 'match' production batch sizes to customer order sizes. Purchasing also works to MRP recommendations and tries to negotiate the best

Table I.3 Examples of customer order types

Order type	Description	Customer examples
Tele-sales	Orders taken daily by telephone for fast-moving goods based on previous day's sales	Mainly retail customers (supermarkets, DIY stores, chain stores)
Schedule	An order is placed for differing quantities to be delivered on specified dates (daily, weekly or monthly)	Customers with planned volume output schedules (automotive, domestic appliances, food/drink or newspapers/magazines)
Repeat	Orders are placed on an as-required basis—there are no schedules, each order being for a single quantity	Non-manufacturers that sell from stock or manufacturers that make to order (builders' merchants, car distributors, bespoke equipment makers)
One-off	Orders placed on an occasional basis by a customer who may never order the same item again	Non-manufacturers that supply bespoke products and manufacturers or specials (architects, power generation equipment)

terms for variable-order quantities, and 'on-time' deliveries. Production works to schedules based on a priority system, which ensures that orders are completed on time.

- *One-off* Since such orders are often for special products, each job should be treated as a project, managed by a project team, and planned from the design stage through to completion. In such cases, completion may involve installation and commissioning, which must be included in the project plan. Where possible, tasks should be overlapped to minimize the overall project time; resource usage (e.g. people and machines) should be planned on a company-wide basis (assuming that several projects share the same resources) to ensure that overloads are not created by conflicting project schedules.

Deliveries to customers

To some extent, the way in which deliveries are made to customers will be governed by the type of orders received (see 'Order types' above), but the way in which delivery is specified on an order can vary, as in the following examples.

- *Direct* In this case the order is likely to be of the schedule type and delivery may be directly to the customer's production area, with no goods-inwards inspection and minimal administrative procedures. Such deliveries are likely to be very tightly scheduled (e.g. daily, at a specific time of day) and therefore it may be advisable for the supplier to use a specialist logistics contractor to make the deliveries.
- *Phased* Certain 'contract-type' orders (e.g. structural steelwork for a supermarket or equipment for a power station) will require deliveries to be phased. These phases should be clearly specified and agreed when the order is placed and design work planned accordingly. Production should then be controlled by splitting the work into a series of works orders, each of which is planned to meet a phase completion date.
- *One-time* Where delivery is to a specified address which is not likely to be repeated (e.g. kitchen units to a domestic user), it is uneconomic to make more than one delivery. It is therefore essential to ensure that all the required items are available from production (or

FUNCTIONS		COMPANY SIZE			
Primary	Secondary	VERY LARGE Over 1500 people	LARGE 501 to 1500 people	MEDIUM 101 to 500 People	SMALL Up to 100 People
ACCOUNTS	Purchase Ledger Sales Ledger Nominal Ledger Payroll Management Accounts	Five Separate Departments	Three Separate Departments	One Department, plus Two Individuals	One Department only
SALES	Sales Orders Estimating Marketing Sales Reps. Customer Relations	Five Separate Departments	Three Separate Departments	Two Departments, plus One Individual	One Department, (incl.Cust.Relations) Outside Agency
PURCHASING	Purchase Orders Vendor Rating Order Progressing Goods Inwards Invoice Validation	Five Separate Departments	Three Separate Departments	One Department, plus Two Individuals	One Department, (incl.Inv.Validation) One Individual
DESIGN & DEVELOPMENT	Concept Design Design Development Production Engineering Product Development Product Testing	Five Separate Departments	Three Separate Departments	Two Departments, plus One Individual	One Department, One Outside Agency
PRODUCTION PLANNING	Production Planning Production Scheduling Production Tracking Work Study Toolmaking & Repair	Five Separate Departments	Three Separate Departments	Two Departments, plus One Individual	One Department, Two Outside Agencies
	etc.	etc.	etc.	etc.	etc.

KEY:	Department (One Function)	Department (Multi-Function)	Individual (One Function)	Outside Agency (One Function)

Figure I.2 Allocation of functions and company size.

stock) on the due date. This requires careful planning if stocks of finished goods are to be minimized.

- *Back-Order* Certain types of customer will not accept 'back-orders' (where a required item is not in stock and is left over until a later date). This applies particularly in the case of tele-sales, where failure to be able to deliver from stock may well result in a lost sale. However, back-orders are often acceptable where regular deliveries are being made. The balance is delicate—should the supplier maintain high stocks (which are costly) or be prepared to lose some sales (also costly)? This is an important policy matter.

CONCLUSIONS

Although all manufacturing companies generally meet the definition given at the beginning of this introduction, it can be seen that there are a great many variations in the way in which the business should be organized and managed. It is essential for a company to be aware of such variations, and to know what type of business it is operating; this may seem obvious, but it is often a fact that the management of a manufacturing company does not meet these essential requirements.

In this book, efforts have been made to cover each topic (chapter) in a general way and also from the viewpoint of the various types of manufacturing business. If some of the text and examples appear to concentrate on a particular business type, this is not deliberate.

It will be noted that references to the size of a business are virtually non-existent. This is because size (in the author's opinion) is not a factor in the application of manufacturing management techniques. The same techniques should apply whether the business employs 15 people or 15 000! The difference is merely one of scale.

For instance, in a large company every function will be covered by at least one person (and possibly several), whereas in a smaller company one person will be responsible for several functions. This is illustrated in Fig. I.2, which shows how functions might be allocated in companies of various sizes.

Figure I.2 splits the functions into primary and secondary types, where all secondaries within a primary work in close cooperation, forming what is sometimes known as an administrative cell. This concept is designed to ensure that all the functions (tasks) involved in a particular operation work together to achieve a common set of objectives (for example, to get a new design into production in the shortest possible time).

CHAPTER

ONE

THE GENERAL PRINCIPLES OF MANAGEMENT

1.1 INTRODUCTION

In this chapter the principles of management as applied to a manufacturing organization are specified, together with the organization required to allow those principles to be applied. This is followed by a more detailed look at the various types of manufacturing organization—in various categories—and the requirements that stem from these classifications.

Because each type of organization requires a different approach to make it operate successfully, this chapter continues with an examination of the various functions that need to be performed and their relationship with one another. It concludes by relating these functions to the various types of organization.

1.2 PRINCIPLES OF MANAGEMENT

Management is a process which, if properly applied, can get the most out of a series of resources. These may be people, money, materials, machines, buildings or land, or most usually a combination of a number of such resources.

Management is said to require the application of the following five activities.

- *Planning* The planning activity involves the preparation of both medium- and long-term plans in order to ensure that all possible eventualities are catered for. These plans are usually in three parts—a marketing plan, a production plan and a business plan.
  - — The *marketing plan* specifies what income can be expected in terms of sales quantities and values and what expenditure this would need in terms of administrative and promotional costs
  - — The *production plan* specifies what will be required in terms of output of products to meet the marketing plan, allowing for the costs of production (labour, materials and overheads) and capital costs for new or replacement equipment
  - — The *business plan* combines the other two plans by specifying what is expected to happen in terms of profit or loss based on income from sales and expenditure on administration and production.
- *Organizing* The organizing or arranging activity is designed to ensure that these plans can be met. This involves determining the specific functions that will need to be performed and providing the necessary facilities (e.g. office and factory space; office, production and storage equipment; machines, tools support services). Organizing requires that clear channels of communication be set up within functions and between functions so that instructions can

flow downwards and information can flow upwards. It also requires the work to be defined in terms of job specifications. The organization should be such that every employee is made aware of his or her responsibilities in terms of work content, quality standards, instruction handling, information reporting and limits of authority.

- *Staffing* The staffing activity is designed to ensure that employees are recruited and trained to carry out all the functions specified during the organizing activity. It also involves such associated activities as motivating and grouping (e.g. into teams) and, by ensuring that working practices and conditions are amenable to change, allows for flexibility in all types of operation.
- *Directing* The directing activity can best be described as the provision of direction by *leadership*. This requires the ability to communicate with subordinates in such a way that instructions are readily accepted and understood. It also requires that reporting back occur honestly and without prompting—a relationship based on mutual respect. This activity cannot operate unless attitudes on both sides are right. For example, if an operator makes a mistake it should be reported at once, but this will not happen if management then conducts a 'witch-hunt' instead of collaborating to solve the ensuing problem.
- *Controlling* The controlling activity is designed to ensure that the plans prepared by management operate effectively. This is achieved by a feedback of information which is then used to modify the plans and implement changes as required. This is shown in Fig. 1.1. This technique is known as the closed loop and ensures that management can react quickly to changes in circumstances—thus the need for flexibility. The application of this control technique means that plans are continuously being confirmed or modified; they must never be 'set in concrete'.

1.3 PRINCIPLES OF ORGANIZATION

In order to operate successfully, a company needs to be organized in such a way that every employee has a clearly defined role in terms both of his or her own work and of work relationships with other employees. However, this does not mean that personal initiative and flexibility should be stifled: in fact, these factors should be positively encouraged.

Most manufacturing organizations start this process by drawing up an *organization chart* similar to that in Fig. 1.2. This defines management responsibilities by function. Such a chart should not be a secret but should be available to all employees. It should not be used to show personal status but as a tool that will allow any employee to identify personal, work group and company responsibilities and communication links.

However, such a chart does not give the complete picture, since it does not show the detailed working relationships between departments and individuals or within departments, nor are the reporting functions made clear. It is therefore essential to back up this chart with two other sets of documents, *job descriptions* for each employee and *input/output charts* for each department or section.

The *job description* is commonly used in many companies and organizations to ensure that employees are made aware of what is expected of them and also to define the limits of their responsibilities. An example is shown in Table 1.1.

This document should clearly show not only the duties and responsibilities, but also the objectives, by defining the way in which the job-holder can contribute towards the overall objective of making the company into a world-class player. World-class organizations generally prefer to consider their business as a 'process' to which functional departments and individuals contribute, and this approach ensures that all individual objectives are geared to meeting the overall business objectives.

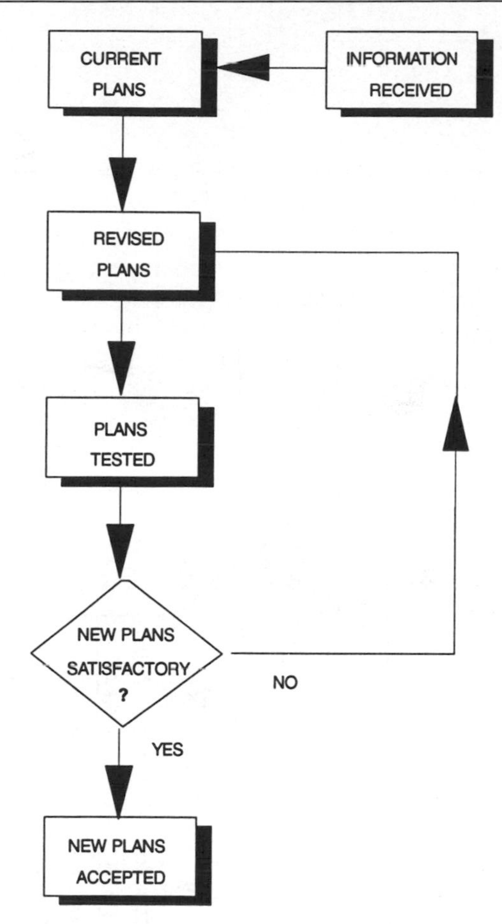

Figure 1.1 The closed-loop technique of control.

However, the job description is unlikely to show the detailed interactions between different people and their functions. It is therefore often helpful to prepare some form of matrix which will define these interactions in 'input' and 'output' terms. This will not only ensure that proper communication links are maintained but also establish the internal customer/supplier relationships.

This concept of such internal relationships is vital in any organization since it enables employees to realize that they receive information, products and services from an internal source (their supplier) and pass these on, in a modified form, to another internal source (their customer). In other words, employees should be aware that quality of goods and services is not just for external customers and that links with internal sources must be on a partnership basis.

In this competitive age, all employees should be aware of the importance of their relationships with their external suppliers and customers (for instance, by the application of TQM — total quality management) but it is perhaps not considered equally important to establish similar relationships internally. This input/output approach should help to destroy such an illusion and an example of an input/output chart is shown in Fig. 1.3. Obviously this figure is incomplete, since it refers only to documents and such a chart should also include information (data) that is given verbally or passed electronically (computer data). It should also specify individuals (by job title, not name) where appropriate.

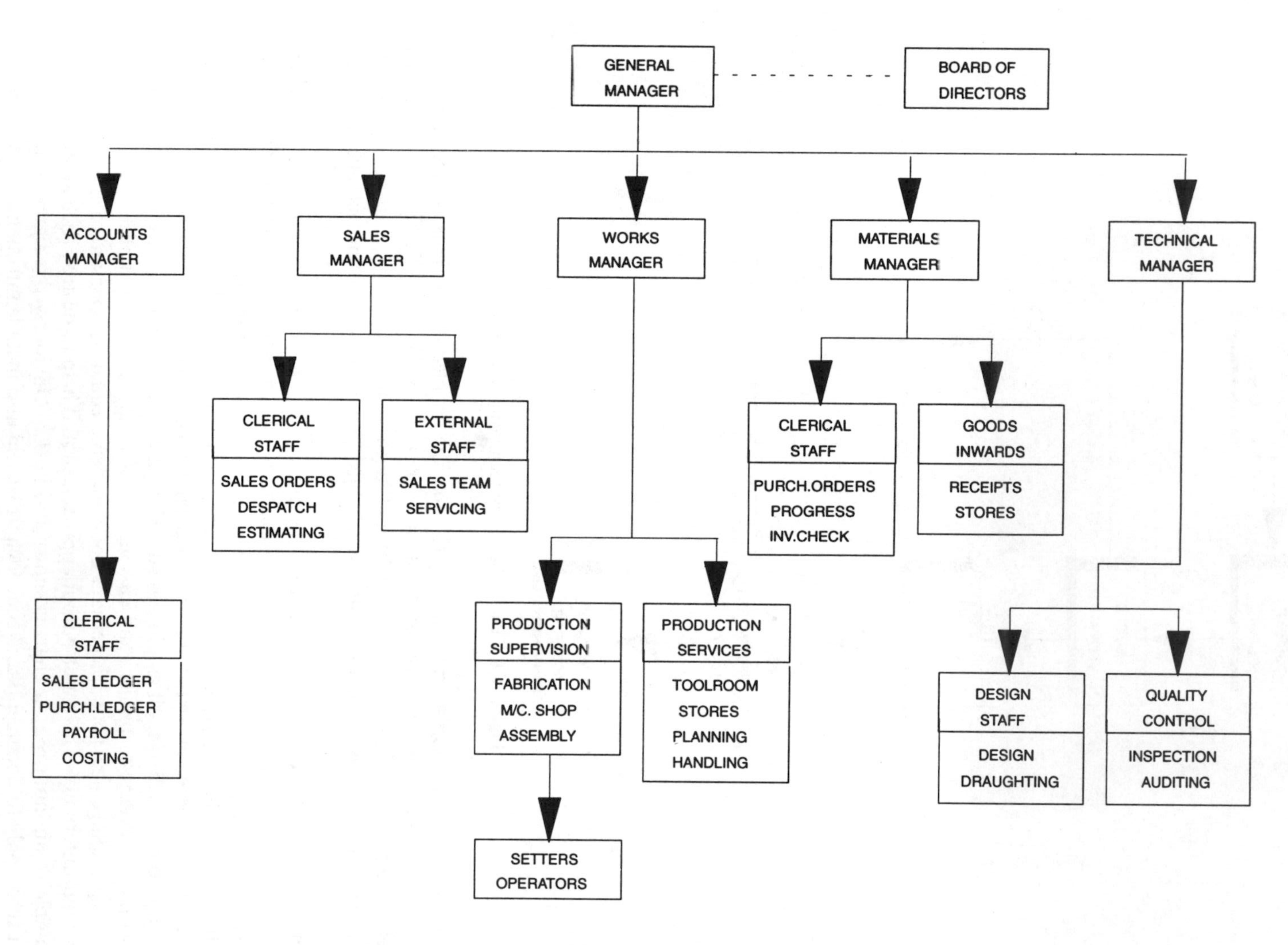

Figure 1.2 Typical organization structure for manufacturing company.

DATA	CUSTOMER/SUPPLIER (Internal or External)									
	EXTERNAL CUSTOMER	EXTERNAL SUPPLIER	SALES OFFICE	PURCHASING DEPARTMENT	ACCOUNTS	DESPATCH DEPARTMENT	GOODS INWARDS	PRODUCTION PLANNING	WORKS OFFICE	QUALITY CONTROL
CUSTOMER ORDERS	FROM		TO					TO		
PURCHASE ORDERS		TO		FROM						
DESPATCH NOTES	TO				TO	FROM				
SALES INVOICES	TO				FROM					
ADVICE NOTES		FROM		TO			TO			
PURCH.INVOICES		FROM		VIA	TO					
WORKS ORDERS								FROM	TO	
PICKING LISTS			FROM			TO				
S/ORDER AMENDMT.	FROM		TO					TO		
TEST CERTS.	TO					VIA				FROM

Figure 1.3 Typical input/output chart for manufacturing company.

Table 1.1 Typical job description

Job title	Sales order clerk
Reporting to	Sales manager
Duties	1. To receive orders from customers by post, fax or phone. 2. To check stock availability and confirm delivery dates from Production. 3. To check with Accounts that customer is not exceeding credit limits and is not on 'stop list'. 4. To enter orders onto the computer as set out in the current user manual and prepare an acknowledgement copy for the customer, etc.
Objectives	1. To ensure that all customers' orders are entered on the day of receipt. 2. To ensure that all customers' queries are answered without delay. 3. To ensure that telephone calls are answered informatively within 15 seconds. 4. To ensure that a good rapport is built with all customers, etc.
Working conditions	Hours: Mon–Fri. 8.30am to 5.00pm Breaks: Lunch 12.30am to 1.15pm Overtime and holidays: Standard—Scale B Salary: Scale 8
Responsibility limits	Cannot, without authority of manager: 1. accept orders for non-standard products 2. quote prices or deliveries for non-standard items 3. amend orders 4. amend prices, etc.
Reporting	Reports to be run weekly as follows: 1. open order value 2. orders by customer 3. orders by product, etc. (copies as specified in current user manual)
Discipline	The company's disciplinary procedures as specified in the current employee handbook must be adhered to unless specific instructions to the contrary are received from the sales manager.

1.4 TYPES OF MANUFACTURING ORGANIZATION

Although, in basic terms, all manufacturing companies operate in the same way, there are many differences in detail due to the markets into which they are selling. Thus, in order to determine the principles under which a company should operate, it is necessary to identify its type of business, and this can best be achieved by locating the company concerned on a scale from 1 to 5.

A *category 1* company is a 'make-to-stock' company, i.e. all its products are made to a series of standards (it does not make specials) and held in stock ready for sale.

A *category 5* company is quite the opposite, i.e. it is a 'make-to-order' company with all its products being made to customers' specifications as specials. In this case the company does not hold any stocks, placing orders on its suppliers to meet orders from its customers.

In fact, most manufacturers are somewhere in between, i.e. in categories 2 to 4, and the main features of each of these categories are shown in the form of a spectrum in Fig. 1.3. As a

Table 1.2 Categories of manufacturing company by product

Category	Typical products
1	Building materials (bricks, cement, tiles, etc.) Hand tools (screwdrivers, hammers, saws, etc.) Office equipment (copy paper, envelopes, pens, etc.) Standard steel sheets and rolled sections
2	Printed forms (standard products overprinted as required) Gift products (pottery: 'A present from Blackpool') Automotive (vehicles finished to suit dealers' orders) Kitchen furniture
3	Ceramic machinery (assemblies allow size/shape options) Professional sports equipment (options to suit user) Limbs for amputees
4	Bespoke tailoring (made up from stocks of material) Conveyors (made up from belts, rollers, frames, etc.) Printed circuit boards
5	Special-purpose machines Injection and blow mouldings Preformed hoses

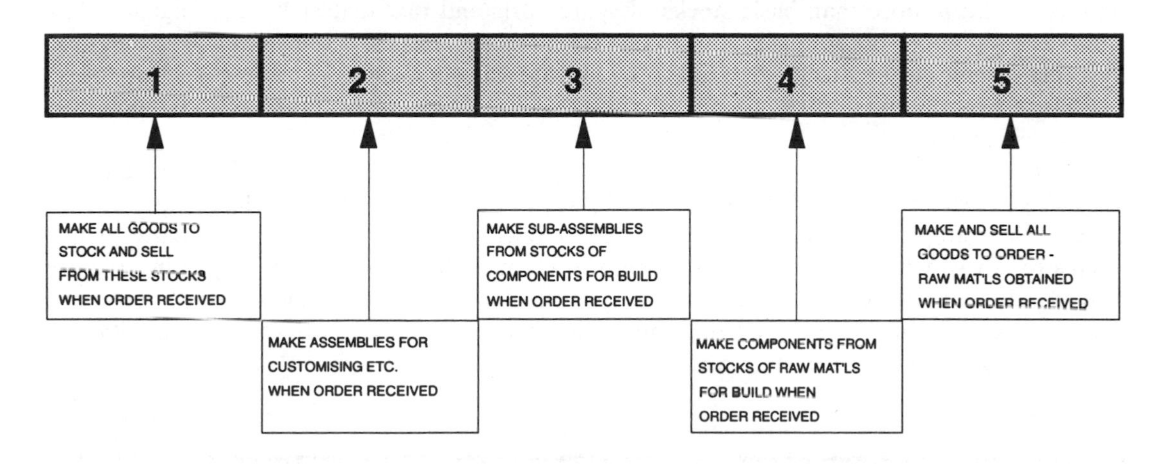

Figure 1.4 The spectrum of company types (by stock policy).

guide, some examples of typical manufacturing companies in the various categories are given in Table 1.2 based on their products.

However, this system of classification is not totally satisfactory since it does not take account of other relevant factors. For example, many suppliers make special products to stock for a particular customer on the basis of scheduled orders (orders that schedule a series of deliveries over a period of several months). It may also be that a manufacturer makes a range of products that fall into different categories. It is probably therefore necessary for a manufacturer to categorize its company type in diagramatic form as shown in Table 1.3, which takes account of such variations and allows such a company to determine its type.

Classification by product may not always be easy since the technique assumes that policies exist on what should be stocked. For example, product group D is classed as scale 2 because stocks of assemblies are held ready for subsequent customizing when orders are received. If such policies do not exist, then it is necessary to create them.

Table 1.3 Diagram used to classify a company type

Product	A	B	C	D	E	F	G	H	Total
Category	1	1	2	2	2	3	4	5	
% By value	10	15	15	10	20	15	10	5	
Ratio (1)	0.10	0.15	0.30	0.20	0.40	0.45	0.40	0.25	2.25

Ratio (1) is the category × the % by value (e.g. 3 × 0.15 for product F)
For this company the average category is 2.25

There is another way of classifying a manufacturing company. This is similar to the method described above (which is based on stock policies), but in this case the basis is types of sales ordering. This second method uses the same scale (1 to 5) as shown in Fig. 1.5, but now Category 1 includes companies that take orders by phone or fax for same-day delivery. Such companies must maintain stocks of all saleable items since any failure to deliver means a loss of business (the customer may well go elsewhere!). At the opposite end of this scale, a Category 5 company usually receives orders on the basis of estimates and quotations and is therefore unlikely to keep more than basic stocks of spare parts and raw materials. Once again, many companies do not fit naturally into one of the five categories and the same approach as that used in Table 1.4 should be used to determine the average category value.

These two methods can be used as alternatives since the results will probably be similar for any given company.

The need to categorize a company in this way may not be immediately apparent, but hopefully it will be revealed when various options and policies are considered later in this book. For example, accurate sales forecasting is essential for a category 1 company (which makes to stock) whereas it is unlikely to be of such importance if the company is in category 5. Similarly, the setting of stock levels to meet particular service levels (the statistical approach which calculates the chances of being out of stock when demand is fluctuating) is vital for category 1 but of little interest to category 5.

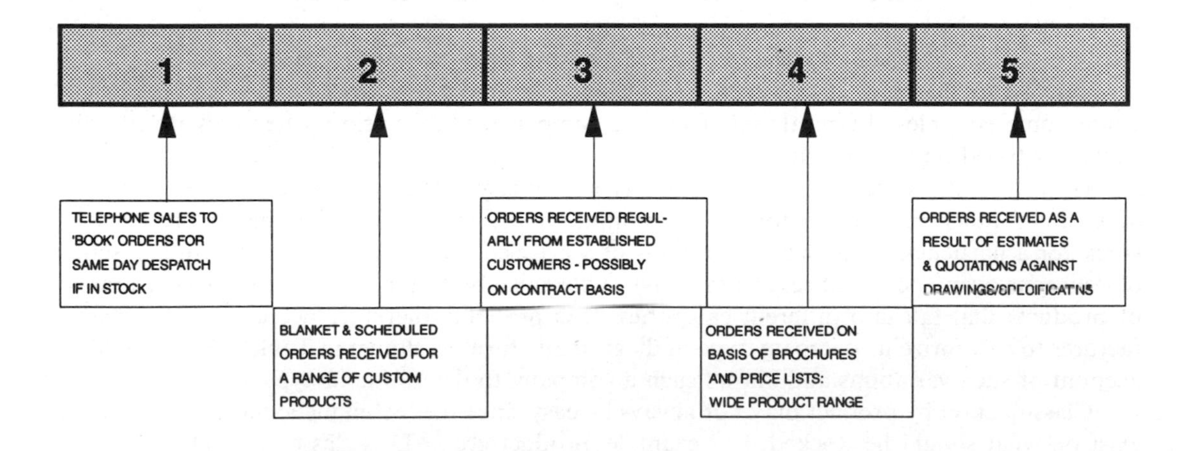

Figure 1.5 The spectrum of company types (by sales ordering).

Table 1.4 Functions and the development cycle

Development cycle stage	Functions
New product design and development	Research and development Design/drawing Production engineering Quality standards
Sales promotion	Customer relations Sales literature Market research Sales forecasting Sales estimates and orders
Employee training	Personnel recruitment and training
Tooling	Tool design Tool making
Plant development	Production engineering Works engineering Systems engineering
Production planned	Production planning/scheduling Capacity planning
Works orders raised	Production control
Purchase orders raised	Purchasing
Materials received	Goods inwards recording Quality control Raw materials stores
Works orders processed	Production control Production supervision Setters and operators Internal transport
Stocks built up	Inter-process/Finished goods stores
Goods despatched	Packing department Despatch department
Accounts prepared	Accounts department
Cash flows calculated	Costing department

1.5 DEFINITION OF FUNCTIONAL AREAS

Although manufacturing companies differ markedly in many respects (size, category, product complexity, etc.), they all share common functions. These can best be defined by examining the sequence of events that must take place if the company is to survive and develop. This sequence can best be shown in the form of a closed loop, as in Fig. 1.6.

As can be seen from this figure, the key to manufacturing success is to develop new products on a regular basis and to sell them profitably. These profits can then be ploughed back for new product development.

The functions that need to be carried out to match this cycle of operations are all based on the various 'stages' in the cycle, as shown in Table 1.4. The work that needs to be carried out within each of these functions will vary slightly depending on company type, but is generally as follows.

Research and development

- To create new prototype products in line with policies agreed with senior management, based on market research and customer needs.

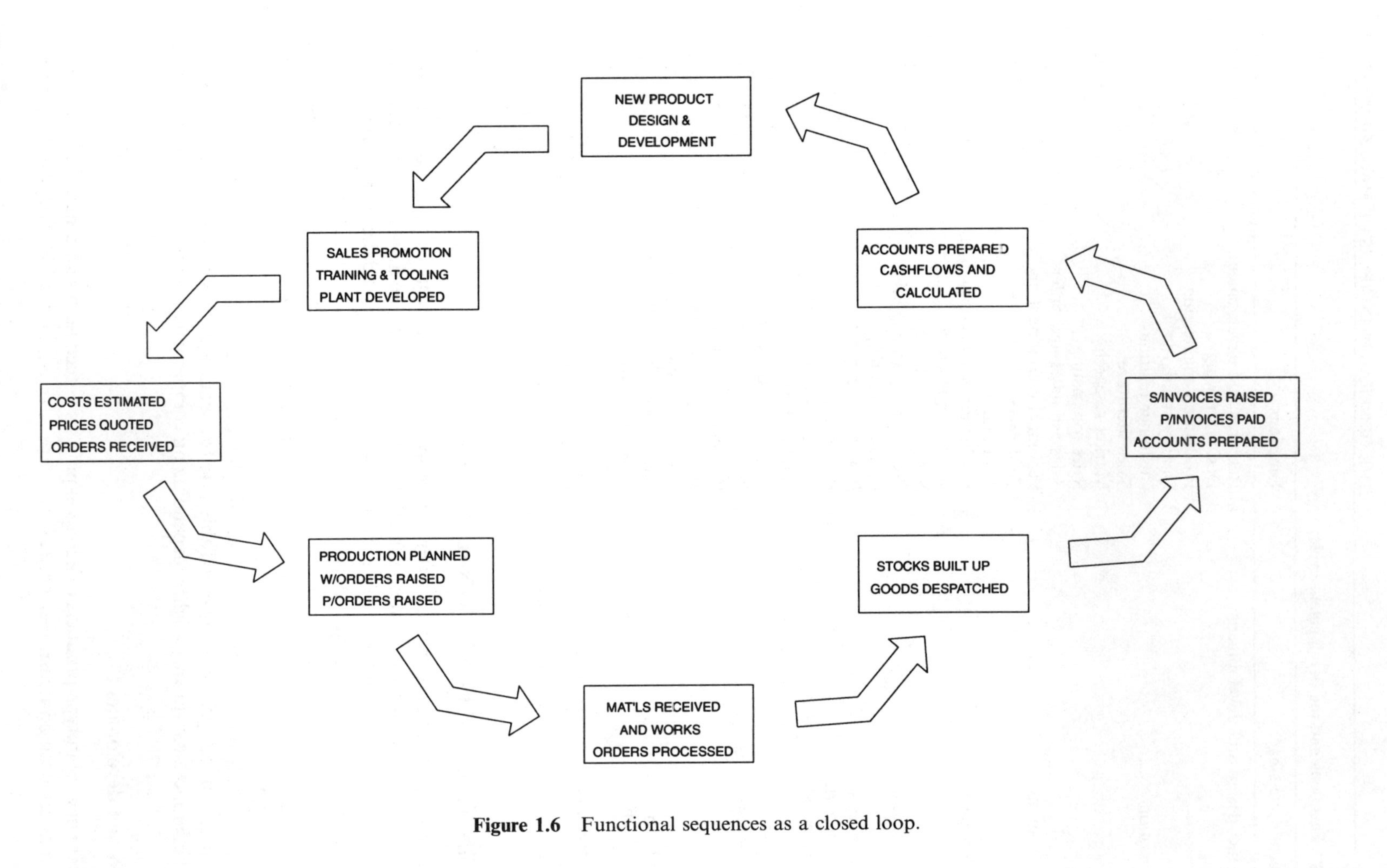

Figure 1.6 Functional sequences as a closed loop.

- To develop modifications to existing products in order to improve their performance.
- To develop testing procedures that can be used to ensure that new and existing products achieve the required standards.
- To maintain a 'library' of information on all the latest relevant techniques, materials and processes.

Design/drawing

- To prepare concept and design drawings for new and modified products from which working drawings can be prepared.
- To prepare working drawings and parts lists which can be used by other departments to order tooling, raw materials and process plant such that the required products can be manufactured.
- To maintain complete records of all drawings and parts lists in such a way that information can be retrieved as required.
- To build up a technical 'library' of information on materials, standard components and services from outside suppliers and to maintain this in such a way that information can be easily and quickly obtained.

Production engineering

- To devise methods of manufacture for products and components based on working drawings and specifications that will minimize costs and wastage and maximize output.
- To evaluate these methods in quantitative terms (output levels and costs) and to prepare detailed work specifications that define the methods, machines, operations, tools and materials to be used.
- To provide—in conjunction with costing department—information that can be used for the preparation of estimates and quotations.

Quality standards

- To specify a series of standards that must be met in order to achieve the quality levels required, e.g. tolerances and test results.
- To specify a series of standard components, materials and methods that should be used wherever possible (to minimize diversity) and to define these in a quality standards manual.
- To maintain a 'library' of national and international standards that are relevant to the product range of the company and to define how these should be applied by other departments.

Customer relations

- To maximize the effectiveness of advertising within the budget in order to present the best possible image of the company and its products. This may involve the use of a public relations company or advertising agency.
- To create a high level of customer loyalty by dealing effectively with all complaints and warranty claims.
- To ensure (in conjunction with other departments) that the highest possible level of customer service is maintained in terms of service agreements, product availability and spares/repairs.

- To maintain effective contact with customers by means of such techniques as mailshots, exhibitions, seminars and sales visits.

Sales literature

- To produce attractive and accurate sales literature. In some cases this will be in the form of technical reference manuals that will assist customers in product selection.
- To ensure that such literature is distributed in the most economic and effective way.

Market research

- To develop market research techniques that will keep the company informed about the latest developments in the market, e.g. new competitive products, changes in customer requirements, revisions to regulations and alternative sales outlets.
- To prepare regular reports containing such information for senior management and to recommend changes in strategy.

Sales forecasting

- To develop the most suitable methods of forecasting that will provide management with data on which production can be planned and so that stocks can be maintained at appropriate levels.
- To ensure that information from which forecasts will be prepared is collected and held in the most suitable form.

Sales estimates/orders

- To set up and maintain a 'library' of information on the basis of which estimates and quotations can be prepared.
- To prepare estimates and quotations against customer enquiries and to progress these to ensure that the maximum number are converted into orders.
- To receive customers' orders and record them in a suitable form for acknowledgement and processing.
- To deal with customers' enquiries on order progress and to record amendments (if these are acceptable).

Recruitment and training

- To work with other departments to prepare a manpower plan (which will be based on the business, marketing and production plans). This plan should specify future manning requirements in terms of numbers and skills.
- To evaluate the skills and age profiles for all existing employees and to compare these with the manpower plan requirements.
- To prepare a series of training and recruitment plans which will ensure that the skills and age profiles of the labour force match the manpower plan.

Tool design and making

- To collaborate with the production engineering function to ensure that suitable tools are designed to meet production requirements.
- To ensure that such tools are made or purchased such that they are available when required.
- To ensure that all existing tools are checked and refurbished as necessary or replaced if refurbishment is not possible.

Works engineering

- To maintain all buildings, plant and machinery in order to minimize production delays and preserve the asset value. This will require the operation of a planned preventive maintenance system.
- To collaborate with production engineering when new plant is being specified/ordered and to assist with installation and commissioning.
- To maintain stocks of engineering materials and spares at the most economic levels.

Systems engineering

- To take responsibility for all production and quality control systems equipment (e.g. machine controllers, PLCs, robots, data capture terminals) in terms of specification and operation.
- To prepare and maintain programmes for all such equipment in collaboration with the production engineering department.
- To maintain a 'library' of information about the equipment and systems available and records of settings for all installed items.

Production planning

- To plan all production operations in order to meet customers' delivery requirements or to maintain agreed stock levels.
- To schedule all such operations in order to meet delivery and stock level requirements in the most economic way by minimizing such factors as subcontracting, overtime working and machine set-ups.
- To reschedule operations when changes in requirements occur (e.g. alterations to customer orders) in such a way as to minimize disruption and costs.

Capacity planning

- To set up a system that will provide information in terms of workloads on each production machine or work centre in relation to the current production plans.
- To assist production planning to plan operations in such a way as to minimize overloads or idle time on each machine or work centre and, where possible, to damp down fluctuations in machine/work centre loading.

Production control

- To create works orders in line with the production planning schedules with full documentation covering operation sequences, tooling, operation specifications, machines, manning, etc.

- To progress these orders through to completion at the date specified and to record this progress so that enquiries can be answered and corrective actions taken when necessary.
- To modify these orders as required (e.g. batch splitting, rescheduling, re-routeing)

Purchasing

- To maintain a 'library' of approved suppliers for all items purchased on a regular basis, including details of prices, discounts and conditions of sale.
- To place orders on suppliers against requisitions or recommendations. These orders must be fully specified in terms of the items/services required, quantities, prices and delivery requirements.
- To progress these orders such that deliveries are made on time and in line with conditions agreed.
- To monitor all suppliers in order to produce 'vendor rating' figures on the basis of which purchasing decisions can be made.

Goods inwards

- To record details of all goods received in order to ensure that deliveries are in line with purchase orders and are covered by fully detailed suppliers' advice notes.
- To ensure that quality control is informed of such deliveries so that goods received can be checked/tested where necessary and put into stock or production as quickly as possible.
- To ensure that all goods inwards are properly identified or labelled and, where necessary, accompanied by certificates of conformance or test.

Quality control

- To prepare a fully detailed quality manual that specifies all QC procedures.
- To specify all inspection and test equipment required and to ensure that this is properly maintained and calibrated at suitable intervals.
- To specify full details of all inspections and tests on all production items (e.g. what, where, how) and methods of recording and analysing results.
- To audit suppliers to ensure that they can maintain the quality standards required.
- To assist customers who wish to carry out quality audits on the company and to investigate any quality problems found by customers.

Raw materials stores

- To maintain records of all raw material and bought-in component stocks and to carry out stock checks to ensure that these records are accurate.
- These records should include details of stock locations, receipts, issues, returns to stock, purchase orders outstanding and allocations such that free and physical stocks can be readily identified.
- To ensure that stocks are issued in the correct sequence (e.g. first in–first out) and are so stocked that batches do not become mixed.

Production supervision

- To ensure that work is carried out in its production area in line with the production schedules or if this is not possible, to inform production planning.
- To ensure that all necessary resources (tools, materials, setters and operators) are available before a works order is authorized to start in its area.
- To ensure that all procedures for safety, quality and work booking are adhered to by employees in its control.
- To maximize the use of all the resources within its control.

Setters and operators

- To ensure that machines are set up and operations carried out as specified on the works documents and associated procedures.
- To book all work carried out on the documents or through the terminals provided, including details of scrap or rectification with reasons where requested.
- To comply with all safety regulations such as machine guards, protective clothing and fault/accident reporting.
- To comply with all procedures for time and attendance recording, overtime working, holidays and sickness.

Internal transport

- To comply with all safety procedures concerning mechanical handling devices such as fork-lifts, pallet trucks, cranes and machine load/unload devices and to ensure that all gangways are kept clear of obstructions.
- To move all goods and materials from stores to production, within production and from production to stores as instructed by the supervisors concerned.
- To ensure that all such items are accompanied by the appropriate documentation and can be identified. Where required the new locations of items moved should be recorded as specified in the procedures.

Finished goods stores

- To maintain records of all items and components in stock and to carry out stock checks to ensure that these records are accurate.
- These records should include details of stock locations, receipts, issues, returns to stock, works orders outstanding and allocations such that free and physical stocks can be readily identified.
- To ensure that stocks are issued in the correct sequence (e.g. first in–first out) and are so stocked that batches and lots do not become mixed.

Packing department

- To pick from finished goods stores all items specified on picking lists, to record any shortages found and to pack these items as specified in the packing documents or procedures. Each package should be clearly labelled and the numbers/weights of packages recorded on the picking list.

- To locate these packages in a suitable area, whose location code should be recorded on the picking list.
- To maintain stock records of packing materials and to requisition when new supplies are required.

Despatch department

- To arrange transport for goods that are to be despatched (details from picking list) and to take the items concerned from their location to the loading bay.
- To prepare despatch notes for the shipment showing what is being despatched, to whom, and by what means.
- To load the transport and ensure that the driver is provided with all the despatch notes required.
- To inform accounts that the despatch has taken place (despatch note copy) so that invoices can be prepared.
- To deal with any customer returns by informing sales department and also to record details of returned stillages or pallets.

Accounts department

- To create sales invoices for shipments to customers, and credit notes to allow for incorrect invoicing or goods shipped.
- To set up an accounting system that will record all financial transactions within the company and will include a nominal ledger which records all items of income and expenditure by type and area.
- To calculate and pay out all wages and salaries and to maintain full records of such payments including deductions, holiday pay, etc.
- To maintain all required statutory records for the purposes of VAT, PAYE, NIC, etc., and to pay these to the appropriate bodies when required.
- To maintain a register of fixed assets for updating annually and to assist the auditors to prepare annual accounts.
- To provide senior management with financial information in the form of monthly accounts and associated reports.

Costing department

- To set up and maintain a costing system that will allow the company to determine the costs of making any product and thus to determine its profitability.
- To set up a system of cost collection with reports that will allow management to identify any area where costs are greater than planned.
- To provide information that can be used by the accounts department for monthly and annual accounts (e.g. stock and work-in-progress values) and also by the sales department for estimating purposes.

1.6 RELATIONSHIPS BETWEEN FUNCTIONAL AREAS

As can be seen in the previous section, it is possible to define almost 30 such areas, and although in smaller companies some of these can be combined it is still necessary to determine the relationships so that communications and responsibilities can be clearly specified. To some extent this can be done by specifying the inputs and outputs for each function (as shown in Fig. 1.3), but this can sometimes lead to omissions and it is better to define the relationships in terms of the functions themselves rather than the workflows only. Such an exercise can be very time-consuming, but is usually well worthwhile, not only for the end results but also for the information (and misconceptions!) that emerge during the progress of the exercise. In many cases employees are not aware of the need to share information or consult with others, and the detailed specification of the relationships allows them to understand how their work fits in to the overall pattern.

In Fig. 1.7 a format is illustrated for such an exercise together with an example of how this should be used, each function being given a code so that the combined codes can then be used to identify and describe the relationship. It is important to try to make these descriptions as full as possible, i.e. not only to include details of inputs and outputs of data but also to consider the levels of cooperation required, sometimes at a personal level.

As an example of this approach, the functional relationship description for code I-N (sales orders to production planning) is given in detail in Table 1.5.

1.7 FUNCTIONS AND ORGANIZATION TYPES

Once the organization type of a manufacturing company has been established (on the scale of 1 to 5), it is possible to define the role of the various functions more specifically. In most companies only a limited number of the functions are directly affected, and Table 1.6 gives some examples of this.

Table 1.5 Functional relationship descriptions

For function	To function	Relationship description
I	N	
Sales orders	Production planning	As soon as orders are received by Sales, details will be discussed with Production Planning in order to agree an acceptable delivery date, taking account of customer requirements and factors affecting production lead time. A copy of the order acknowledgement will be passed to Production Planning so that it can then plan the order into production and arrange for resources to be made available. Production Planning will inform Sales Orders of any problems that arise either before production starts or during production, and if a delay is unavoidable agree a new delivery date. Production Planning will assist Sales to answer any queries from the customer; it will not deal with the customer directly. Any changes initiated by the customer will be checked by Sales with Production Planning and a suitable course of action agreed for transmission to the customer. *At no time will sales agree delivery dates or order amendments without prior discussion.*

		R & D	DES'N /DRG.	PROD. ENG.	QUAL. STDS.	CUST. REL.	SALES LIT.	MKT. RES.	SALES FCST.	SALES ORDS.	REC.& TRNG.	TOOL DESN.	WKS. ENG.	SYST. ENG.	PROD. PLAN	etc
		A	B	C	D	E	F	G	H	I	J	K	L	M	N	
RES.& DEVELOPMENT.	A															
DESIGN/DRAWING	B															
PROD.ENGINEERING	C															
QUALITY STANDARDS	D															
CUSTOMER RELATIONS	E															
SALES LITERATURE	F															
MARKET RESEARCH	G															
SALES FORECASTING	H															
SALES ORDERS	I															
RECRUIT & TRAINING	J															
TOOL DESIGN/MAKE	K															
WORKS ENGINEERING	L															

RELATIONSHIP		RELATIONSHIP DESCRIPTION
FOR FUNCTION	TO FUNCTION	
A RES.& DEVELOPMENT	B DESIGN & DRAWING	xxxxxxxxxxxxxxxxxxxxxxxxxxxxxxxxxxxxxxxxxxxxxxxxxxxxxxxxxxxxxx xxxxxxxxxxxxxxxxxxxxxxxxxxxxxxxxxxxxxxxxxxxxxxxxxxxxxxxxxxxxxx xxxxxxxxxxxxxxxxxxxxxxxxxxxxxxxxxxxxxxxxxxxxxxxxxxxxxxxxxxxxxx xxxxxxxxxxxxxxxxxxxxxxxxxxxxxxxxxxxxxxxxxxxxxxxxxxxxxxxxxxxxxx xxxxxxxxxxxxxxxxxxxxxxxxxxxxxxxxxxxxxxxxxxxxxxxxxxxxxxxxxxxxxx
A RES.& DEVELOPMENT	C PROD.ENGINEERING	xxxxxxxxxxxxxxxxxxxxxxxxxxxxxxxxxxxxxxxxxxxxxxxxxxxxxxxxxxxxxx xxxxxxxxxxxxxxxxxxxxxxxxxxxxxxxxxxxxxxxxxxxxxxxxxxxxxxxxxxxxxx

Figure 1.7 Relationships between functional areas.

Table 1.6 Examples of functions and the effects of organization type

Function	Effect of organization type
Research and development	In a *type 1* company, with a standard product range, R&D will be mainly concerned with the long-term development of new products and this development will involve collaboration with other departments such as Market Research and Customer Relations. In a *type 5* company, making 'specials' to order, R&D will be involved with Design and Production Engineering to develop such products within a much shorter timescale based on a series of standard ideas and concepts.
Sales forecasting	In a *type 1* company, which is selling from stock, sales forecasting is critical in deciding what levels of stock should be held. If stocks are excessive then costs will be too high, but if stocks are too low then orders and customers will be lost. The sales forecasting function must therefore have a high profile and use sophisticated techniques. In a *type 5* company, which is making to order, sales forecasting is necessary only to develop the long-term management plans and to determine what resources will be required.
Tool design and tool making	In a *type 1* company the tooling will be designed and made to maximize output and minimize costs on the standard product range. This can be time-consuming and costly in terms of sophisticated tooling, but is worth while when the life of a product (how long it can retain market share) may be several years. In a *type 5* company tooling will probably have to be designed and made for many products that may be 'one-offs'. Thus the approach will be totally different, tooling cost being minimized at the cost of longer operation times.
Production planning	In a *type 1* company production planning is very short-term, often requiring production to be planned to very exact schedules, for example, on a shift-by-shift basis (e.g. fresh food production). Schedules may need to be prepared only a few hours before the shift starts because demands change rapidly and the output from previous shifts must be taken into account. In a *type 5* company planning is on a much longer timescale, each project possibly spanning several months, or even years (e.g. Concorde or the Channel Tunnel). In such companies some form of project planning is often employed.
Quality control	In a *type 1* company, quality is usually maintained by the use of some form of sample checking using SPC (statistical process control) to ensure that operations or processes remain within specified limits of tolerance. In a *type 5* company every component is checked and every assembly tested to ensure that the final product matches design and performance specifications.

PROBLEMS

1. Prepare an outline marketing plan for a company that manufactures a range of tinned food products. Last year's results gave a turnover of £2.5m with a gross profit margin of 35 per cent. Your answer should include the estimated sales quantities for each product with selling prices and costs in each market (UK, rest of Europe, etc.), together with last year's actual figures. If you feel that there are likely to be seasonal variations in sales on a month-by-month basis these should be shown in tabular format. You should also show the planned marketing expenditure (sales team, advertising, promotions, etc.) compared with last year.
2. Prepare an outline job description for the production planner of a company manufacturing a variety of central heating boilers, some of which are in a standard range (domestic) and others made to order (industrial). The company uses a computer to determine its material

requirements based on a combination of orders and forecasts and maintains stocks of standard domestic boilers and standard components for industrial boilers.

3. Prepare a functional relationship description showing the workflows and other areas of cooperation between the following functions of a company that manufactures kitchen units (in kit form). The company buys timber and hardware (screws, hinges, handles etc.).
   (a) Purchasing and quality control
   (b) Quality control and production
4. Prepare a list of typical industries that are likely to be in the following organizational categories:
   (a) category 2
   (b) category 4.
5. Compare the following functions as carried out in (a) a type 1 company and (b) a type 5 company.
   (i) Design and drawing
   (ii) Warehousing and despatch

FURTHER READING

Dilworth, J. B., *Production and Operations Management: Manufacturing and Non-manufacturing,* 4th Edn, McGraw-Hill, New York, 1989.

DTI, *Managing into the 90s*—A series of short publications on various topics from the Department of Trade and Industry, available from Media Scope (Tel. 01443 821877).

Handy, C. B., *Understanding Organisations*, Penguin, London, 1985.

Freeman-Bell, G. and Balkwill, J., *Management in Engineering: Principles and Practice*, Prentice-Hall, London, 1993.

Hill, T., *Production and Operations Management*, 2nd Edn, Prentice-Hall, Englewood Cliffs, NJ, 1991.

Lanigan, M. *Engineers in Business: The Principles of Management and Product Design*, Addison Wesley, Wokingham, 1992.

Lucas Engineering and Systems, *The Lucas Manufacturing Systems Engineering Handbook*, Lucas Engineering and Systems, Solihull, 1988.

Muhlemann, A., Oakland, J. S. and Lockyer, K. G., *Production and Operations Management*, 6th Edn, Pitman, London, 1992.

Pfeifer, T., Eversheim, W., Konig, W. and Weck, M., *Production Engineering—The Competitive Edge*, Chapman and Hall, London, 1993.

Robey, D., *Designing Organisations*, Irwin, Boston, MA, 1991.

Slack, N., *The Manufacturing Advantage*, Mercury Books, London, 1991.

Todd, J., *World-class Manufacturing*, McGraw-Hill, London, 1994.

CHAPTER

TWO

MANUFACTURING MANAGEMENT BY COMPUTER

THE LINK

Having established the main techniques and principles of manufacturing management in chapter 1, it is now necessary to examine the role of the computer in a manufacturing company.

It must be stressed that the computer is not a substitute for good management, but is rather a tool that management can use to make good management better. It has been said of certain people that they are 'strong in the arm and weak in the head', and this is true of computers.

Computer systems currently available cannot think for themselves (although development work is under way to change this situation, for example to make a computer learn from its mistakes, using an approach known as artificial intelligence) but are excellent 'workhorses' which can process data at a speed and in quantities beyond the capacity of any human.

It is therefore essential to ensure that the computer is used to do what it does best, thus relieving management and other employees of routine tasks so that they can concentrate on the 'thinking' parts of their jobs. To do this it is necessary to make sure that the right type of system is installed, that its users are properly trained and that it is aimed in the right direction.

2.1 INTRODUCTION

In this chapter the place of the computer in manufacturing is examined in relation to the management principles considered in chapter 1. This is followed by an examination of the range of computer systems available. Methods are suggested that should enable a manufacturing company to approach the problems of selection and specification in a logical way.

Once a system has been selected it is important to justify the project in order to ensure that expectations of benefits are clearly compared with likely costs. Various possible methods of justification are therefore considered and compared.

Computer systems that fail to meet expectations generally do so because the implementation phase has not been properly planned and executed. This chapter therefore ends by suggesting how an implementation should be carried out to minimize the chances of failure.

2.2 THE PLACE OF THE COMPUTER IN MANUFACTURING

In chapter 1 management was defined as the process that gets the most out of a set of resources. The five activities that make up this process were listed as planning, organizing, staffing, directing and controlling. A computer system, suitably chosen and properly implemented, can assist management in all these activities.

However, it is a fallacy to believe that computer systems are a universal solution to management problems; the truth is that computers can make bad worse and good better. The role of the computer system in management is as a useful tool—not as a replacement for the manager!

An examination of how the computer can assist management in the five activities should help to define the computer's role.

Planning

The computer is an ideal tool for the planner because of its ability to carry out a large number of operations in a very short time. To prepare a plan it is necessary to set up a series of 'statements' and then test these in combination against a set of objectives.

If this test does not meet the plan objective, one or more of the statements must be changed and the plan re-tested, and this process continued until the plan is satisfactory. This process can be illustrated by the example in Fig. 2.1.

In this example a marketing plan is being prepared in order to meet objectives which are defined in terms of sales value (turnover) and gross profit. The statements are that for each product the selling price, sales volume and cost will be at stated monetary values or quantities.

These figures are then used to calculate the overall profit and turnover figures, and plan 1 falls short of both the objectives. Plan 2 is then developed by increasing the selling prices and volumes of some of the products. The test shows that this is much better, the turnover being within 1 per cent of the objective.

However, the profit margin is still 12 per cent below requirement and therefore a third plan is prepared (not shown in Fig. 2.1). This reduces the costs of some products to increase the profit margin. *Note: A number of ways of reducing costs are examined in chapter 14.*

It can be seen from this example that this process (known as iteration) can be very time-consuming if carried out manually. However, if a computerized spreadsheet is set up, it is very easy to examine the effects of changes in the plan instantly.

In any manufacturing company there are numerous applications of this planning technique, all of which can benefit from the use of computer systems. Table 2.1 shows how some of these can operate.

In all these examples, the same basic techniques are applied, i.e. objectives are defined and plans developed to meet these objectives by varying particular elements of the plan. Without computers these planning applications would be virtually impossible, due to the need to carry out the process quickly and accurately.

Organizing

In chapter 1 the organizing activity was defined in terms of the provision of a framework that would allow plans to be carried out effectively. This framework should include:

- trained staff
- facilities and equipment
- information.

MARKETING OBJECTIVE The 1994 marketing objective is to increase sales by 10% over the 1993 value (from 2.40m to 2.64m) with a gross profit margin of 30% (609,230)

MARKETING PLAN # 1

PRODUCT	SELLING PRICE	SALES VOLUME	SALES VALUE	UNIT COST OF SALES	TOTAL COST OF SALES	PROFIT
	A	B	A × B	C	B × C	A × B − B × C
A214	3.45	45,000	153,000	2.85	128,250	24,750
B906	4.25	75,500	320,875	3.74	282,370	38,505
C103	6.40	52,000	332,800	5.25	273,000	59,800
	↓	↓	↓	↓	↓	↓
K632	1.95	125,000	243,750	1.35	168,750	75,000
ALL			2,450,000		2,014,950	435,050
OBJECTIVE			2,640,000		2,030,770	609,230

THIS PLAN FALLS SHORT OF THE OBJECTIVE BOTH IN TERMS OF SALES VALUE AND PROFIT

THUS: ... SALES VALUE IS ONLY 86% OF OBJECTIVE

... PROFIT IS ONLY 71% OF OBJECTIVE

MARKETING PLAN # 2

PRODUCT	SELLING PRICE	SALES VOLUME	SALES VALUE	UNIT COST OF SALES	TOTAL COST OF SALES	PROFIT
	A	B	A × B	C	B × C	A × B − B × C
A214	3.75	55,000	206,250	2.85	156,750	49,500
B906	4.25	75,500	320,875	3.74	282,370	38,505
C103	6.50	57,000	370,500	5.25	299,250	71,250
	↓	↓	↓	↓	↓	↓
K632	2.15	140,000	301,000	1.35	189,000	112,000
ALL			2,625,000		2,089,800	535,200
OBJECTIVE			2,640,000		2,030,770	609,230

THIS PLAN FALLS SHORT OF THE OBJECTIVE IN TERMS OF THE PROFIT

THUS: ... SALES VALUE IS 99% OF OBJECTIVE

... PROFIT IS ONLY 88% OF OBJECTIVE

Figure 2.1 Modifying a plan to meet an objective.

Table 2.1 Planning applications in a manufacturing company

The marketing plan	This plan would normally be prepared before the start of a financial year by defining sales targets to meet a set of objectives. It would then be monitored by comparing it with monthly 'actuals'. This is an ideal spreadsheet application on computer.
The production plan	This plan would also be prepared at the start of the financial year, and would give details of how production was to be scheduled (with planned costings) to meet the marketing plan. It should also be monitored monthly, on a spreadsheet.
The business plan	This plan combines the marketing and production plans from a financial viewpoint in order to determine such key factors as cash flow, commitments and profits. It should also be monitored monthly on a spreadsheet.
Production planning	Production should be planned ahead in detail on a daily or weekly basis in order to ensure that customer demands are met. This plan will almost certainly be subject to frequent changes as demands alter and available capacity varies. Special computer software can be obtained commercially for this purpose which allows the user to test alternative plans on a 'what-if' basis.
Materials planning	Materials requirements planning (MRP I) is now a well-established technique using computer-based software. Material requirements are established to meet sales demands by predicting the shortfall between what will be required and what is likely to be available on the basis of stocks and orders already placed on suppliers.
Resource planning	Manufacturing resource planning (MRP II) is a development of MRP I whereby all resource requirements are calculated (labour, machines, materials, tools, etc.) against sales demands. Computer-based software is available for this type of planning.

Thus, management should ensure that the people who are to operate the planning functions are suitably trained in the principles of planning and in the use of the facilities and equipment provided.

Staff involved in planning should know the sources of the information they require and clearly understand how it is obtained. They should also be aware of the way in which their plans are used to manage the company's operations.

As previously stated, an effective planning function relies on computer software systems and these should therefore be suitable for the particular applications. Such systems rely completely on accurate information, and much of this is also computer-based. Management must therefore ensure that proper procedures are in place to ensure such accuracy.

As an example of this, materials requirements planning uses stock figures to predict potential shortages. Stocks should therefore be checked regularly to ensure that the physical quantities are equal to those on the computer, and any discrepancies should be investigated to determine the reason so that it can be eliminated. (Some typical reasons for such stock discrepancies are listed in Table 2.2.)

However, stock is only one of a number of areas where inaccurate data can cause plans to be adversely affected. Others include the following.

Table 2.2 Typical reasons for stock discrepancies

Paperwork	When stock transactions take place (e.g. issues, receipts, returns or transfers) it is usual for them to be authorized by a document (e.g. stock requisition, goods inwards note, returns note). These documents are then used to update the computer record. If such notes are mislaid, incorrectly or illegibly written or incorrectly entered, or if stock is moved without such documents, then the computer stock will be inaccurate.
Stock checks	When physical stocks are being checked, it is advisable to ensure that the checkers are familiar with the items concerned so that they are able to identify such items correctly (by computer code). The checkers should ensure that their stock-check sheets are legible and that all locations for an item have been checked.
Free stocks	In many companies stocks are allocated (reserved) against sales orders (finished goods) or works orders (raw materials and components). During a stock check, allocated stocks should not be counted as free stock and vice versa. It is also essential to ensure that all such allocations are linked to a specific order so that, if that order is amended or cancelled, the allocation can also be amended or reversed.
Units of measure	In many computer-based systems it is possible to set up different units of measure (UOMs) for purchasing and stocking. Thus, the UOM for buying a type of screw could be 'box of 100' whereas the stock UOM could be 'each'. If such UOMs are confused during a stock transaction or stock check, then stock figures will be inaccurate.

- *Bills of materials* These define the quantities of raw materials, components and sub-assemblies required to make one unit of a particular product. Inaccuracies will affect the material requirements plan.
- *Routeings* These define the production sequence to be followed when making a component, sub-assembly or assembly by specifying operations, machines, labour, tools and materials and set-up/operation times. Inaccuracies will affect the production and capacity requirements plans.
- *Master production schedules* These define what is planned for despatch during a series of future periods of time. Inaccuracies can affect production and materials requirements plans and may also lead to late deliveries or overstocking.

Staffing

The importance of having fully trained employees has already been stressed, but in areas where computers are being used such a policy is vital. Most people are aware of the term 'garbage in – garbage out', but they are not always aware of all the implications.

In fact, where computers are concerned, training is not enough; users must also be motivated by understanding the crucial nature of their own role. They must know where the information they use comes from, and how it is derived. They must also know what happens to the information they provide and be aware of its importance to the success of the company.

In the area of computer operation users should understand how errors can arise and, more importantly, what damage such errors can cause. They should also be able to identify errors because of their knowledge and experience (the 'that doesn't look right' factor).

Directing

In chapter 1, directing was defined as the provision of leadership whereby the courses of action to be followed in specific circumstances are clearly set out. However, individual initiative should not be stifled by oppressive interference from above, and employees should be encouraged to work together in teams to identify and analyse potential problem areas. This approach is particularly relevant in the field of quality control, where employees can often suggest method improvements that will improve quality performance.

In companies where computer systems are in operation, senior management relies almost totally on the data such systems can provide. In such cases a high standard of direction depends on a flow of accurate and timely information in both directions.

Controlling

The 'closed-loop' technique of control shown in Fig. 1.1 (chapter 1) is the basis on which a computer system should operate. Figure 2.2 shows how this technique has been developed to control production using a computer with two levels of test being applied. The first of these is to ensure that the plan is workable in terms of production capacity, materials and tooling; the second, after implementation, is to monitor it against the work-in-progress situation.

2.3 SELECTING THE RIGHT COMPUTER SUPPLIER

In order to ensure that a company gets the most out of its computer system, great care must be taken in the original selection of supplier. The procedures suggested for making this selection may seem laborious, but there is no doubt that they are worth while.

The techniques employed in this exercise, although being used in this case to select a computer system, can and should also be applied whenever expenditure is being considered on a capital project, e.g. new or replacement machines, a new factory or a new fleet of vehicles.

In the past, many companies have purchased systems that are not compatible with their type of industry and methods of working. This has led to frustration and a great deal of time being wasted or lost. It has also been very costly.

There are three factors to be considered in choosing a suitable supplier.

- *Hardware* The supplier should be able to supply a range of hardware options, types and configurations, not just one.
- *Software* The supplier should be able to supply a full range of software modules that can interface with one another to form an integrated system. It should also be possible to integrate other suppliers' software if necessary.
- *Service* The supplier should be able to offer a complete package of services including installation, training, helplines, hardware and software maintenance and software upgrades.

The sequence of events that should be followed in choosing the right supplier is not dissimilar to those used by any efficient purchasing department; this sequence should comprise the seven stages given in Table 2.3.

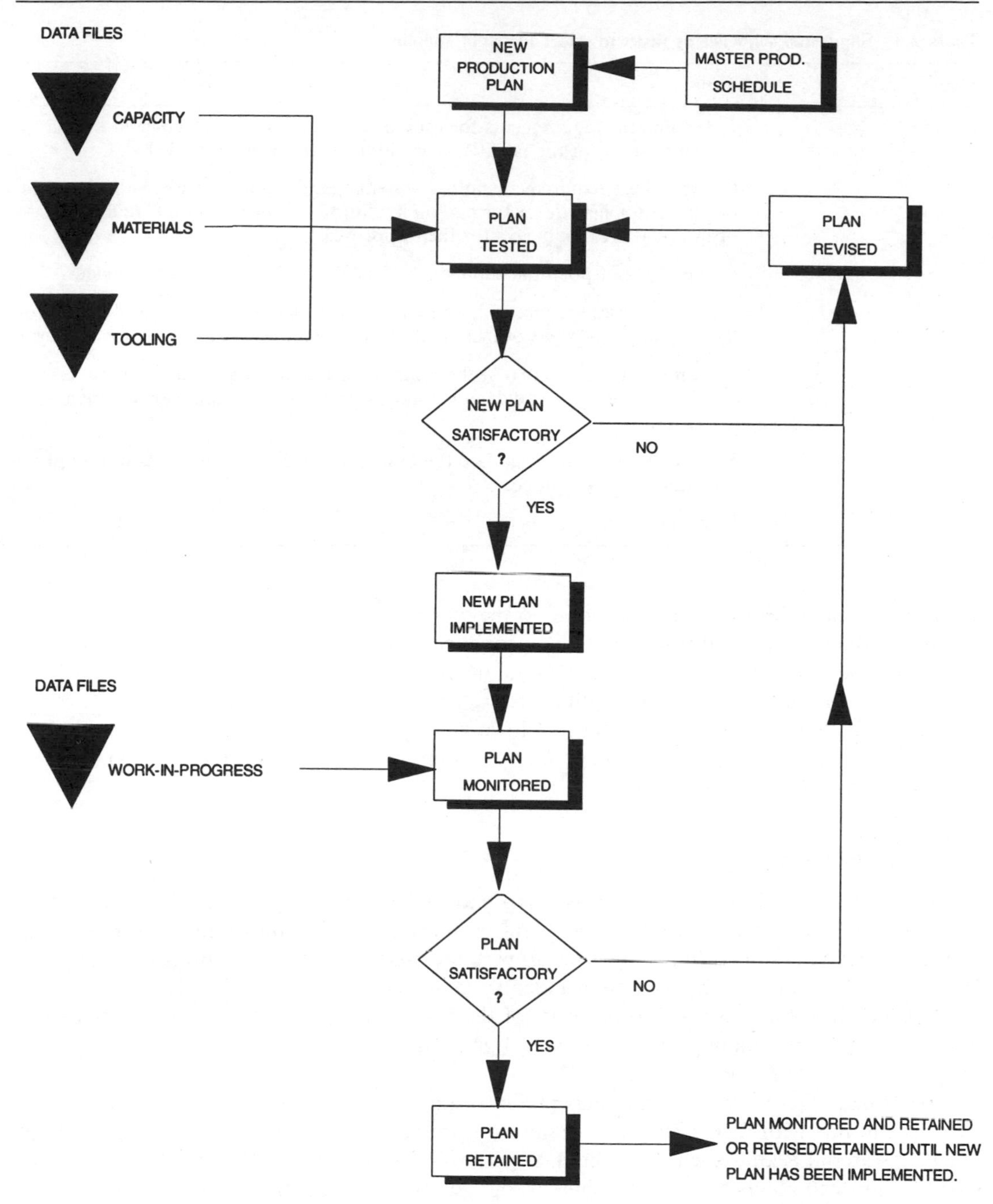

Figure 2.2 Controlling production by computer.

2.4 SPECIFYING THE RIGHT SYSTEM

In order to ensure that the system specification covers all requirements completely, it is necessary to examine seven specific areas in order to:

Table 2.3 Suggested sequence of tasks to select the right supplier

Stage	Task
1	Determine the requirements for each operational area to be covered by the system in terms of the facilities and modules that will be needed.
2	Convert these requirements into a system specification that can be sent out to potential suppliers in the form of an invitation to tender (ITT). Circulate this to operational area managers for their approval.
3	Obtain a list of potential suppliers and send out the invitations to tender.
4	When responses are received, assess each proposal against the specification for 'fit' (identify any areas of incompatibility or weakness) and value for money.
5	Prepare a short list of two or three suppliers and investigate further by means of third-party enquiries and demonstrations. Involve operational area managers in the latter.
6	Select a supplier and negotiate the best terms, including the costs of any programming work required.
7	Place the order.

1. select the modules that are appropriate to the company
2. specify the facilities and features required for each module
3. specify the required interactions between modules
4. specify the required interactions with other systems
5. specify the required size of the system, allowing for growth
6. specify any special requirements (e.g. long part numbers or hardware preferences)
7. specify any adverse conditions.

Select the modules

Most computer-based manufacturing systems are available in modular form, which means that each module can be purchased separately and in many cases can run on its own. Figure 2.3 shows a typical manufacturing system set up with the links between the various modules; Fig. 2.4 shows other modules that can be included in some systems.

It is important at this stage to be aware of the basic functions of the various modules and also the facilities normally provided. Table 2.4 gives this information in summary form; fuller details are given in subsequent chapters.

Some manufacturing companies may prefer to start with a basic system and then develop it by stages in line with their needs. For example, a production system can be operated without MRP I (Material Requirements Planning). Works orders would then be planned on the basis of stocks of finished goods or order schedules from customers, and purchase orders on the basis of stock replenishment policies.

Select the facilities and features

At this stage it is advisable to prepare a list of possible suppliers and to obtain literature giving details of their products, perhaps by visiting an exhibition. This literature will almost always give a list of the modules available and the features/facilities in each module. This information

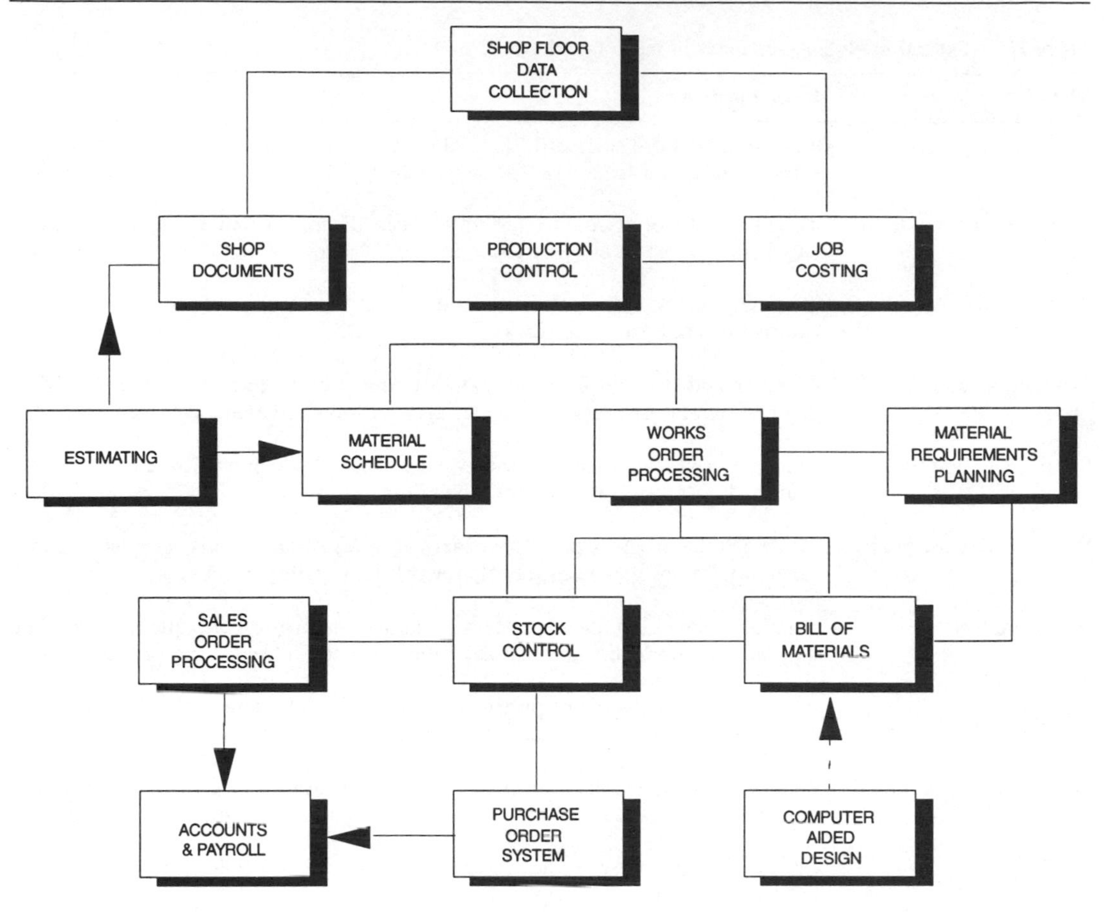

Figure 2.3 Typical computer system showing modules available (courtesy of Kewill Systems plc).

can then be checked against the list of modules already prepared and used to determine whether a supplier's system is likely to be suitable. The specification of the required facilities and features within a module is often difficult (what to put in and what to leave out). However, the following approach is generally satisfactory.

Facilities and features can be classified into three types:

1. those included by almost every supplier
2. those included by some suppliers, but not all
3. those that are specific to the company seeking to purchase a system.

There should be no need to specify type (1) facilities other than in general terms, for example, for the SOP module:

This module should contain all the standard features and facilities necessary to enter and amend sales orders, to allocate from stock against these orders, to print picking lists and despatch documents and to create sales invoices.

Table 2.4 Typical modules—summary of basic functions

Module	Basic functions
Sales order processing	To enter and store sales orders, to allocate stock, to despatch goods, to create sales invoices and to update the sales ledger
Purchase order system	To enter and store purchase orders, to record supplier deliveries, to validate suppliers' invoices and update the purchase ledger
Stock control	To hold details of all stock items and policies, to record details of all stock transactions and to value stocks
Production control	To enter and store works orders, on a planned basis, to meet due dates for completion and to check that materials and tools are available when needed
Shop documents	Program used to design and print works order documents—route card, operation cards, requisitions, identity labels, etc.
Works order processing	To record the progress of works orders in terms of operations completed, times taken, quantities produced, rejects/rework and quality checks
Bill of materials	A database with details of product structures in terms of the quantities of raw materials, components and sub-assemblies needed to make one product unit
Routeings	A database with details of production routes for all manufactured items with operation sequences and descriptions, times, tools and work locations
Work centres	A database with details of locations where operations are performed (machines, machine groups, etc.) and data on capacities, costs and hours worked
MPS	(master production schedule) Series of lists of what it is planned to make, each list covering a future time period
MRP I	(material requirements planning) Program that calculates what should be bought/made to meet the MPS allowing for free stocks and current orders
CRP	(capacity requirements planning) Program that calculates percentage loads in each work centre on the basis of production plans and work-in-progress.

The inclusion of type (2) facilities in the specification is recommended and the inclusion of type (3) is essential. In fact, the general rule should be 'If in doubt don't leave it out'. There are several good reasons for this policy:

1. to allow the potential purchaser to eliminate suppliers that are unable to meet the requirements at an early stage
2. to allow the potential purchaser to assess the real cost of the proposed system by adding the costs of any 'bespoking' that may be needed to meet the specified facilities and features
3. to force the potential purchaser to carry out a critical examination of its business in order to determine what its special requirements really are (are they really necessary or just traditional?)
4. to safeguard the potential purchaser in the event of a dispute between the supplier and purchaser if the system fails to meet the specified requirements.

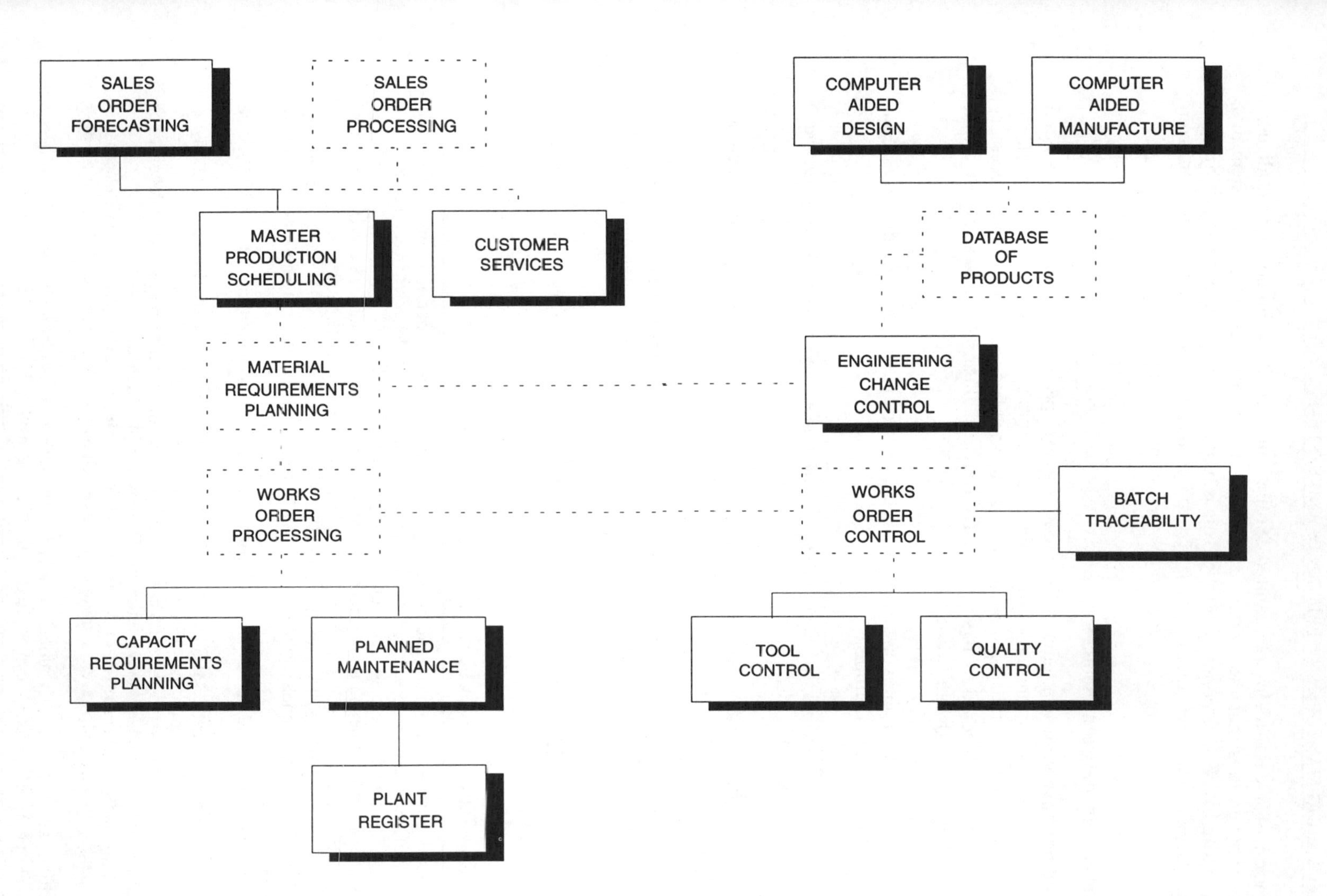

Figure 2.4 Typical additional software modules available.

The preparation of a detailed list of required features and facilities should involve all user departments. This ensures that these users have an input into the 'design' of the system and are therefore more likely to be committed to it. This is in fact a good management principle—the users should be made to feel that it is *their* system and has not been imposed on them from above.

Specify interactions between modules

Once the modules have been selected and their requirements specified in terms of features and facilities, it is necessary to consider the way in which they should work together. Once again, the problem here is to decide what needs to be included in the specification, since it is not necessary to include links that are standard to all suppliers. To overcome this problem it is suggested that links be specified at the following two levels.

1. A general statement that links are required between certain specified modules, without detailed specification, for example:

   The purchase and sales ledger modules should be linked to the nominal ledger module to allow automatic transfer of transaction data.

2. A more detailed statement which defines the programs that may be required to achieve specific objectives, for example:

   The SOP module should be linked to the stock control and sales ledger modules so that stock availability and credit status can be viewed in a window on the order entry screen and used to allocate stock or 'hold' an order for accounts clearance.

It is not possible to provide a complete list of such linkage statements, but some additional examples are listed in Table 2.5.

Table 2.5 Examples of specified linkages between modules

1. The estimating/quotation module should be linked to the SOP module so that, if an order is received against a quotation, it can be copied over automatically, given a sales order reference, amended if necessary, and this reference copied back automatically to the quotation.
2. The tool management module should be linked to the production control module so that when a works order is being planned it is possible to check tooling availability and indicate shortages.
3. The shop-floor data capture (SFDC) module should be linked to the works order processing, costing and payroll modules so that bookings are copied over to record works order progress, costs incurred and piecework earned.
4. The SOP and quotations modules should be linked to the master production schedule (MPS) module such that
   (a) orders are copied over to the MPS automatically unless they have been derived from quotations, in which case the quotation on the MPS must be replaced by the order
   (b) quotations are copied over to the MPS automatically and retained until replaced by a sales order or deleted automatically when the quotation 'valid until date' is reached.
5. The MRP I module should be linked to the MPS module in order to obtain data on scheduled production and purchases. However, it should only take details of orders from the MPS, not quotations.
6. The quotations and SOP modules should be linked to the sales forecasting module such that orders received and quotations prepared can be used to update forecasts.

Specify interactions with other systems

Most manufacturing systems include modules for financial accounting and payroll. If not, they should include links to systems that can perform these functions.

However, other systems may be required and details of these should be given by the specifier to potential suppliers. These details should include the make and version of the system (for example AutoCad version 12) and the functional links that need to be established. Table 2.6 gives examples of how such details might be written.

Other systems that might be covered in this way include computer-aided manufacture (CAM), computer-aided process planning (CAPP), Computer-aided engineering (CAE) and computer-based simulation.

Table 2.6 Examples of specified links with other systems

Other system	System name/version	Functional links
Computer-aided design	AutoCad v.12.0	To copy component drawing details to the parts master and bill of materials files, to enquire on these files and to initiate engineering change procedures
SPC (quality control)	Brankamp v.5.03	To copy process and inspection specifications from the routeings file and data from stock and work-in-progress file for traceability
Planned maintenance	Matrix Impact v.5.0	To share codes and details of plant with asset register, to copy purchase requisitions to POP module and to link with costing module.
Customer liaison	Wang Quest	To share customer and order details with SOP module and product details with parts master files.
Spreadsheet	Excel for Windows	To allow data from any file in the manufacturing system to be copied into Excel on a system-linked PC for processing.
Word processing	WordPerfect	To allow data to be taken from the relevant manufacturing files for processing quotations, mail shots, price lists, certificates of conformance, etc.

Specifying the system size

It is not necessary for the specifier to understand or use computer jargon when preparing a system specification, in fact such terminology should be avoided. The objective should be to define what the system should be able to do in terms of average quantities. Two categories are involved.

1. *Storage volumes* This will be a measure of the amount of data that will have to be stored on the system in terms of the number of records on each file. For example, how many products are made? How many customers are there? How many production routes? How many completed sales orders in the history file? It is not necessary for the specifier to quantify the size of a record in bytes — the supplier knows that already.

2. *Process volumes* This is a measure of the number of transactions that will have to be carried out by the system in a stated period of time. For example, how many purchase orders are placed in a month? How many sales orders? How many issues from stock? How many shop floor bookings?

Figure 2.5 shows a typical sizing document, but if there are any special features in the records or transactions these should be included as part of the document, e.g. large numbers of items on an order or operations in a routeing sequence.

From this information a supplier should be able to determine the system requirements in terms of disk size and processing speed. However, to assist it further, some basic details of the company should also be given as an introduction to the specification or in the invitation to tender (ITT) letter, e.g. products, employees, turnover and process types.

The question of an allowance for extra system capacity should also be dealt with in this part of the specification. This should be considered at three levels.

- *Short term* Allowance should be made for 'freak' periods of activity where volumes could be double the average or even more. This should be stated by quoting examples of maximum and minimum values where appropriate.
- *Medium term* Allowance should also be made for normal growth in the size of the business over the next two to three years. This should be stated as an annual percentage.
- *Long term* Allowance should not be made for long-term growth (up to seven years), but a statement should be included that will ensure that the system can be expanded (upgraded) on site without the need to alter files or operating procedures.

Specify special requirements

It is particularly important to include in the specification any requirements that may not be catered for in a standard 'off-the shelf' system, since the costs of such requirements may be prohibitive or impossible to achieve. Before such requirements are specified it is usually advisable to look at them objectively. Are they really essential or just fads? Is there another way? These requirements can be classified as company requirements, software requirements and hardware requirements.

1. *Company requirements* These can include such factors as long part numbers (over 16 characters) or descriptions containing characters not normally included in the standard set (e.g. foreign words). Other possibilities could include very long descriptions, very complex products and more than three suppliers for a purchased item.
2. *Software requirements* These can include such factors as special search or browse facilities (for example search on a partial code), complex reporting facilities where several files need to be opened and special requirements in terms of security levels, 'windows' facilities, built-in calculations or report-writing software.
3. *Hardware requirements* These can include such factors as a particular make of equipment, a network system of PCs, a mix of PCs and terminals, a series of networks and remote terminals or printers with modems or multiplexers.

Consideration should also be given to screens, printers and keyboards. Are any high-definition or colour screens required? Is there a need for centralized system printing or can all printers be

DEPT.	ITEM	QTY.OF ITEM PER MONTH	No. OF RECORDS TO BE HELD
SALES	SALES ORDERS PER MONTH	450	---
	AVERAGE ITEMS PER ORDER	---	3 (MAX 50)
	OPEN SALES ORDERS	---	1700
	INVOICES PER MONTH	1000	---
	AVERAGE ITEMS PER INVOICE	---	2 (MAX 10)
	MONTHS OF INVOICE HISTORY	---	12 ROLLING
STORES	TRANSACTIONS PER MONTH	6500	---
	NUMBER OF STOCK ITEMS	---	8000
	AVGE. SUPPLIERS PER ITEM	---	1.3 (MAX 6)
	MONTHS OF TRANS. HISTORY	---	12 ROLLING
PLANNING	NO. OF PRODUCTION ROUTES	---	8500
	NO. OF WORK CENTRES	---	85
	AVGE. OPERATIONS PER ROUTE	---	6 (MAX 15)
	OPEN WORKS ORDERS	---	1200
	WKS.ORDERS CREATED/MONTH	350	---
DESIGN	NO. OF B.O.Ms ON FILE	---	10,500
	AVGE. NO. OF ITEMS/B.O.M	---	8 (MAX 40)
COSTING	NO. OF LABOUR RATES HELD	---	15
	NO. OF MAT'L. RATES/ITEM	---	2.2 (AVGE)
	NO. OF COST CENTRES	---	10
	NO. OF COSTS HELD ON FILE	---	15,000
BUYING	PURCH. ORDERS PER MONTH	320	---
	AVERAGE ITEMS PER ORDER	---	2.5 (MAX 10)
	OPEN PURCHASE ORDERS	---	900
	MONTHS OF PURCH. HISTORY	---	12 ROLLING
ACCOUNTS (PURCHASE)	NO. OF SUPPLIERS	---	300
	PURCH. INVOICES PER MONTH	450	---
	PAYMENTS PER MONTH	300	---
ACCOUNTS (SALES)	NO. OF CUSTOMERS	---	110
	SALES INVOICES PER MONTH	see sales	---
	PAYMENTS PER MONTH	400	---
ACCOUNTS (NOMINAL)	NO. OF ACCOUNTS	---	450
	TRANSACTIONS PER MONTH	3750	---
	NO. OF COMPANIES	---	3
	NO. OF FIXED ASSETS	---	650
	MONTHS OF TRANS. HISTORY	---	12 ROLLING
QUOTES	ESTIMATES PER MONTH	30	---
	QUOTATIONS PER MONTH	200	---
	MONTHS OF HISTORY	---	12 ROLLING
	AVGE. ITEMS PER ESTIMATE	---	2.4 (MAX 12)
	AVGE. ITEMS PER QUOTE	---	3.2 (MAX 9)

Figure 2.5 Typical system sizing document.

local? What type(s) of printer? Are special keyboards required (large keys, dust/moisture protection)? What about barcode readers?

Specify adverse conditions

Most PCs and terminals can cope with normal office or factory conditions, but if there is any doubt regarding adverse conditions it is safer to check with the supplier before an order is placed. Factors that fall into this category include dust, dirt, corrosive fumes, vibration, high temperatures and moisture. It is also advisable to inform the potential suppliers of special conditions such as clean room areas, high magnetic fields and the likelihood of power surges or failures.

2.5 THE TENDER AND PROPOSAL PHASES

Preparing the invitation to tender (ITT)

Once the specification has been prepared and agreed, it can be converted into an ITT for sending to potential suppliers. Often, all that is required is to prepare a letter that can accompany the specification. This letter should include the following.

1. Background information on the company (if not included in the specification), e.g. products (a brochure or catalogue?), summary of previous year's accounts, processes used, organization structure, number of employees, major markets.
2. Details of any existing systems or computer equipment that is to be retained and integrated into the new system.
3. Any conditions that may be required for proposal submissions, e.g.
   (a) your proposal should be received not later than [date] and be in triplicate
   (b) your proposal should list any requirement in the specification that cannot be met by your standard software with suggested means of overcoming such problems
   (c) your proposal should include contact names at reference sites who can be approached for an independent view of your system
   (d) your proposal should include fully itemized prices for all software, hardware and services (maintenance, installation, training, etc.) and suggest possible arrangements for an on-site demonstration.

Selecting suitable suppliers

The task of selecting suitable suppliers to whom the ITT can be sent is sometimes difficult, but can often be achieved by reference to one of the trade directories or by examination of some of the regular computer publications which (from time to time) contain tables with details of suppliers of manufacturing systems. It may also be possible to obtain advice from a trade association or employers' federation.

If possible no fewer than four and not more than eight suppliers should be initially selected for inclusion in the list of those invited to tender. After evaluation of the proposals submitted this should be reduced to a short list of two or three.

Evaluating the proposals

It is generally advisable to approach this task in a planned and formal manner, although this can be somewhat time-consuming.

The objective of such an evaluation should be to obtain a 'value for money' score for each proposal so that the best of these can be shortlisted for further investigation (demonstrations and discussions with other users).

To carry out such an evaluation it is necessary to prepare — in advance — three documents. These are:

- *specification* To evaluate 'fit' between specification and proposal
- *costs* To list proposal costs
- *summary* To summarize proposals in value for money terms.

Specification
A typical format is shown in Fig. 2.6 — one per proposal. In this example, points are to be given for the various features and facilities listed in the specification in terms of the degree with which these are met in the proposal.

For example, an important facility could carry a maximum of five points; points might then be awarded as follows:

specification met in full	5
specification met partially	3
specification poorly met	1

On completion, all points would be added together and compared with maximum total available to give the percentage fit factor.

Costs
A typical format is shown in Fig. 2.7 (one to be used per proposal). Cost details are entered from the proposal and added to give software, hardware and other costs. All costs are classified as 'one-off' (hardware purchase, installation, etc.) or 'annual' (maintenance, software licences, etc.). Points from the specification sheet can also be entered for reference purposes.

Summary
A typical format is shown in Fig. 2.8; this summarizes the costs and fit factors for all proposals with space for entry of key specifications from each proposal. The VFM (value for money) factor is calculated as shown in Table 2.7 and shows that

- the proposal from supplier C offers the best value for money
- the weighted average VFM factor for the five suppliers is 0.65
- the budget is probably too optimistic and/or the specification too demanding.

2.6 JUSTIFYING THE PROJECT

Once a system supplier has been selected, the details of the expenditure and facilities available will be known. It will therefore be possible to prepare a project justification report in order to obtain approval from the Board. This report should contain full details of the costs and benefits

SUPPLIER		SOFTWARE NAME		PAGE 1

MODULE & FACTORS	POINTS AVAILABLE	POINTS AWARDED
1. DATABASE MANAGEMENT SYSTEM		
Allow for on-line entry & maintenance of:		
- Item master records	3	
- Bills of materials (product structures)	5	
- Work centre information	1	
- Manufacturing routings (Operations)	3	
Provide product costing functions:		
- Level by level costs roll-up	3	
- Simulation of cost changes	3	
Reporting facilities	1	
Interface with MRP	5	
DATABASE MANAGEMENT TOTALS	24	
2. MATERIAL REQUIREMENTS PLANNING		
Support net change/regenerative MRP	3	
Requirem'ts to be manual/machine generated	3	
Produce planned orders (Mfg./Purch)	5	
System Parameter range	3	
Interface with IM. POP. SOP. & CRP.	5	
Provide Kitting Instructions	5	
MRP I TOTALS	24	
3. INVENTORY MANAGEMENT (STOCK CONTROL)		
PAGE 1 - CARRIED FORWARD	62	

Figure 2.6 Typical proposal evaluation sheet format.

of the project, and these can best be presented by using one of the three recognized investment analysis techniques:

- payback
- rate of return on investment (ROI)
- discounted cash flow (DCF).

There may be two objectives of applying such analysis techniques. The first is to check that the projected investment will be worth while; the second is to compare various possible alternative projects. In the first case, many companies specify a minimum acceptable level (e.g. 20 per cent per annum) and in the second case, this is the only fair method of comparison.

Table 2.7 Calculation of value-for-money (VFM) factors

The VFM factor shows how a proposal compares with others and is calculated by comparing each proposal with a target. This target is calculated by assuming a software percentage fit factor of 100% on a budgeted five year cost (£70 000) as shown.

Supplier	One-off cost	Annual cost x 5	Total 5 year cost	Software fit, %	VFM factor
A	50 000	35 000	85 000	70	0.58
B	45 000	32 000	77 000	72	0.65
C	65 000	25 000	90 000	92	0.71
D	60 000	25 000	85 000	83	0.69
E	70 000	30 000	100 000	92	0.58
Target	45 000	25 000	70 000	100	1.00

The VFM factor for the target is calculated thus: (5 year cost)/(percentage fit) for target = 700. This is then assumed to be 1.00. The same formula is then used for the five suppliers and compared with the target ratio to give the other VFM factors, for example: for supplier B, 77 000/72 = 1069. Thus VFM factor is 700/1069 = 0.65.

The payback method

This is the method most commonly used, not because it is the most accurate, but because it is easy to apply and understand. It is based on answering the question 'How long before I get my money back?', and four sets of figures are normally used:

- capital expenditure (costs of purchase and installation)
- capital receipts (sales value of surplus equipment)
- revenue expenditure (annual costs of operating the new installation)
- revenue receipts (savings in annual operating costs).

This method is useful for quick comparisons between projects, but is inaccurate because it does not take account of such factors as depreciation, taxation and inflation. An example of the use of this method is given in Table 2.8.

The rate of ROI

This is probably the method most easily understood. It can allow for depreciation and gives the rate of return as an annual percentage; thus, a comparison can be made with any other type of investment, for example money invested in a building society savings account.

As previously stated, many companies use this method to set a standard that any proposed project must attain—for instance a minimum return of 20 per cent. An example of the use of this method is given in Table 2.9.

SUPPLIER:

ITEM	SPECIFICATIONS		COSTS		POINTS
	ORIGINAL	SUPPLIER	ONE-OFF	ANNUAL	VALUATION
FILE SERVER					
RAM (Mb)					
DISC (Mb)					
BACK-UP					
TERMINALS	12 Off				
PCs	2 x '386'				
DOT MAT.PRINTER	4 x 200cps				
etc.	etc.				
DELIVERY					
INSTALLATION					
TRAINING					
SUPPORT					
OPERATING SOFTWARE					
etc. etc.					
SOFTWARE MODULES					
ESTIMATING					
SALES ORDERS/INVOICING					
SALES ANALYSIS					
MAT'L REQ'MENTS PLANNING					
INVENTORY MANAGEMENT					
etc. etc.					
TOTAL HW.COSTS & POINTS					
TOTAL SW.COSTS & POINTS					
TOTAL OTHER COSTS					
GRAND TOTALS					

Figure 2.7 Typical tender summary sheet.

SUMMARY OF ALL TENDERS					
ITEM NB. COSTS ARE IN 'OOO	SUPPLIER A	SUPPLIER B	SUPPLIER C	SUPPLIER D	SUPPLIER E
INITIAL COSTS SUMMARY TOTAL INITIAL COSTS HARDWARE ONLY SOFTWARE ONLY OTHER COSTS					
ANNUAL COSTS SUMMARY TOTAL ANNUAL COSTS HARDWARE ONLY SOFTWARE ONLY OTHER COSTS					
POINTS & FIT (320 MAX) HARDWARE POINTS APPLICATION. S'WARE POINTS TOTAL POINTS PERCENTAGE FIT					
SYSTEM SPECIFICATIONS CPU. RAM (Mb) DISC (Mb) BACKUP DEVICE OPERATING SYSTEM NETWORKING SYSTEM					
VALUE FOR MONEY TOTAL INITIAL COSTS TOTAL ANNUAL COSTS x 5 TOTAL 5 YEAR COST PERCENTAGE FIT VFM FACTOR (VFM TARGET = 850)					

Figure 2.8 Typical summary of all tenders.

Table 2.8 Use of the payback method of investment analysis

Cost/revenue type	Cost/revenue item	£k	Total £k
Capital costs	Hardware	25	
	Software	15	
	Installation	3	
	Training	5	
	Others	3	51
Capital revenue	Hardware sold	5	
	Other items sold	3	8
Annual costs	Hardware maintenance	3	
	Software licences	2	
	Software maintenance	3	
	Other costs	4	12
Annual savings	Staff reductions	12	
	Increased output	8	
	Reduced stocks	12	32

Payback display

Year	Outflows	Inflows	Balance
0	(51)	8	(43)
1	(12)	32	(23)
2	(12)	32	(3)
3	(12)	32	17

Thus payback occurs early in year 3. Alternatively the payback formula can be used to calculate when payback occurs.

$$\frac{\text{Capital costs} - \text{capital revenue}}{\text{Annual savings} - \text{annual costs}} = \frac{51 - 8}{32 - 12} = 2.15^{*}$$

* i.e. early in year 3

Table 2.9 Use of the ROI method of investment analysis
In this example the same figures have been used as in Table 2.8, i.e. capital costs = £51 000, annual costs = £12000, capital revenue = £8000, annual savings = £32 000.

1. The net capital cost at the start of the project is 51 000 − 8000 = £43 000

2. The average annual profit over the life of the project, allowing for depreciation (but excluding taxation) can be calculated as:

   (Annual savings − Annual costs) × depreciation factor *
   (32 000 − 12 000) × 0.6 = £12 000

3. Thus, the rate of return is (12 000 × 100)/43 000 = 27.9%

*This factor depends on the life of the project in years and a percentage rate of depreciation, and can be otained from a standard depreciation chart.

The DCF method

This is probably the most accurate method, since it is designed to calculate the true return on the investment as an equivalent annual investment rate, based on the values and timing of the cash flows. It can therefore take account of both depreciation and taxation over the life of an asset and allow for the residual asset value to be taken into account if it can be sold at the end of its life. An example of the use of this method is shown in Fig. 2.9.

As with many exercises of this sort, this is an iterative process, i.e. various values are tested until a suitable value is found. In this example, the value being varied is the discount rate. The first test was at 20 per cent and gave a present value total of +£13 410; this rate is too low. Sixty per cent was then tested and gave a value of −£8540; this rate is too high. The third test was therefore on 40 per cent which gave a value of −£70. This is close enough to zero to be accepted.

Companies that use this method may also set a standard that any proposed project must attain, e.g. a minimum DCF rate of 25 per cent.

Computerizing the investment analysis process

These three methods can be time-consuming if carried out manually where there is a large number of cost and savings elements. However, low-cost PC-based software is available (for example IVAN from Applied Technology) that not only simplifies the process but also allows easy application of alternative scenarios (what ifs), for example changes in asset life or rate of depreciation.

2.7 USING STANDARD PACKAGES IN MANUFACTURING

Three standard computer-based packages can be used to assist in the management of manufacture. These are

1. spreadsheet packages
2. database packages
3. project management packages.

Spreadsheet packages

A spreadsheet is the computer equivalent of a very large sheet of blank paper ruled into vertical columns and horizontal rows, each rectangle thus formed being known as a cell. Each cell has a unique name formed by its position, thus a cell in column C and row 10 is called C10. The width of a column can be adjusted to suit the application.

A cell can hold data (text or a number) and a formula. Thus it is possible to set up a spreadsheet that will display information and calculated values for many applications. The example in Fig. 2.10 has been prepared for stock control and valuation purposes in a small company, and the way in which the entries have been set up is given in Table 2.10.

Most spreadsheet packages include facilities for display of data in graphical form and allow the user to manipulate data by copying and sorting. For instance, in the example, it would be possible to prepare a new sheet each week by copying and adjusting the previous one or to sort the parts into a sequence that put the highest 'total value' item into row 3 and the lowest into row 254 or vice versa.

DISCOUNTED CASH FLOW (DCF) METHOD

GIVEN THE FOLLOWING DATA: £ K

... Capital Expenditure (Year 0) = 40

... Profit before Depreciation and Tax = 24

... Life of Investment = 3 Years

... Residual Value at end Year 3 = 5

YEAR	INVESTMENT	PROFIT	NET CASH FLOW	TEST DISCOUNTED @			PRESENT VALUE @		
				20% p.a.	60% p.a.	40% p.a.	20% p.a.	60% p.a.	40% p.a.
	£ K	£ K	£ K				£ K	£ K	£ K
A	B	C	D	E	F	G	H	I	J
0	−40	NIL	−40	1.00	1.00	1.00	−40	−40	40
1		+24	+24	0.833	0.625	0.714	+19.99	+15	+17.13
2		+24	+24	0.694	0.391	0.510	+16.66	+9.38	+12.24
3	+5	+24	+29	0.578	0.244	0.364	+16.76	+7.08	+10.56
TOTAL	−35	+72	+37				+13.41	−8.54	−0.07

NOTES: (the letters refer to the columns in the table)

(B) Original Investment in Year 0 = 40,000; after 3 years asset sold for 5,000

(C) The Profit on the Investment is 24,000 per Year, after first year

(D) The Net Cash Flow is the Difference Between Columns B and C.

(E to G) The Discount Factors For 20%, 60% and 40% p.a.

(H) Column D Multiplied by Column E.

(I) Column D Multiplied by Column F.

(J) Column D Multiplied by Column G.

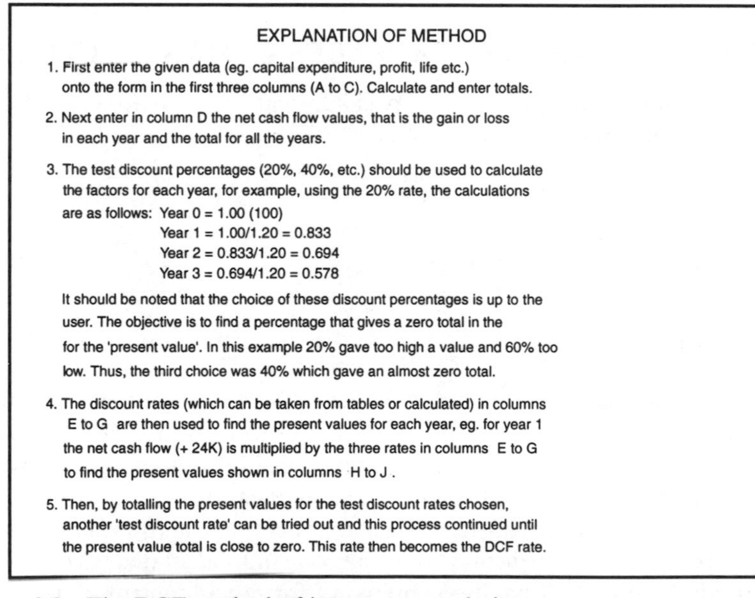

EXPLANATION OF METHOD

1. First enter the given data (eg. capital expenditure, profit, life etc.) onto the form in the first three columns (A to C). Calculate and enter totals.
2. Next enter in column D the net cash flow values, that is the gain or loss in each year and the total for all the years.
3. The test discount percentages (20%, 40%, etc.) should be used to calculate the factors for each year, for example, using the 20% rate, the calculations are as follows: Year 0 = 1.00 (100)
   Year 1 = 1.00/1.20 = 0.833
   Year 2 = 0.833/1.20 = 0.694
   Year 3 = 0.694/1.20 = 0.578

   It should be noted that the choice of these discount percentages is up to the user. The objective is to find a percentage that gives a zero total in the for the 'present value'. In this example 20% gave too high a value and 60% too low. Thus, the third choice was 40% which gave an almost zero total.
4. The discount rates (which can be taken from tables or calculated) in columns E to G are then used to find the present values for each year, eg. for year 1 the net cash flow (+ 24K) is multiplied by the three rates in columns E to G to find the present values shown in columns H to J .
5. Then, by totalling the present values for the test discount rates chosen, another 'test discount rate' can be tried out and this process continued until the present value total is close to zero. This rate then becomes the DCF rate.

Figure 2.9 The DCF method of investment analysis.

A spreadsheet can often be linked to a database package (see 'Database packages' below) or to a word processor so that reports, with graphics, can be easily prepared. There are many possible manufacturing applications for the spreadsheet, e.g. cost estimating, costing, financial accounts, capacity planning, sales analysis, stock recording and production recording.

COLUMNS

ROWS	A	B	C	D	E	F	G	H	I	J
1	PART NO.	DESCRIPTION	PREVIOUS QTY	UOM	RECEIPTS	ISSUES	CURRENT QTY	UNIT COST	TOTAL VALUE	COMMENTS
2	==	==	==	==	==	==	==	==	==	==
3	10010	75mm DIA EN3 ROD	50	METRE		4	46	7.50	345.00	
4	10025	90mm DIA EN3 ROD	40	METRE		3	37	8.60	318.20	
5	10030	R/125 ROLLER BEARING	80	EACH			80	4.25	340.00	
6	10040	M10 UNF STUD x 25	10	BOX x 20		1	9	2.50	22.50	
7	10045	M20 UNF NUT	4	BOX x 100	4	1	7	1.50	10.50	
8	10050	25MM WASHER	2	BOX x 100	6		8	0.75	6.00	
9	10060	1OMM PIN X 25	3	BOX x 200			3	1.25	3.75	
254	41070	P124 CASTING	24	EACH		2	22	15.55	342.10	
255		WEEK 12 - 1993						TOTAL	57,558.00	
256										

NOTES: 1. FOR DETAILS OF CALCULATIONS SEE TABLE 2.10

2. IF MORE THAN ONE ISSUE OR RECEIPT IN A WEEK ADD NEW FIGURE TO EXISTING

Figure 2.10 Weekly stock control using a spreadsheet.

Table 2.10 Calculations and entries for spreadsheet in Fig. 2.10

Text entries	Rows 1 and 2	Column headings with underlining
	Columns A and B	Part numbers and descriptions
	Column D	Units of measure
	Column J	Comments
	Cells B255 and H255	Week no./year and total
Numeric entries	Column C	Previous quantity
	Column E	Receipt quantities
	Column F	Issue quantities
	Column H	Unit cost
Calculations	Column G	(C3 + E3 – F3) . . . (C254 + E254 – F254)
	Column I	(G3*H3) . . . (G254*H254)
	Cell I255	SUM (I3:I254)

Database packages

A database is basically a very large filing cabinet containing numerous files that can be viewed or processed as required. For example, it could contain a file of saleable product records, each record containing a number of fields of data as shown in Table 2.11. This table gives the record structure in terms of the fields included, the field types and the field sizes. These terms can be defined as follows.

- *File* A file contains a group of records, all in the same format, for example a customer file would contain the records for a number of customers.
- *Record* A record contains a number of fields, each field containing a piece of information, for example, a customer's record could include fields for name, phone number, address, business type and turnover (year-to-date or previous year).
- *Field* Each field is set up to contain a specific type of information or data (see field types below) with a unique field name (e.g. CUSTNAME). Its type depends on the data it is to contain.

Table 2.11 Record and fields in a database file (saleable parts file record)

Field	Field name	Type	Size
1	Product code	Character	12
2	Description	Character	40
3	Drawing no.	Character	10
4	Selling price	Numeric	6
5	Quantity sold last year	Numeric	5
6	Unit cost	Numeric	6
7	Planned sales quantity	Numeric	5
8	Sales quantity YTD	Numeric	5
9	Date of last sale	Date	8
10	Stock item (Y/N)	Logical	1
11	Specification	Memo	*
12	Market (H/E/O)	Character	1
	etc.	etc.	

* See below for explanation of memo-type fields

There are five field types in this example, as follows.

- *Character* Used to store words, for example names, locations, or numbers that are treated as text, e.g. part numbers.
- *Numeric* Used to store numbers, with or without decimals, and signs (+ or −). These fields can be used for calculations.
- *Date* Used to store calendar dates. These can be used to calculate time differences in days, for example, between two dates or between a field date and today's date.
- *Logical* Used to record whether a statement is true or false, for example, 'This is a stock item'—Yes or No.
- *Memo* Used to record free-form text. The advantage of such fields is that they can be very large but occupy only the space of the text entered (unlike character fields which are of limited size and occupy the space allocated even if it is not all used). The disadvantage is that memo fields cannot be searched or indexed.

Records on a database file can be searched to find a particular record or a group of records with a common factor, for example, all home sale products (H) in field 12. Indexes can also be set up so that by calling up a particular index the records can be seen in a sequence set by a given factor or a combination of factors.

For instance, a 'description' index would display records in an alphabetic sequence of descriptions instead of in part number sequence. It is also possible to copy a record or a file which can then be amended as required and to specify calculations on numeric fields.

Like the spreadsheet, a database package has many uses. For instance, by setting up a series of files, a workable sales invoicing system can be developed as shown in Fig. 2.11. The same principles could be applied to a production system with files containing details of parts, materials, processes and machines being used to set up works order file records and to print works order documents. Database packages usually include very full facilities for setting up reports which can then be viewed or printed and, as previously suggested, can be linked to a spreadsheet so that, for example, a database file of prices can be used for spreadsheet estimating.

Project management packages

Project management can be defined as a technique for planning and managing a project, and a project can be defined as a series of tasks and activities which start at a specific point in time and terminate at a later point. Some manufacturing companies use project management as a form of production management, usually because they are in the 'make to order' business where each order can be defined as a project.

There are numerous packages available, but all operate on the same set of basic principles which can be summarized by considering the work sequences that should be carried out for a successful project, as shown in Table 2.12. As with all planning functions, iteration is necessary to ensure that the plan achieves its target objectives, and most packages include facilities for this. The plan can be displayed on the computer screen either as a Gantt chart or as a CPM (Critical Path Method) network. Examples of both types are shown in Fig. 2.12.

The following points about project management procedures should be noted.

- *Criticality* When the details of all tasks have been set up on the computer, a number will be classed as critical and highlighted accordingly. This means that these tasks have no margin of error in terms of time—if they overrun then subsequent critical tasks will also overrun.

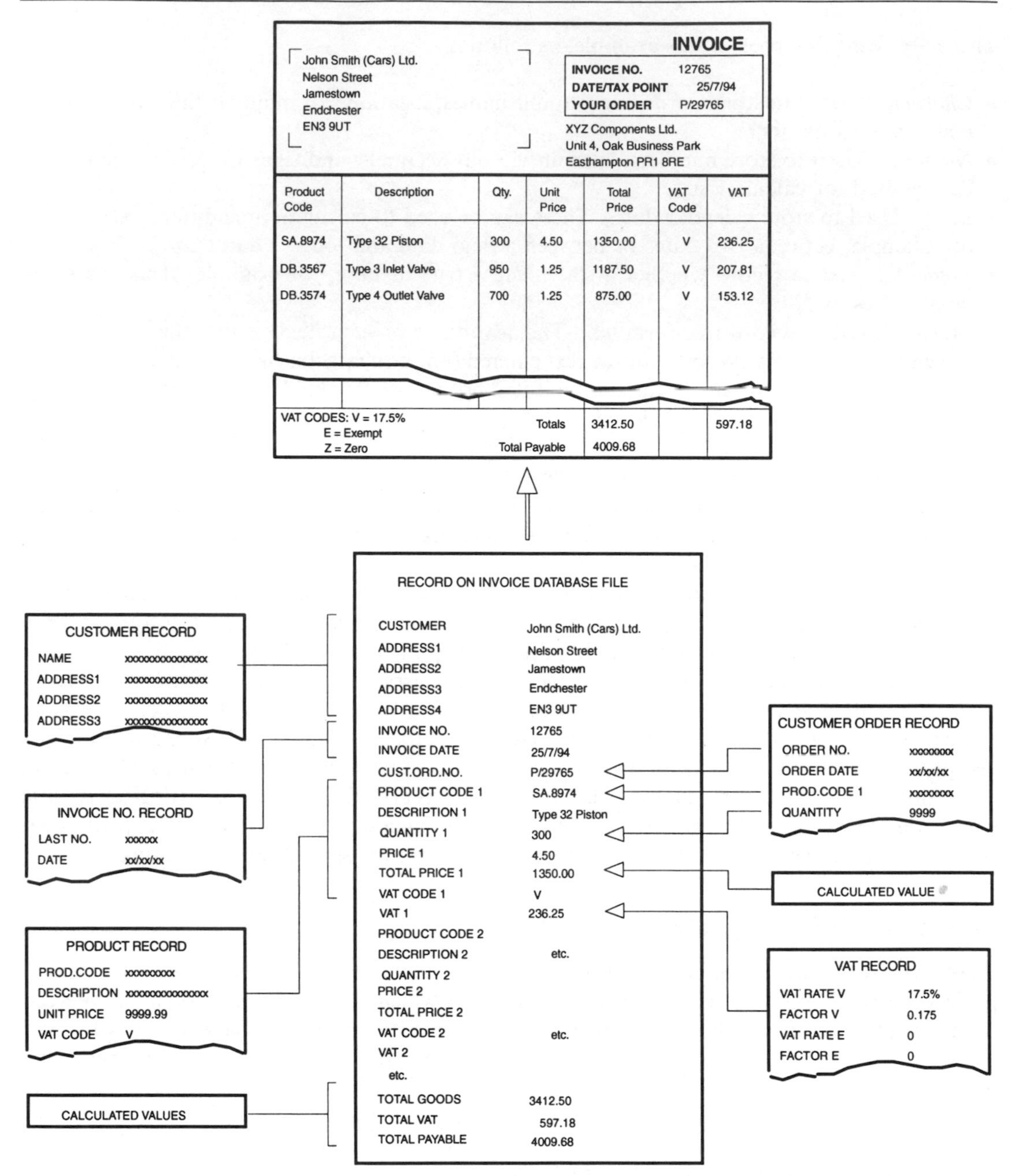

Figure 2.11 Example of use of database package.

However, non-critical tasks are said to have a degree of 'float'. For example, a task with a planned duration of 10 days does not have to be completed until 14 days have elapsed; it therefore has 4 days of float and could start 4 days late (or finish 4 days late). The sequence containing these critical tasks is known as the critical path.

- *Resources* Resources can be people, machines or services. They should be defined in terms of their cost per time unit and their availability, for example 15 hours per week. The computer

Table 2.12 Work sequence for a project management operation

Phase	Job	Job detail
Planning	Define tasks	List all tasks that must be carried out
	Estimate durations	Determine time that task should take
	Allocate resources	Resources available to carry out tasks
	Define sequences	Sequence in which tasks should be performed
	Set up plan	Enter details to computer system
	Test plan	Test to determine if resources overloaded or due date exceeded
	Adjust plan	Adjust if test fails by adding resources or rescheduling tasks
	Cost plan	Check that costs are within agreed limits
	Adjust plan	Adjust if costs exceed agreed limits or report level of excess
	Iterate	Continue to test/check/adjust until a satisfactory plan is evolved
Control	Monitor progress	Based on reports enter progress and note any 'slippage'
	Monitor costs	Based on job bookings and invoices, calculate costs to date
	Report progress	Report to superiors any slippages in time or cost
	Adjust plan	Adjust plan if slippage has occurred and note effects
	Complete project	Continue to monitor/report/adjust until project is completed

will add all demands on a resource in a given time period and report on overloading. Such overloads can sometimes be eliminated by 'profiling', i.e. by allocating the resource in a different way as shown in Fig. 2.13.

- *Progressing* (or tracking) A computerized project management system will allow a user to record progress on a task by entering a percentage completion at regular intervals. This allows the user to identify where an overrun or overspend is likely. An example is given in Table 2.13. This shows that, by comparing what has been booked with what has been planned, theoretical completion percentages for time and cost can be calculated. These can then be compared with the actual completion percentage estimates to show whether a task is running to plan or not.

Table 2.13 Identifying problems from progress reports

Task code	Time planned	Cost planned	Time booked	Cost booked	Percentage completed	Time ratio	Cost ratio	Comments
A100	24	£480	15	£320	50%	62%	67%	Overrunning on time and cost
A110	60	£1800	15	£600	25%	25%	33%	Overrunning on cost only
B130	100	£2500	20	£550	25%	20%	22%	Underrunning on time and cost
B170	20	£750	2	£75	10%	10%	10%	On time and on cost

* Time and cost ratios are calculated as percentage (booked)/(planned)

PROBLEMS

1. A small (120 employees) company has decided to purchase its first manufacturing computer system. It makes to stock a wide range of wall tiles and also makes to order 'specials' for contracts (e.g. hotels). Write a fairly brief specification (up to 200 words) for the inventory control module covering links required with other modules, special requirements, stock records and policies, etc.

XYZ Components Ltd. Page 1

GANTT CHART OF PROJECT IMPLEMENTATION PLAN

TASKS/OPNS.	WHO?	01	02	03	04	05	06	07	08	09	10	11	12	13	14	15	16	17	18	19
INSTALLATION																				
M/C DELIVERY	JONES	▒																		
M/C TEST	JONES		▒																	
SYSTEM TRNG.	LIST A		▒	▒																
PROGRESS MTG.1	LIST B				▒															
PREP.DATABASES																				
PRODUCTS	PETERS			▒	▒	▒														
B.O.M's	FOX				▒	▒	▒													
ROUTINGS	DAVIES						▒	▒												
STOCKS	PETERS							▒	▒	▒										
WORK CENTRES	DAVIES				▒	▒														
ENTER DATA																				
PRODUCTS	HUGHES						▒	▒												
B.O.M's	WILLIAMS							▒	▒											
ROUTINGS	HUGHES									▒	▒									
STOCKS	WILLIAMS											▒	▒							
WORK CENTRES	DAVIES							▒												
etc. etc. etc.																				

(Column headings 01–19: WEEK NO.)

XYZ Components Ltd. Page 1

CPM NETWORK OF PROJECT IMPLEMENTATION PLAN

MACHINE DELIVERY

24/3/94 - 30/3/94

JONES - FL. Od

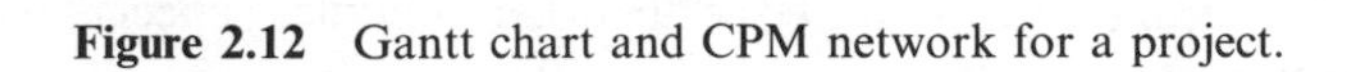

Figure 2.12 Gantt chart and CPM network for a project.

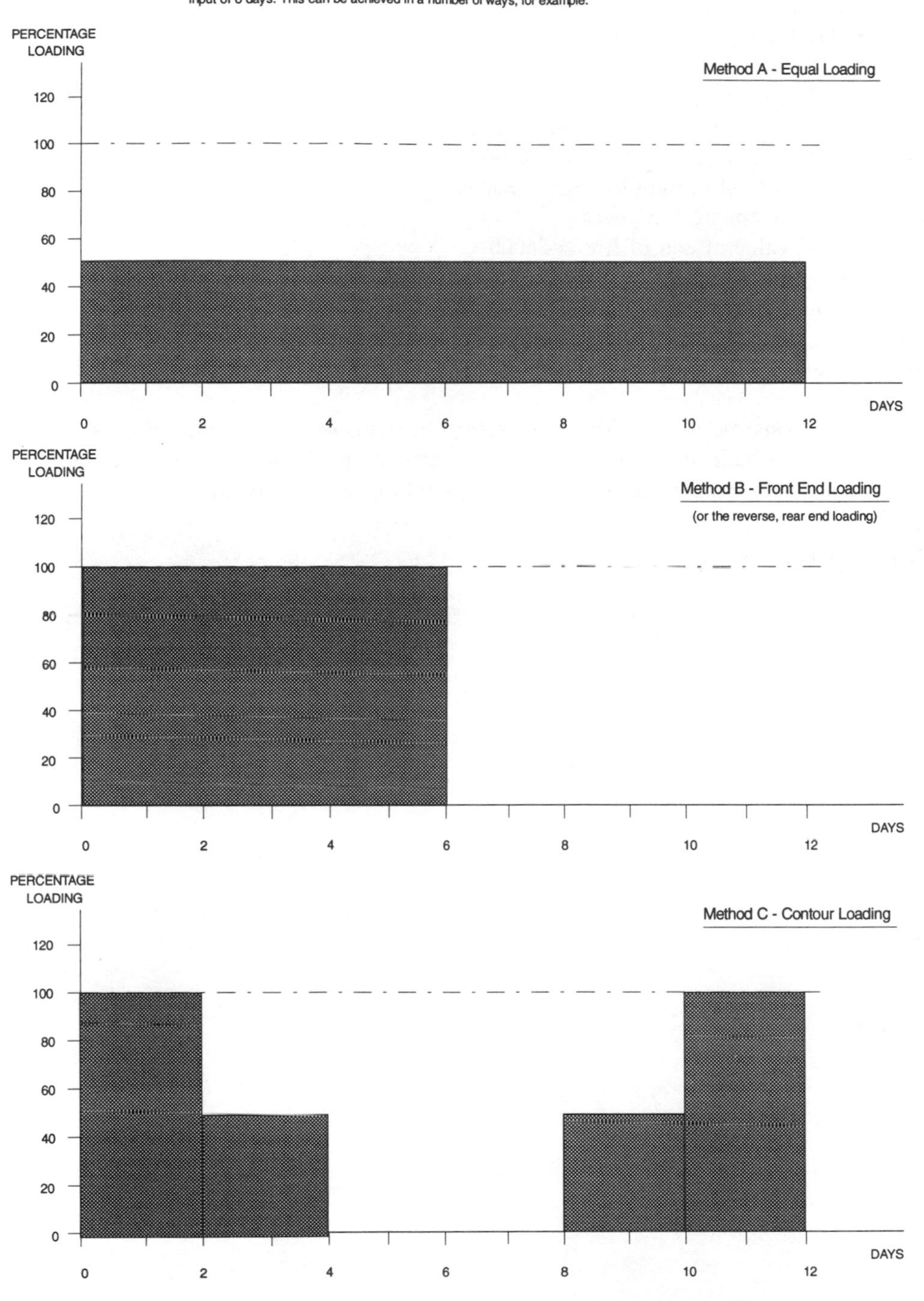

Figure 2.13 Methods of scheduling resources.

2. Prepare a draft 'invitation-to-tender' letter for the above company to send with its specification to potential suppliers.
3. Use the DCF method of investment analysis to determine the discount rate for a project for which the following data are given (solution in appendix A):
   (a) capital expenditure on new project = £85 000
   (b) capital recovered by sale of old equipment = £15 000
   (c) annual costs of new project = £30 000
   (d) annual costs of current system = £60 000
   (e) life of investment = 5 years
   (f) residual value at end of life = £5000
4. Devise a spreadsheet that can be used to determine capacity loadings on a number of production machines. The forward visibility should be four weeks, with loadings shown on a weekly basis. The spreadsheet should show, for each machine, the works orders making up the load, the loads in hours for each operation and the percentage loading on each machine.
5. Specify the construction of a database record in terms of fields, field sizes and types, which can be used to build up a file of production operations. This file is intended to be used to provide information for the preparation of works order documents on the database.

FURTHER READING

Bedworth, D. D. and Bailey, J. E., *Integrated Production Control Systems: Management Analysis and Design*, Wiley, New York, 1992.

CHAPTER

THREE

FORECASTING IN MANUFACTURING

THE LINK

Managing a company without information is like flying a plane without instruments, on a dark and stormy night — highly dangerous! The pilot needs to know where the plane is, in which direction it is travelling, and what lies ahead — so does the manager!

Forecasting is a vital instrument for the manufacturing manager; it can show what future demand is likely to be and thus enable plans to be made and policies established to meet this demand. Thus, given that the forecasts are scientifically based (statistically sound), it is possible for management to determine what products should be made and in what quantities, and from this to decide what should be purchased, what should be stocked and what grades and numbers of people should be employed.

Forecasts can also be used to look at possible trends in other areas, that may affect the business, for example, changes in foreign currency exchange rates, raw material availability (feast or famine) and economic indicators that may affect costs (interest rates and inflation). It is therefore essential for the effective manager to understand how forecasts are prepared and presented in order to be able to use the resulting data with confidence. For example, a knowledge of the laws of probability (all forecasts should include probability figures) is needed.

This chapter is designed to help the manager achieve this objective.

3.1 INTRODUCTION

In this chapter, forecasting, the first of the various tools used to operate and manage a manufacturing company, is studied, and the need for some form of forecasting is examined. The possible application areas are listed and considered in detail in the context of the different types of manufacturing organization. This is followed by consideration of the various techniques available, with examples of how they can be applied. The chapter concludes by looking at the relationship between forecasting and other manufacturing functions.

3.2 THE PLACE OF FORECASTING IN MANUFACTURING

The success or failure of any enterprise depends on the ability of its management to look into the future with a reasonable degree of confidence. This usually means trying to find answers to questions of the 'what will happen if' type, e.g. 'What will happen if'

- we apply a 10 per cent price increase?
- there is a 10 per cent rise in the price of our basic raw material?
- we make a 10 per cent cut in our stocks?
- we buy a new production machine?
- we install a new computer system?
- we change our methods of costing?'

In other words, without information about what is likely to happen as a result of a decision, the chances of that decision being right are considerably reduced. The effectiveness of this 'what-if' approach depends on two factors — the availability of accurate past data and the nature of the forecasting technique — and if the latter is not based on properly applied forecasting principles the results may be misleading.

For example, if the answer to the 'what-if' question suggests that sales will increase as a result of a price reduction, stocks may be built up in anticipation. However, if the suggested increase in volume does not take place the company could face considerable losses, especially if the products are perishable.

In many companies forecasts are based on hunches and 'guesstimates', some of which are likely to be of dubious value — perhaps because they are influenced by personal opinions. For example, sales representatives are by nature optimists and may therefore overstate their 'sales forecasts'. Similarly, production managers are pessimists, so their 'production output forecasts' will probably be understated.

It can be seen from the 'what ifs' listed above that forecasting is a tool not just of sales management but of general management (since decision making is one of the primary management functions). In chapter 1 the need for forward planning was emphasized in terms of plans for marketing, production and finance. These plans can only be based on what is expected to happen, and these expectations must depend on some form of forecasting.

The marketing plan is the starting point since, by defining what is likely to be sold, other factors can be defined, for example:

- what is to be made?
- what is to be stocked?
- what changes are needed in the production and stores areas?
- what are the design and development implications?
- what are the financial implications?

The answers to these questions can often only be found by applying knowledge obtained from previous experience, this being expressed in the form of a time series which is a collection of observations made at different times, for example monthly sales of a given product or prices for a raw material. By deriving forecasts based on these observations, it is possible to plan ahead and to increase/reduce production or to build up stocks at the lowest price.

Thus, forecasting, in its true sense, can be defined as:

> *a technique that is used to predict, with some measure of probability, what is likely to happen to a time series in the future, on the basis of past experience or past relationships with other time series.*

Many people in industry believe that a scientific approach to forecasting requires a knowledge of advanced mathematics that is beyond them, but this is no longer the case and, since the development of PC-based systems, the expertise required has been reduced to an average level of numeracy.

3.3 GENERAL FORECASTING METHODS

The main essential for a good forecast is an accurate record of what has happened in the past. It is generally impossible to have too much information.

This information should include not only what might be called 'direct data', for example monthly sales or production figures broken down into product groups and markets, but also 'indirect data' which can be used to explain (and possibly discount) values that do not fit into the general pattern. Typical examples of indirect data could include dates of

- strikes or shutdowns
- price increases by product group
- serious disruptions due to bad weather
- government legislation, etc.

There are also certain general principles that should apply whichever forecasting method is being used, for instance:

- the future period should not be markedly different from that on which the forecast is based
- enough data should be used to ensure that minor variations do not carry undue weight
- different forecasting methods and factors should be tested on a representative sample of historical data and the results compared with the actual data to find the most suitable method.

Various methods of forecasting could be considered suitable for use in the manufacturing area. These are

- the moving averages method
- the exponential smoothing method
- time series models
- econometric models.

The moving averages method is not a true forecasting technique since the user is required to make the forecasts from an examination of the graph. The other three are forms of statistical forecasting, although exponential smoothing is a very simple method, which produces a level forecast several periods ahead, and is therefore not suitable for time series with a trend or seasonality.

3.4 THE MOVING AVERAGES METHOD

This method is simple to operate and is very commonly used, the data from each period being given the same importance (weight). The first step is to plot the data to reveal the average period of significant fluctuation. By averaging blocks of data, fluctuations are levelled out. Each of these averages is taken to represent the trend at its middle point, the moving average being plotted graphically. However, care should be taken when selecting the length of the fluctuation period (over which averages will be taken), since too short a period may not give reliable future trends and too long a period may hide seasonal fluctuations.

Most manufacturing companies prepare monthly sales figures, and these figures can be used to take simple averages for the fluctuation period chosen and to prepare a graph. It should then be possible to estimate, from an examination of the apparent trend, what is likely to happen in the next few periods.

This technique is illustrated in Fig. 3.1, which shows the effects of selecting a three-monthly average (graph A) and a five monthly average (graph B). In this case the better choice is three-monthly, since this eliminates the fluctuations and shows the trend clearly, whereas the five-monthly line gives a more out-of-date picture (since the last trend point is further from the end of the series). It should be noted that this is essentially a judgemental method because the user, having plotted the data, must then project the trend into the future.

The selection of the most suitable period for averaging depends on the amount of fluctuation in the raw data, and it is always advisable to try two or three alternatives before a decision is made. If the averaging period is too long, the latest trend point is out of date, making it difficult to make a forecast. On the other hand, if the period is too short the graph will fluctuate too much and no clear trend will be discernible.

If the series appears to have a seasonal pattern (which will not be apparent until at least three years' data are available), the method is more complicated:

1. a 12-month moving average will provide an initial estimate of the trend, free of seasonality
2. this can be subtracted from the original series, to give deviations from the trend
3. these deviations can be averaged, for corresponding months in different years, to create 'seasonal factors'
4. these factors should be subtracted from the original series, which should be smoothed by a shorter moving average, to give a more up-to-date trend
5. finally, the forecast can be made as a combination of the projected trend and the seasonal factors.

As stated above, this method is easy to understand and to operate and is made even easier if a spreadsheet can be set up, because the calculations and graphs can be processed automatically. However, it is not particularly accurate, due to the equal weighting given to all readings in an average value. In fact, less weight should be given to a reading as it becomes older, as happens with exponential smoothing.

3.5 TIME SERIES

The derivation of a forecast from a time series can sometimes be judgemental, by plotting the graph and estimating its extension into the future. However, such an approach should be carefully considered, since it is easy to misinterpret trends if the values are erratic. It is therefore suggested that the plot be defined in terms of its features, thus:

- is it smooth with few or no turning points?
- is there any sustained trend, upwards or downwards?
- if the series is monthly or quarterly is there an obvious seasonal pattern that is broadly repeated every year?

MONTH	JAN	FEB	MAR	APR	MAY	JUN	JUL	AUG	SEP	OCT	NOV	DEC
SALES	1000	1200	1300	1100	1200	1000	900	700	800	1100	900	1300
3 MONTH TOTALS	---	3500	3600	3600	3300	3100	2600	2400	2600	2800	3300	
3 MONTH MV'G AVGE.	---	1167	1200	1200	1100	1033	867	800	867	933	1100	
5 MONTH TOTALS	---	---	5800	5800	5500	4900	4600	4500	4400	4800		
5 MONTH MV'G AVGE.	---	---	1160	1160	1100	980	920	900	880	960		

NOTE: The average figures are 'centred', that is on month 2 for the 3 month av'ge and on month 3 for the 5 month av'ge.

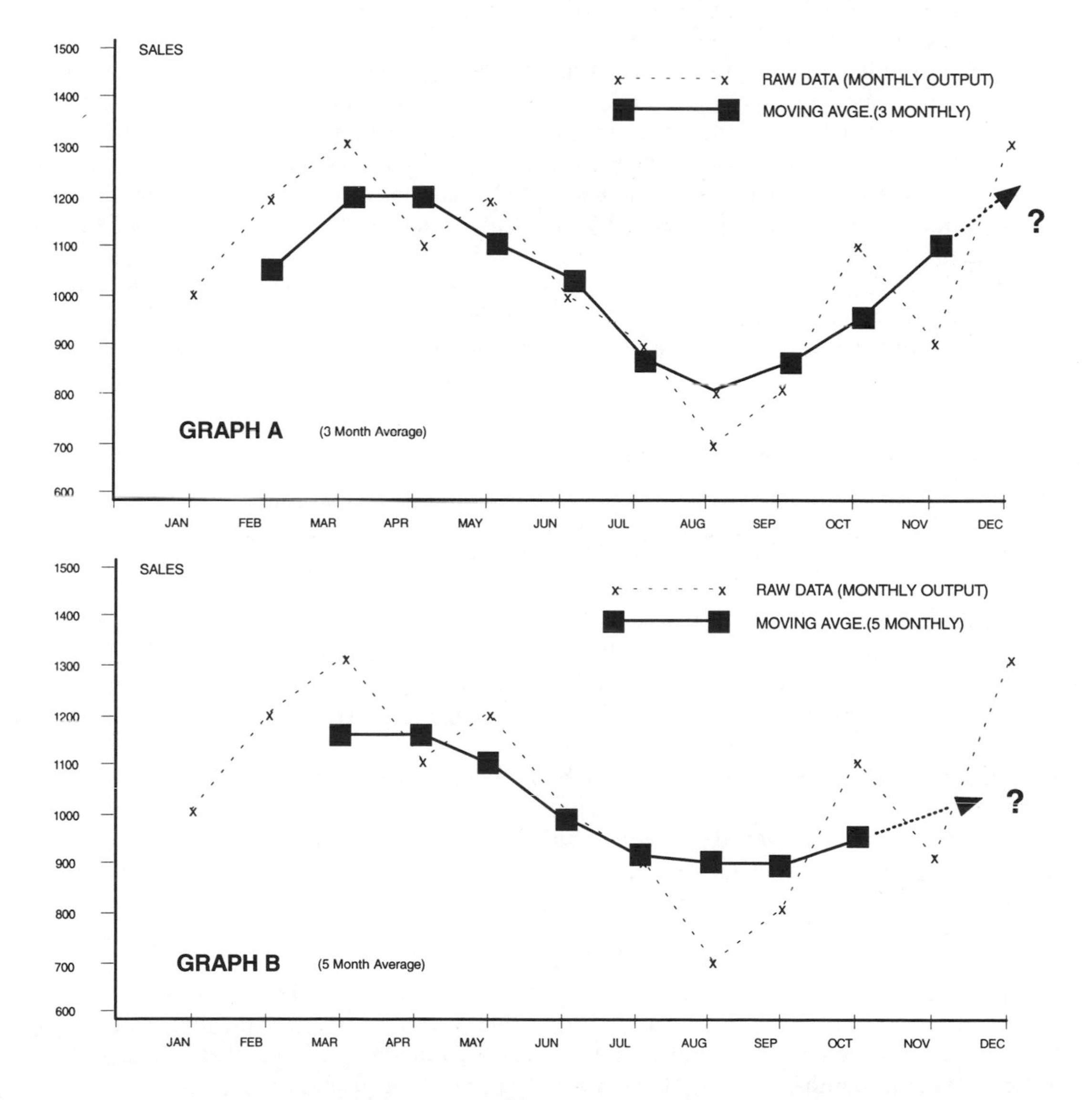

Figure 3.1 Forecasting—the method of moving averages.

- are there any obvious 'outliers' (points that appear out of line with the local trend and seasonal pattern) and, if there are, can these be explained by external causes (e.g. extreme weather conditions)?

If the answers to these questions suggest that the series is in some way irregular or seasonal, then the judgemental approach may be difficult and will probably produce inaccurate forecasts. In such circumstances one of the statistical forecasting methods should be used, for the following reasons.

- If the series is irregular or seasonal, the true trend may not be apparent and seasonality may be incorrectly assessed.
- Although the human eye can read a graph quickly, the brain cannot absorb a long run of data, and yet such a long run is essential for accurate forecasting. Many companies use only limited amounts of data, because they are easier to read or because it is believed that old data are irrelevant. They perhaps do not realize that they are ignoring valuable information. For example, if a monthly time series without seasonal variations is to produce accurate forecasts, a minimum of three years' data is required, and this should be extended to five years if there is seasonality.
- Judgemental forecasting provides no rational means of assessing the inevitable margins or bounds of error, which widen as forecasts go further into the future. However, if statistical methods are used, these bounds can be defined in terms of the probability of their being exceeded.

3.6 THE EXPONENTIAL SMOOTHING METHOD

This method is fairly simple to use, but can be laborious unless a spreadsheet has been set up. The calculations required involve comparisons between forecasts and actuals for previous periods and the current period, with a forecast for the first period being assumed. This allows a forecast for the second period to be derived, and so on.

This is achieved by using the formula: $\text{NF} = \text{OF} + \alpha\,(\text{AV} - \text{OF})$, where NF is the new forecast (the forecast for the next period), OF is the old forecast (for the current period just ended), α is a smoothing factor and AV is the actual value (for the current period just ended). A suitable value for α must be found by iteration, to give the forecasts that most nearly match the actuals.

Figure 3.2 shows how the forecasts have been calculated using an α factor of 0.1, the actual values being the same as those used in the example on moving averages for monthly sales (Fig. 3.1). The forecasts thus derived are plotted on the accompanying graph. This shows a flattish line because, when the value of α is small, the forecast responds slowly to a change of level.

Figure 3.3 shows similar graphs using α values of 0.2 and 0.3, and the forecast can be seen to be responding more quickly to level changes. However, there is now a greater risk of it responding to what are merely random variations about the true level.

The selection of the most suitable value for α can be looked at in various ways: first by comparing the annual total value of sales with the three forecast totals (for the three values of α) to see which has the lowest difference, and secondly by examining the cumulative totals of the differences on a monthly basis. Both methods of selection are given in Table 3.1.

- If $\alpha = 0.1$ The annual deficit is 560 units and in the worst month (June) it is 872 units; there is never a surplus.

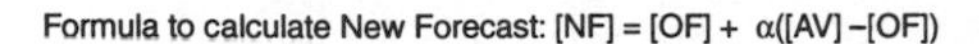
Formula to calculate New Forecast: [NF] = [OF] + α([AV] −[OF])

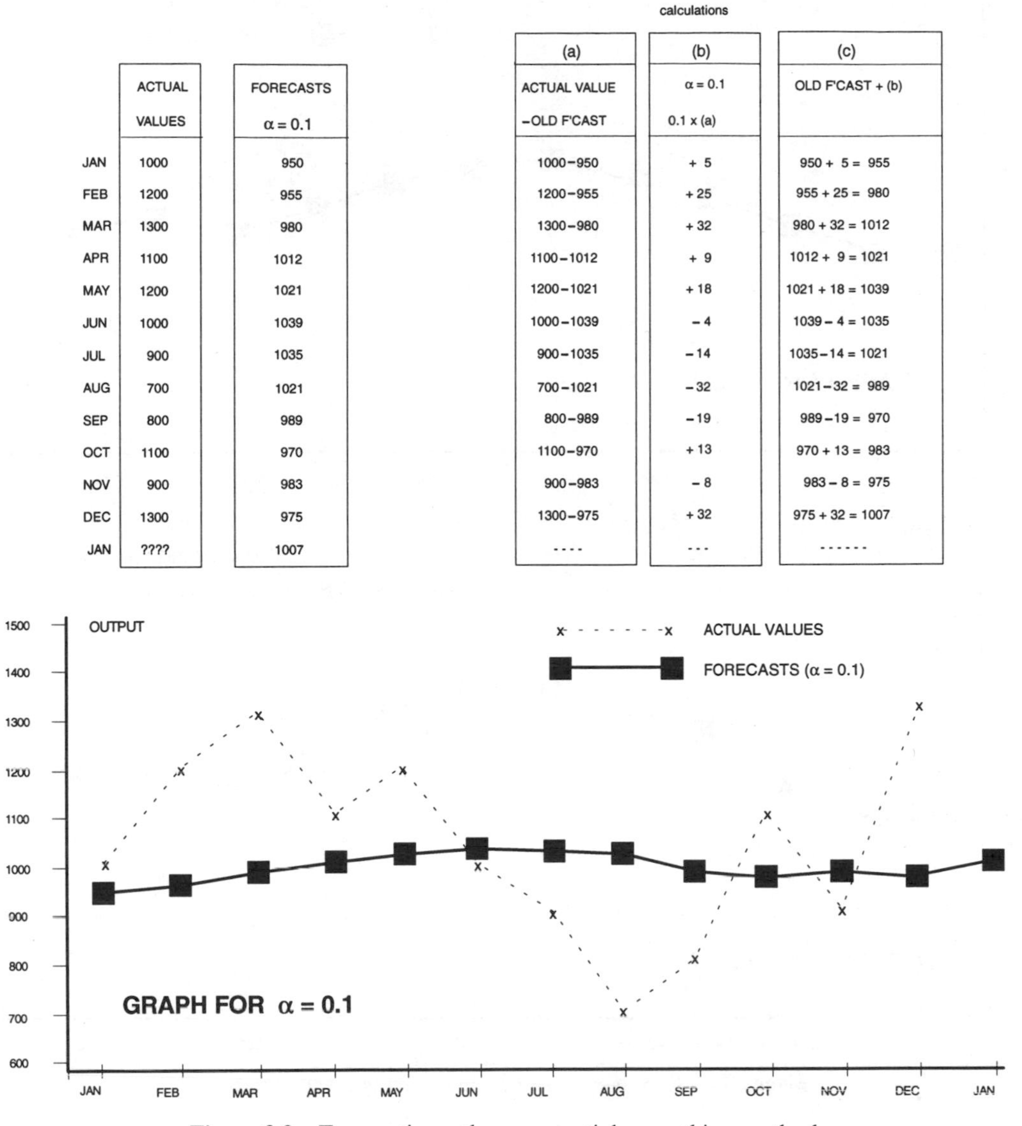

calculations

	ACTUAL VALUES	FORECASTS α = 0.1	(a) ACTUAL VALUE −OLD F'CAST	(b) α = 0.1 0.1 x (a)	(c) OLD F'CAST + (b)
JAN	1000	950	1000−950	+ 5	950 + 5 = 955
FEB	1200	955	1200−955	+ 25	955 + 25 = 980
MAR	1300	980	1300−980	+ 32	980 + 32 = 1012
APR	1100	1012	1100−1012	+ 9	1012 + 9 = 1021
MAY	1200	1021	1200−1021	+ 18	1021 + 18 = 1039
JUN	1000	1039	1000−1039	− 4	1039 − 4 = 1035
JUL	900	1035	900−1035	− 14	1035−14 = 1021
AUG	700	1021	700−1021	− 32	1021−32 = 989
SEP	800	989	800−989	− 19	989−19 = 970
OCT	1100	970	1100−970	+ 13	970 + 13 = 983
NOV	900	983	900−983	− 8	983 − 8 = 975
DEC	1300	975	1300−975	+ 32	975 + 32 = 1007
JAN	????	1007	----	---	------

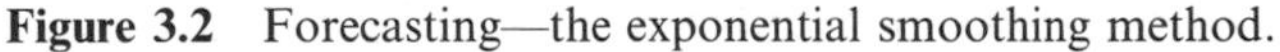
Figure 3.2 Forecasting—the exponential smoothing method.

- If $\alpha = 0.2$ The annual deficit is 28 units and in the worst month (May) it is 713 units; the worst surplus is 271 in November.
- If $\alpha = 0.3$ The annual surplus is 476 units and in the worst months the deficit is 550 (May) and the surplus 753 (November).

Thus, the best value for α in this case would appear to be a little over 0.2, which gives zero annual error, and it must be accepted that safety stocks will have to be built up to counteract the deficits, which would otherwise affect sales in the first six months of the year. There appears to be a case here for looking at the seasonal factors.

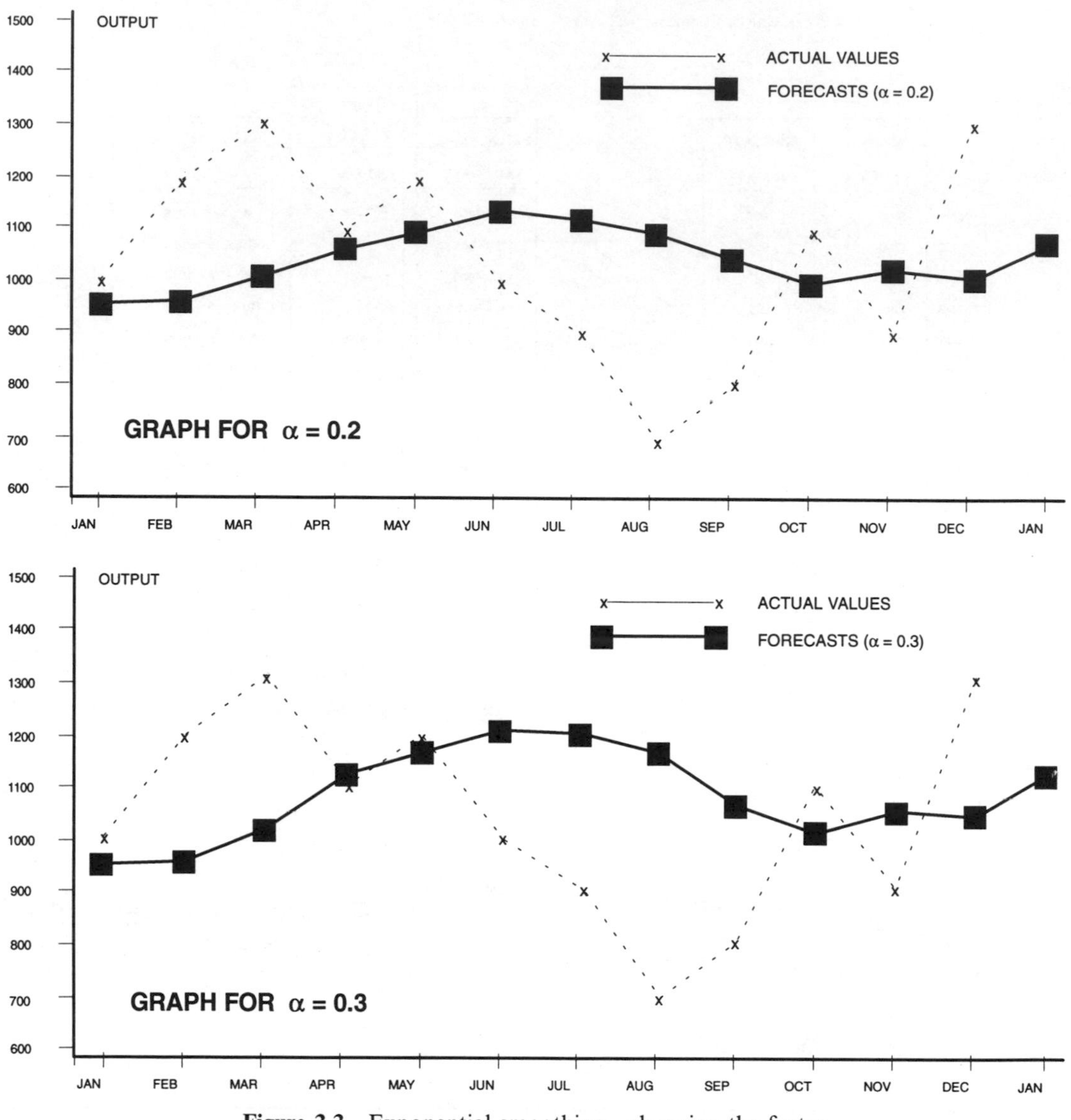

Figure 3.3 Exponential smoothing—changing the factor.

It should be noted that this fairly simplistic method is not valid when the series has a trend (upwards or downwards) as opposed to shifts in the level. For this a more complex method, e.g. Holt's 2-parameter method (which is included in some commercially available spreadsheets), should be used.

3.7 BASIC STATISTICAL FORECASTING

This book is not designed to be read by those who wish to obtain an in-depth knowledge of statistical forecasting. If this is required, Cryer (1986) could be helpful.

The simplest and most basic example of statistical forecasting — exponential smoothing — has already been described, but if a more accurate and scientific approach is required, a number

Table 3.1 Selecting the best value for the factor α

Month	Actual sales	Forecasts $\alpha = 0.1$	Forecasts $\alpha = 0.2$	Forecasts $\alpha = 0.3$
Jan.	1000	950 (−50)	950 (−50)	950 (−50)
Feb.	1200	965 (−285)	960 (−290)	965 (−285)
Mar.	1300	980 (−605)	1009 (−581)	1039 (−546)
Apr.	1100	1012 (−693)	1075 (−606)	1135 (−511)
May	1200	1021 (−872)	1093 (−713)	1161 (−550)
Jun.	1000	1039 (−833)	1129 (−584)	1215 (−335)
Jul.	900	1035 (−698)	1121 (−363)	1203 (−32)
Aug.	700	1021 (−377)	1094 (+31)	1162 (+430)
Sep.	800	989 (−188)	1030 (+261)	1066 (+696)
Oct.	1100	970 (−318)	992 (+153)	1009 (+605)
Nov.	900	983 (−235)	1018 (+271)	1048 (+753)
Dec.	1300	975 (−560)	1001 (−28)	1023 (+476)
TOTALS	12 500	11 940	12 472	12 976

In this example the forecast for January of 950 has been derived from the forecasts from the previous year. In other words, this is only a part of the data derived from a much longer time series. The values in brackets are the cumulative differences between the forecasts and the actuals.

of commercially available computer-based packages can produce statistical forecasts without requiring the user to understand the finer points of the processes and calculations involved. However, for those who feel that they need a basic knowledge of the terms and techniques in order to understand the way in which such systems operate, the following sections should prove useful.

3.8 MODELLING METHODS

There are two techniques that rely on the use of a process known as 'modelling', which can be defined as the derivation of a mathematical formula (or a set of formulae) that will fit the data concerned and will allow the user to calculate the value at the next point in the future—the forecast value. The two techniques both have a contribution to make in industry and commerce. They are:

1. econometric—where a correlation is established between the original series and one or more others that are easier to forecast
2. time series—where a close correlation is established between the series and its own past.

In manufacturing terms, the basis of an econometric forecasting system for, say, demand for a product is the ability of the system creator to select the factors that will influence that demand. A mathematical model must then be built which, when fed with data relating to these factors, will calculate the product demand forecasts.

A time series system also uses the technique of mathematical modelling, but in a different way. In this case, the historical data for the product series are used, as in the case of exponential smoothing, to prepare a forecast that suggests the likely trends for the future.

Judgemental forecasting provides no rational means of assessing the inevitable margins or bounds of error which widen as the prediction goes further into the future, but, if statistical modelling is used, these bounds can be defined in terms of probabilities. These are measured on a scale of 0 to 1, where a value of 0 means that something will never happen (person will swim the Atlantic) and a value of 1 means that something is certain to happen (the sun will rise tomorrow).

As an example, assuming that the forecast errors follow the principles of normal distribution (see Appendix B), the bounds can be calculated so that the probability of a particular forecast lying outside the interval between them is any selected number. A common choice for the probability is 1 in 20 (0.05) but this usually gives a rather wide interval. For a probability of 1 in 3 (0.33), the interval is only half the size.

3.9 ECONOMETRIC MODEL FORECASTING

In order to illustrate how the econometric technique operates, it is probably best to consider an example, in this case a manufacturer of vehicle batteries which wished to develop a forecasting system for its products. In order to do this it decided to build an econometric model which would provide this information based on the sales of passenger vehicles and commercial vehicles. The way in which this model was developed is shown in Fig. 3.4.

This shows that data has been obtained for quarterly sales over a five year period for batteries (z_t), cars (x_t) and trucks (y_t), the subscript t being used to indicate a time from the start. Thus z_1 is the quantity of batteries sold in the first quarter of the first year, z_2 is the second quarter quantity and so on.

The objective in building the model is to find correlations (or relationships) between the three sets of data, and these can be either positive or negative.

- A positive correlation between two sets of data means that they tend to move up and down together. Thus, in the example, if the sales of cars and trucks are above average for a given period, then the sales of batteries would usually also be above average.

CONSIDERING THE ECONOMETRIC EXAMPLE OF THE AUTOMOTIVE BATTERY MANUFACTURER:

LET z_1, z_2 ....z_{20} BE FIVE YEARS OF QUARTERLY SALES OF VEHICLE BATTERIES (MEASURED IN SOME STANDARD UNITS).

LET x_1, x_2 ....x_{20} BE FIVE YEARS OF QUARTERLY SALES OF CARS.

LET y_1, y_2 ....y_{20} BE FIVE YEARS OF QUARTERLY SALES OF TRUCKS.

THEN A RELATIONSHIP IS SOUGHT OF THE FORM:

$$z_t = c_0 + c_1 x_t + c_2 y_t + a_t \qquad (1)$$

WHERE t STANDS FOR TIME (t = 1, 2, ... 20)

IN OTHER WORDS, GIVEN z_1, x_1 AND y_1, THE ERROR IN THE RELATIONSHIP a_t CAN BE CALCULATED.

SIMILARLY FOR t = 2, t = 3, AND SO ON.

c_0 AND c_1 AND c_2 ARE CONSTANTS WHICH HAVE TO BE ESTIMATED

Figure 3.4 Example of econometric methods.

- A negative correlation is also possible in some cases, for example, an increase in taxation on a manufacturing company could lead to a reduction in investment in new plant.

The strength of a correlation is measured in terms of a 'correlation coefficient' which always lies between +1 and −1.

The possibility of 'lag' must also be considered: for instance, in the example an increase in car and truck sales in period 5 might not show in battery sales until period 6, due to the existence of buffer stocks at the vehicle manufacturer's.

The constants c_0, c_1 and c_2 in equation (1) in Fig. 3.4 are known as the model parameters, and are estimated by using a technique known as linear regression. For any set of values of the parameters, predictions are made compared with the actuals and the *sum of squares* of the prediction errors is obtained, the values of the parameters being varied in such a way that the minimum sum of squares is found. The corresponding values of c_0, c_1 and c_2 are the best estimates of the parameters.

From these, forecasts can be made about future values. The average size of the prediction errors in the past gives one element in setting the bounds of the forecasts; the others are uncertainty about the forecasts of the explanatory variables (x_t and y_t) and uncertainty about the true values of c_0, c_1 and c_2.

3.10 TIME SERIES MODEL FORECASTING

In time series models the error bounds also depend on the average size of the past prediction errors and on uncertainty about the true parameter values. The use of statistical methods does not remove the need to exercise judgement, but it gives a firmer base. For example, time series modelling will give an objective estimate of the downward trend during a recession, but will not indicate when it is going to end; for this, market assessments are required.

3.11 TYPES OF TIME SERIES MODEL FORECASTING

The remainder of this chapter is concerned with the various types and features of time series models, which can be listed as follows:

1. auto-regressive models
2. moving average and non-stationary models
3. ARIMA models
4. forecasting steps
5. model identification and estimation
6. seasonality
7. transformation
8. outliers
9. PC-based systems.

Auto-regressive models

The oldest and most obvious of the time series models is the weighted average. This determines the current observation as a weighted average of past observations plus a forecast error, for example:

$$z_1 = c + \phi_1 z_{t-1} + \phi_2 z_{t-2} + a_1 \quad (2)$$

By analogy with the econometric model (Fig. 3.4), this is known as an auto-regression of order 2 or AR(2). In general, if the model contains p past observations, then it is described as an auto-regression of order p or AR(p). The success of this type of model (as with econometric models) depends on the strength of the correlations, except that here they are known as auto-correlations.

For example, if there are 20 observations then the correlation between z_t and z_{t-1} is estimated from the correlation between $z_1, z_2, \ldots z_{19}$ and $z_2, z_3 \ldots z_{20}$. This is called the auto-correlation at lag 1. Similarly, the correlation between $z_1, z_2 \ldots z_{18}$ and $z_3, z_4, \ldots z_{20}$ is the auto-correlation at lag 2, and so on. In general, if the auto-correlations are high and positive at low lags, the series will appear smooth, whereas if they are low or negative the series will appear to be irregular.

An experienced statistician can often decide what order of model to choose by looking at the pattern of the auto-correlations. However, there are many series for which the AR model is unsuitable or which require a very large value of p (number of past observations) to make a good forecast. This is particularly true when there is a persistent upward or downward trend or a drift in the mean level (as often happens to stock market prices, for example). In such cases the series is said to be non-stationary.

Moving average and non-stationary models

Another group of forecasting methods, which developed out of the needs of quality control, is exponential smoothing, the simplest form of which was described and illustrated in section 3.6. An example of this technique is shown in Fig. 3.5, where the original equation in terms of forecasts has been changed to refer only to the series and the errors.

It should be noted that the left-hand side of equation (3) in this figure ($z_t - z_{t-1}$) is the change in the original series from one period to the next. This is known as the 'first difference' of the series. The right-hand side of the same equation ($a_t - \theta a_{t-1}$) is called a moving average of order 1 or MA(1). In an extension of this method, the model includes the previous forecast errors at times $t-2, t-3$, etc. If the modification depends on q previous forecast errors, with parameters $\theta_1, \theta_2, \ldots \theta_q$, the method is described as moving average of order q or MA(q).

ARIMA models

The great contribution of the statisticians Box and Jenkins (1976) was to show that a combination of auto-regressive and moving average models was more parsimonious in terms of the number of parameters than either taken alone. They called their new method 'auto-regressive moving average' or ARMA. Thus, the term ARMA(p, q) describes a model in which p previous observations and q previous errors have been used in the forecast.

Expressing a time series model in terms of previous observations is equivalent to expressing it in terms of previous errors, but it is a question of which, or what mixture of the two, is the most parsimonious in terms of model parameters. Box and Jenkins also showed that, if the series is non-stationary, it should be differenced—as in equation (3) in Fig. 3.5—before an ARMA model is fitted. If the new series is still non-stationary, the differencing should be repeated. When a stationary series has been obtained, its auto-correlations should be examined to determine possible values of p and q.

BASED ON A FORECAST FOR TIME (t −1) A FORECAST FOR TIME t

IS OBTAINED, THUS:

$$\hat{z}_t = \hat{z}_{t-1} + ka_{t-1}$$

WHERE THE OVERHEAD 'CAPS' INDICATE FORECASTS. HOWEVER, SINCE

$z_{t-1} = \hat{z}_{t-1} + a_{t-1}$ THEN $\hat{z}_{t-1}$ CAN BE REPLACED IN THIS EQUATION

TO OBTAIN:

$$\hat{z}_t = z_{t-1} - \theta a_{t-1}$$

WHERE $\theta = (1 - k)$. THEN ADDING a_t TO BOTH SIDES, THE SERIES

FOLLOWS THE MODEL:

$$z_t - z_{t-1} = a_t - \theta a_{t-1} \qquad (3)$$

Figure 3.5 Example of adaptive forecasting.

If the series is differenced d times and an ARMA(p, q) model is fitted to the resulting series, the overall model is known as 'auto-regressive integrated moving average' or ARIMA(p, d, q), where p is the number of previous observations, d is the number of times the series is differenced and q is the number of previous errors used in the forecast. The word 'integrated' is used to indicate that the model for the differenced series has to be summed or integrated to get back to the original series. The simple exponential smoothing technique is just the ARIMA(0,1,1) model of equation (3) in Fig. 3.5. which is applicable only to series having no linear trend. Holt's more general adaptive forecasting method for series with a trend is in fact a restricted version of the ARIMA(0,2,2) model.

An example of the use of the ARIMA modelling technique is shown in Fig. 3.6.

Forecasting steps

A forecast can be made several steps (periods) ahead if the following method is used.

1. The forecast one step ahead should be treated as the next observation, the forecast error should be assumed to be zero and the model applied to forecast two steps ahead.
2. The process should then be repeated to obtain the forecast three steps ahead, and so on.

TO ILLUSTRATE THE (1,2,2) INTEGRATED MOVING AVERAGE MODEL:

$$w_t = \theta_1 w_{t-1} + a_t - \theta_1 a_{t-1} - \theta_2 a_{t-2} \qquad (4)$$

WHERE $w_t = z_t - 2z_{t-1} + z_{t-2}$ (THE SECOND DIFFERENCE)

AND a_t IS THE ERROR IN FORECASTING w_t (OR z_t) AT TIME (t − 1)

$\hat{w}_t$ CAN BE OBTAINED FROM EQUATION (4) BY SIMPLY SETTING a_t TO ZERO.

AND THEN $\hat{z}_t$ FOLLOWS FROM $\hat{w}_t$, z_{t-1} AND z_{t-2}

THE NUMBER OF PAST VALUES OF THE SERIES THAT ARE NEEDED

TO MAKE A FORECAST IS NOW (p + d). IN PRACTICE,

p AND q CAN BE QUITE SMALL AND d IS EITHER 1 OR 2.

Figure 3.6 Example of ARIMA modelling.

The pattern of the forecasts depends on the type of model. For a pure auto-regressive (AR) type the forecasts are influenced by recent observations, but tend towards the mean of the series, as the forecast period moves further away from the present into the future. However, for an integrated moving average (ARIMA) model, when the series has a trend, the forecasts follow the latest estimate of this trend — usually a straight line — after the first few periods, which are influenced by recent forecast errors. If there is no consistent trend, but the mean level of the series seems to drift up or down, the forecasts will be at the estimated current level.

Model identification and estimation

The order of differencing needed to eliminate any trend or drift can be judged by eye, although this is more difficult when the series is seasonal, but it can be more reliably determined by a simple statistical test derived from the auto-correlations. The appropriate values of p and q will be suggested after an examination of the auto-correlations of the differenced series, but this requires some expertise. Fortunately it is possible to automate the choice of model, or at least to restrict that choice to a small set of alternatives. Within this set, the selection will be made that involves the smallest number of parameters and has properties (such as a smooth trend) that accord with the user's intuitions.

As in the case of linear regression, the values of the ϕs and θs are found that minimize the *sum of squares* of the 1-step forecast errors within the data. When the model is pure auto-regressive (after differencing if necessary), estimation is straightforward, but when moving average terms are involved, it is more complicated: in such cases the errors can be calculated on a recursive, or step-by-step, basis, as in exponential smoothing, provided that some assumptions are made about the errors at the beginning of the series.

Figure 3.7 illustrates this recursive process, the estimation program varying the values of θ_1 and θ_2 until it has minimized the sum of squares of the errors. In practice a more complicated, but more accurate, method of estimation is used.

Seasonality

For seasonal series, comparisons of actuals and forecasts should, as before, be made for monitoring purposes. But the forecast is now a combination of trend and seasonal components; the seasonal component normally changes only slowly, so the user may well accept this, but wish to

RECURSIVE (STEP-BY-STEP) ESTIMATION CAN BE CARRIED OUT PROVIDED SOME ASSUMPTIONS ARE MADE ABOUT THE ERRORS AT THE BEGINNING OF THE SERIES.

FOR THE MODEL (0,2,2):

$$w_t = a_t - \theta_1 a_{t-1} - \theta_2 a_{t-2} \qquad (5)$$

WHERE $w_t = z_t - 2z_{t-1} + z_{t-2}$

THE RECURSION STARTS WITH w_3 SINCE w_1 AND w_2 CANNOT BE CALCULATED. IF IT IS ASSUMED THAT a_1 AND a_2 ARE ZERO a_3 CAN BE ESTIMATED FROM EQUATION (5), AND GIVEN w_4 THEN a_4 CAN BE ESTIMATED AND SO ON.

Figure 3.7 Example of recursive estimation.

modify the trend forecasts in the light of personal knowledge of future events (e.g. an advertising campaign). In the example figures in appendix C, the forecasts of the series are monitored and the forecasts of the trend are compared with the estimates from the past series.

Many series, for example monthly sales figures, show seasonal variations that manifest in high auto-correlations at multiples of the seasonal period, for example, in a monthly series, in periods 12, 24, 36, etc. Box and Jenkins proposed an extension of the ARIMA model to deal with this, which combines two models:

- *non-seasonal* this model is described in 'ARIMA models' above
- *seasonal* this model is expressed in terms of the change in the series since a year ago and the forecast error a year ago.

To be specific, the changing pattern is represented by an intermediate series u_t, thus $u_t - u_{t-s} = a_t - \theta_s a_{t-s}$ and the original series z_t follows a non-seasonal ARIMA model in terms of u_t. The forecasts several steps ahead for a seasonal model will, as before, show a continuation of the latest trend and, superimposed on this, an estimate of the latest seasonal variation.

Transformation

In the previous sections it has been assumed that the forecasting errors remain of the same average size throughout the series. Before a model is fitted, these errors are unknown, but their general character can be inferred by estimating a simple trend (such as a 12-month moving average) and observing the variation about this.

In practice, where the series grows two- or three-fold or more in size (for example where sales are by value rather than quantity terms), the variation usually grows in line with the trend. It is then necessary to transform the data, perhaps by taking logarithms or square roots before fitting a model.

Outliers

Sometimes a freak observation appears, leading to an exceptionally large forecasting error, for instance a negative error in a production series due to a strike or shutdown. If the strike ends quickly and production returns to normal, this negative error will be followed by decreasing positive errors, because the original outlier has upset the later forecasts. It should be noted that normal cold weather (affecting outside work such as construction) will be included in the seasonal variation.

PC-based systems

There are a number of PC-based statistical forecasting systems on the market, but care should be taken to ensure that the chosen package includes such facilities as the ability to identify and measure outliers (see above). Details of a package called Prophet, which has been developed at the University of Kent, are included in appendix C as an example of what is now available.

Year	Month												
	JAN	FEB	MAR	APR	MAY	JUN	JUL	AUG	SEP	OCT	NOV	DEC	TOTAL
1	30	18	36	72	78	84	78	72	60	24	24	32	602
2	30	36	36	108	112	120	120	118	106	36	40	36	898
3	52	46	52	142	160	172	150	154	148	64	58	52	1248
4	60	59	66	165	174	190							

Figure 3.8 Monthly sales volumes (problem 1).

3.12 LINKS WITH OTHER SYSTEM MODULES

If a computer-based statistical forecasting package is to be used in a manufacturing environment, which also uses other computer systems, it is advisable to ensure that facilities to import data to the package and export data from it are available. These facilities do not have to be via a direct link — they could be by disk copy and transfer — but the data formats should be compatible. For example, data may be held in the sales order processing files or archives that could be copied over monthly to run a new sales forecast, and this forecast could then be copied into the production planning module to update the master production schedule (MPS).

Similarly, data from an econometric modelling exercise could be used to assist in the calculation of resource requirements for a series of future periods which would allow long-term plans to be related to forecasts taken from national or international statistics.

PROBLEMS

1. Figure 3.8 shows a sequence of observations in a time series of sales for a manufacturing company. These should be plotted on a graph and examined.
   (a) Are there any observable trends in this series and, if there are, what conclusions can be drawn from them?
   (b) Is there any seasonal pattern? If there is, it should be defined.
   (c) Try to predict the sales volumes for the last six months of year 4.
2. Use the sales volumes in Fig. 3.8 to draw a three month moving average graph. Superimpose this on your original raw data graph.
3. Revise the predictions made on the raw data (problem 1(c)). Which do you think are the more accurate, and why?
4. What types of manufactured products are likely to have seasonality? Define what the patterns might be, and explain the probable causes.
5. Suggest which functional areas in a manufacturing company would benefit from the application of the following statistical forecasting techniques:
   (a) econometric methods
   (b) time series methods.

REFERENCES AND FURTHER READING

Anderson, E. J., *The Management of Manufacturing*, Addison-Wesley, Wokingham, 1994.

Box, G. E. P. and Jenkins, G. M., *Time Series Analysis: Forecasting and Control*, Holden Day, San Francisco, USA, 1976.

Cryer, J. D., *Time Series Analysis*, Duxbury Press, Boston, MA, 1986.

Lewis, C. D., *Industrial and Business Forecasting Methods: a Practical Guide to Exponential Smoothing and Curve Fitting*, Butterworth Scientific, London, 1982.

Mair, G., *Mastering Manufacturing*, Macmillan, London, 1993.

Moroney, M. J., *Facts from Figures*, Penguin, London, 1960.

Smith, S. B., *Computer-based Production and Inventory Control*, Prentice-Hall, Englewood Cliffs, NJ, 1989.

Vollmann, T. E., Berry, W. L. and Whybark, D. C., *Manufacturing Planning and Control Systems*, 3rd Edn, Irwin, Boston, MA, 1988.

CHAPTER

FOUR

THE MANAGEMENT OF SALES ORDERS

THE LINK

Success in business depends on the customers — look after them and they will look after you — and there are three ways in which this 'caring' relationship can be established.

The first, which is covered in this chapter, is to provide a sales management system that ensures that customer's enquiries and orders are processed quickly and efficiently with accurate paperwork and deliveries when promised. The second is to ensure that the products are designed and made to the highest possible standards and meet the customer's expectations. This is dealt with in the later chapters on design and quality (chapters 5 and 13). The third is to ensure that an effective 'after-sales' service system is operating such that any problems are dealt with without delay; this is covered in the chapter on customer services (chapter 15).

This question of relationships between customers and suppliers is crucial to any business. All manufacturing companies have both suppliers and customers, and must try to build their relationships on the basis of mutual trust and respect with both. This 'partnership' concept is a recurring theme whenever the modern management approach is being discussed, and is rapidly replacing outdated confrontational methods.

4.1 INTRODUCTION

In this chapter the sales order processing functions are considered in detail, starting with ways in which orders can be received, the order types and the associated company types. Next, the methods that should be used to process these orders to ensure that customer requirements are met are considered, including the records that should be kept and the information that should be communicated to other departments and the customers. The packing and despatch requirements are then examined, including the provision of suitable documentation and the procedures needed to deal with customer returns. Finally, methods are suggested for dealing with the financial aspects of sales, i.e. pricing, discounts, invoices and credit notes.

4.2 THE SALES ORDER PROCESSING FUNCTIONS

The functions that should be performed by the sales and other departments can be defined as follows.

1. *Estimates and quotations* In companies that do not sell from stock or from a standard range, it is necessary to provide customers with quotations specifying what will be provided and at what price. These are prepared on the basis of estimates that are built up from known or estimated costs plus a profit margin.
2. *Sales order entry* When an order is received it must be entered into the company system (which may be manual or computer-based). In either case full details should be entered so that a record is available for the sales and other departments and also, in some cases, for the customer in the form of an order acknowledgement.
3. *Sales order processing* When orders have been entered, they must be processed into production (for make-to-order) or the despatch area (for make-to-stock). If a computer-based system is in operation this will be simply a matter of confirming the order details, which will automatically update the master schedule or allocate stock. However, if the system is manual then order copies or tabulations must be sent to production or despatch.
4. *Sales order despatch* The despatch department will normally be responsible for picking goods from stock or receiving from production, packing to customer specification, organizing transport, creating despatch notes and despatching. A despatch note copy must be passed to the accounts department if a manual system is in operation; despatch will also be responsible for receiving and recording customer returns.
5. *Sales order invoicing* The accounts department will be responsible for raising invoices and credit notes against despatch notes and customer returns notes.
6. *Sales order analysis* Records should be kept of all sales orders so that the figures can be analysed as required, preferably by computer. Table 4.1 shows how the information that such records can contain could be analysed—the number of options is vast!

In the chapter on forecasting (chapter 3) it was suggested that data of this type should be retained for as long as possible in order to ensure maximum possible forecast accuracy. This applies equally to associated data files: for example, even if a product has been superseded it

Table 4.1 Sales analysis records and options

Order header	Order items	Other related files
Order no.	Order no.	Products
Customer code	Item no.	Customers
Delivery code	Due date	
Order date	Product code	
Market code	Despatch date	
Customer order no.	Quantity	
Agent code	Unit price	

Sales analysis reports can be generated using almost any combination of the above data, e.g.

- sales by value and/or quantity for a customer for a period
- sales by value and/or quantity for a market for a period
- sales by value and/or quantity for a product for a period
- sales by value and/or quantity for a product group for a period
- sales by value and/or quantity for an agent for a period
- order items late delivered, by product
- order items late delivered, by customer
- order items late delivered, by market
- order items late delivered, by agent

should not be deleted from the file but merely 'flagged' (marked with a code) so that it can still be read for analysis purposes.

4.3 SALES ORDER AND COMPANY TYPES

In chapter 1, section 1.4, the classification of manufacturing companies into types was considered on the basis of stock policies and, as an alternative, on the basis of sales order types. The various sales order types that were listed defined, to some extent, the company type, as follows.

- *Type 1* Telephone sales to 'book' orders for same-day despatch if in stock. This defines a company that relies on meeting customers' orders from stock with a minimal response time. If the item is out of stock it probably means a lost sale (*example:* supply of perishable foods to a supermarket chain).
- *Type 2* Blanket and scheduled orders received for a range of custom products. This defines a company whose products are generally specific to a customer, with items being made in batches against order schedules and stocked where necessary (*example:* supply of components to an automotive manufacturer)
- *Type 3* Orders received regularly from established customers—often on a contract basis. This defines a company that provides a service as a subcontractor (*example:* supply of plating, fabrication or machining on customer's materials or components).
- *Type 4* Orders received on the basis of brochures and price lists: wide product range. This defines a company that does not hold stocks of finished goods but can supply fairly quickly from stocks of components and sub-assemblies (*example:* supply of cranes or conveyors that have to be made to suit a specific application).
- *Type 5* Orders received as a result of estimates and quotations against drawings or specifications. This defines a company that makes purely to order and holds minimal stocks (*example:* supply of structural steel fabrications for industrial and commercial buildings).

4.4 SALES ORDER PROCESSING

Although there are a number of different sales order and customer types, the sequence of operations needed to process any type of order can be illustrated in one flow chart, as shown in Fig. 4.1. From this it can be seen that the only real variation is between orders that are to be delivered from stock and those that are make-to-order. However, there are some minor variations in the earlier stages of the sequence, which can be described by examining each order type in detail.

1. telephone sales
2. blanket and schedule sales
3. regular and contract sales
4. brochure- and catalogue-based sales
5. quotation-based sales.

Electronic data interchange (EDI) can be used for order types 1 and 2.

Telephone sales

Sales are generated by the company's 'tele-sales' team, who phone customers on a regular basis to obtain their orders. The operation sequence for this type of sale on a computer-based system

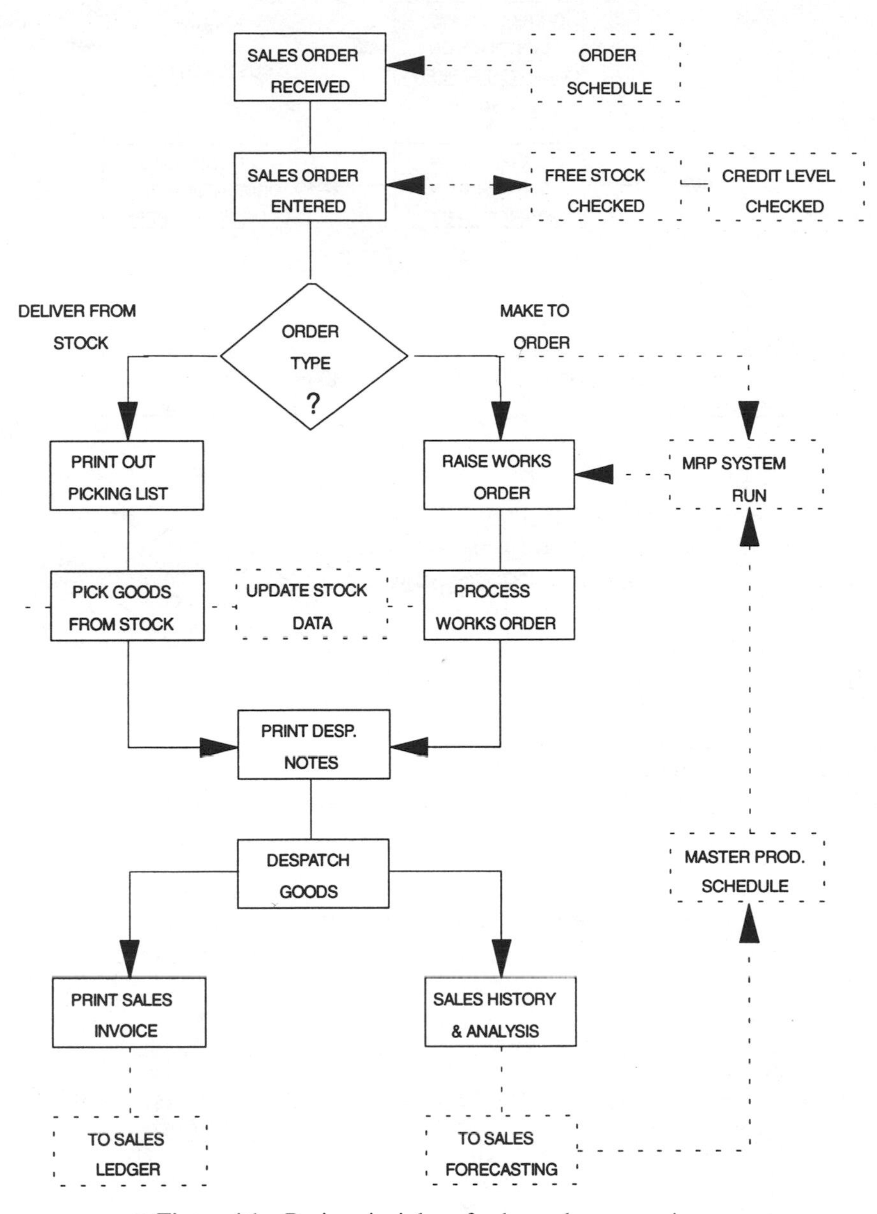

Figure 4.1 Basic principles of sales order processing.

is shown in Fig. 4.2. This sequence, which is difficult to operate without a computer, relies on speed of response to customers and their requirements.

A member of the tele-sales team sets up a customer record on a computer terminal; this displays such information as contact name, products normally ordered, net prices (including discounts) and credit status. The system then allows a telephone call to be made to the contact via the terminal, and if an order is to be placed an order form is called onto the screen (Fig. 4.3).

The customer contact then specifies the items and quantities required, and after each item the sales clerk enters the quantity to the screen (quantity-required box) and either confirms that

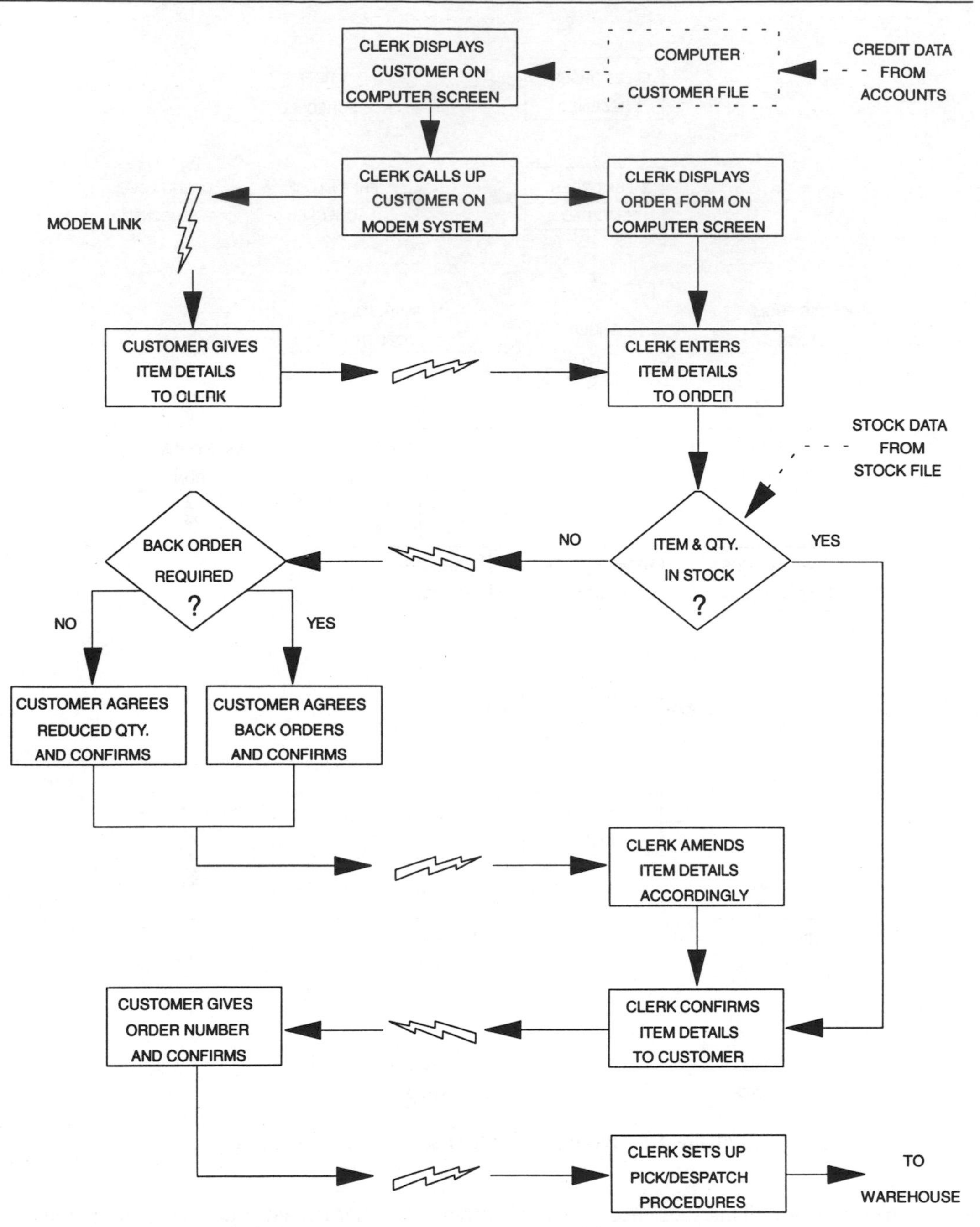

Figure 4.2 Telephone sales order system.

stock is available or reports the stock position. If stock is available a key is pressed which copies the quantity required entry into the quantity-ordered box; if it is not a reduced quantity is entered into that box and the contact is asked if a 'back order' is acceptable, i.e. should the balance be delivered when available, and a (Y)es or (N)o is entered.

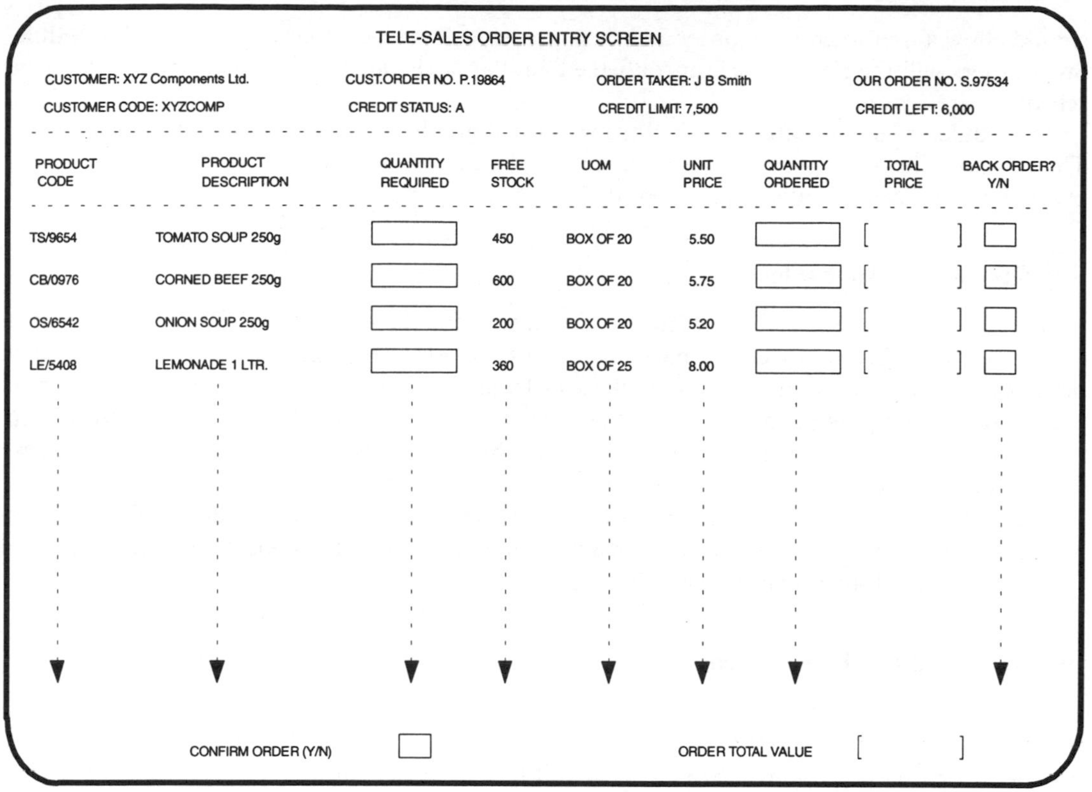

TELE-SALES ORDER ENTRY SCREEN

CUSTOMER: XYZ Components Ltd. | CUST.ORDER NO. P.19864 | ORDER TAKER: J B Smith | OUR ORDER NO. S.97534

CUSTOMER CODE: XYZCOMP | CREDIT STATUS: A | CREDIT LIMIT: 7,500 | CREDIT LEFT: 6,000

PRODUCT CODE	PRODUCT DESCRIPTION	QUANTITY REQUIRED	FREE STOCK	UOM	UNIT PRICE	QUANTITY ORDERED	TOTAL PRICE	BACK ORDER? Y/N
TS/9654	TOMATO SOUP 250g		450	BOX OF 20	5.50		[ ]	
CB/0976	CORNED BEEF 250g		600	BOX OF 20	5.75		[ ]	
OS/6542	ONION SOUP 250g		200	BOX OF 20	5.20		[ ]	
LE/5408	LEMONADE 1 LTR.		360	BOX OF 25	8.00		[ ]	

CONFIRM ORDER (Y/N) | ORDER TOTAL VALUE [ ]

Figure 4.3 A tele-sales order entry screen.

As each entry is made in the quantity-ordered box an item price is calculated and displayed by the system and a running total is displayed at the foot of the screen. This allows the sales clerk to check that the credit limit is not being exceeded. On completion of the order, the details are called back to the contact for confirmation and the order is then accepted and processed for packing and despatch.

Blanket and schedule sales

These are very similar types of order, which can be defined as follows.

1. *Blanket order* An order from a customer that does not specify in detail what is required but merely lists a range of products and prices and states that order quantities will be notified from time to time. In effect it is a form of contract to purchase.
2. *Schedule order* An order from a customer that specifies what is required in detail, but with quantities phased over a period of time, on a regular basis which could be monthly, weekly, or even daily. It is generally accepted that such schedules are approximate and subject to frequent changes.

In the case of blanket orders, the supplier usually creates a new order for each notification of a requirement by the customer referenced to the blanket order number, and therefore a conventional order-processing system can be used. However, in the case of schedule orders the system should allow a user to enter an order and then reproduce it several times, once for each schedule. Such orders will then be held as 'unconfirmed' and as such can be amended as changes to the schedules are received.

Schedule orders are often supplied from finished goods stocks or from consignment stock. This variant allows a customer to hold stocks on the suppliers' premises on an agreed payment basis and these can be called off as required.

Regular and contract sales

Where the supplier is a subcontractor to its customers there may well be an arrangement (contract) with agreed prices for particular types of work, e.g. plating of components at a price per thousand. In such cases no purchase order is placed and no formal sales order entered, each batch of components being sent to the subcontractor with an advice note and returned in the same way. The customer is then invoiced monthly against these notes at the agreed prices, these procedures being used to eliminate unnecessary paprwork.

However, in some cases a sales order (contract) is created by the contractor and used as an internal control document. His customer's advice notes (sent with the goods to be processed) are then treated as call-offs against this sales order.

Brochure- and catalogue-based sales

Manufacturers of a standard range of products will issue brochures or catalogues of their product range, with associated price lists. Customers can then order such items quoting a product code and price, and the sales order will be processed normally.

It is necessary to allow for at least four general price lists to be in use at any given time, plus a number of lists for specific customers or markets. For example, as a minimum:

- home and EU—previous
- home and EU—new list
- other export—previous
- other export—new list.

This will allow either previous or new prices to be used on a sales order during a price change period because, during such a period, certain customers may be allowed to continue to buy at old prices for policy reasons.

If a computer system is in use the facility should be available to produce a new list by copying and amending the current list. Thus, the following facilities could be required:

- to make an across-the-board percentage change
- to make percentage changes for different product groups or markets
- to make a percentage change to 'flagged' products (previously selected)
- to change prices of flagged products (previously selected)
- to change the price of an individual product
- to change the prices of a product group or market.

Percentage changes can of course be plus or minus, and prices may be subject to discounts (see section 4.9 below).

In some companies the standard range of products is very wide, with a variety of 'features and options' against each basic item (for example, a wide range of colour combinations, sizes

CUSTOMISED SALES ORDER SCREEN

Enter Customer Code [CA1207] Order and Line References [S.23578] [010] Number of Units Required [25]

[ABC BUILDERS & PLUMBERS MERCHANTS] Product Code [ALBANY.WB] [ALBANY WASHBASIN]

COLOURS		SIZES (mm)		WASTES		OPTIONS	
CANARY YELLOW		400 x 250		CHAIN WASTE 1.25"		WITH PEDESTAL	Y
DONKEY GREY		600 x 300	Y	CHAIN WASTE 1.5"		WITH TAPS	Y
GRASS GREEN	Y	700 x 350		POP-UP WASTE 1.25"	Y	WITH SOAP TRAY	
HYACINTH		750 x 400		POP-UP WASTE 1.5"		WITH MIRROR	
JONQUIL		800 x 450		**OVERFLOWS**		WITH TOWEL RAIL	
K'FISHER BLUE				NO OVERFLOW	Y	WITH MONOGRAM	
MIDNIGHT BLUE		**TAPHOLES**		SLOT OVERFLOW		WITH SPLASHBACK	
ROYAL PURPLE		2 TAPHOLES		CIRCULAR OVERFLOW			
SNOWFLAKE		100 mm MIXER					
VICTORY RED		200 mm MIXER	Y				
WESTMINSTER		1 CR. TAPHOLE					

Figure 4.4 Sales order entry screen (features and options).

and inlet/outlet variations). To simplify the task of order entry and to reduce the number of products in the product file, a special facility may be used as shown in Fig. 4.4. This shows a washbasin which is available in about 650 feature variations and a number of options. Use of this facility ensures that all features are entered (by entering Y in the appropriate boxes) and that all entries are valid (the system will reject any unacceptable combination, e.g. a size/colour that is not available). Only one feature can be accepted in any group but all options could be accepted. The system will then convert the entries made/validated into order lines, using the selections made, as in Table 4.2.

Quotation-based sales

Many 'make-to-order' companies are expected to quote against drawings and specifications before an order is placed by their customers, and therefore require an estimating procedure or system that will facilitate this. Figure 4.5 illustrates such a system, showing how an estimate can be built up in a series of 'steps', using as an example the structural steelwork for an architect-designed industrial building, as follows.

1. The raw materials are standard rolled-steel sections for which prices and other details already exist. The first step therefore is to prepare a detailed 'bill of quantities' (bill of materials, BOM) defining what items and quantities are required.

Table 4.2 Order lines converted from features and options entries

010	25	Albany washbasin	Grass green, 600 × 300 mm, 200 mm mixer, pop-up waste, no overflow
011	25	Albany pedestal	Grass green
012	25	CP mixer tap	200 mm with pop-up waste

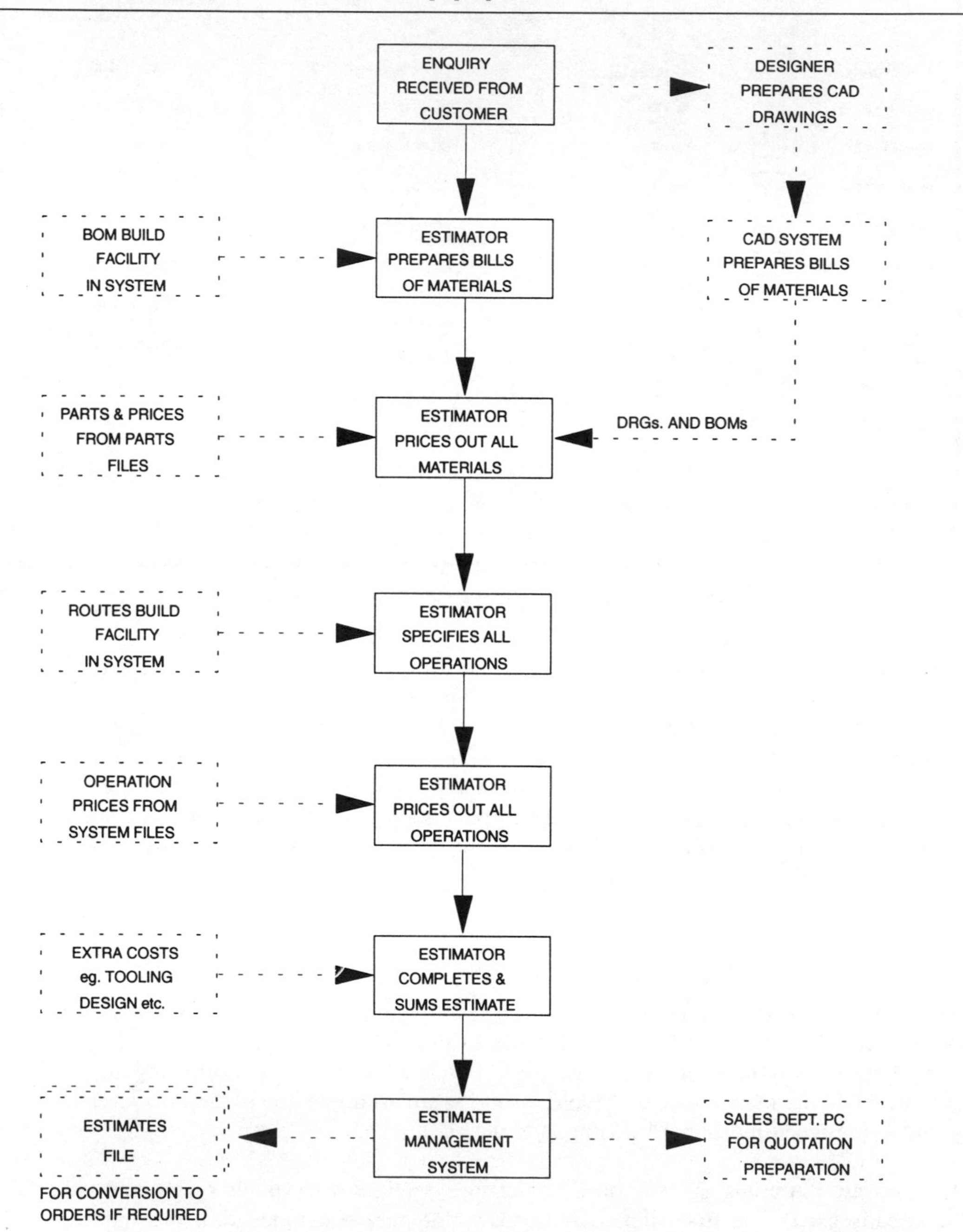

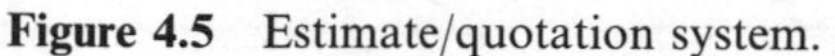

Figure 4.5 Estimate/quotation system.

2. These are priced by the estimator using the current price lists or computer-based records. In this case the BOM should include not only the steel sections but also quantities of the various types of standard connections, stanchion feet, etc., for which prices have already been calculated.
3. The estimator must then prepare a series of 'routeings', i.e. a list of the operations to be carried out to make the various structural units. For example, a roof truss could be made by cutting a number of steel sections to length, dressing the cut ends, welding the parts together, drilling holes and painting. Each of these operations should be given an estimated time and the grades of labour and machines to be used should be specified.
4. Prices for the various operations can be obtained from the files or computer records and used to estimate the cost of each operation.
5. Additional costs should be included for such items as the preparation of drawings and bills of materials, together with the costs of such items as special jigs and fixtures for work holding while drilling or welding.
6. Finally, the estimate is completed by adding the profit margin required and the quotation can then be prepared on this basis.

All estimates should be dated and numbered, and filed in such a way that they can easily be retrieved (preferably on computer). This allows the estimate to be easily converted into a sales order if successful.

The quotation is usually a summary of the estimate, with a specification and a price breakdown by major units. For example, the price breakdown for the structural steel building could be by construction stages, the format often being set by the customer. However, a quotation should always include payment terms (e.g. 10 per cent with order, 30 per cent on delivery of phase I, 50 per cent on delivery of phase II and 10 per cent on completion). It should also quote delivery dates and specify the validity date (e.g. 'this quotation is valid until 30 June 1995').

4.5 SALES ORDER ENTRY

A sales order is a record containing a series of data and is usually divided into three parts:

1. header data
2. product data
3. additional data.

Header data

The header contains information common to all the items on the order. Typical data that should be recorded are given in Table 4.3.

Product data

Information is required for each product on the order. Typical data that should be recorded are given in Table 4.4. The item number can be set by the supplier or generated by computer during order entry. However, if the customer's order includes item numbers then these should be used.

The sales unit of measurement can be defined by the user, e.g. 'each', 'per 1000', 'per kilo', 'per gross', 'per box of 25', and will normally be a part of the product record.

Table 4.3 Sales order header data

Internal order ref.	Manually obtained from a register or computer-generated
Customer order no.	Given by customer; if verbal order then must be confirmed
Customer code	Account code from which other details can be obtained
Order date	Date on customer's order
Order receipt date	Date received by supplier
Order entry date	Date entered by supplier
Currency code	Used to convert prices for invoicing
Delivery code	Used to define a delivery method

Table 4.4 Sales order product data

Item number	Each item should be given an item or line number
Product code	Defines the product from which other details can be obtained
or	
Free text	To define if a non-standard item, including UOM
Quantity	In terms of the sales unit of measurement
Price	In terms of the sales unit of measurement
Delivery required	Date required by customer
Delivery promise	Date promised by supplier
Special instructions	Any special requirements (free text), e.g. packing method or inspection procedure

Additional data

These will normally be a free text section that will allow the user to enter any special instructions or requirements. For example, this part of the order could be used to specify any delivery instructions ('Deliver to Gate 6 between 9am and 11am') or to specify maximum vehicle height. Equally it could be used to give shipping instructions for an export order.

A useful facility that is available or some computer systems is standard text, which may be available for the entry of both 'additional data' and product 'free text'. This allows the user to set up and code wording that is frequently used so that entry of the code calls off the text to save typing. Such text can then be amended or expanded if required.

4.6 WORKS ORDER CREATION IN AN SOP SYSTEM

If a company can be classified as a 'make-to-order' type, then the sales order processing system may be required to include a facility for the preparation of works orders. This may be achieved by passing a computerized sales order via the MPS (master production schedule) module to the MRP (material requirements planning) module or, if MPS is not being used, directly to MRP which will break the order down into gross requirements (i.e. requirements that ignore free stocks and outstanding purchase or works orders) for parts and sub-assemblies, and then to net requirements. The MRP module will then produce recommendations for new works and purchase orders to cater for the sales order.

However, in a pure make-to-order situation, MRP is not likely to be suitable since many of the parts and sub-assemblies will not be stock items. In such a case, the sales order must be broken down to give a 'schedule of requirements' or 'planning bill' which will be, in effect, a bill of materials (BOM) usually obtainable from the assembly drawings or the CAD (computer-aided design) system. The items on this schedule can then be converted into a series of works and purchase orders which will be 'pegged' to the sales order, i.e. they will include the sales order reference in the works or purchase order header. This will allow enquiries to be made by Sales or Purchasing when they wish to check progress against the sales order concerned.

4.7 ALLOCATIONS AND PICKING LISTS IN AN SOP SYSTEM

If a company can be classified as a 'make-to-stock' type, then the sales order processing system may be required to include a facility for the allocation of stock and the preparation of picking lists. Allocation is a method of reserving stock against a demand (usually a sales order) so that although the stock is visible it is not available (free). Allocation does not normally involve a stock movement but merely an entry on the stock record, and if necessary (perhaps because of an order cancellation) the facility to deallocate should also be provided.

A picking list is a printed document which lists items that should be taken out of stock for packing and despatch. A typical format is shown in Fig. 4.6. This shows that a number of orders for a customer have been listed with details of quantities required and stock locations for each item. The list also has spaces for the order picker to enter quantities actually picked with packing and despatch details. It will be noted that the order items are shown separately even when they are identical (for example, the Avon 10″ plate). This is to help customers to check off receipts against their orders, which would be difficult if identical items on different orders were shown as a single item.

The procedure for creating picking lists will vary depending on the system being used and the customers' needs, but generally operates as follows.

1. Stock is allocated at the order entry stage, or later if the delivery is some way ahead.
2. Selections are made on the basis of a customer, a range of orders, a batch of orders or a single order. These selections will attach a picking list number to an order or orders.

CUSTOMER: xxxxxxxxxxxxxxxxxxxxxxxxxxxxxxxxxxxx
xxxxxxxxxxxxxxxxxxxxxxxxxxxxxxxxxxxx
xxxxxxxxxxxxxxxxxxxxxxxxx
xxxxxxxxxxxxxxxxxxxxxx

PICKING LIST

DESPATCH METHOD: xxx

PICKING LIST NO. xxxxxxx
DESPATCH DATE: DD/MM/YY
PAGE NO. 1 of 1

ORDER NO.	ORDER LINE NO.	PRODUCT CODE	DESCRIPTION	QTY.	UNITS	W'HOUSE	LOCATION (RACK)	QTY. PICKED	CARTONS OR PALLETS	PICKED BY
AB.27502	10	234/56	AVON 10" PLATE	15	EACH	FG.6	C.23			
	20	234/72	AVON 14" DISH	24	EACH	FG.6	C.30			
	40	500/45	TEST CREAM JUG	5	EACH	FG.7	D.14			
	50	275/86	TAMAR 10" PLATE	20	EACH	FG.7	D.19			
BD.3579	20	234/56	AVON 10" PLATE	30	EACH	FG.6	C.23			
	30	234/42	AVON 7" PLATE	40	EACH	FG.6	C.32			
	50	726/80	TYNE COFFEE POT	16	EACH	FG.6	E.32			
	70	435/78	TEES CUTLERY SET	12	SET	FG.7	F.21			
BD.3585	30	234/56	AVON 10" PLATE	35	EACH	FG.6	C.23			
	10	234/42	AVON 7" PLATE	5	EACH	FG.6	C.32			
			UNITS LISTED	202				CARRIER		
								WEIGHT		
			UNITS PICKED					CARTONS		

Figure 4.6 Typical picking list format.

Table 4.5 Picking list sequences

Sort	Order	Line	Product	Quantity	Warehouse	Rack
Location	AB.27502	10	234/56	15	FG.6	C.23
	BD.3579	20	234/56	30	FG.6	C.23
	BD.3585	30	234/56	35	FG.6	C.23
	AB.27502	20	234/72	24	FG.6	C.30
	BD.3579	30	234/42	40	FG.6	C.32
	BD.3585	10	234/42	5	FG.6	C.32
		etc.			etc.	

3. The picking lists can then be printed, with a choice of sequence to suit the user, e.g. by location within order number, by location only (see Table 4.5) or by product.

4.8 PACKING AND DESPATCH IN AN SOP SYSTEM

All despatches should be fully covered by documentation, normally a printed three-part despatch note set. This set has the following functions.

1. The top copy provides the customer with information about the delivery so that it can be identified and checked against the customer's purchase order. It is retained by the customer.
2. The second copy provides the supplier with evidence that the delivery has been made and accepted. This is achieved by the carrier obtaining a signature from the customer on this copy, which is then returned to the supplier. This is known as the PoD (proof of delivery) copy.
3. The third copy is retained by the supplier and held in the despatch office. Once the PoD copy has been returned this third copy can be passed to accounts for an invoice to be raised and sent out.

A copy of a typical despatch note (sometimes known as a delivery or advice note) is shown in Fig. 4.7. It can be seen that this is very similar to the picking list (Fig. 4.6).

In fact, despatch notes are usually developed on a computer system by the entry of data taken from picking lists. This method is shown in the example in Fig. 4.8, as follows.

- *Step 1* Enter a picking list number; screen will display the customer name for confirmation. Enter a delivery address code or leave blank if default address is required. Screen will display address for confirmation.
- *Step 2* Screen will display product details from the original picking list; enter quantity if different from original or leave blank to default to original. Also enter numbers and details of cartons and pallets; confirm.
- *Step 3* Enter other details that will be used for invoicing purposes, using rules set up in the system, e.g. stillages charges at £10 each.

When an item is marked with a delivered quantity during the despatch note creation phase, the quantity on the original sales order will be reduced by that quantity automatically. If this leaves a balance outstanding this becomes a 'back order' and will be delivered at a later stage when

DESPATCH NOTE

ABC POTTERY LIMITED

CUSTOMER
xxxxxxxxxxxxxxxxxxxxxxxxxxxxxxxxxxxxxxxx

DELIVERY ADDRESS
xxxxxxxxxxxxxxxxxxxxxxxxxxxxxxxxxxxxxxxx
xxxxxxxxxxxxxxxxxxxxxxxxxxxxxxxxxxxxxxxx
xxxxxxxxxxxxxxxxxxxxxxxxxxxx
xxxxxxxxxxxxxxxxxxxxxxxx

DESPATCH NOTE NO. xxxxxxx
TAX POINT: DD/MM/YY
PAGE 1 of 1

ORDER NO.	LINE NO.	PRODUCT CODE	DESCRIPTION	QTY.	UNITS	PACKAGES	PACKAGE TYPE	CUSTOMER COMMENTS
SO.2736	10	234/56	AVON 10" PLATE	15	EACH	3	CARTON A	
	20	346/50	TYNE 10" PLATE	20	EACH	3	CARTON A	
	30	147/92	TEES 10" PLATE	150	EACH	15	CARTON A	
SO.2942	10	932/70	TRENT 12" DISH	30	EACH	6	CARTON C	
	20	502/07	SEVERN 7" PLATE	50	EACH	4	CARTON B	
	30	800/43	CLYDE 7" PLATE	120	EACH	10	CARTON B	
	40	280/21	TWEED 7" PLATE	40	EACH	4	CARTON B	

DELIVERY INSTRUCTIONS: xxxxxxxxxxxxxxxxxxxxxxxxxxxxxxxxxxxxxxxxxxxxxxxxxxxxxxxx
xxxxxxxxxxxxxxxxxxxxxxxxxxxxxxxxxxxxxx

SIGNED FOR: ....
DATE: ....

Figure 4.7 Typical despatch note layout.

DESPATCH NOTE CREATION SCREEN

Enter Picking List No. [PL17543] Customer Name Displayed — Is Name Correct ? (Y/N) [Y]

Enter Delivery Address Code. [DY1420] Delivery Address Displayed — Is Address Correct ? (Y/N) [Y]

ORDER NO.	ORDER LINE NO.	PRODUCT CODE	DESCRIPTION	QTY.	UNITS	QTY. PICKED	CARTONS OR PALLETS	CARTON/ PALLET TYPE	DATA CORRECT ? Y/N
AB.27502	10	234/56	AVON 10" PLATE	15	EACH				
	20	234/72	AVON 14" DISH	24	EACH				
	40	500/45	TEST CREAM JUG	5	EACH				
	50	275/86	TAMAR 10" PLATE	20	EACH				
BD.3579	20	234/56	AVON 10" PLATE	30	EACH				
	30	234/42	AVON 7" PLATE	40	EACH				
	50	726/80	TYNE COFFEE POT	16	EACH				
	70	435/78	TEES CUTLERY SET	12	SET				
BD.3585	30	234/56	AVON 10" PLATE	35	EACH				
	10	234/42	AVON 7" PLATE	5	EACH				

Method of Despatch (Text) [ ]
Packing Charges ? £ [ ]
Carriage Charges ? £ [ ]
Carrier (Text) [ ]

Insurance Charges ? £ [ ]
Number of Cartons/Pallets — cartons [ ] pallets [ ]

Figure 4.8 Despatch note creation—user entries and responses.

stock becomes available, unless cancelled by the customer. The quantity despatched will also be used to adjust the finished goods stock record in terms of allocated and total stock.

4.9 INVOICING AND DISCOUNTS IN AN SOP SYSTEM

Invoices are created from despatch notes, and often carry the same reference number for easy identification. Most companies apply unit prices to an order at the order entry stage and these prices are then applied to the item quantities on the despatch note to give gross item prices (e.g. 60 at £1.50 = £90.00).

An example of a typical invoice is shown in Fig. 4.9 with products, quantities, unit prices and gross prices (before discount and sales tax). The remaining four columns are used for discounts and VAT.

Discounts can come in a variety of types but all have the same objective, to improve sales turnover. Discounts can be applied at item or invoice level and are almost always quoted in percentage terms.

- *Customer discounts* Discounts given to specific customers in order to retain their loyalty. These are normally applied at the foot of the invoice.
- *Market discounts* Discounts given in specific markets to retain or expand business in the face of competition. These can also be applied at the foot of an invoice or at item level.
- *Product discounts:* Discounts given for purchases of specific products in order to boost flagging sales or to launch a new product. These are always applied at item level.
- *Break discounts:* Discounts given for purchases in larger quantities to encourage customers to order less frequently by stocking more. These can be at invoice, product group, or item level (see Table 4.6).

A problem that may be encountered when discounts are being applied is the combination of several discount rates which reduce the net price to an uneconomic level. It is therefore necessary to build a series of rules into the system, e.g.

INVOICE

CUSTOMER: xxxxxxxxxxxxxxxxxxxxxxxxxxxxxxxxxx
xxxxxxxxxxxxxxxxxxxxxxxxxxxxxxxxxx
xxxxxxxxxxxxxxxxxxxxxx
xxxxxxxxxxxxxxxxxxxx

INVOICE NO. xxxxxxx
TAX POINT: xx/xx/xx
ORDER NO. xxxxxxxxx

LINE NO.	PRODUCT CODE	DESCRIPTION	QTY.	UNITS	PRICE PER UNIT	GROSS PRICE	DISCOUNT TYPE	DISCOUNT %	NET PRICE	VAT RATE
10	234/56	AVON 10" PLATE	15	EACH	2.50	37.50		NIL	37.50	17.5%
20	346/50	TYNE 10" PLATE	20	EACH	2.00	40.00	PRD	5.00	38.00	17.5%
30	147/92	TEES 10" PLATE	150	EACH	1.50	225.00	BRK2	5.00	213.75	17.5%
40	932/70	TRENT 12" DISH	30	EACH	4.00	120.00	BRK1	2.50	117.00	17.5%
					GROSS	422.50	NET	3.85 %	406.25	
							LESS 2.5% CUST.DISC.		396.09	
							PLUS 17.5% VAT		69.32	
							INVOICE TOTAL		465.41	

DISCOUNT CODES			
MKT	Market	BRK1	Break Level 1
PRD	Product	BRK2	Break Level 2
CUST	Customer	BRK3	Break Level 3

Figure 4.9 Typical invoice with various discounts.

Table 4.6 A break discount system

No. of units	Discount %
Up to 50	Nil
51 to 100	2.50
101 to 200	5.00
Above 200	7.50

1. if both market and customer discounts are applicable on a particular invoice, then only the larger of the two should be applied, or if they are the same then the customer discount only should be applied
2. if both product and break discounts are applicable on a particular invoice, then only the larger of the two should be applied, or if they are the same then the break discount only should be applied
3. break discounts are applicable only at the delivery stage, not at the order stage.

4.10 CUSTOMER RETURNS IN AN SOP SYSTEM

Regardless of the type of SOP system being used, it should include facilities for dealing with returns from customers. A number of reasons can be identified for a return to be made, and a number of procedures for ensuring that the return is correctly managed. In fact, there are three types of return (covering seven reasons) as given in Table 4.7. The procedures for dealing with each type can be summarized as follows.

- *Type A* The returned goods should be inspected and then returned to stock or, if substandard, to a quarantine area to be reprocessed, scrapped or regraded. The old order should be cancelled, a new order raised, and a credit note (Fig. 4.10) created to cancel out the invoice that was produced when the shipment was made. The correct goods should then be despatched and invoiced against the new order using normal despatch procedures.
- *Type B* The goods should be delivered to the correct address or as required by the delivery instructions. This will require the creation of a new despatch note set, which will result in a new invoice being created. The old invoice must therefore be 'cancelled' by means of a credit

Table 4.7 Types of customer return

Type A	Correct goods ordered, wrong goods supplied Wrong goods ordered and supplied Goods below required quality standard
Type B	Wrong delivery address Delivery instructions not observed
Type C	Quantity supplied in excess of order Order cancelled by customer but not actioned

CUSTOMER: xxxxxxxxxxxxxxxxxxxxxxxxxxxxxxxxxxxxxxx
xxxxxxxxxxxxxxxxxxxxxxxxxxxxxxxxxxxxxxx
xxxxxxxxxxxxxxxxxxxxxxxxxx
xxxxxxxxxxxxxxxxxxxxxxx

CREDIT NOTE

CUSTOMER CODE: WA.2356

CREDIT NOTE NO. 99999
TAX POINT: DD/MM/YY
DESPATCH NOTE NO. 11111

ORDER NO.	LINE NO.	PRODUCT CODE	DESCRIPTION	QTY.	UNITS	PRICE PER UNIT	GROSS PRICE	DISCOUNT %	NET PRICE	VAT RATE
S0.13754	10	234/56	AVON 10" PLATE	15	EACH	2.50	37.50	2.50	36.56	17.5%
	20	346/50	TYNE 10" PLATE	20	EACH	2.00	40.00	2.50	39.00	17.5%
S0.13825	20	147/92	TEES 10" PLATE	150	EACH	1.50	225.00	5.00	213.75	17.5%
	30	932/70	TRENT 12" DISH	30	EACH	4.00	120.00	2.50	117.00	17.5%
	40	750/21	SOAR 12" DISH	15	EACH	4.50	67.50	2.50	65.81	17.5%

COMMENT:
This Credit Note refers to our invoice M.009654 for goods returned by you on your Debit Note DN.67402

NET CREDIT	472.12
LESS 2.5% CUST.DISC.	460.32
PLUS VAT	80.56
CREDIT NOTE TOTAL	540.88

Figure 4.10 Typical credit note layout.

note, and it will also be necessary to cancel or amend the original order so that it carries the correct delivery address or instructions.

- *Type C* The returned goods should be inspected and then returned to stock or, if substandard, to a quarantine area to be reprocessed, scrapped or regraded. A credit note should be raised and the original order amended or cancelled.

In all cases of a customer return it is important to ensure that all records are corrected so that data in the system are accurate. The records concerned are as follows.

- *Sales ledger* This can normally be corrected by creating a credit note and then a new invoice. It is incorrect procedure to cancel or amend an invoice.
- *Sales order analysis* The importance of sales order analysis for forecasting and planning has already been stressed. It is therefore essential to cancel and re-enter or amend sales orders when a return has been received from a customer.
- *Stock records* The stock records can be corrected by a 'return-to-stock' transaction if the goods returned are not substandard. However, it is advisable to set up a quarantined stock location so that a record is visible for suspect returns.

4.11 SALES ORDERING THROUGH EDI

Electronic Data Interchange (EDI) allows a customer to place orders on a supplier directly by accessing the supplier's computer system. This saves time and possible transcription errors; it also allows the customer to amend an order and in some cases to enquire about stock availability.

Such a system can operate only between companies which recognize that their supplier or customer is a partner and that certain conventions must be observed. For example, it is probably not acceptable to cancel an order that has already left the supplier's factory, except in very exceptional circumstances.

EDI operates best where the supplier is making regular deliveries to the customer against schedules, since it allows goods to be called off and delivered 'today' from the supplier, in line with what was used or sold 'yesterday'. This is particularly relevant if the customer is operating a JIT (just-in-time) system or is a large retailer (e.g. a supermarket chain).

An EDI link can be used as part of a tele-sales system by allowing the customer to enter the tele-sales form (Fig. 4.3) and to see the responses in terms of stock availability. This will allow the customer to make decisions about back orders.

EDI can also be used for 'self-billing', the creation of self-addressed sales invoices by a customer on behalf of the supplier, charging for goods taken from consignment stock. In other words, goods are supplied to the customer but not invoiced since they are held in a special stock location. Then, when a quantity is withdrawn from this location for use by the customer, EDI is used to notify the supplier that the withdrawal has taken place and that the customer must pay the supplier on an agreed basis (e.g. monthly).

This facility is examined in more detail in chapter 8. The various EDI facilities are also examined in chapter 8, from the point of view of the purchaser.

PROBLEMS

1. Prepare print layouts for two sales analysis reports that will illustrate the following factors:
   (a) sales by value and quantity for a customer for a period
   (b) order items that are late delivered for a customer and period.
2. Write a brief memo to the sales director of your company commenting on and explaining the figures in the two reports prepared in problem 1.
3. Prepare a print layout for a sales order form in terms of the header, product data and additional data for a company that manufactures a wide range of confectionary products with typical data entries.
4. Prepare a picking list sequence (similar to that in Table 4.5) on the following basis:
   (a) sorted by location within product, assuming that a product can be stocked in several locations
   (b) sorted by product only, assuming that a product can be stocked in several locations.
5. Prepare a procedure to be used for dealing with customer returns.

FURTHER READING

Anderson, E. J., *The Management of Manufacturing*, Addison-Wesley, Wokingham, 1994.
Mair, G., *Mastering Manufacturing*, Macmillan, London, 1993.
Smith, S. B., *Computer-based Production and Inventory Control*, Prentice-Hall, Englewood Cliffs, NJ, 1989.
Vollmann, T. E., Berry, W. L. and Whybark, D. C. *Manufacturing Planning and Control Systems*, 3rd Edn, Irwin, Boston, MA, 1988.

CHAPTER
FIVE
THE MANAGEMENT OF DESIGN

THE LINK

The need for a company to provide a world-class service to its customers was established in the previous chapter, together with suggested means of achieving this. However, this objective is unlikely to be met if product design is not up to standard in terms of the required degrees of 'fitness for function', 'process capability' and 'useful life'.

It is therefore a primary design function to consider the various aspects of product quality, bearing in mind the needs of the market at world-class level, and the capability of the existing production facilities. In addition, design must recognize that 'time-to-market' is often a critical factor. In other words, a new product must be available for sale before the competition can 'steal' its potential customers with its new product.

To achieve these two major objectives, the design department should be organized as a team, or a series of coordinated teams, so that the 'buck-passing' situation does not arise. For example, it would be totally unsatisfactory if a design were prepared that was then rejected by production engineering, marketing, or purchasing because of technical manufacturing problems, likely customer reaction, or unavailability of supplies.

This teamworking is the basis of concurrent engineering, which is described in this chapter, and which ensures that everyone in the company who needs to be involved during the design phase is involved, at the earliest possible time.

5.1 INTRODUCTION

In the previous chapter, references were made to the role of design, and it was suggested that a design department must endeavour to match the desires of the customer with the capabilities of the production resources. In this chapter, the way in which a design department should operate to meet this requirement is considered in detail, by suggesting answers to such questions as:

- how can the time taken to get a new product onto the market be reduced?
- how can the value of a product be increased?
- how can quality be built into the product at the design stage?
- how can the designer ensure that the product fits the manufacturing resources?
- how can the designer meet the desires of the customer?

Various 'tools' and techniques are now available to help to solve the problems that these questions raise, and these will be examined in this chapter, which will also consider how design must integrate with other departments if it is to be successful.

5.2 DESIGNING FOR MINIMUM TIME-TO-MARKET

For many years the design function was considered to be related more to art than to science, and it was believed that artists could not be expected to work to a timetable when inspiration was the key. After all, it took Michelangelo seven years to paint the ceiling of the Sistine Chapel!

Thus, so the theory went, the designer should not be put under pressure to meet a deadline or inspiration would be lost. However, this theory is based on the fallacy that all products must be artistic masterpieces rather than commercial successes, and without commercial success there is no job for the designer.

Commercial success is based on a variety of factors, and one of these is that a company cannot stand still but must operate a policy of continuous new product development (NPD), since no product has an infinite life. This is illustrated in Fig. 5.1, which shows how a product should be monitored in order to determine when an upgrade or replacement is required.

The importance of a short time-to-market for the replacement can be seen, since any delay will cause a considerable loss of sales income, and can also be costly in terms of an unwanted extra period of investment without any return. In order to minimize the time it takes to bring a new product onto the market, it is necessary to examine the sequence of events involved and then to try to 'compress' the time by overlapping, or parallel working. This process is known as concurrent (or simultaneous) engineering, and the sequence is shown in Fig. 5.2 firstly in 'series' format (method A) and then 'in parallel' (method B), showing the reduction in time.

Concurrent engineering means that a new product is developed by a team or a number of teams, each containing several 'function specialists' such as

- concept and detail designers
- production engineers
- quality engineers
- materials specifiers.

To make concurrent engineering work, teamwork is essential, and it is not enough merely to assign people to a group without leadership or direction; true teamwork exists only when various conditions have been satisfied, as given in Table 5.1.

It is also important to ensure that members of the team communicate closely with one another, as shown in Fig. 5.3, and this communication is greatly enhanced if the team can all work in the same office, with shared facilities. The use of milestones and overlapping is illustrated in Fig. 5.4, and the four 'phases' shown can be defined as follows.

Concept phase

The phase starts with a marketing study to determine customer and market needs and to set the basic parameters, e.g. is it 'top of the range' or 'low-cost economy'?

Next, a feasibility review is carried out to determine whether these design parameters are viable, and this will probably include a QFD (quality function deployment) exercise as described in chapter 13, section 13.8. At this stage it may also be necessary to prepare prototype models and/or computerized simulations or visualizations for market testing (see section 5.7 below).

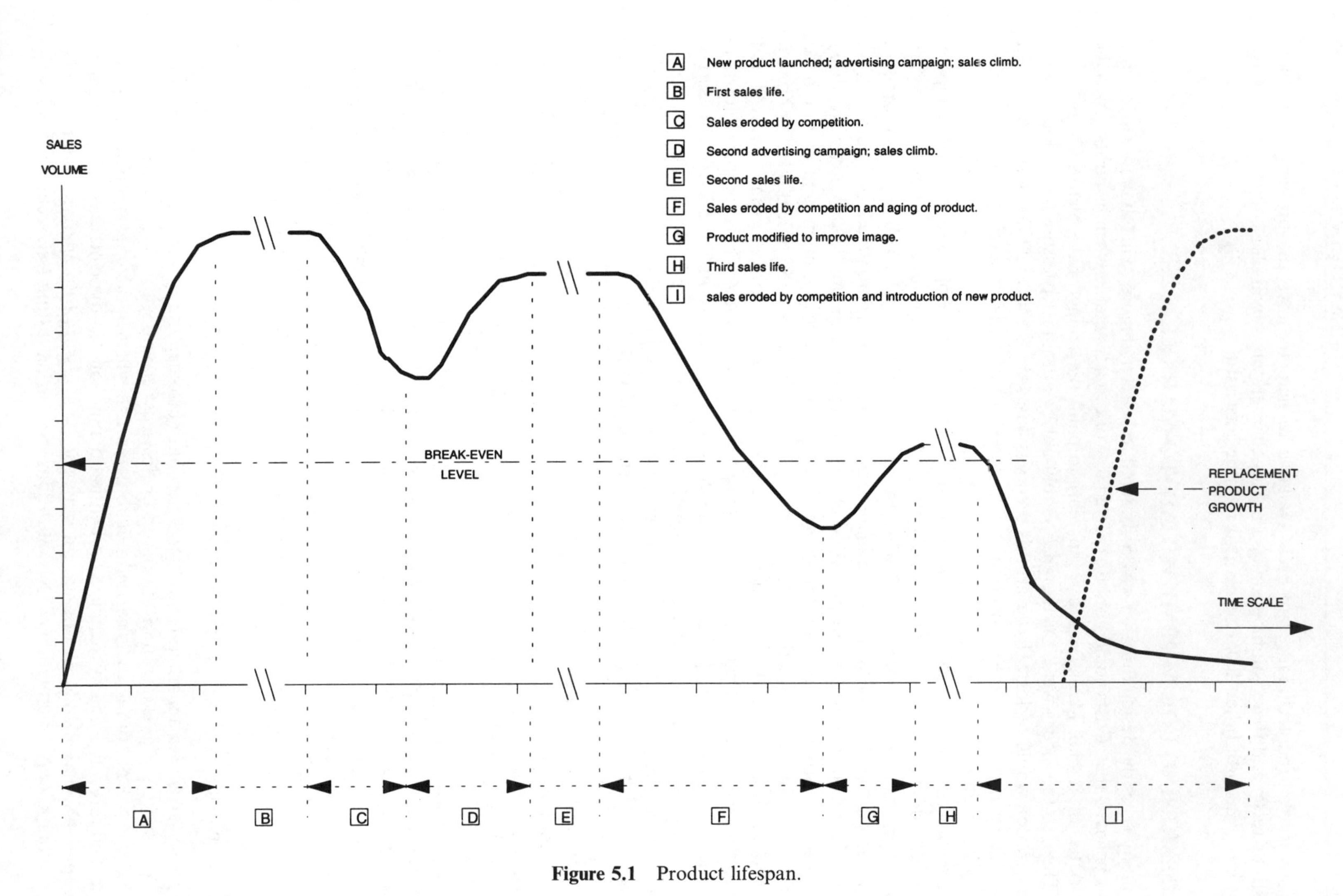

Figure 5.1 Product lifespan.

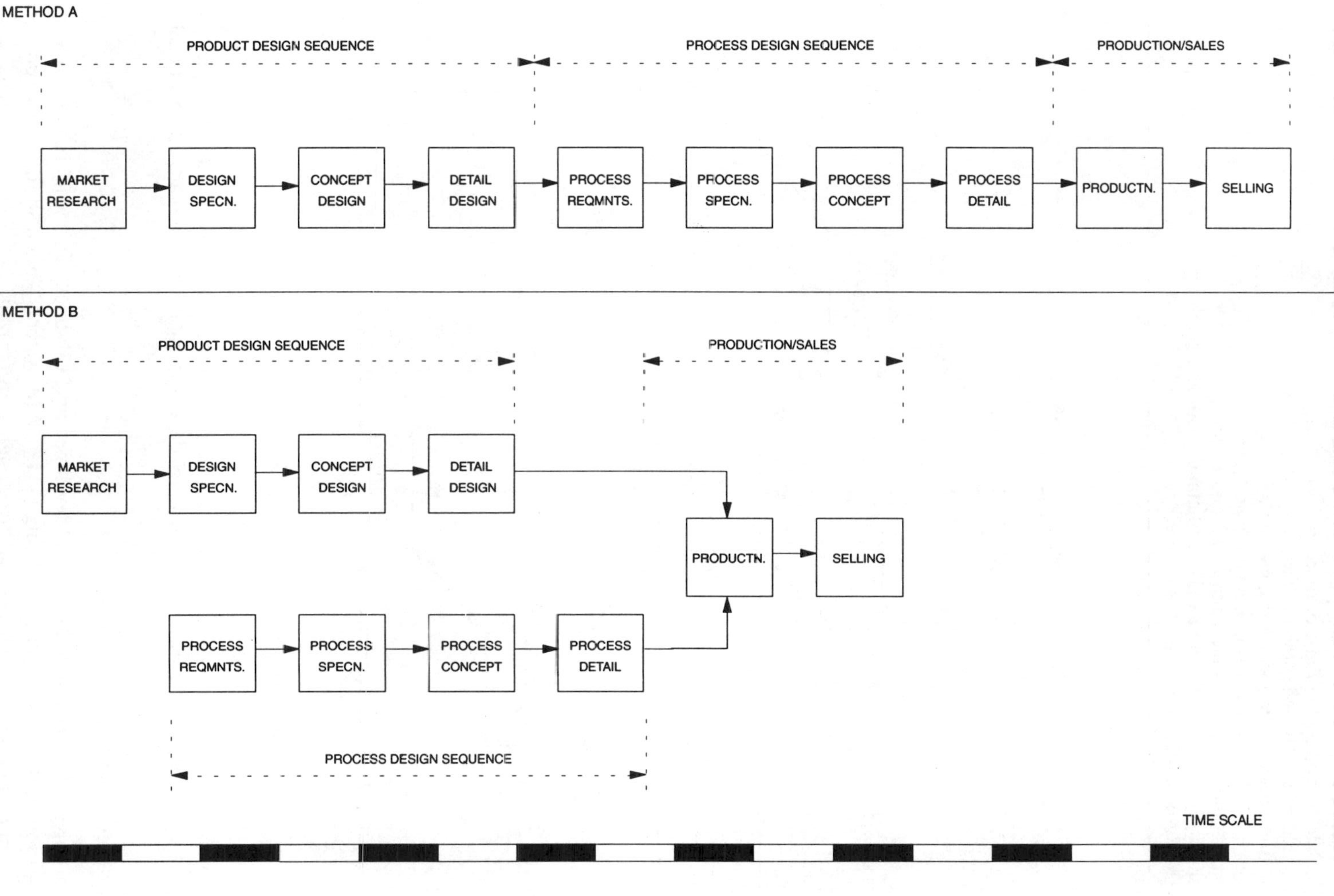

Figure 5.2 Sequence of activities for new product development.

Table 5.1 The right conditions for teamwork

- There should be a shared purpose or objective.
- There should be clearly understood priorities.
- Participation should be balanced and consistent.
- Tasks should be linked to a set of common 'milestones'.
- Decisions should be based on facts and data, not opinions.
- There should be a leader who can handle the human issues and resolve problems.
- There should be authority to act.

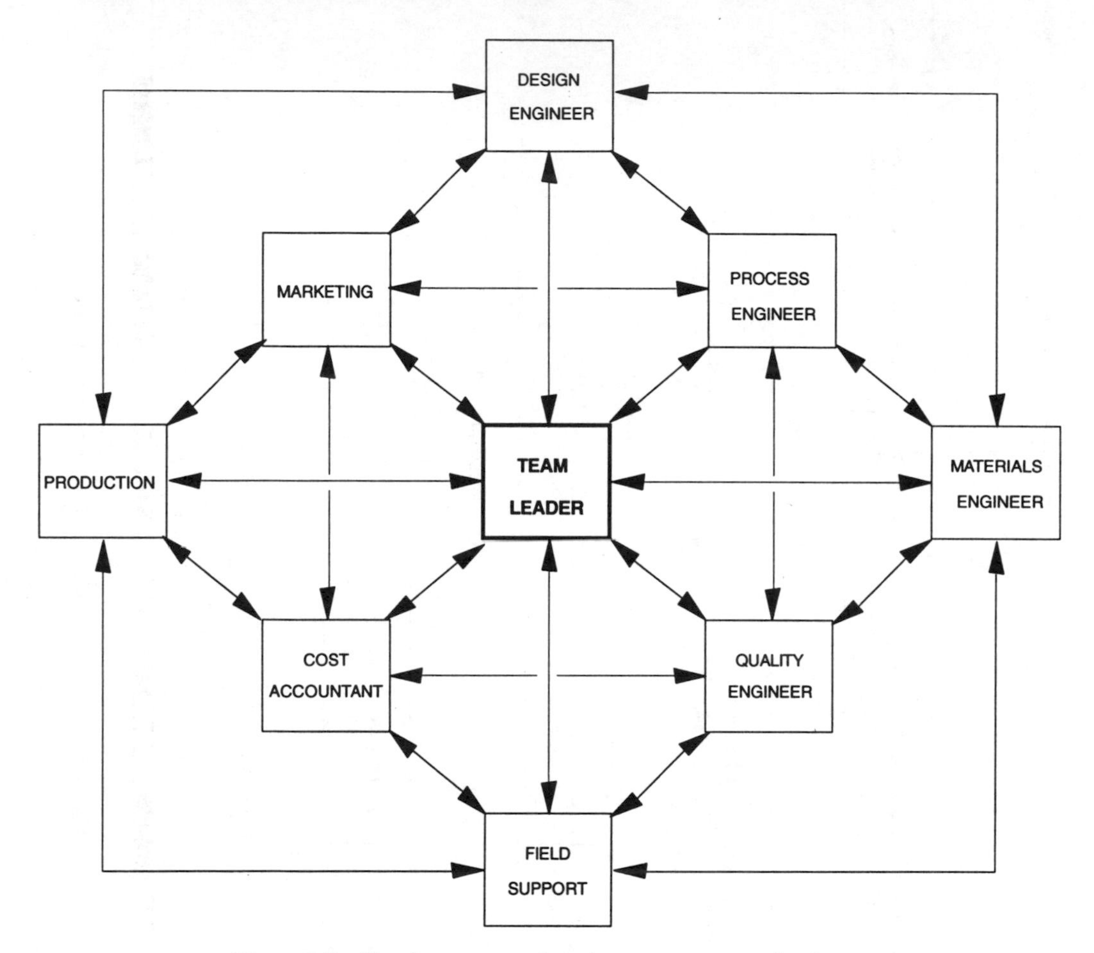

Figure 5.3 Simultaneous engineering team communications.

The milestone at the end of this phase will be a decision to proceed, based on a design review, although work on the next phase (product development) should already have started.

Product development

During this phase, the overall design will be developed and validated. This will involve not only the preparation of drawings (preferably using CAD/CAM systems), but also the application of various techniques designed to ensure that the design is optimized.

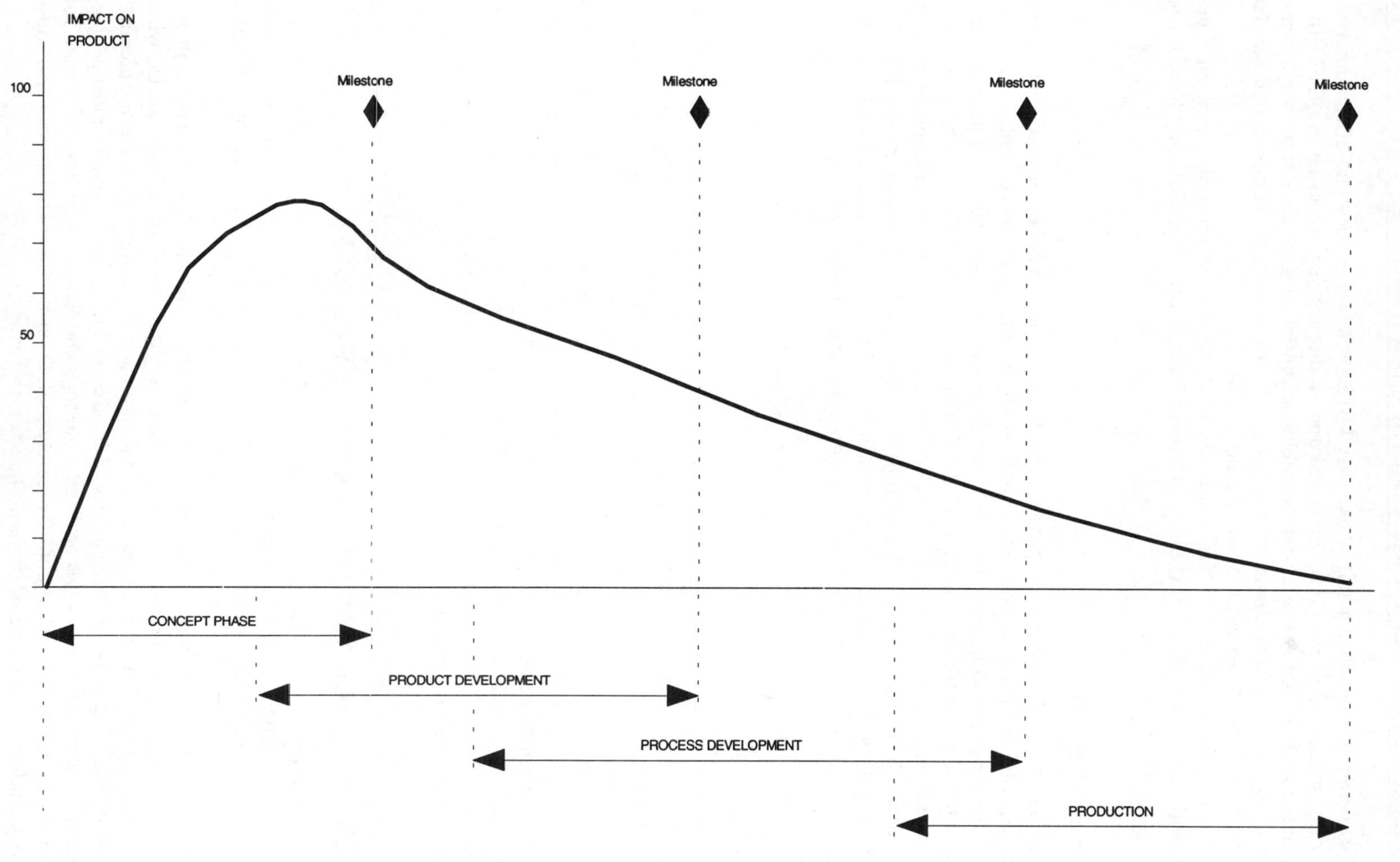

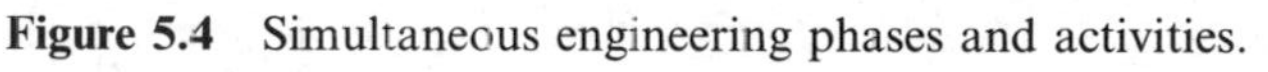

Figure 5.4 Simultaneous engineering phases and activities.

For example, it is during this phase that FMEA (failure modes and effects analysis—chapter 13, section 13.8) should be employed and studies carried out using the DFM (design for manufacture) and DFA (design for assembly) methodologies. Testing of prototypes should be planned using Taguchi omtimization methods and, where available, computerized simulation and CAE (computer-aided engineering) should be used.

The milestone at the end of this phase will be a decision to proceed, based on a design review, although work on the next phase (process development) should already have started.

Process development

During this phase, the production processes that will be used will be designed and validated. This will involve the preparation of flow charts to define the operation sequences, the development of detailed operating procedures (e.g. machine settings), the specification and design of tooling, and the preparation of programs for NC (numerically controlled) machines, robots and computerized measuring devices. If a CAPP (computer-aided process planning) package is available, this will reduce the time taken to develop the detailed operating procedures.

The milestone at the end of this phase will be a decision to proceed, based on a design review, although work on the next phase (production) should already have started.

Production

During this phase, pilot production will be used to resolve initial problems, although if simulation has been used effectively, these should be minimal. The team should then be in a position to increase production rates rapidly to the target volumes and to ensure that all the required process controls (such as SPC—statistical process control) are operating satisfactorily.

Once production has reached full volume, the team can reduce the frequency of their involvement, but should continue to monitor the output so that a series of continuous improvements can be implemented.

The milestone at the end of this phase will be the attainment of the original project objectives and the preparation of a final report that highlights not only the achievements, but also the failures and the lessons learned.

5.3 DESIGNING FOR MAXIMUM PRODUCT VALUE

Value engineering (sometimes known as value analysis) should not be confused with cost reduction. Value engineering is designed to enhance the value of a component or assembly by improving its functionality for the same or a reduced cost, whereas cost reduction is an attempt to reduce the cost of an item without any change in its functions.

To illustrate this, it is necessary to understand the difference between cost and value. Generally speaking, cost is a fairly constant factor regardless of the circumstances, whereas value varies markedly between individuals and situations. For example, the cost of widening the entrance to a building is the same regardless of the use to which it is put, and for most people the increase in value of this alteration is negligible, but to a wheelchair user it has a very high value. In other words, while the accountant may merely look at costs, the value engineer will consider both costs and usage, and the relationship between the two.

The FAST technique

One of the tools used by value engineers is a technique called FAST (function analysis system technique) which allows the engineer or designer to view the design picture at any stage in the product development process. The FAST technique requires that every design function for a product or part be identified and described by a noun–verb phrase and that this be used to link the functions in a network. This is done by asking two questions for each function: 'why?' and 'how?'.

The network is then drawn in such a way that 'why' answers are placed to the right and 'how' answers to the left. This method can be seen by examining the FAST network in Fig. 5.5, which has been drawn for a light bulb (an example from Fox, 1993).

Starting on the left and moving to the right, the 'why' questions are answered or, by reversing the process and reading from right to left, the 'how' questions are answered, e.g.

Q. Why provide a gas-filled bulb? **A**. To provide an oxygen-free atmosphere
Q. How to create a very high temperature? **A**. By providing a filament and a flow of current.

Once a FAST diagram has been prepared it can be used as the basis for establishing the critical parameters of a design as described in chapter 13, section 13.8 (critical parameter management (CPM)).

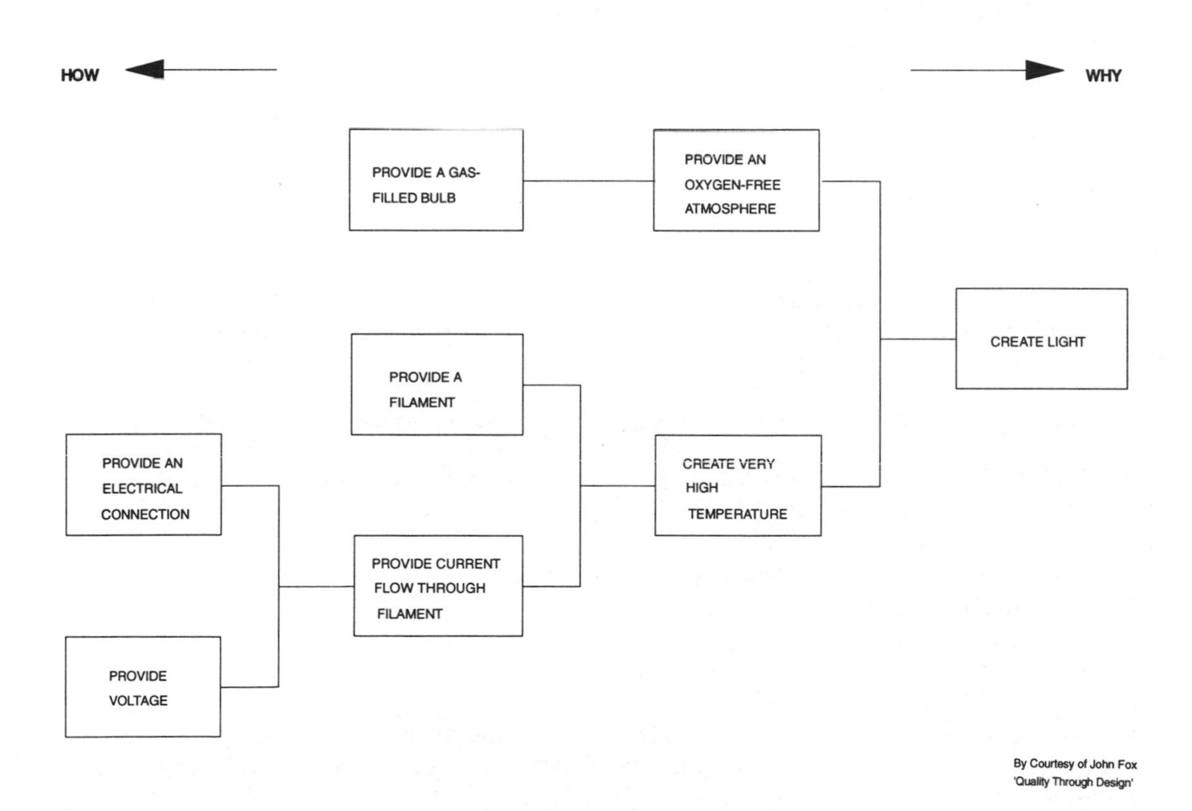

Figure 5.5 FAST diagram for a light bulb.

Value engineering

Once a design has been established (using techniques such as FMEA, FAST and CPM), it is possible to carry out a value engineering exercise since all the functions, and their relationships, should be known. The first step is to identify the parts that enable a particular function to be performed and to determine the cost of each such part. This can best be achieved by using the FAST diagram, since the parts are generally on the left of the diagram and the functions on the right. These costs can then be 'rolled up' across the diagram so that the total cost of delivering each function is known; this is then the basis for evaluation.

The evaluation starts by the application of judgement to the value and cost of each function. Next, any function that appears to have an unreasonable cost should be identified and these should be the targets for value engineering.

Various techniques are available to the engineer to improve the functional values, e.g. *brainstorming* within a team or group can result in a long list of ideas, many of which may be totally unpractical but some of which may form the basis of a solution that could increase functionality, reduce cost, or both.

Alternatively, a technique known as the *problem-solving cycle* could be used. This process involves six steps which, if followed correctly, will eliminate the pitfalls of a less structured approach. These are:

1. identify the problem
2. analyse the problem
3. generate potential solutions
4. select and plan a solution
5. implement the solution
6. evaluate the solution.

Value engineering solutions generally fall into one or more of three categories:

- *part reduction* eliminating unnecessary parts or combining two parts into one
- *part modification* modifying a part to reduce its cost or allow it to perform multifunctions
- *function improvement* changing the way in which a part performs its function.

5.4 DESIGNING FOR MAXIMUM PRODUCT QUALITY

In chapter 13, considerable coverage is given to the various techniques that can be used to ensure that the design meets two criteria, i.e. that it

1. matches the customer requirements
2. is matched by the process capability.

These techniques include:

- FMEA—failure modes and effects analysis to minimize the effects of failure in service
- QFD—quality function deployment to ensure that market requirements for quality are built into the design
- CPM—critical parameter management to ensure that the factors that can adversely affect product performance are identified and controlled.

However, two aspects of the design/development/quality relationship that have not yet been covered — first, Taguchi's design of experiments; and second, his work on the 'quality loss function'.

As a part of the design and development function it is usually necessary to set up tests or experiments in order to determine how a product behaves under a variety of conditions. For example, to test a car radiator hose connector, it is necessary to apply variations in temperature, pressure and vibration, in different combinations. This requires the use of a designed experiment, since the 'one-variable-at-a-time' approach would take too long and be too costly. Designed experiments have the ability to:

- evaluate the interactions of two or more variables
- eliminate confusing or confounding effects of two or more variables
- establish cause–effect relationships between several variables and their outcomes
- identify the variation from planned changes while isolating it from the remaining or random variation.

The Taguchi approach to DoE (design of experiments)

Dr Genichi Taguchi introduced a new approach to DoE using standard orthogonal arrays and linear graphs, and he also made the distinction between controllable and non-controllable factors, treating each type differently in his experiments.

This approach can best be demonstrated by an example, which is given in appendix D. This shows how the arrays are used to set up a series of test runs, each run using a different combination of controllable and non-controllable factors. The results of these can then be studied in order to identify which combination gives the most satisfactory product design. For a more 'in-depth' view, Bergman and Klefsjö (1994) is a valuable source.

As a result of Taguchi's work, it is now possible to minimize the number of changes required to each variable when searching for the *optimum* combination. Standard factorial tables are used to replace the traditional one-at-a-time experiments. This minimizes effort, and finds the performance plateau which avoids high sensitivity of performance to small changes in a variable or variables. This results in what Taguchi called a *robust* product or process design being developed economically. However, the method does require multiple experiments in which changes in variables and performance must be tabulated.

The terms used by Taguchi may require definition: some of these are given in Table 5.2.

The Taguchi approach to quality

Taguchi did not accept the traditional view that quality is an absolute — that if an item is within tolerance it is 'good' but outside tolerance it is 'bad', and that quality loss occurs only when limits are exceeded.

Instead he believed that quality is continuous, with the loss beginning as soon as a product or process starts to deviate from its target value. He therefore defined his *loss function* in mathematical terms, whereby the loss increased quadratically in proportion to the amount of deviation from the target.

These two approaches (traditional and Taguchi) can be seen in Figs 5.6 and 5.7. In the latter, the shaded areas, although within the specification limits, are actually hidden losses; Taguchi's loss function can identify and quantify these. These hidden losses, although not visible

Table 5.2 Definitions of terms used in Taguchi method

Confounding	The mixing of different terms or items such that they cannot be separated
Factor	An independent variable that can be varied from trial to trial
Factorial design	An experiment that measures all possible combinations of factors (fractional factorial design is a subset of this)
Inner array	The array that holds the controllable factors and their interactions
L_N	Standard for denoting a Taguchi orthogonal array, where N is the number of test combinations required
Level	A setting of variable—the actual value to which a factor is set
Linear graph	A combination of points and lines used to depicit interactions between factors when they are assigned to an orthogonal array
Loss function	A quadratic, continuous curve used to illustrate that poor quality, or loss, occurs when output deviates from nominal values, and not just from non-conformance to specifications
Noise factor	A factor that is too difficult or expensive or impossible to control
Orthogonal array	A factorial or fractional factorial design
Outer array	An array for holding noise factors
Response variable	The output of a process or product—a variable you are interested in measuring for the purpose of optimizing it
Robust	Insensitive to the effects of noise—a process is said to be robust if its output is consistent under variable conditions

through a quality check, can be seen in terms of customer dissatisfaction, customer returns and poor product life.

The way in which the loss function is calculated can be seen from the example in Table 5.3. In this case a part is assumed to have a nominal dimension of 200 mm, with a tolerance of ±1 mm. If the part measures less than 199 mm, or more than 201 mm, it is returned by the customer as unusable and has to be replaced at a cost of £75. The hidden losses can then be calculated using Taguchi's quadratic loss function formula and drawn as a graph, as shown in Fig. 5.8.

Table 5.3 Calculating a Taguchi loss function

The quadratic formula used is:

$$L = k(y - m)^2$$

where L is the loss (£75), k is the constant, y is the measured value (199 or 201 mm) and m is the target value (200 mm).

Step 1 Calculate the constant k: $75 = k(201 - 200)^2$ or $k(199 - 200)^2$, thus $k = 75$.

Step 2 Use this k value to draw an L–y axis graph based on the formula $L = 75(y - 200)^2$ to plot the points on the line as shown in Fig. 12.8

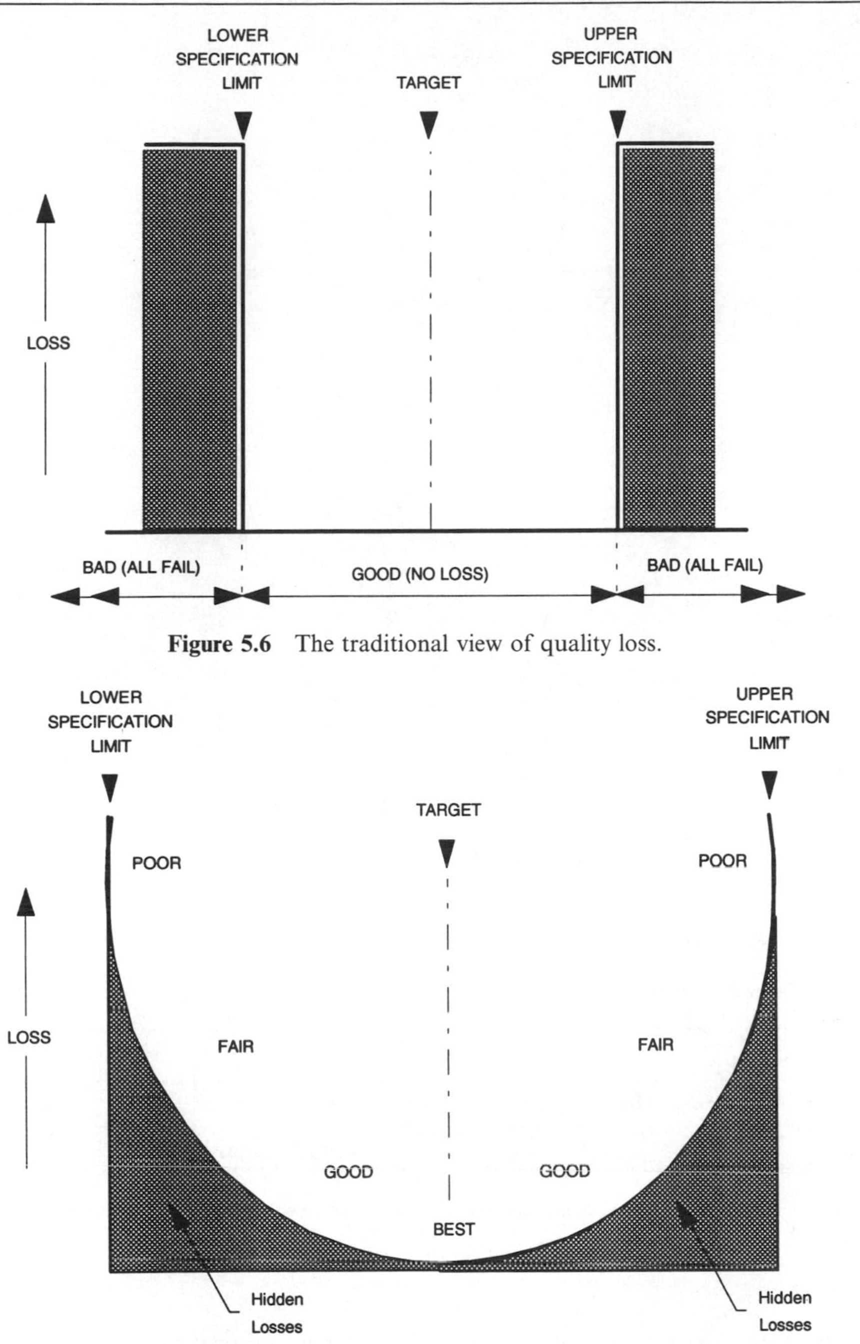

Figure 5.6 The traditional view of quality loss.

Figure 5.7 The Taguchi view of quality loss.

Taguchi's loss function can also be used to set production specification limits and this calculation can be seen in Table 5.4, using the same example as Table 5.3. These limits are much tighter than those originally specified, and it can also be seen that the costs of quality losses can be calculated for any mix of measurements. However, this would be an onerous task unless a computer system (such as SPC) could be used to make the calculations.

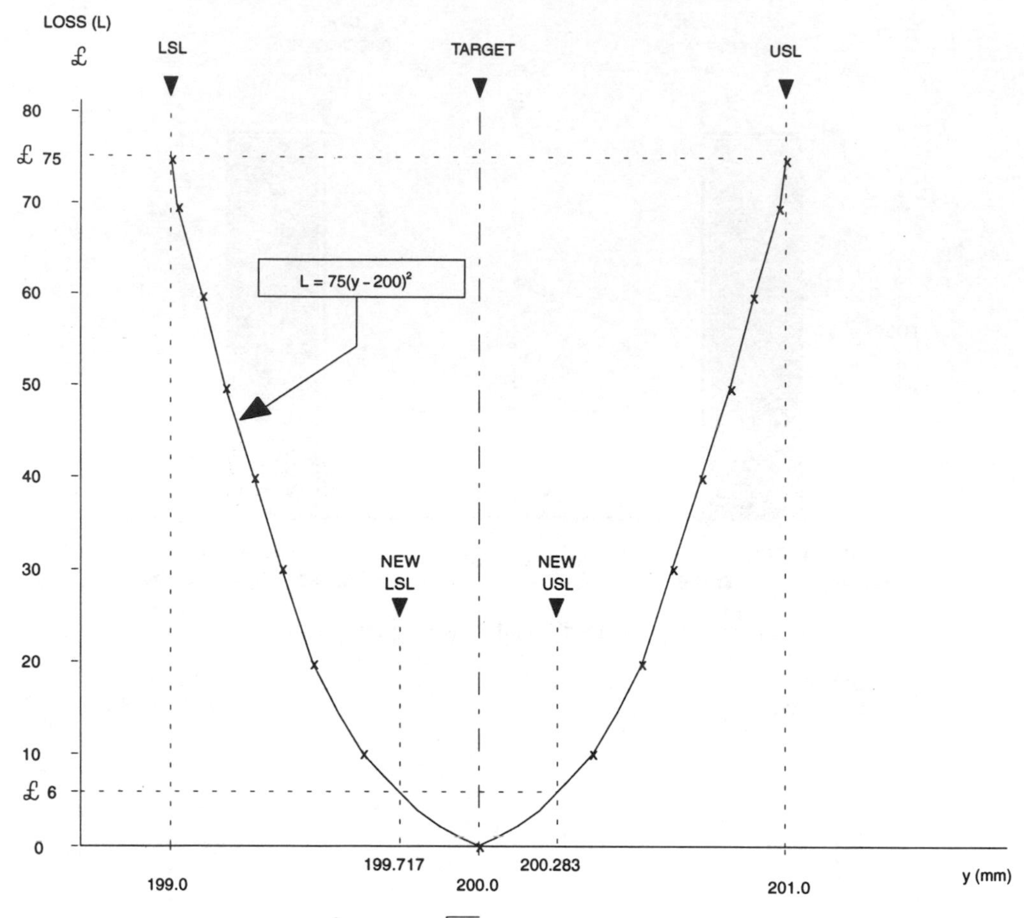

From the Taguchi equation, if $L = k(y - m)^2$ then $y = m \pm \sqrt{L/k}$

Given that k = 75 and m = 200, then:

Then, if L =	0	10	20	30	40	50	60	70	75
y =	200.000	200.365	200.517	200.632	200.730	200.817	200.894	200.966	201.000
or y =	200.000	199.635	199.483	199.368	199.270	199.183	199.106	199.034	199.000

NOTE: This data has also been used to calculate the new upper and lower specification limits, details of the calculations being shown in table 12.4.

Figure 5.8 Calculation of Taguchi quality loss graph.

5.5 DESIGNING TO FIT MANUFACTURING RESOURCES

The need for design specifications to be met by production capability is fully covered in chapter 13 (section 13.10). This generally tends to look at the problem from one side only, i.e. how can production ensure that the design needs are met?

However, in many companies it is important for the design team to understand the strengths and weaknesses of the production facilities, and to try to balance these against the quality and efficiency requirements. This objective can often be achieved by applying a strategy known as

Table 5.4 Using the loss function to calculate specification limits

By means of a costing exercise, it was found that the replacement cost of the item—if caught during production—was £6. The loss L in the equation previously used therefore becomes £6 and the specification limits can then be calculated as follows:

If $L = 75(y - 200)^2$, then $6 = 75(y - 200)^2$.

Let X be $(y - 200)$, which will be the ± specification limits. Then, from the equation: $6 = 75X^2$ and thus $X = \sqrt{(6/75)} = \pm 0.283$. Thus the limits are 200 mm ± 0.283.

DFM (design for manufacture) and its derivatives, e.g. DFA (design for assembly) or DFF (design for fabrication).

DFM, in its broadest sense, is concerned with understanding how product design interacts with the other component parts of a manufacturing system, and defining product design alternatives to optimize all aspects of that system. In order to look at DFM in a more specific way, it is necessary to consider certain principles, which can be defined as follows.

Minimize the number of parts

Every part in an assembly should be critically examined in terms of its function. Can the functions of two parts be combined? Can two parts be combined to make one part? Can parts be fitted together with fewer fastening devices? Can a part be eliminated by a change in design?

Develop a modular design

Modular design can be defined as a form of standardization whereby a module can be used in a variety of products. It is thus possible to develop a range of products using a series of standard modules, leading to a lower number of parts and sub-assemblies overall.

Use standard components

Where possible, standard components should be used, whether purchased or manufactured. Designers should therefore be able to access a standards manual that classifies all commonly used components in terms of degrees of acceptability, e.g.

A. should be used wherever possible
B. should be used only where special circumstances pertain
C. should be used only as a last resort.

Design parts to be multifunctional/multi-use

The designer should try to use a part to carry out more than one function, e.g. as an electrical conductor and a structural member, and also to think in terms of multi-use, e.g. by providing extra holes, a mounting plate could be used to locate different components or in different sub-assemblies.

Avoid separate fasteners

Where possible, fasteners such as screws and rivets should be replaced in the design by integral clips or tabs. However, if fasteners must be used, they should be chosen carefully to minimize the number and variety of types, and if automatic feeding devices are being used in the assembly process, the fastener type should be compatible with the device.

A number of methods are now available for rating an assembly in terms of the success (or otherwise) of the application of DFM techniques. These are generally based on the costs of making the parts and assembling them. Thus, a theoretical standard cost might be set up and given a rating of 100 per cent and the actual compared with this.

For example, Boothroyd and Dewhurst (1987) consider a spindle housing assembly as shown in Fig. 5.9. In the original design there are 10 parts and the assembly efficiency rating is 7 per cent. However, by applying DFM techniques, the number of parts has been reduced to two and the efficiency rating increased to 93 per cent. This has been achieved by changing the housing from stainless steel to nylon, which allows the spindle to be clipped directly into the housing without the need for nylon bushes and brass screws.

The costs quoted for this example by Boothroyd and Dewhurst, are summarized in Table 5.5. This shows that the costs of the product have been reduced by almost 50 per cent by using DFM. The improvement in rating from 7 to 93 per cent has been calculated using the Boothroyd–Dewhurst Method which is described in their *DFA Handbook* (Boothroyd and Dewhurst, 1987).

It has been established that DFM, and in particular DFA, is most effective if applied during the concept phase of a project, because it is at this time that 70 per cent of a product's life cycle impact is determined, with only 5 per cent of the resources allocated to the project having been spent. This can be seen in Fig. 5.10.

Table 5.5 Costings for the spindle housing DFM example

Part	Assembly	Material	Manufacture	Tooling	Total cost
Original design					
Housing	0.02	1.74	1.56	7.83	11.15
Bush (2)	0.09	0.01	0.06	9.03	9.19
Screw (6)	0.35	0.72	—	—	1.07
Spindle	0.04	0.26	1.29	—	1.59
Total	0.50	2.73	2.91	16.86	23.00
New design					
Housing	0.02	0.14	0.24	10.05	10.45
Spindle	0.02	0.26	1.29	—	1.57
Total	0.04	0.40	1.53	10.05	12.02

5.6 DESIGNING TO FIT CUSTOMER REQUIREMENTS

In a recently published paper (Institution of Mechanical Engineers, 1992) entitled 'A vehicle characterization process', the author introduced his topic thus.

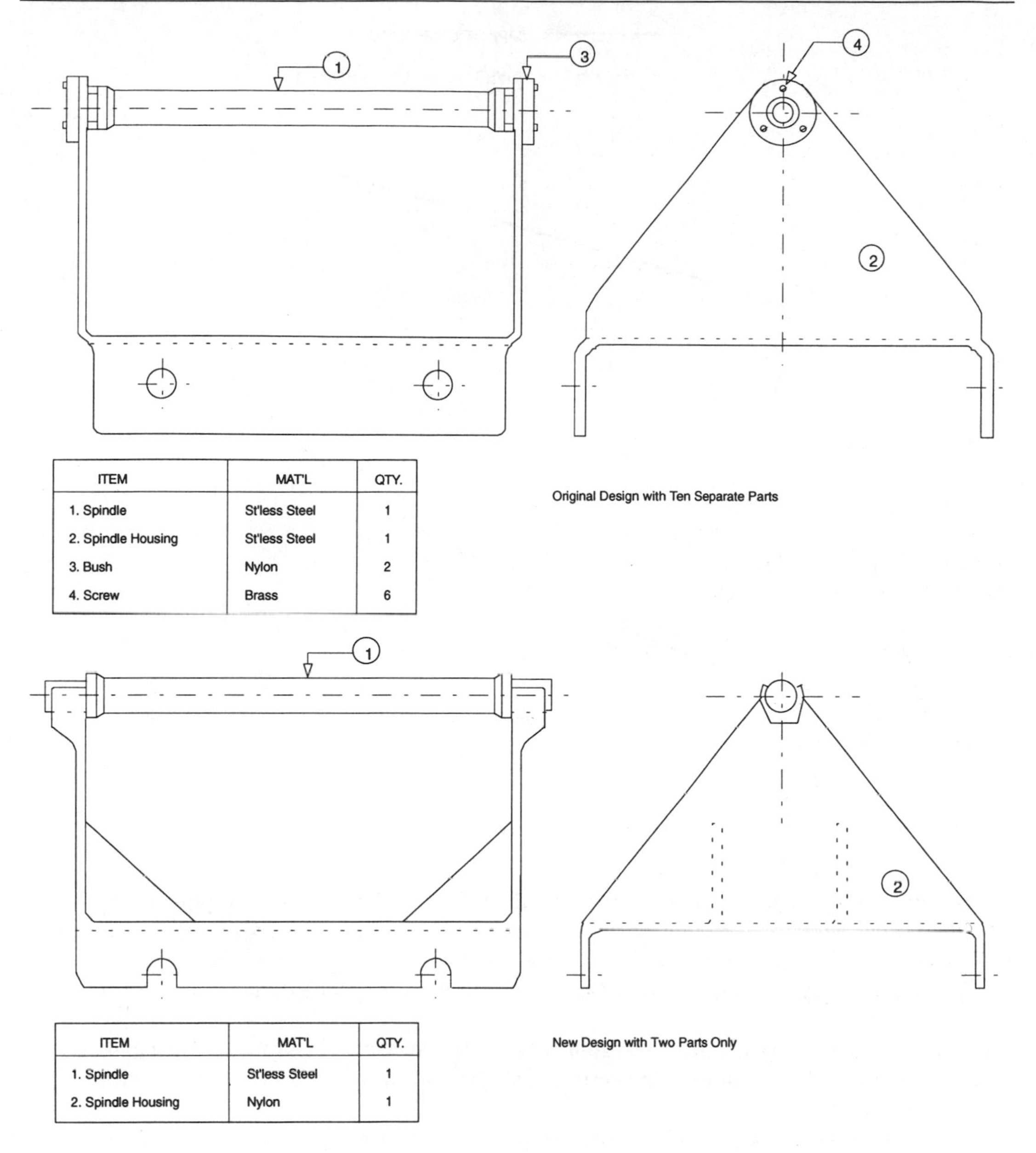

ITEM	MAT'L	QTY.
1. Spindle	St'less Steel	1
2. Spindle Housing	St'less Steel	1
3. Bush	Nylon	2
4. Screw	Brass	6

ITEM	MAT'L	QTY.
1. Spindle	St'less Steel	1
2. Spindle Housing	Nylon	1

Figure 5.9 Application of design for assembly (DFA).

> *A comprehensive process has been developed to characterize a new passenger car* [in this case a Jaguar] *with the objective of ensuring that the appropriate emphasis is given to those areas which are important to the targeted customers. Vehicle characterization is used to set the targets for a programme during the definition phase and design solutions are measured against the targets during the implementation.*

This process could equally well be applied to any product, and it is worth noting that the manufacturer not only identified the areas that were important to the customers, but also identified the customers who were most likely to buy the product.

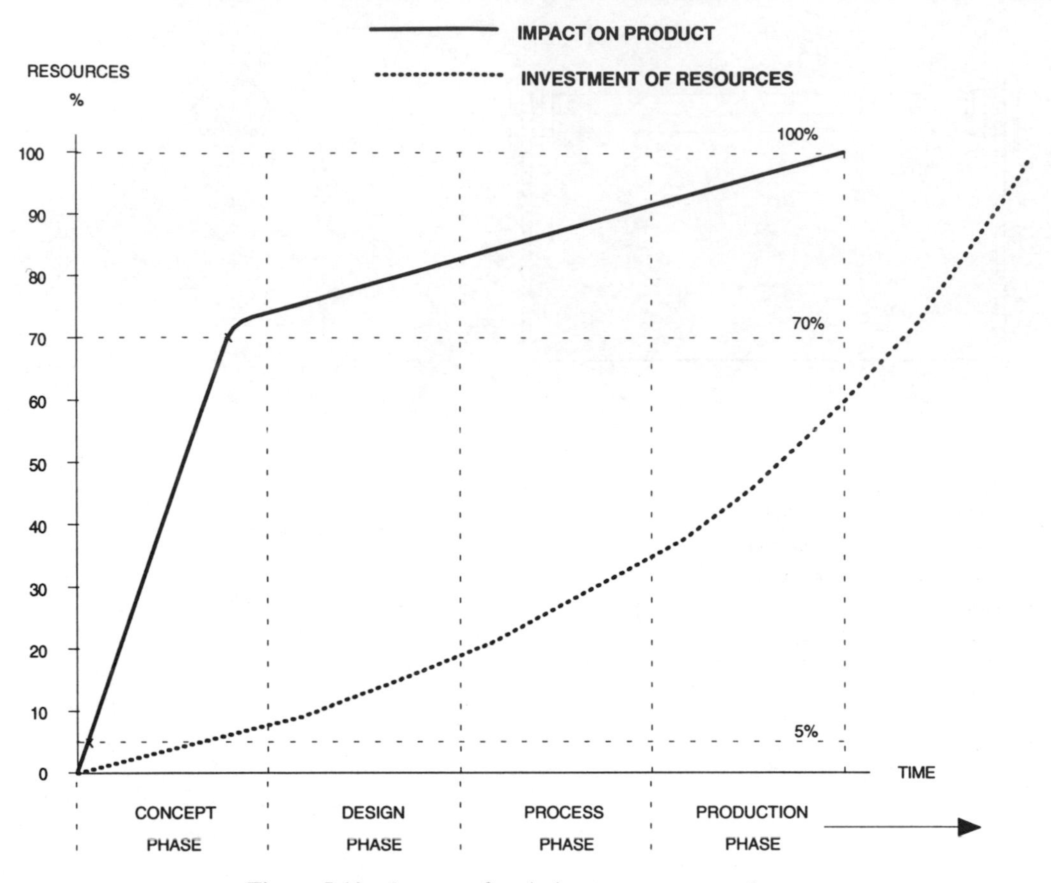

Figure 5.10 Impact of early investment on product.

In fact, the approach used was similar to the QFD (quality function deployment) process described in chapter 13 (section 13.8, where a 'how–what' chart is prepared as shown in Fig. 5.11. (compare this with the QFD chart in Fig. 13.8).

In order to carry out this exercise a number of 'rules' were specified, including the following.

- Because the project was large, it was split between five specialist groups, each group (exterior, occupant space, driver controls, powertrain and dynamics) running its own chart with a sixth chart for the total car.
- All targets set had to be numerical and realistic, since it was recognized that it is impossible to be best at everything.
- Parameters were defined in terms of changes from the baseline rather than in complete detail; primary parameters were identified as those that were customer-perceived (maximum speed, fuel consumption, etc.), whereas sub-parameters were probably imperceptible (weight, aerodynamic drag, etc.).
- The same priority or weighting system had to be given to a feature regardless of the chart in which it appeared.

As a basis for comparisons, the characterization process required subjective judgements on various features from competitors' vehicles and this was achieved by allowing existing or poten-

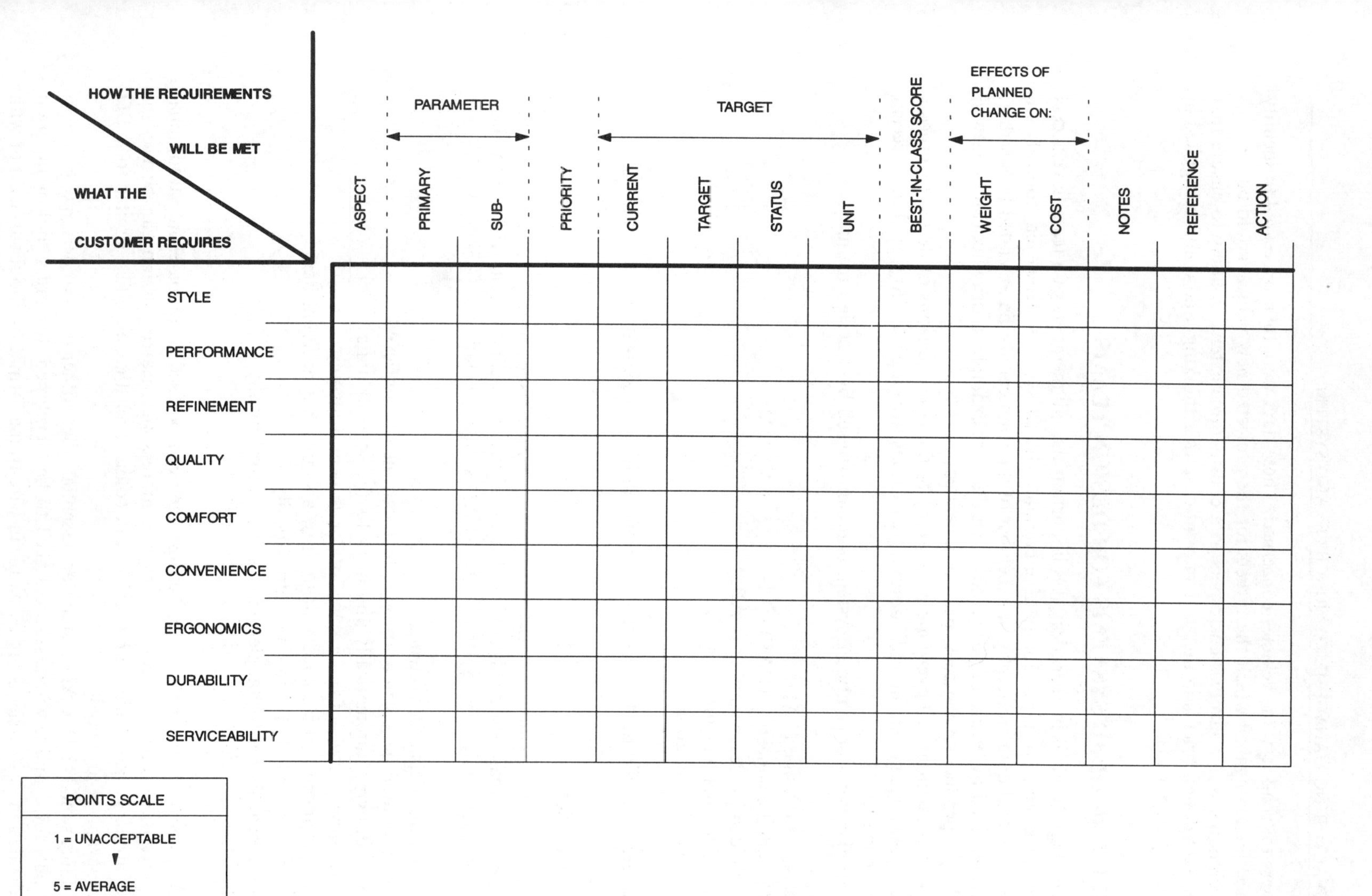

Figure 5.11 Chart for vehicle characterization process.

tial customers to 'road test' the vehicles concerned. These data were then processed to identify 'best in class' factors against which the targets for the projected new vehicle could be set.

This technique, which involves customers to a considerable extent, is likely to ensure that the new product design brief is in line with current customer thinking, and could be applied in many manufacturing areas.

5.7 COMPUTER-AIDED SYSTEMS FOR DESIGN TEAMS

If concurrent engineering is in operation, it is important that all members of a team have access to the same data, and are using compatible computer-based systems. However, controls are needed to ensure that data created by one member of the team cannot be amended by another. This can be achieved by means of passwords such that, for example, a drawing can be viewed and copied by a production engineer, but not amended.

Equally, it is essential that everyone be working to the latest versions of design or process plans. Thus, a message system should inform a user who is working on a copy, if a more recent design or plan exists.

There are five types of computer system that can be used by a design team:

1. computer-aided design (CAD)
2. computer-aided engineering (CAE)
3. computer-aided process planning (CAPP)
4. computer-aided manufacture (CAM)
5. computer-based simulation

These packages naturally need to operate with one another, as shown in Fig. 5.12.

Computer-aided design

CAD systems vary greatly in their scope, complexity and sophistication, but all have the same objectives, i.e. to increase the productivity of the designer, to increase accuracy, and to allow data to be exchanged with other systems. Most CAD systems include the facility to:

- copy or replicate parts of a drawing
- store frequently used details in a library which can then be retrieved
- dimension a drawing automatically and work to any scale or units
- build an assembly drawing from a series of component drawings
- change the parameters of a component to draw a similar but different item
- create a parts list or bill of materials automatically
- draw geometrical shapes (rectangles, circles, etc.) automatically
- display another drawing in a window for reference.

Basic CAD systems are known as 2-D, because they can be used to prepare only traditional multi-view, two-dimensional images. However, a more sophisticated version can be used to 'model' the surfaces with shading and textures in colour such that an effect similar to a 2-D colour photograph is produced.

Three dimensional (3-D) CAD systems are essential where other types of design work are to be carried out, for example where surfaces need to be 'developed' for press tool or injection moulding die design. 3-D systems are available in wire frame format (without surfaces) or with

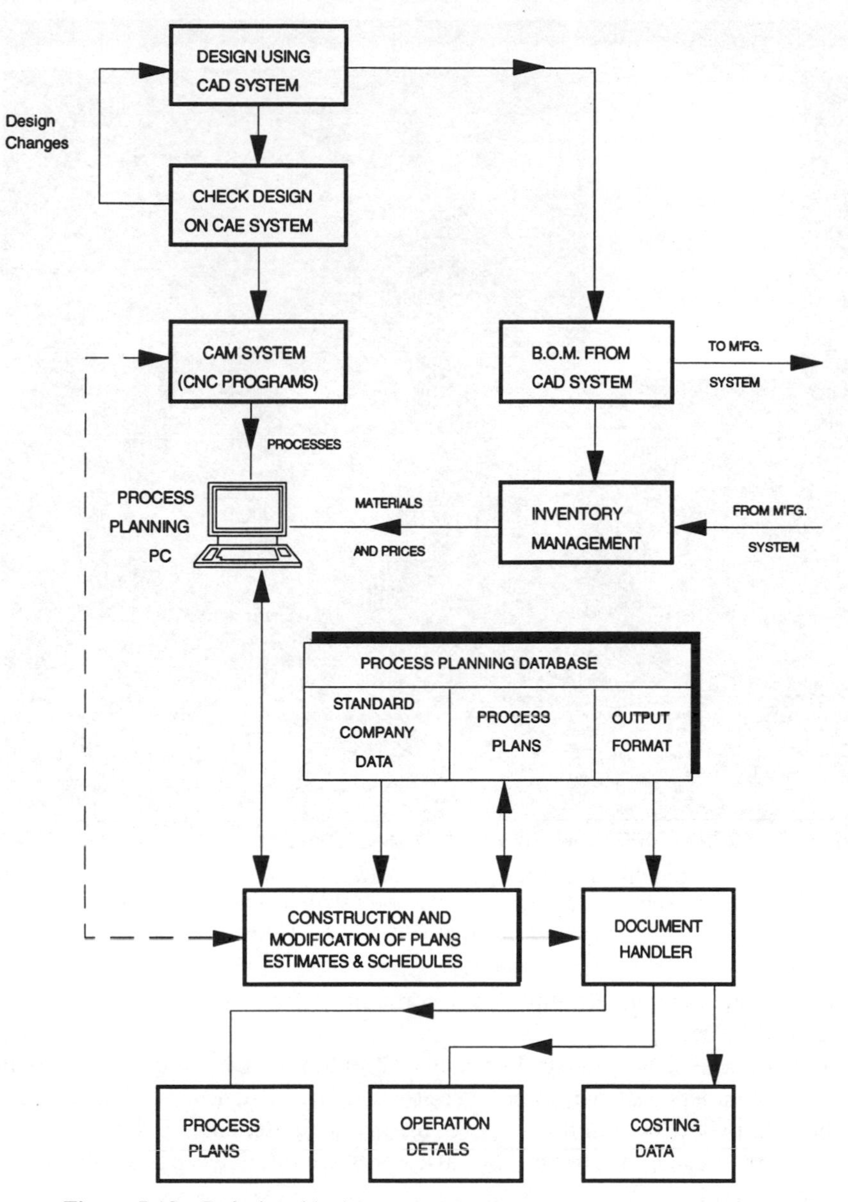

Figure 5.12 Relationship between CAD, CAE, CAM and CAPP.

full surface/solid modelling capability. Fig. 5.13 shows two types of CAD software from the wide range of options available.

Computer-aided engineering

Once a design has been produced on a CAD system, it may be necessary to check it, to ensure that it meets the design specification in terms of the functions it will be asked to perform; typical of this 'checking' software is FEA (finite element analysis), which allows a user to carry out a

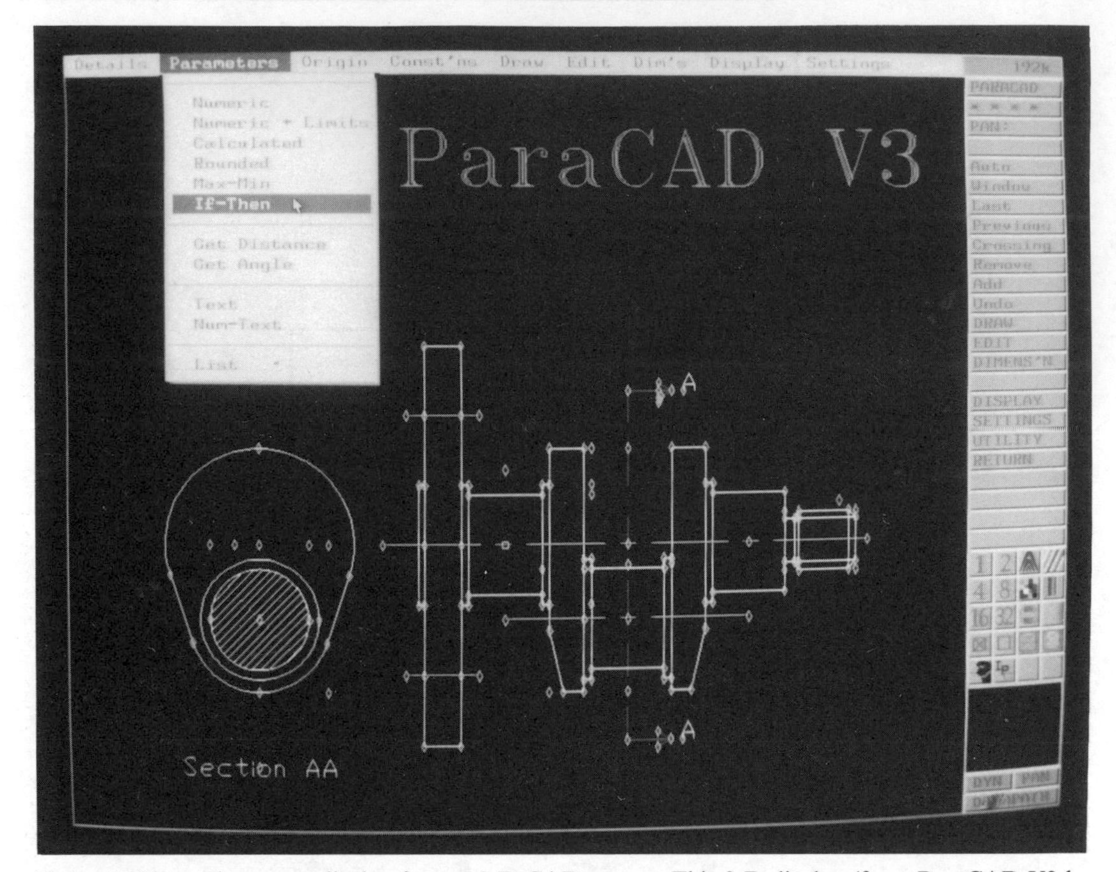

Figure 5.13(a) The screen display from a 2-D CAD system. This 2-D display (from ParaCAD V3 by CADlogic) is being used to alter the parameters on a crankshaft so that the designer can test out a series of 'what-if' design changes. This system is designed for AutoCAD users.

series of static and dynamic loadings on a component or assembly in order to calculate the stresses that will be set up.

For example, the structural steelwork of a building can be subjected to wind and snow loads, and the members that would be over-stressed strengthened. Similarly, redundant members can be eliminated and under-stressed members lightened. FEA will not only allow a designer to test the design for stresses, but also show where unacceptable deflection or bending will occur under load.

Another CAE facility is known as CFD (computational fluid dynamics). In this case, any type of flow through a component or assembly can be simulated and used to improve the design. There are numerous applications for this type of package, some of which are described in Table 5.6.

CFD packages usually include a library of the 'properties' of gases, fluids, and solids so that various material combinations can be tested and the effects readily assessed. Figure 5.14 shows a typical example.

Computer-aided process planning

There is some degree of overlap between CAPP and CAM (computer-aided manufacture), and the differences can best be explained by defining them both.

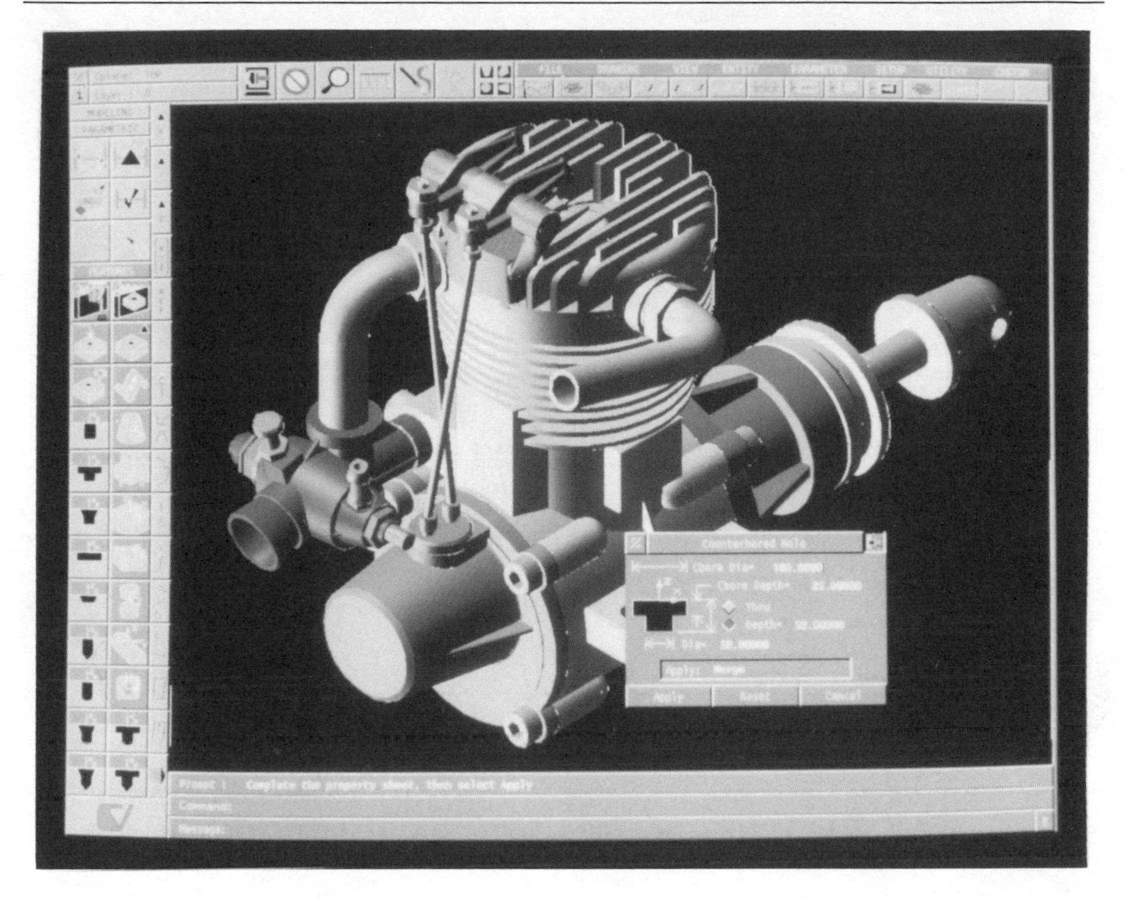

Figure 5.13(b) The screen display from a 3-D CAD system. This 3-D display (from CVdesign by Computervision) is being used to model and visualize using surface shading, textures and colour. The system also has parametrics and can 'window' a detail as shown.

Table 5.6 Typical applications of CFD (computational fluid dynamics)

Smoke	Smoke concentration and air velocity vectors at a specified time after the start of a house fire in a specific room of a particular type of house
Air-flow	Air-flow patterns and velocities over a vehicle or aircraft under specified conditions
Pressure	Pressure distribution in a car engine valve passage (e.g. an exhaust valve): CFD could also be used to study the temperature distribution in the valve, valve guides and passage
Particle flow	Flow of particles in a spray dryer or a paint spraying system
Moulding	Flow of metal in a die casting process or of plastic in an injection moulding process for die design
Press work	Flow of sheet metal in a press to form a component for press tool design and pressure required

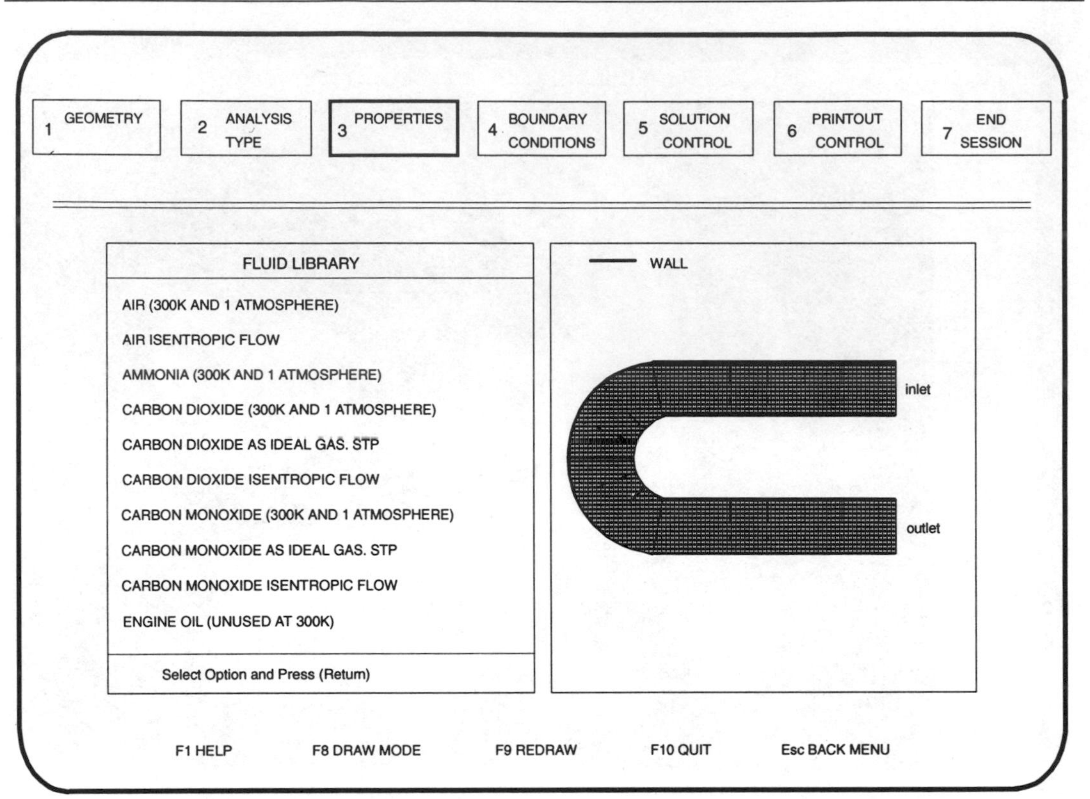

Figure 5.14 Part of a typical Fluid Properties Library. By Courtesy of CHAM 'Easyflow' Huntsville, AL.

- *CAPP* Most CAPP packages are designed for defining and estimating the work to be done in manufacturing a component, or in some cases an assembly. These packages allow the user to set up a series of synthetic data (machine feeds and speeds, handling and gauging times, materials, cutting tools, etc.) so that, by entry of the required details, a standard method can be specified and a standard time calculated. CAPP packages can include modules for machining, fabrication and assembly work and are most commonly used for estimating. They are not normally associated with a CAD package since the user works directly from a hard-copy drawing.
- *CAM* CAM packages are almost always associated with a CAD system and are used to determine how a component can best be manufactured. In this case, the user works on the computer screen and can develop and simulate tool paths to ensure that there are no problems (for example, tool collisions with a work-holding device). For CNC (computer numerical control) machines, a CAM system can be used to generate the machine programs and to optimize process times and methods. CAM packages can include modules for machining, sheet metal work cutting and robot handling, and are most commonly used for specifying how a process should be carried out.

As an example of how a CAPP system operates, part of a process planning procedure is illustrated in Fig. 5.15, showing how questions are posed in a logical, step-by-step approach such that the user, in answering, develops both the process plan and the standard times. In this case, the operation was on a machining part on a lathe, but it could equally have been fabrication work or the assembly of parts.

COMPUTER QUESTION	TYPICAL ANSWER
USER' NAME (16 CHARs MAX)	M. STEVENSON
PART/DRG.NUMBER	SP.19564
SELECT UNITS OF INPUT - (1) METRIC - (2) IMPERIAL	1
SELECT MATERIAL FROM LIST (1) EN1 M.STEEL (2) EN2 M.STEEL (3) EN3 M.STEEL (4) EN8 CBN.STEEL (5) EN26 CBN.STEEL (6)STAINLESS STEEL	1
GIVE MATERIAL DETAILS/DIMENSIONS (24 CHARs MAX)	125mm DIA B.D.MS,EN1
GIVE PART NAME (20 CHARs MAX)	PIVOT PIN
GIVE OPERATION No.	10
GIVE SHORT ELEMENT DESCRIPTION	FACE AND TURN DIAS.
GIVE OPERATION TEXT (2 LINES x 70 CHARs)	FACE OPEN END AND TURN DOWN TO 112mm x 150mm LG.
SELECT ELEMENT TYPE (1)C.DRILL (2)DRILL (3)C.BORE (4)REAM (5)TURN (6)BORE (7)FACE (8)GROOVE (9)RECESS (10)PART (11)FORM (12)SCREW (13)TAP (14)D.BOX (15)CSNK	7
SELECT MACHINE TOOL (PRESS RETN FOR MORE) (1)HOLBROOK MODEL D.18 (2)CVA MODEL 1A. (3)SPRINT (4)CHURCHILL 290	1
CAPACITIES OF M/C SELECTED ARE GIVEN BELOW. OK? CHUCK DIA. 355mm BED SWING 508mm SADDLE SWING 280mm CENTRES 915mm	Y
GIVE NUMBER OF ELEMENTS	1
GIVE START OR STOCK DIA.	125
GIVE LIST OF DIAS.	O
GIVE LIST OF LENGTHS	1
SELECT TOOL MATERIAL: (1)HS STEEL (2)TU.CARBIDE (3)CERAMIC (4)DIAMOND	2
SELECT FINISH REQUIRED	3

NO.	MIN.TOL.	MAX.TOL.	MIN.FIN.	MAX.FIN.
(1)	0.0127	0.0381	0.40	0.77
(2)	O.0406	0.0736	0.80	1.57
(3)	0.0762	0.1499	1.60	3.18
		etc.		etc.

CUTTING TIME = 0.91 SMs

NO.OF CUTS=1; DEPTH OF CUT=1.000mm; SPEED=286rpm;
LNGTH OF CUT=62.50mm; FEED=0.3175mm/Rev.

ADDITIONAL TIME ELEMENTS FOR THIS WORK ELEMENT ARE:

APPLY COOLANT (CODE 14) = 0.08 SMs
START MACHINE (CODE 15) = 0.05 SMs

ELEMENT 1 (FACE): TOTAL TIME = 1.04 SMs

Figure 5.15 Developing a computer-aided process plan. By courtesy of Marlow Microplan Limited (Capes).

Computer-aided manufacture

CAM systems are generally used to develop programs for numerically controlled (NC) machines working from a CAD drawing. There are various 'levels' of such machines, but in most cases the programming is carried out off-line, i.e. remotely rather than on the machine controller. CAM systems allow the user to simulate the tool path that a machine will follow in order to carry out the metal removal process, and this path is then converted into computer code by means of a 'post-processor' to form the NC program, which can then be loaded into the machine controller. However, CAM can also indicate to the user how to optimize the process, for example to reduce tool wear by ensuring that cutting is carried out at the ideal rate, to reduce time taken by eliminating all unecessary movements, to reduce metal wastage by using the correct size of raw stock material, and to suggest the best locations for work holders.

CAM systems are available for most types of process application, e.g. turning, milling, drilling, punching, cutting (laser, flame or plasma) and forming. However, to illustrate the process, a CNC sheet metal punch has been chosen. In this example, the following two facilities are available in the package.

- *Nesting* Given that a number of different sheet metal components are required to be cut from a sheet of a given size, the system will lay these out in such a way as to minimize the wastage (scrap). This can be seen in Fig. 5.16, where 38 pieces in eight varieties have been laid out by the nesting program on a standard size of sheet.
- *Punching* The punching program is then used to optimize the tool paths to give minimum process time, including tool changing as required (the machine includes a magazine that can

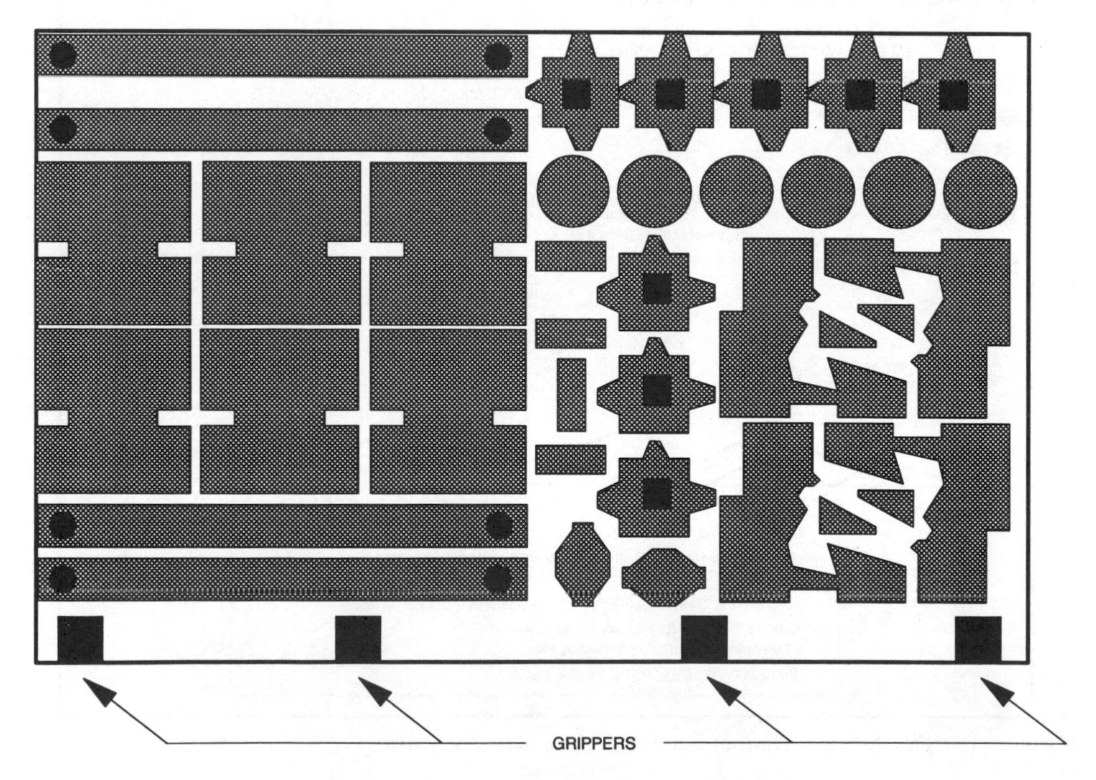

Figure 5.16 Example of nesting using CAM.

contain up to 24 different types of punch). The machine also has a programmable load/unload facility.

Most modern CNC machines also include facilities for programming tool changing and devices to measure and warn if a tool has been damaged or broken.

Computer-based simulation

Many CAM systems include the facility to run a simulation of a process so that it can be checked for time and practicality, but in addition there are specialized simulation packages that can be used to run a combination of processes or to test the operation of a complex electronic or mechanical system. As an example, Fig. 5.17 shows a simulated work cell layout with a combination of machines and handling devices. This can be 'run', with each unit operating on its own program, to ensure complete integration within the cell. This simulation could include feasibility checks on all control equipment to make sure that the possibility of an accidental collision or impact was minimized.

The automotive industry uses simulation extensively, the best-known example being crash simulation, but other applications include design checks on suspension systems under a variety of conditions, testing of the on-board computer systems, and verification of steering geometry.

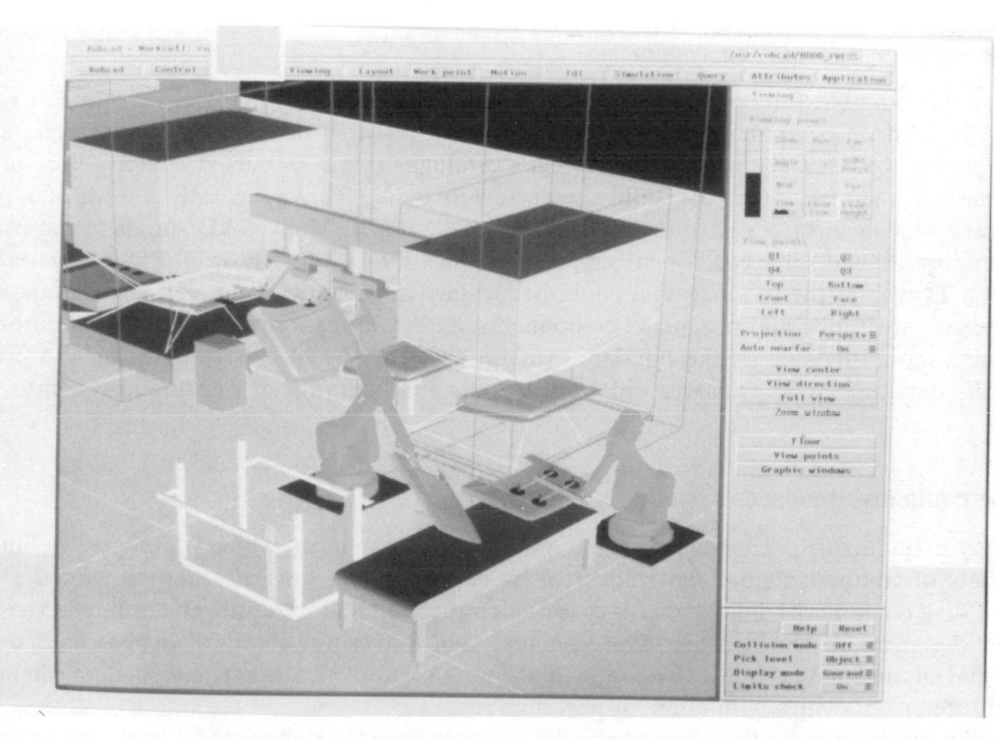

Figure 5.17 The screen display for a work cell simulation. This display (from ROBCAD by Technomatix Technologies) shows how 3-D graphical simulation and off-line programming facilities can be used to verify the operation of an automated manufacturing cell.

5.8 INFORMATION MANAGEMENT IN A DESIGN DEPARTMENT

A design department cannot operate without an information management system. There are three categories of such systems to be considered:

1. the drawing and data management system
2. the supplier information system
3. the company standards system.

The drawing management system

A fully cross-indexed register of all drawings and associated data (calculations, specifications, etc.) is needed so that any authorized user can gain access by means of a logical search procedure. This can best be achieved by using standard computer software, such as a database, or one of the proprietary systems generally available.

Such information is usually required at short notice to deal with a customer complaint, a design change, a production problem or a new product development, and ideally the information should be displayed on a computer screen so that the user can work directly from a data copy, without affecting the original, which will still be available for other users. All the usual controls should be applied to eliminate unauthorized access and to ensure that amendments cannot be made to the original data without proper authority.

The supplier information system

Wherever possible, fully detailed information about all products available from suppliers should also be held on computer, with easy access by logical search or browse. For example, a user should be able view the pages of a supplier's catalogue in a screen window and select an item which can then be copied over into the current drawing. This is, in fact, a form of symbols library which can save a great deal of time for the designer. Many CAD suppliers can provide such libraries and some component suppliers provide library data (in various formats) for CAD users. Typical of the products that could be included are electrical and electronic components, fasteners, hydraulic and pneumatic components, and transmission equipment. In addition, the library should include all the standard symbols that may be used by the designer to prepare wiring diagrams, process charts, pipework layouts and hydraulic or pneumatic systems.

The company standards system

Many manufacturing companies maintain a library of standards, which is designed to limit the variety of components and materials used by designers, and such information should also be held on a computerized database for ease of identification and retrieval. This will tend to ensure that designers all work to the same standards and that a non-standard item is used only in special circumstances. It will also eliminate the not unknown situation where an item can appear on different drawings with different part numbers.

The company standards system should cover not only the standards for parts and materials, but also such matters as quality standards, drawing standards (sizes, dimensioning, tolerance notations, etc.) and technical specification standards.

Data translation

An international industry standard format known as STEP (Standard for the Transfer of Engineering Product Data) has been formulated by ISO (International Standards Organization) to facilitate the computerized storage and transmission of all product-related data. This will allow users to exchange data even though they are using different computer systems. Thus, a supplier could, by using a 'translator', produce a catalogue on computer tape or disk which could be sent to all its customers who work with a CAD system.

Warning!

Out-of-date information can seriously damage a company's health. It is therefore essential to ensure that the information systems are updated on a regular basis.

5.9 THE MANAGEMENT OF ENGINEERING CHANGES

Engineering changes can arise in two ways: from a need to change the product or a need to change the process. However, in both cases the procedures should be the same, i.e. a formal sequence of events using an ECN (engineering change note) in order to identify:

1. the reason(s) for the required change
2. the effects of the change and the areas involved
3. the action that should be taken
4. the communication requirement.

An example of a typical ECN is shown in Fig. 5.18, which shows how this formal procedure would operate. First, a note is created (this could be by any employee) and approved by the departmental manager; copies of this note are then circulated to various other departments for their comments, evaluations, and suggestions before the change is finally approved and implemented.

Reason(s) for the change

The reasons for the required change must be included on the ECN in specific terms. For example, a generalized statement such as 'to reduce scrap' or 'to improve product appearance' will not do, since it is not possible from such statements to identify the effects of the proposed change in quantitative terms. The example in Fig. 5.18 shows the sort of information that should be provided.

Effects of the change

First, the costs of the change should be calculated and these could include such factors as existing stock that can be made obsolete, redesign costs, new tooling, machine modifications, or an extra cost on a purchased item. Second, the financial savings should be calculated and these two figures compared using a method such as payback for a low cost change or 'rate of return' if the change is more costly (see chapter 2, section 2.6 for details of these methods). These calculations can then be shown on the form as comments and suggested action, as shown in Table 5.7.

ENGINEERING CHANGE NOTE		DATE	dd/mm/yy	ECN No.	ECN.2347
ORIGINATOR	C. Blackwood	DEPT.	Work Study	MANAGER	PWK

Description of Proposed Change

It is suggested that the 15mm diameter holes (marked A on the attached drawing) on part number H.27342 (hinge block for the Commodore range of cubicles) be altered to include a countersink 18mm diameter by 5mm deep on the outer face.

Reasons for Proposed Change

The holes are used to insert the two hinge pins and by countersinking them in the manner suggested, the automatic insertion device which is used as a part of the assembly process will work more effectively with less down time when a pin wedges in the hole. I estimate that this downtime is currently 0.5 hours/day.

Comments on Proposed Change

Suggested Action on Proposed Change

Agreed Action on Proposed Change

CIRCULATION (Please Tick)	
ACCOUNTS MANAGER	
DESIGN MANAGER	
MATERIALS MANAGEMENT	
PROD.CONTROL MANAGER	
Q.A MANAGER	
PURCHASING MANAGER	
SALES MANAGER	
WORKS MANAGER	

Figure 5.18 Typical engineering change note (ECN).

Other factors may be difficult to evaluate but could include the effect on customers, on quality, on purchasing, or on associated production processes. However, all such factors should be noted on copies of the ECN by the departments concerned.

Action to be taken

Once the copies of the ECN have been circulated and returned with their comments and suggested actions by the departments concerned, a decision to accept or reject must be taken. This could be at various levels, depending on the expenditure involved, and the company should already have a policy on this for dealing with other types of project (for example, a maximum payback period or a minimum rate of return). If the change is approved then various actions will

Table 5.7 Example of ECN comments on a proposed change

ECN No. 2347 'Comments on proposed change'	
Costs	Current stock of parts = 1200 at a value of £1.00 each = £1200.00 Redesign cost, say £100.00 Purchased item: cost £1.00 Extra cost for countersinking £0.05/piece Annual usage = 12 000 (or 50/day × 240 days). Thus costs of change are £100 (one-off) and £(0.05 × 12 000) = £600 (annual)
Savings	0.50 hours per day downtime @ £16/hour = £8/day, × 240 days per year = £1920
Payback	If current stock is used before change is implemented the annual saving is £(1920 – 600) = £1320. Thus payback is £100/1320 = 0.076 years or 18 days
ECN No. 2347 'Suggested action on proposed change'	
Proposal	Change is justified; use up the current stock before making the change. The current usage rate is 50/day and thus the stock of 1200 will be exhausted in 24 days. This will allow time for the change to be implemented.

need to be taken, and these should already have been defined in detail in the quality manual, for example:

- drawing revision procedure; numbering and recording
- recall of superseded drawings procedure; register of copies issued to be kept
- computer records revision procedure; bills of materials, stock records, routeings, etc.
- work-in-progress procedure; method of dealing with items in production
- purchase orders procedure; method of dealing with outstanding P/Os
- marketing procedure; methods of informing customers/agents, revising catalogues and service manuals, etc.

Effectivity dates

As part of the agreed action, a date should be set for the change to take place (the effectivity date) which will be based on the time for implementation plus any additional time for policy decisions to be effected (e.g. to use up existing stocks). In some computer systems the facility exists to set up effectivity dates in the database files (e.g. stock records, bills of materials and routeings) such that the appropriate records are changed automatically when that date is reached, as shown in Table 5.8 for a BOM.

Priorities for changes

In some companies, priority codes are allocated to engineering changes in order to indicate the importance (or otherwise) of the change, for example as follows.

- *Priority A* This is the highest priority, requiring immediate action, since the reasons for the change involve health or safety or a serious product malfunction. In this case, all stocks will be withdrawn and all outstanding works and purchase orders cancelled.

Table 5.8 The use of effectivity dates

Level	Part no.	Quantity	UOM	Effectivity date (end)	Effectivity date (start)
0	H.23567	1	Each		
1	H.27342	1	Each	27/10/94	
1	H.27342A	1	Each		28/10/94
2	E.40615	0.2	kg		
1	P.35335	2	Each		
2	E.13487	0.1	m		
1	W.95344	2	Each		
	etc.		etc.		

In this example part H.27342 (hinge block) is to be replace by part H.27342A after 27/10/1994. No other parts are affected.

- *Priority B* This priority is less serious but still requires urgent action, for example if a supplier goes out of business and a different component from another supplier has to be used. In this case, existing stocks can be used up provided that there are sufficient stocks of other affected items.
- *Priority C* This priority is important but not crucial, since the reason for the change is purely financial (cost reduction). Some delay can therefore be accepted, to run out existing stocks and existing orders, unless the cost reduction justifies stock write-off and possible order cancellation charges from the supplier.
- *Priority D* This is the lowest priority, since the proposed change is purely cosmetic and can therefore be delayed until a convenient time in the future (for instance so that several 'priority D' changes can be implemented together).

These priority settings are particularly relevant when considering finished goods stocks, either ready for despatch at the manufacturer or recently delivered to a customer.

Change coding of drawings, documents and computer records

Any document or record that is subject to an engineering change should carry not only its own unique code but also a version, issue, or revision number and a date showing when the change took place. The conventional method is to code the first version with a 'zero', the second with a 'one' and so on. Thus, drawing number C.17432 v.0 would be the first version, and C.17432 v.1 the second version.

However, in some companies a double coding system is used for the version number whereby minor changes are shown by adding 'one' to the secondary code, and major changes by adding 'one' to the primary code and reducing the secondary code to 'zero', thus:

C.17432 v.0.0 original version
C.17432 v.0.1 first minor change
C.17432 v.0.2 second minor change
C.17432 v.1.0 first major change
C.17432 v.1.1 third minor change
etc. etc.

Warning Whichever system is used (and some companies use neither but create a totally new document number), it is essential to ensure that complete records are kept of all hard-copy issues so that these can be replaced when a new issue is made. Failure to do this can be very costly if an item is made or supplied to the wrong version!

PROBLEMS

1. In Table 5.1, the right conditions for teamwork are defined. Prepare a similar table showing the *wrong* conditions for teamwork.
2. You are part of a design team involved in the development of a new ball-point pen, and have been asked to prepare a FAST diagram for the product functions. Starting with a 'box' on the right of the diagram labelled 'enable writing', work to the left by asking 'how' this can be achieved (see section 5.3 for more details of the method; When you have completed your diagram, examine the suggested solution in appendix A and compare the two, noting any functions you have missed.
3. Calculate the Taguchi loss function, and thus the specification limits, and plot the values on a graph, given the following information (see section 5.4 for details of the method; see appendix A for solution). A component has a nominal length of 75 mm with a tolerance of ±0.8 mm. If it falls outside these limits but reaches the customer, the replacement cost is £20. However, if the faulty part is found during production, the cost of rectification or replacement is £3.50.
4. Suggest at least four suitable applications for FEA (finite element analysis), which is described in section 5.7, in a similar manner to the applications listed for CFD (computational fluid dynamics) in Table 5.6.
5. Prepare a draft procedure for inclusion in the quality manual covering the recall of superseded drawings when an engineering change has taken place. Your draft should also define the procedures for maintaining the drawing register and what that register should contain.

REFERENCES AND FURTHER READING

Bedworth, D. D., Henderson, M. R. and Wolfe, P. M. *Computer Integrated Design and Manufacture*, McGraw-Hill, New York, 1991.

Bergman, B. and Klefsjö, B., *Quality: from Customer Needs to Customer Satisfaction*, McGraw-Hill, London, 1994.

Boothroyd, G. and Dewhurst, P., *Product Design for Assembly Handbook*, Boothroyd–Dewhurst Inc., Wakefield, RI, USA, 1987.

Bralla, J. G., *Handbook of Product Design for Manufacturing*, McGraw-Hill, New York, 1986.

Corbett, J., Meleka, J., Pym, C. and Dooner, M., *Design for Manufacture*, Addison-Wesley, Wokingham, 1991.

Cross, N., *Engineering Design Methods: Strategies and Tactics for Product Design*, 2nd Edn, Wiley, Chichester, 1993.

FISITA 92, *Engineering for the Customer*, 'A Vehicle Characterisation Process', by Bradley, I. P., Jaguar Cars Ltd, Institute of Mechanical Engineers, London, 1992.

Fox, J., *Quality through Design*, McGraw-Hill, London, 1993.

Pugh, S., *Total Design*, Addison-Wesley, Wokingham, 1990.

Turino, J., *Managing Concurrent Engineering*, Van Nostrand Reinhold, New York, 1992.

CHAPTER

SIX

THE MANAGEMENT OF PRODUCTION

THE LINK

In a make-to-order company, when sales orders have been received and entered, Production must be immediately informed so that it can plan and execute works orders. However, in a make-to-stock company the sales orders do not directly trigger works orders, which are only required when stock needs to be replenished.

In either case, production is related directly or indirectly to sales and this relationship must be close, in fact a partnership, in the same way as Production and Sales are related to Design. Sales must therefore be aware of the problems of production and vice versa. For example, the question of delivery dates is often controversial. Sales may require a short delivery promise to keep a particular customer happy, whereas production must take account of all the factors (stock shortages, supplier lead times, machine and labour availability, etc.) in setting priorities.

Conflicting demands on sales and production must therefore always be resolved by discussion and compromise, and never by confrontation. It is always better to give a longer delivery promise, which can be met, than to make an unachievable promise. In fact, if a supplier fails to meet a due date, it is more likely to upset the customer and risk losing future orders than if a realistic date had been given which was a little longer than the customer wanted.

6.1 INTRODUCTION

In this chapter the various operating functions that need to be carried out in the production area are examined together with the variations that occur when they are applied in different types of manufacturing organization. The stages that are needed to convert demand into finished product are then considered with particular reference to the departmental control and planning aspects—the issue of instructions and the feedback of information. Finally, the relationships between the production staff and other departments are discussed, in terms of two-way communication and cooperation.

6.2 DEMAND AND PRODUCTION

Production takes place because demand has to be met and demand arises, because either

- stocks are insufficient to meet expected orders (make-to-stock), or
- new orders are received (make-to-order).

If an MRP (material requirements planning) system is in operation this will suggest that works orders be created to meet the demands that are currently contained in the master production schedule (MPS), which is a combination of sales orders and forecasts. The MRP system will compare these demands with what is available in stock and—taking account of stock policies—recommend that works and purchase orders be placed to make up the shortfalls. If MRP is not in operation then sales orders drive the demand in a make-to-order company or stock policies in a make-to-stock company.

These are both known as 'push' systems, where production is initiated at a stage before production. Alternatively, 'pull' systems are initiated as production takes place, the best-known example being JIT (just-in-time), where demand at the end of a production sequence triggers the operations that are necessary to meet the immediate demand—but no more.

These techniques—MRP, JIT, etc.—are considered in greater detail later in this book.

6.3 THE ELEMENTS OF A PRODUCTION SYSTEM

A production system can be considered as a series of associated functions, each function providing information to the others. This principle is illustrated in Fig. 6.1, the functions being defined as:

- works order creation
- works order planning and scheduling
- works order issue
- works order kitting
- works order progressing
- works order completion.

As stated in the previous section, not all production systems are based on works orders (for example JIT is not). However, at this stage it is easier to assume that this is not the case in order to establish the basic principles.

6.4 WORKS ORDER CREATION

A works order is created in response to a demand which can come from one of two sources (or both), depending on the type of business. In a make-to-order company the demand comes from sales orders received, whereas in a make-to-stock company it comes from stock policies which may be influenced by forecasts. To create a works order it is necessary to have access to three information databases which contain details of what is needed to make a product and how it is made. These are the parts, the bill of materials and the routeings files.

The database files

The *parts file* is often a combination file that is used to hold details of all raw materials, components, sub-assemblies and finished goods items. Each record on this file can have 100 or more data fields covering full details of the item (e.g. part number, description, analysis

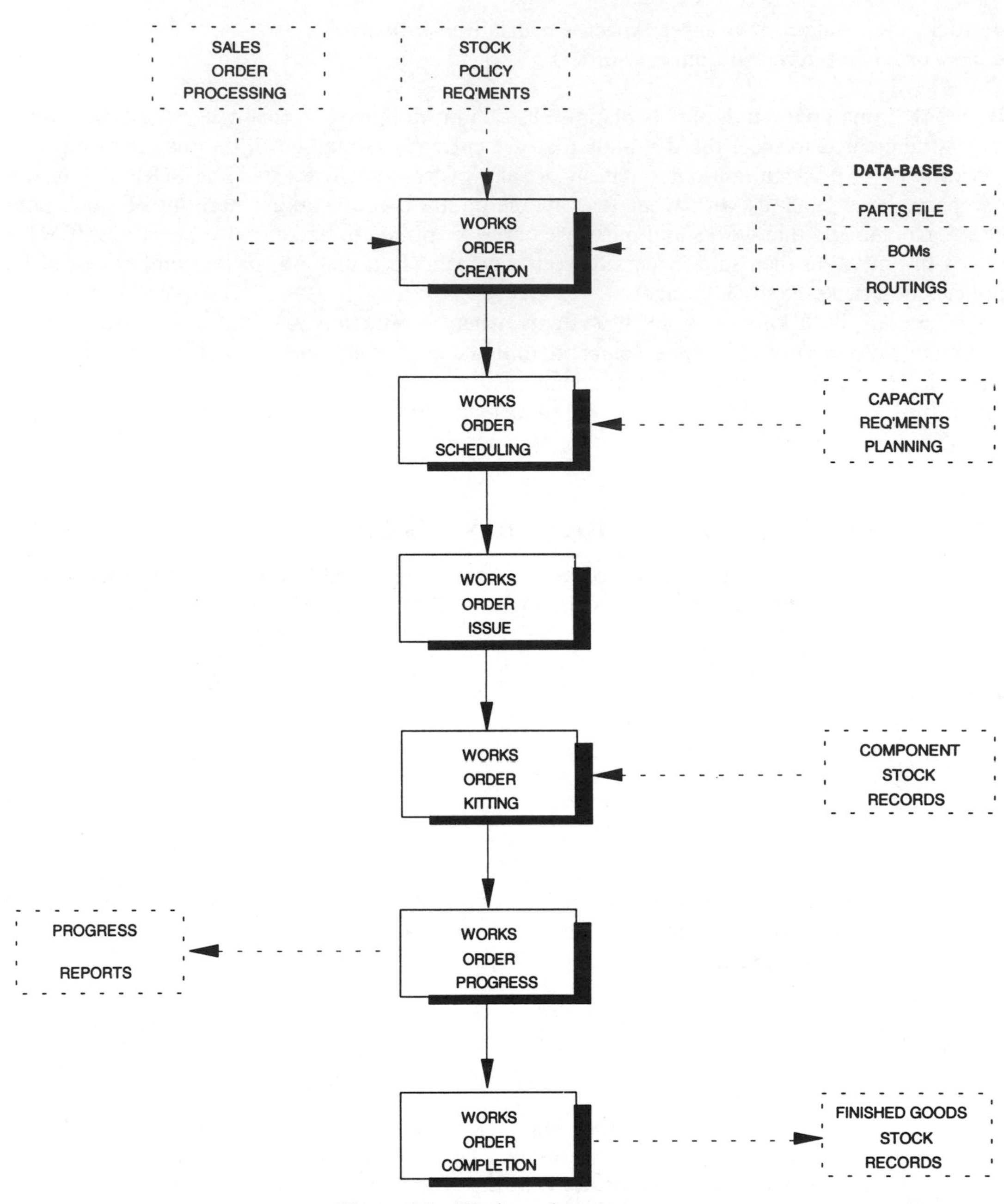

Figure 6.1 Works order sequence.

codes, stocks, stock locations, costs, and stock policies). If an item is not on this file it cannot be ordered or stocked.

The *bill of materials (BOM) file* defines the way in which the item is built up in stages from its basic components and raw material through to the complete final product. Figure 6.2 illustrates the principle: in this example there are four 'levels', the final product by convention being at level 0 (zero). This figure shows two presentations of the same data: the 'indented bill',

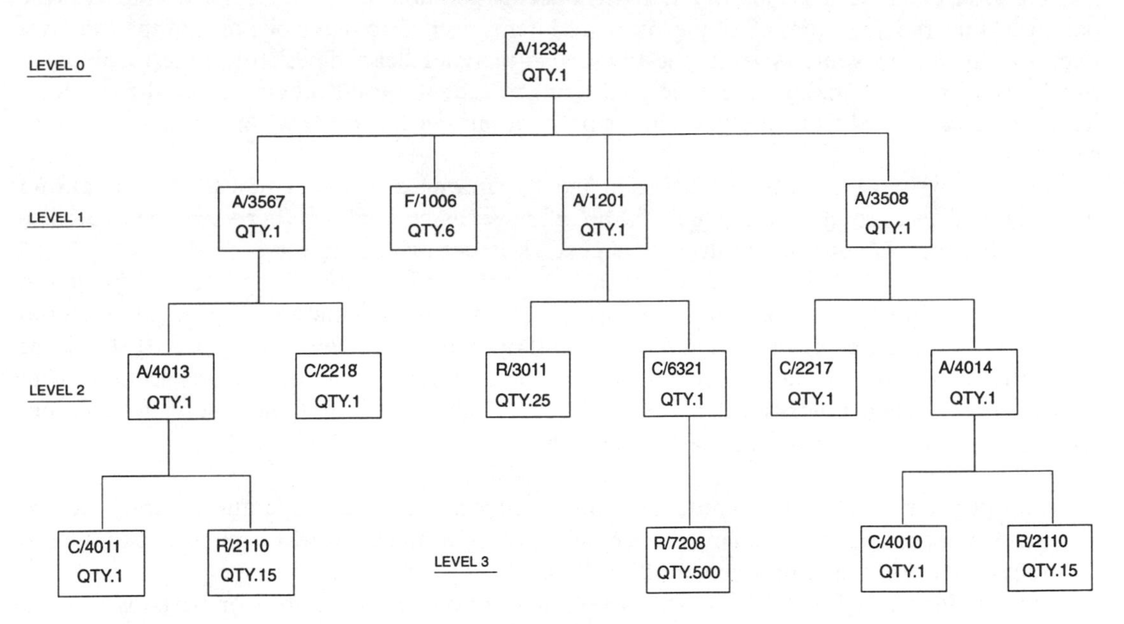

INDENTED BILL

BILL OF MATERIALS				
PRODUCT CODE: A/1234			LT = 2 days	
LVL	ITEM CODE	QTY	UoM	L/T Days
1	A/3567	1	EACH	2
.2	.A/4013	1	EACH	3
..3	..C/4011	1	EACH	1
..3	..R/2110	15	mL	1
.2	.C/2218	1	EACH	1
1	F/1006	6	EACH	2
1	A/1201	1	EACH	1
.2	.R/3011	25	mL	3
.2	.C/6321	1	EACH	2
..3	..R/7208	500	Gms	2
1	A/3508	1	EACH	1
.2	.C/2217	1	EACH	1
.2	.A/4014	1	EACH	2
..3	..C/4010	1	EACH	1
..3	..R/2110	15	mL	2

Figure 6.2 Bill of materials (product structure).

which is a typical computer screen display, and the 'product structure', which is easier to visualize.

In this case the final product (A/1234) is an assembly of three sub-assemblies plus six screws (F/1006), each sub-assembly consisting of a component plus a sub-sub-assembly, the latter being made from raw material and/or a component.

The BOM includes the quantity of an item needed to make the item on the level above (the parent). Thus, the quantities of all the lower level items needed to make one unit of the top-level item are known. In some systems, the BOM also includes 'lead times' (the time to obtain a purchased item or to make a manufactured item) so that the total time to make the top level item can be calculated from any point in the production cycle. The BOM can be also be used to calculate costs.

In many systems, the facility exists to include 'phantoms assemblies' in a BOM, a phantom being defined as a group of parts that—although never assembled together—can be treated as an assembly for administrative purposes (e.g. stock issues or costing).

The *routeings file* defines the way in which an item is manufactured in terms of production stages. For example, part no. C/6321 in Fig. 6.2 is made by machining 500 g of raw material R/7208, the routeing being defined in a sequence of operations as shown in Fig. 6.3. It should be noted that this is the master routeing which will be used as a basis for the preparation of an actual routeing when a works order is created. Fig. 6.3 also shows how the following two other files may be used to prepare the master routeing.

1. The work centres file, which contains details of all locations where operations can be carried out. A work centre can be a single machine, a group of machines with similar capabilities, a process, a work bench or a group of benches.
2. The standard text file, which contains text descriptions of operations or perhaps tooling. Each is identified by a unique code and, by using this code, time and effort can be saved when master routeings are being set up.

Most computer systems include the facility to copy and amend a BOM or a routeing, thus saving work when data for a family of products are being set up. Such systems also often include a 'find and replace' facility which allows a part or a machine that is being superseded to be replaced on any BOM or routeing by a new part or machine.

Setting up a works order

If an MRP (materials requirements planning) module is installed, recommended works orders will be set up automatically in response to demand (allowing for stock availability); this aspect is dealt with in a later chapter. However, without MRP, works orders must be created by the planner on the basis of known plus forecast demand and available stocks. This process is considerably easier to carry out on a computer system than manually, but essentially the procedure is the same.

The planner will enter the details of what is to be made onto a works order creation screen or a form that will carry a unique works order number, the actual entries being a part number (item code) and a quantity (or a default to the standard batch quantity). If a computer system is being used this will add to the order the correct quantities of materials or components needed to make that quantity (using the BOM file record) and also the routeing (operation sequence) from the master routeing. Otherwise this will have to be done manually.

The order has now been created but has not been planned; it is therefore flagged with a status of 'unplanned' or 'provisional' and as such is not available for issue to production. At this stage, if a computer system is in use, a facility known as 'trial kitting' can be used to determine what materials or components needed to make the item quantity on the order are not available (shortages). This operates on the basis of the bill of materials by looking one level down from the ordered item. This is shown in Table 6.1, based on the BOM in Fig. 6.2.

PRODUCT ROUTEING RECORD

PRODUCT CODE: C/.6321 STD.BATCH = 10
ADDITIONAL DESCRIPTION: xxxxxxxxxxxxxxxxxxxxxxxxxxxxxx

OPN. NO.	WK. CR.	STD. TEXT	SU. TIME	OPN. TIME	MOVE TIME	QUEUE TIME	OPERATION DESCRIPTION	QTY.	TOOLING
10	RMST	100	1.50	0.12	2	4			
20	LCLA		2.50	0.40	2	8	xxxxxxxxxxxxxxxxxxxxx xxxxxxxxxxxxxxxxxxxxx		xxxxxxxxxxxx
30	VMIL		5.00	0.20	1	6	xxxxxxxxxxxxxxxxxxxxx xxxxxxxxxxxxxxxxxxxxx		xxxxxxxxxxxx
40	GRND		4.50	0.25	1	4	xxxxxxxxxxxxxxxxxxxxx xxxxxxxxxxxxxxxxxxxxx		xxxxxxxxxxxx
50	FGST	330	4.50	0.25	1	4			

WORK CENTRES FILE

W/C. CODE	DESCRIPTION	M/C. CODES	CAP'Y. Hrs.	EFFY. %	RATE Per Hr.
RMST	RAW MAT'L STORES	010	80	85	15.00
LCLA	LIGHT CR. LATHES	020-029	240	80	20.00
VMIL	VERTICAL MILLERS	030-039	120	80	25.00
GRND	GRINDING	040-049	80	85	25.00
ASS1	ASSY.BENCH 1	050	40	80	18.50
etc.		etc.		etc.	

STANDARD TEXT FILE

CODE	OPERATION DESCRIPTION	TOOLING
100	DRAW RAW MATERIAL FROM STORE RM/2	
110	xxxxxxxxxxxxxxxxxxx xxxxxxxxxxxxxxxxxxx	xxxxxxxxxxxxxxxx
120	xxxxxxxxxxxxxxxxxxx xxxxxxxxxxxxxxxxxxx	xxxxxxxxxxxxxxxx
	etc.	etc.
330	RETURN TO COMPONENT STORE CM/1	

Figure 6.3 Product routeing.

Table 6.1 Trial kitting example

Works order W.98765 requires a quantity of 50 of part A/3567 and a trial kitting enquiry shows, one level down:

Component	UOM	Required	Free stock	Shortage
A/4013	Each	50	10	40
C/2218	Each	50	120	Nil
A further trial kitting is then carried out on component A/4013, again one level down:				
C/4011	Each	40	75	Nil
R/2110	ml	600	2000	Nil

In some computer systems the facility to trial kit at multi-level is available, but in this example it has been assumed that this is not the case. This trial kitting exercise shows that, before works order W.98765 can be issued, another order (W.98766) must be created and run in order to provide stock of component A/4013. It may then be decided to confirm these two works orders, and at this point stock of raw materials and components may be allocated (reserved) against the orders.

6.5 WORKS ORDER PLANNING AND SCHEDULING

Once a works order has been created it must be processed through the planning system in order to determine when it should be issued to production. Four factors can affect this start date:

1. the due date for completion
2. the time to complete the work
3. the availability of production capacity
4. the availability of lower level components.

Due date for completion

The due date for a finished product may be directly set by a delivery promise to a customer or by a date by which it must be in stock to avoid a stock-out situation. This date then sets the due dates for any lower level items that have to be produced to make that finished product, as shown in Fig. 6.4, which uses the lead times from the BOM structure to calculate planned start and finish dates backwards from the due date of the finished product (assembly A). This is called 'Back-scheduling'.

Time to complete the work

In the case of purchased items, lead times are set up which define how long it should take to obtain the components or materials required. As shown in Fig. 6.4, the same approach can be used for manufactured items. However, this method does not take account of batch size (unless standard batches are always used) and may therefore be inaccurate. It may be better to use the data that are normally available on the master routeing. This gives three, or in some systems, four sets of time values (usually in hours) for each operation:

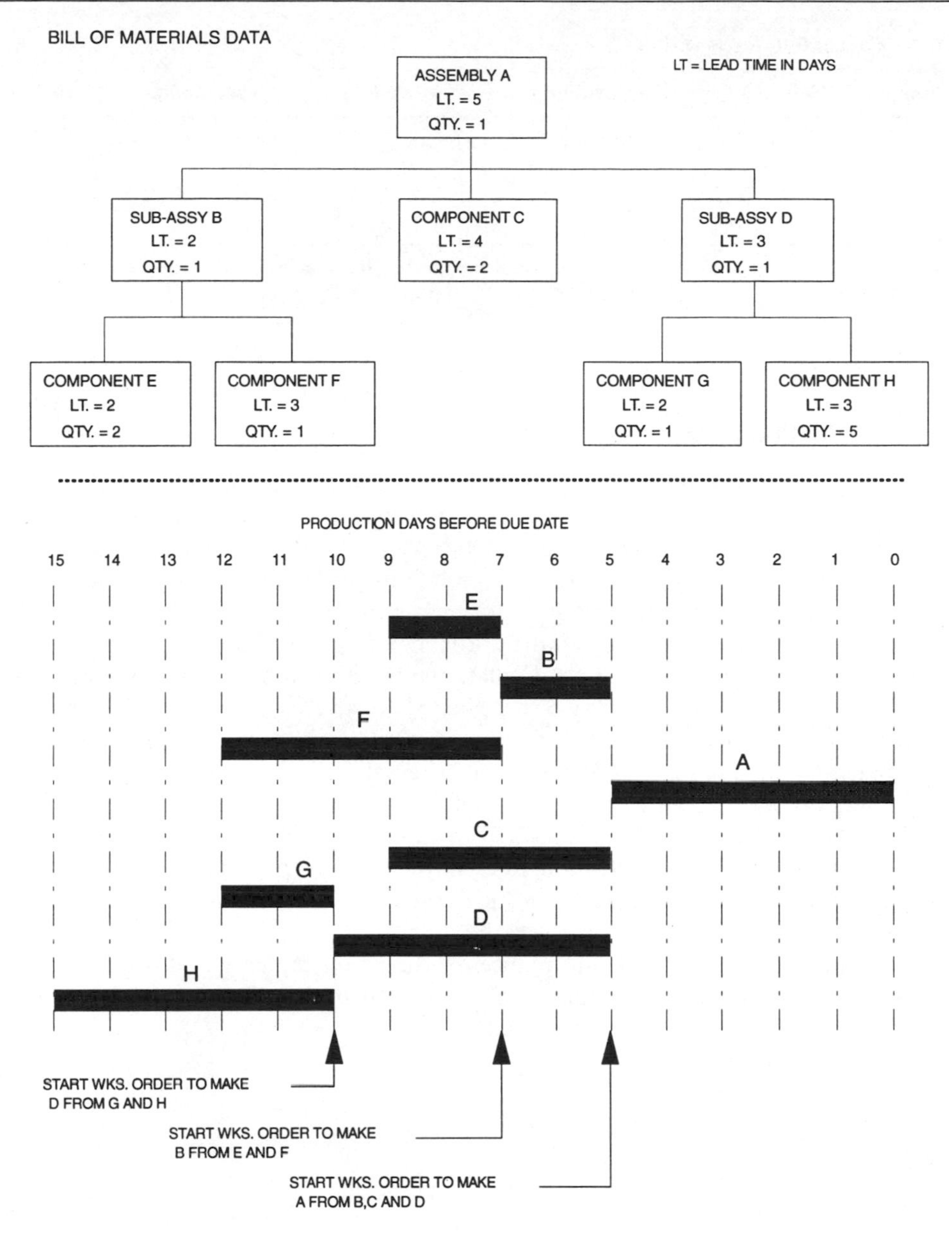

Figure 6.4 Scheduling method for associated orders.

- *set-up time* a time to set up a machine for an operation
- *operation time* a time to process one unit of product
- *move time* a time to move the batch to the next operation
- *queue time* an average time waiting to start an operation.

From these data it is possible for the system to calculate the production lead time for a given batch size, as shown in Table 6.2.

Table 6.2 Calculating the lead time

Operation no.	Set-up hours	Operation hours	Move hours	Queue hours	Comments
10	4.00	0.25	6.00	5.00	
20	2.00	0.15	4.00	4.00	
30	1.00	0.20	4.00	8.00	
40	15.00	Nil	2.00	3.00	See note

This is a batch operation, not quantity-related, e.g. heat treatment. Thus the lead time for a batch of 100 would be 22.00 + 100(0.60) + 16.00 + 20.00 = 118 hours or, say, 15 days

Available production capacity

The planned start and finish dates for a series of works orders may be theoretically perfect but can fail to be met if there is insufficient production capacity. It is therefore necessary to examine the effect of a new batch of works orders on the production load, and this can be done in a computer system by running the capacity requirements planning (CRP) module. In a manual system this can be very time-consuming, if not impossible!

The CRP module looks at all open works orders, both planned and in progress, and uses the production times from the routeings to calculate the load on each machine or work centre in a given time period. At this stage it may be assumed that capacity is unlimited (infinite capacity) or limited to a fixed number of hours for the period (finite capacity). In general, infinite-capacity planning is used as an initial check and then, when the production plan has been partially refined, finite planning is used to make the final adjustments.

To illustrate the process, screen displays from an infinite capacity planning run are shown in Fig. 6.5. At this stage no prioritization rules have been set up and therefore works orders are taken as planned, i.e. as scheduled to start and finish with intermediate operations given start/finish dates to suit.

The refining process is designed to reduce overloads on capacity and can be achieved in a number of ways, each adjustment being followed by a further CRP run to measure the impact. Typical adjustments that could be made are to:

- increase capacity by overtime or weekend working
- add capacity through an extra shift
- subcontract some of the work
- re-route some works orders to use alternative operations.

Availability of lower level components

If a works order cannot start on its planned start date because of component or material shortages it must be considered as an indication that the planning function has failed. It is accepted that such failures are always possible due to the totally unforeseen, but are rare. For example, if a supplier is unreliable then good planning would suggest a change in supplier or, if this is not possible, a safety stock should be held. Similarly, if a machine is liable to break down then it should be subjected to more intensive planned maintenance and an alternative process defined and set up (internal or subcontract). In the last resort consideration should be given to machine replacement.

TABULAR DISPLAY OF A WORK CENTRE WORK LOAD

WORK CENTRE: GRND GRINDING   CAPACITY: 80 HRS/WK   EFFICIENCY: 0.85

WK.No 24			WK.No 25			WK.No 26			
ORD.No.	HRS.	CUM.	ORD.No.	HRS.	CUM.	ORD.No.	HRS.	CUM.	
W.1274	15.50	15.50	W.1322	12.00	12.00	W.1346	11.75	11.75	→ MORE
W.1283	1.25	16.75	W.1326	6.25	18.25	W.1347	5.50	17.25	
W.1290	1.75	18.50	W.1327	3.50	21.75	W.1349	4.00	21.25	
W.1291	2.50	21.00	W.1329	4.25	26.00	W.1352	3.50	24.75	
W.1294	2.25	23.25	W.1333	2.75	28.75	W.1354	5.25	30.00	
W.1298	4.75	28.00	W.1335	8.30	37.05	W.1355	9.20	39.20	→ MORE
W.1302	9.35	37.35	W.1338	7.50	44.55	W.1357	12.25	51.45	
W.1309	7.25	44.60	W.1340	16.80	61.35				
W.1311	3.40	48.00							
W.1314	5.75	53.75							
W.1315	8.25	62.00							→ MORE
W.1319	7.35	69.35							
W.1321	7.25	76.60							
CAPACITY = 68 HRS.			CAPACITY = 68 HRS.			CAPACITY = 68 HRS.			
LOAD = 112.6 %			LOAD = 90.2 %			LOAD = 75.7 %			

GRAPHICAL DISPLAY OF WORK CENTRE WORK LOADS

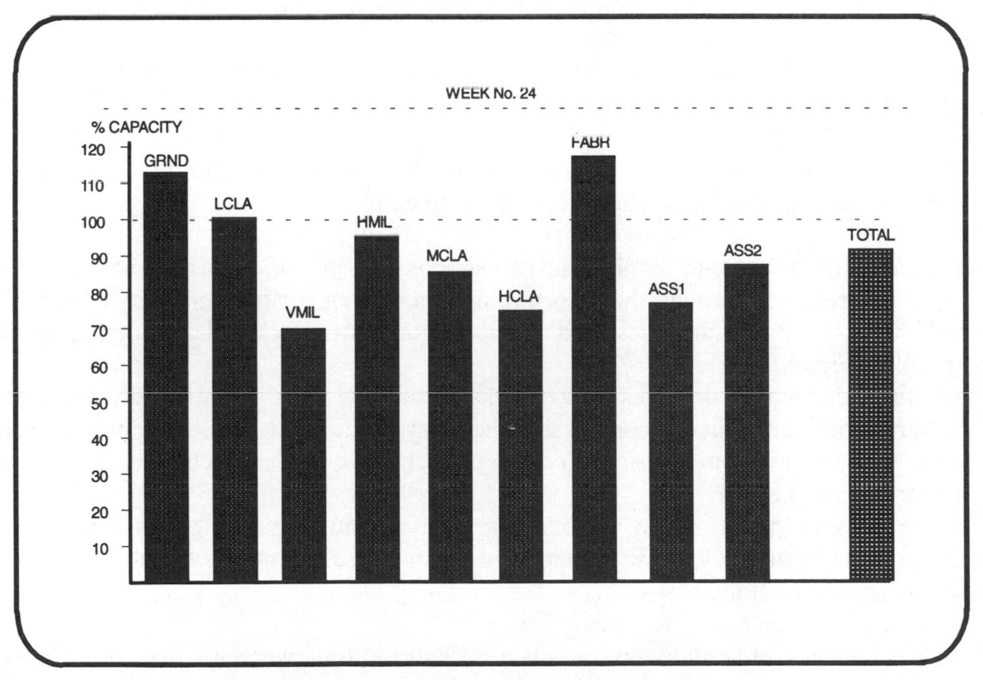

Figure 6.5 Typical capacity planning reports.

Some of these suggestions may seem excessive, but it is simply a matter of economics. What is the cost of holding safety stock compared with the cost of a late delivery with the possible loss of a customer or a penalty clause payment?

Author's note: A form of production planning known as 'line of balance' may be appropriate where delivery schedules are specified by the customer for products that are assembled from a number of components. An example of how this technique is used is included in appendix E.

6.6 WORKS ORDER ISSUE

Once a works order has been confirmed and planned, it can be issued. This generally means that the status is altered to 'issued' to indicate that work is now about to start. The works documentation can now be printed and issued to the production department. A typical document set could include the following.

- *Operation route card* This is sometimes known as the 'traveller'. It gives full details of what is to be made and the operation sequences involved. This card can also be used to record quality checks, and travels with the job (Fig. 6.6).
- *Material requisition* This card gives details of materials, components and possibly tools to be drawn from stores, and is the authority to make the withdrawal. This card is then used to update the computerized or manual stock records (Fig. 6.7).
- *Operation cards* One card is printed for each operation, giving full details of the work to be done. It is then used by the operator to enter quantities produced (and in some cases time booked). This card is then used to update the computerized or manual works order progress records (Fig. 6.7).
- *Stores receipt* This card gives details of components or assemblies being passed into the stores, and is the authority for this. The card is then used to update the computerized or manual stock records.
- *Identity labels* If the batch is in a series of stillages or otherwise split, then identity labels may be required in addition to the travelling route card.

In some companies a 'work-to list' is issued to supervisors, giving details of the works orders that should be processed through their work centres in a given period (shift, day or week). The format of a work-to list is usually designed to suit the company concerned, but a typical example is illustrated in Fig. 6.8.

From this information the supervisor can ensure that the work in his or her area is carried out—as far as possible—in the correct sequence and at the right time. This process is often assisted by applying a system of priority coding to each works order; such codes are shown on the example in Fig. 6.8.

Various possible prioritization systems can be used, but they can be classified into two types, 'static' and 'dynamic'. Static systems can be operated manually, but dynamic systems are practical only if computer-based.

1. *Static* A static prioritization system sets a code at the time when a works order is issued, and this code is changed only if a decision is made by management. For example, such a code could be based on the due date as shown in Table 6.3.
2. *Dynamic* A dynamic prioritization system sets a code at the time when a works order is issued, but this code changes automatically as the order progresses. For example, such a code

OPERATION ROUTE CARD

PRODUCT CODE: C/.6321
DESCRIPTION: xxxxxxxxxxxxxxxxxxxxxxxxxxxxxxxx
ADDITIONAL DESCRIPTION: xxxxxxxxxxxxxxxxxxxxxxxxxxxxxxxx

WORKS ORDER: W.24653
ORDER QTY.: 10
DUE DATE: DD/MM/YY

OPN. NO.	WK. CR.	WORK CENTRE DESCRIPTION	SU. TIME	OPN. TIME	MOVE TIME	QUEUE TIME	OPERATION DESCRIPTION	QTY.	TOOLING	START DATE	FINISH DATE
10	RMST	RAW MAT'L STORES	1.50	1.20	2.00	4.00	DRAW RAW MATERIAL FROM STORE RM/2	5 Kg		DD/MM/YY	DD/MM/YY
20	LCLA	LIGHT CR. LATHES	2.50	4.00	2.00	8.00	xxxxxxxxxxxxxxxxxxxxxx xxxxxxxxxxxxxxxxxxxxxx		xxxxxxxxxxxx	DD/MM/YY	DD/MM/YY
30	VMIL	VERTICAL MILLERS	5.00	2.00	1.00	6.00	xxxxxxxxxxxxxxxxxxxxxx xxxxxxxxxxxxxxxxxxxxxx		xxxxxxxxxxxx	DD/MM/YY	DD/MM/YY
40	GRND	GRINDING	4.50	2.50	1.00	4.00	xxxxxxxxxxxxxxxxxxxxxx xxxxxxxxxxxxxxxxxxxxxx		xxxxxxxxxxxx	DD/MM/YY	DD/MM/YY
50	FGST	FINISHED GOOS STORES	4.50	2.50	1.00	4.00	RETURN TO COMPONENT STORE CM/1			DD/MM/YY	DD/MM/YY

Figure 6.6 Typical operation route card (traveller).

MATERIAL REQUISITION CARD

PRODUCT CODE: C/.6321
DESCRIPTION: xxxxxxxxxxxxxxxxxxxxxxxxxxxxxx
ADDITIONAL DESCRIPTION: xxxxxxxxxxxxxxxxxxxxxxxxxxxxxx

WORKS ORDER: W.24653
ORDER QTY.: 10
DRG.NO. FR/86437

OPN. NO.	WORK CENTRE DESCRIPTION	SU. TIME	OPN. TIME	OPERATION DESCRIPTION	QTY.	START DATE	FINISH DATE
10	RAW MAT'L STORES	1.50	1.20	DRAW RAW MATERIAL FROM STORE RM/2	5 Kg	DD/MM/YY	DD/MM/YY

STOREMAN		COMPUTER		DATE STARTED	
QUANTITY ISSUED		STOCK LOCATION		DATE COMPLETED	

MATERIAL REQUIRED: xxxxxxxxxxxxxxxxxxxxxxxxxxxxxxxxxxxxxxxx
xxxxxxxxxxxxxxxxxxxxxxxxxxxxxxxxxxxxxxxx
xxxxxxxxxxxxxxxxxxxxxxxxxxxxxxxxxxxxxxxx

OPERATION CARDS

PRODUCT CODE: C/.6321
DESCRIPTION: xxxxxxxxxxxxxxxxxxxxxxxxxxxxxx
ADDITIONAL DESCRIPTION: xxxxxxxxxxxxxxxxxxxxxxxxxxxxxx

WORKS ORDER: W.24653
ORDER QTY.: 10
DRG.NO. FR/86437

OPN. NO.	WORK CENTRE DESCRIPTION	SU. TIME	OPN. TIME	OPERATION DESCRIPTION	TOOLING	START DATE	FINISH DATE
20	LIGHT CR. LATHES	2.50	4.00	TURN DOWN END 'A' TO DRG. AND ROLL THREAD	TOOL KIT TK/1054	DD/MM/YY	DD/MM/YY

OPERATOR		COMPUTER		DATE STARTED	
QUANTITY GOOD		QUANTITY SCRAP		DATE COMPLETED	

COMMENTS:

Figure 6.7 Typical operation and material requisition cards.

Table 6.3 Example of static prioritization

The priority code is to be based on a week and year in which delivery is due. Thus:

Due date	Week no.	Year no.	Code
22/10/93	42	3	342
06/12/93	49	3	349
13/01/94	02	4	402
08/03/94	10	4	410

could be based on the number of days remaining before the due date or some similar principle, as shown in Table 6.4.

This shows that the top dynamic priority is works order C, followed by B, D and A. If the static system had been used the sequence would be A, C, D and B.

Table 6.4 Example of dynamic prioritization

One commonly used form is known as the 'critical ratio', which is based on the time remaining before the due date and the time to complete the work. This is expressed as a formula

$$CR = \frac{(\text{Due date day no}) - (\text{Time now day no}) \times 100}{\text{Lead time remaining in days}}$$

Works order	Due date (day)	Time now (day)	LT remaining (days)	Priority (code)
A	120	100	15	133
B	140	100	40	100
C	125	100	30	83
D	130	100	25	120

SUPERVISOR: ALAN BROWN

WORK-TO LIST

DATE FROM: DD/MM/YY
DATE TO: DD/MM/YY
PAGE: 1 of 1

WKS.ORDER NO.	OPN. NO.	START	FINISH	PRODUCT CODE	QTY.	PREV. WORK CENTRE	NEXT WORK CENTRE	SETUP TIME	OPN. TIME	PRIORITY
LCLA	LIGHT CENTRE LATHES									
W.17685	20	DD/MM/YY	DD/MM/YY	P/17532	150	RMST	GRND	1.50	4.40	50
W.17690	20	DD/MM/YY	DD/MM/YY	P/18670	60	RMST	GRND	2.00	6.25	80
W.16489	30	DD/MM/YY	DD/MM/YY	R/89654	45	VMIL	DRIL	2.00	9.00	105
W.20976	25	DD/MM/YY	DD/MM/YY	S/34268	100	RMST	BORE	3.00	7.80	120
GRND	GRINDERS - C'LESS									

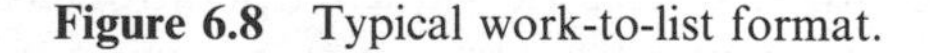

TOTAL = 15 ORDERS

PLANNED BY: ....

Figure 6.8 Typical work-to-list format.

6.7 WORKS ORDER KITTING

In some cases the number of items that need to be issued from the stores against a works order is too great to be readily handled using the materials requisition alone. This is particularly true at the assembly level which can involve many components and sub-assemblies. In such cases a kitting list can be printed as shown in Fig. 6.9. This can be given to the stores in advance of works order documentation issue, and the materials requisition will then refer to this list.

In certain types of work (again, usually on assemblies) some low-cost items are stocked as 'bulk-issue' (or free-issue) in the assembly area. This means that such items are issued against a requisition in a practical quantity rather than against an individual order (e.g. a box of screws or a tin of adhesive). These items are therefore excluded from the kitting list, although they are included in the bill of materials and flagged as such. The stock control procedures for this situation are described in chapter 9.

PRODUCT CODE: S.14367
DESCRIPTION: xxxxxxxxxxxxxxxx
QUANTITY: 100

KITTING LIST

WORKS ORDER No. W.13578
DATE: DD/MM/YY
PAGE: 1 of 1

ITEM CODE	DESCRIPTION	U.O.M.	QTY. REQD.	STORES	BIN/ RACK	FULL ISSUE?	PARTIAL ISSUE	COMMENT
12/345	xxxxxxxxxxxxxxxxxxxxxxxxxxx	EACH	100	STORE A	23/1			
13/444	xxxxxxxxxxxxxxxxxxxxxxxxxxx	EACH	200	STORE A	24/2			
11/907	xxxxxxxxxxxxxxxxxxxxxxxxxxx	LTRS	5	STORE A	30/5			
275608	xxxxxxxxxxxxxxxxxxxxxxxxxxx	KGS	50	STORE A	33/3			
P.43356	xxxxxxxxxxxxxxxxxxxxxxxxxxx	EACH	100	STORE A	34/1			
17/0987	xxxxxxxxxxxxxxxxxxxxxxxxxxx	EACH	400	STORE B	62/2			
12/0978	xxxxxxxxxxxxxxxxxxxxxxxxxxx	EACH	100	STORE B	73/2			
23/4689	xxxxxxxxxxxxxxxxxxxxxxxxxxx	MTRS	600	STORE C	92/9			

TOTAL = 8 ITEMS ISSUED BY: .... ENTERED BY: ....

Figure 6.9 Typical kitting list printout.

The kitting list can be considered as a stores requisition, and Fig. 6.9 shows that space has been left for entries by the stores staff. In this example, there are three columns. Full issue is ticked if appropriate, or the actual quantity issued should be entered into partial issue. The comment column can be used to record other details, for example, bin location used or lot number, for traceability.

The 'kit' that is to be used to make the item on the order does not include raw materials or components only; tools may also be required and, in some processes, drawings and NC (numerically controlled) machine tool or robot programs (if these are not stored on the machine or loaded directly). Such items should be included on the bill of materials and itemized on the kitting list.

It may also be necessary or desirable to 'phase' the kitting, for example to make some issues before the first operation and others when a subsequent operation is due to start. This procedure requires careful control but can prove advantageous by reducing the chances of items that are not immediately required becoming lost or damaged in the production area. If a computer system is being used, facilities are probably available to call up the list onto a screen and to enter the issues, as shown in Fig. 6.10.

In this example, if all the items on the list are issued 'complete' then only the bottom box (*Is kitting list issued complete?*) need be entered unless stock has been taken from a different location from that specified. The use of default values should also be noted as a time-saver, a default value being defined as an entry that will be made automatically by the computer system if no other entry is made.

6.8 WORKS ORDER PROGRESSING

When a works order has been 'kitted' it is considered to be a part of the work-in-progress (until then the materials or components were stock), and a system of recording progress should start

KITTING LIST

WORKS ORDER No. W.13578
DATE: DD/MM/YY

Recording Stock Issues

LINE	ITEM CODE	DESCRIPTION	U.O.M.	QTY. REQD.	STORES	BIN/ RACK	CORRECT STORE/BIN? [Y/N]	COMPLETE ISSUE ? [Y]	PARTIAL ISSUE ? [QTY]
01	12/345	xxxxxxxxxxxxxxxxxxxxxxxxxxxxx	EACH	100	STORE A	23/1			
02	13/444	xxxxxxxxxxxxxxxxxxxxxxxxxxxxx	EACH	200	STORE A	24/2			
03	11/907	xxxxxxxxxxxxxxxxxxxxxxxxxxxxx	LTRS	5	STORE A	30/5			
04	275608	xxxxxxxxxxxxxxxxxxxxxxxxxxxxx	KGS	50	STORE A	33/3			
05	P.43356	xxxxxxxxxxxxxxxxxxxxxxxxxxxxx	EACH	100	STORE A	34/1			
06	17/0987	xxxxxxxxxxxxxxxxxxxxxxxxxxxxx	EACH	400	STORE B	62/2			
07	12/0978	xxxxxxxxxxxxxxxxxxxxxxxxxxxxx	EACH	100	STORE B	73/2			
08	23/4689	xxxxxxxxxxxxxxxxxxxxxxxxxxxxx	MTRS	600	STORE C	92/9			
							Default = Y	Default = Y	

To use Default Values [SKIP] to next Box.

Is Kitting List Issued Complete? [Y/N]

Is any Store/Bin Incorrect? If so Enter Line No. & New Store/Bin — Line No. — Store — Bin

If [Y] in 'Complete' no entry allowed in 'Partial'

If [N] in 'Complete' entry must be made in 'Partial'

Figure 6.10 Stock issues recording from a kitting list.

to operate. This recording process (sometimes known as tracking) can be manual, semi-manual or fully computerized, but in all cases the data will generally be the same, as shown in a typical work sheet in Fig. 6.11 (which also shows the computerized version). Most companies that use this type of record operate on a daily or shift basis, and expect all time spent to be entered even if it is non-productive.

- *Manual* In this case the operative enters details onto a work sheet or on an operation card which is then passed to the planning office to update its records and then to the costing department. If a PBR (payment-by-results) system is in operation, details will also be required by the wages department.
- *Semi-manual* This procedure will be the same as *manual* except that data from the work sheets or operation cards will be entered to a computer terminal which will result in automatic updating of the progress records and costs and allow access for Wages to calculate bonus payments.
- *Computerized* This procedure is often called shop-floor data capture (SFDC) and requires a series of computer terminals (often specially adapted to suit the conditions) to be located throughout the production area. The operator enters details of the work done into the nearest terminal, daily or at the end of an operation. A typical SFDC screen format is shown in Fig. 6.11; it can be seen that the supervisor's identity code (Id) and password are needed to enter rework or non-productive work.

DAILY WORK SHEET

DATE .... SHIFT CODE ....

OPERATOR: .... DEPARTMENT: .... CLOCK NO: ....

REF.(1)	OP.NO.(2)	DESCRIPTION (Machine No. or NP Work)	TIME ON	TIME OFF	QTY.GOOD	QTY.SCRAP	SCRAP CODE(3)

NOTES:
1. Works Order Number or Non-Productive Time Code
2. Operation Number from Route Card
3. As List in Standard Procedure Manual

SUPERVISOR: ....

SFDC - OPERATOR WORK BOOKING SCREEN

CLOCK NO. [ ] NAME OK? [Y/N]

WKS.ORDER NO. OPERATION NO. MACHINE NO.

QTY.GOOD QTY.SCRAP SCRAP CODE

IS THIS OPERATION COMPLETED? [Y/N]

SUPERVISOR ID SUPERVISOR PASSWORD

NON-PRODUCTIVE WORK CODE

NON-PRODUCTIVE WORK DESCRIPTION

REWORK CODE

WKS.ORDER NO. OPERATION NO. MACHINE NO.

RE-WORK WORK DESCRIPTION

Figure 6.11 Daily work sheet and SFDC work booking screen.

Some systems use barcodes to assist with SFDC entries. For instance, an operator or supervisor can use a bar coded card to enter his or her Id, other codes (works order, operation, scrap and non-productive work) being picked up from the route card or a data sheet.

As each booking is made against a works order its status changes to indicate its progress towards completion, and if a computer system is in use (and up to date!) a screen enquiry will show the latest position. Figure 6.12 shows a fairly typical screen with the order status at 'started' and the operations showing flag settings for status (ST):

N not started
S started
C completed
H held.

The held flag is set to indicate that there is a problem, and no further bookings can be made until it is unset.

A secondary level of progress is given by the quantity columns. Thus, on operation 20, which is started, 20 out of the batch of 75 have been completed and 1 scrapped. Thus the exact position of the order can be determined, as shown in the example in Fig. 6.12, including, in this case, a critical ratio (priority) of 85, indicating that it is behind schedule.

WORKS ORDER STATUS

PRODUCT CODE: C.2218
DESCRIPTION: xxxxxxxxxxxxxxxxxxxxxxxxxx
QUANTITY REQD. 75
DUE DATE: WEEK 24/3

WORKS ORDER No. W.13142
ORDER STATUS: STARTED
DATE: WEEK 21/5

OPN. NO.	WK. CR.	OPERATION DESCRIPTION	SU. TIME	OPN. TIME	MOVE QUEUE	STRT DATE	FIN DATE	QUANTITY OK	QUANTITY SCR	ST	%age C'PLETE
10	RMST	xxxxxxxxxxxxxxxxxxxx xxxxxxxxxxxxxxxxxxxx	0.15	9.00	6.00	22/2	22/3	75	0	C	100
20	LCLA	xxxxxxxxxxxxxxxxxxxx xxxxxxxxxxxxxxxxxxxx	0.25	30.00	12.00	22/4	23/2	20	1	S	28
30	VMIL	xxxxxxxxxxxxxxxxxxxx xxxxxxxxxxxxxxxxxxxx	0.50	15.00	12.00	23/3	23/5	0	0	N	NIL
40	GRND	xxxxxxxxxxxxxxxxxxxx xxxxxxxxxxxxxxxxxxxx	0.45	18.75	10.00	24/1	24/3	0	0	N	NIL
50	INSP	xxxxxxxxxxxxxxxxxxxx xxxxxxxxxxxxxxxxxxxx	0.25	12.00	15.00	24/4	25/1	0	0	N	NIL

BASED ON 7.5 HOURS PER DAY:
LEAD TIME AT START: 141.35 HOURS OR 18.8 DAYS
CURRENT LEAD TIME: 114.19 HOURS OR 15.2 DAYS

WORKS ORDER PROGRESS:
WORKS ORDER IS 19.2% COMPLETED
PRIORITY VALUE (CR) = 85

Figure 6.12 Typical works order status screen.

6.9 WORKS ORDER COMPLETION

When the last operation on a works order has been 'flagged' as completed it could be closed, but in many company systems it is not. This is to ensure that it is still available for modification or amendment if a booking error has been made and also to make it more visible for enquiries. The order status is therefore set to 'completed', and may remain in this state for several weeks or even months. However, at some point it is closed and archived, which means that it is no longer immediately visible but can be viewed in the archive files if required.

When an order is completed, the final items, which may well be needed to make a higher level item, are put into stock which means that they no longer form a part of the work-in-progress.

6.10 SCRAP AND REWORK

Any production system should include the facility to record full details of any item that is scrapped together with reasons in the form of codes and text. This information can then be analysed to determine whether changes should be made, for example in design, process or material used. A typical list of scrap codes and descriptions is given in Table 6.5.

The production system should also facilitate the control and recording of all rework on a works order, such work being carried out only on the authority of the supervisor. All time spent should be recorded on the daily work sheet or a time card with a reason code and description. These should be similar to those used for recording scrap.

Another form of unplanned work, which should also be authorized by the supervisor, may be known as 'excess' or 'extra' work. In this case, for reasons beyond the control of the operator, the operating time allowed is insufficient and extra time must be booked. Such booking should be recorded seperately, and once again should be accompanied by a reason code and description.

Table 6.5 Examples of scrap codes and descriptions

S100	Faulty material—wrongly specified
S110	Faulty material—wrongly supplied
S120	Faulty material—damaged in transit
S200	Incorrectly made—design error
S210	Incorrectly made—wrong drawing issue
S220	Incorrectly made—operator error
S230	Incorrectly made—setting error
S300	Damaged in production—tool breakage
S310	Damaged in production—bad handling
S320	Damaged in production—machine fault
S400	Damaged in transit—packing fault

6.11 LOST AND NON-PRODUCTIVE TIME

In order to control production efficiency and to provide accurate information for costing purposes, it is necessary to ensure that all lost and non-productive time is properly controlled and recorded. All such time should therefore be booked as before, with bookings authorized by the supervisor. In companies where shop-floor data capture has been installed, or where work sheets are in use, the system can be adapted to allow such bookings; alternatively a special clock card can be used.

Lost time and non-productive work can be defined as any time that cannot be booked to a works order. In some companies the two terms mean the same, but in others non-productive work is defined as work not directly associated with production (e.g. attending meetings or training courses) whereas lost time is idle time (e.g. waiting for work or machine breakdown). Some examples are given in Table 6.6.

Table 6.6 Examples of lost time codes and descriptions

Code	Description
L100	Waiting—machine breakdown
L110	Waiting—for work
L120	Waiting—for setter
L130	Waiting—for material
L140	Waiting—for crane
L150	Waiting—power failure
N200	Union meeting
N210	Training course
N230	Machine cleaning
N240	Other department

6.12 ORDER AND BATCH SPLITTING

There may be reasons why a works order, which started in production at a certain batch size, needs to be split into two or more smaller batches. For example, a part of the original order may be urgently required to meet an unplanned customer requirement (e.g. to replace faulty goods). In such a case it is important to be able to split the batch as a standard procedure within the system by calling up the order on a computer screen and specifying the split (and other details) thus:

Order W/1356 for 300 units for delivery in 15 days' time is to be split into two orders:
Order W/1356/1 new order for 100 units for delivery in 5 days' time
Order W/1356 original order amended for 200 units for delivery in 15 days' time.

This is generally easier to do between between operations, and care must be exercised to ensure that the original documents are amended, new documents are issued for the new order and traceability (see section 6.13 below) is maintained on both. If, however, it is necessary to split the batch before the order has been issued or kitted it may be easier to cancel it and issue two or more new orders.

This batch-splitting procedure, which is designed to overcome specific problems, should not be confused with the deliberate policy of 'overlapping'. In Fig. 6.13 three different methods of order scheduling are shown.

- *Gap* This is the most common method. A gap is allowed between each operation by specifying positive move and queue times. The lead time in the example is 21 days.

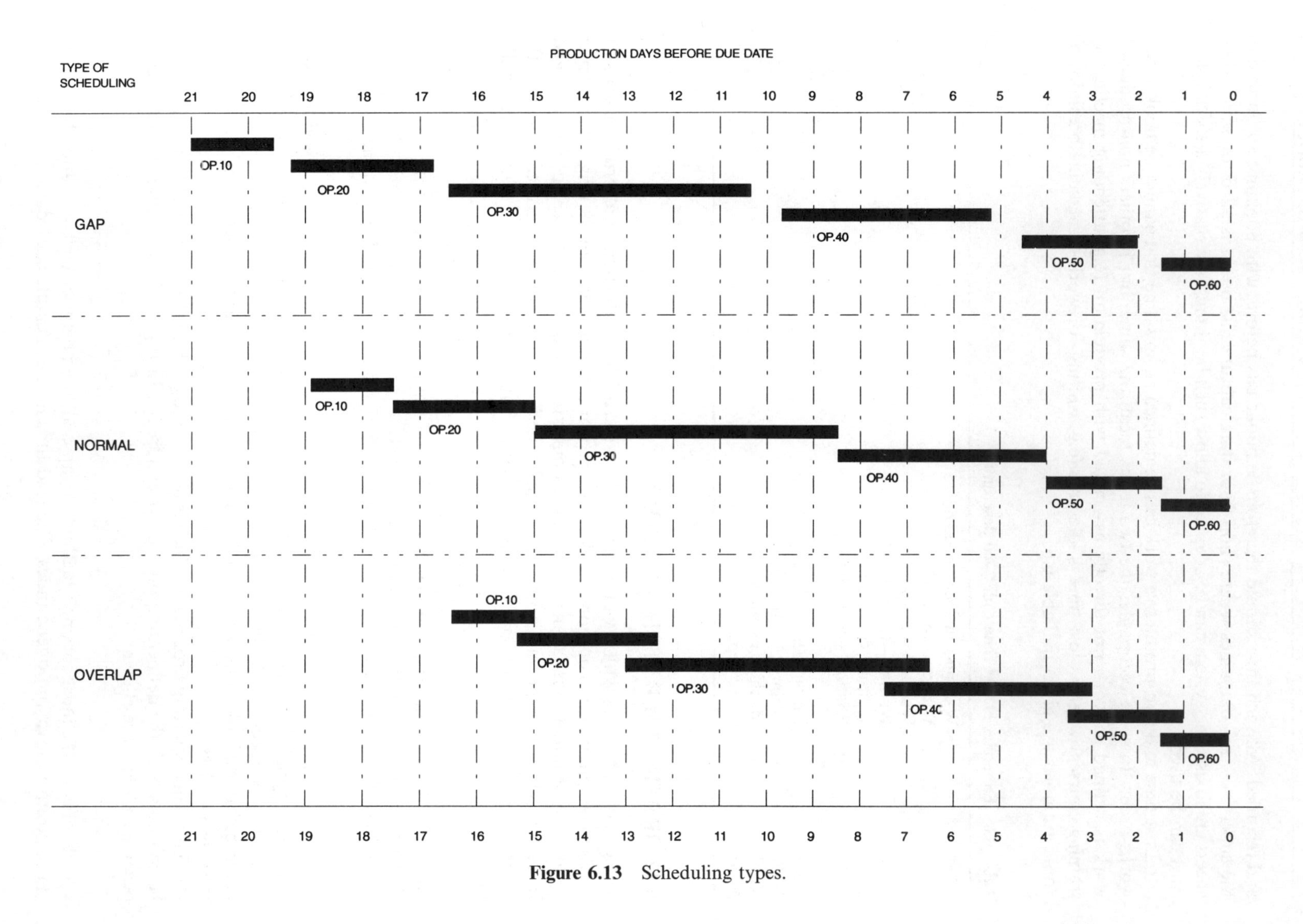

Figure 6.13 Scheduling types.

- *Normal* This is a faster method but requires tighter control. Move and queue times are set to zero. The lead time is reduced to 19 days.
- *Overlap* This is the 'fast-track' method but requires very close control. Move and queue times are specified as negative values. The lead time is reduced to $16\frac{1}{2}$ days.

6.13 TRACEABILITY

This topic is covered more fully in chapter 13, but the principles should be observed within a production management and control system where traceability is required. Full traceability can be defined as the ability to determine, for any batch of product, where it came from and where it goes to. For example, if a component fails unexpectedly on a car, the manufacturer should be able to identify all the components from that production batch and (if manufactured) the batch of raw material used to produce those components. It can then contact all the customers whose cars contain these components so that a replacement part can be fitted.

In terms of production this means that raw materials and bought-in component stocks must be clearly associated with a specific batch from a supplier, for example, for steel wire, a cast number and cast analysis or a certificate of conformance (C of C) for a batch of components. Stocks from different batches should be clearly labelled or held in non-adjacent locations, and issues for a works order should not contain any mixed lots. Lot numbers should always be recorded on a works order record.

6.14 COMMUNICATION WITH OTHER DEPARTMENTS

The production department of any company is the 'engine room', and as such must maintain constant communication with other departments to ensure that production is maintaining the right speed and travelling in the right direction. Some examples of this two-way communication are given in Table 6.7. In addition to these, production should maintain close links with the service departments such as tool room, tool stores, internal transport and maintenance.

Table 6.7 Production communication links

Department	From production	To production
Sales	Any factor that could affect delivery dates and progress	Any change in orders or new orders received
Purchasing	Any requirement to expedite or delay purchase orders	Any problems that could delay deliveries of ordered items
Costing	Any information that could explain cost variances	Any cost variances that could affect budgets or improve efficiency
Stores	Any information on unplanned issues or returns to stores	Any likely delays in making issues to production
Quality	Any production problems that could affect quality levels	Any changes in quality requirements or operating procedures

6.15 OPERATOR TRAINING AND DEVELOPMENT

The importance of well-trained and motivated operators in the production area cannot be overemphasized. Thus, the objective should be to make all operators multi-skilled for maximum flexibility and to develop the concept of team working.

This concept is based on the fact that each operator has strengths and weaknesses but, by working as a team, these are cancelled out so that the strength of the team is greater that the combined strengths of the individuals working alone. This is known as synergy, and can be seen on the football field when a team, working together and for each other, defeats a team of 'all-stars' working for themselves.

The current trend for multi-skilling can be seen in action in the following examples.

- *Self-inspecting* an operator carries out the inspection function on work he or she has produced. This eliminates delays (waiting for an inspector) and allows the operator to see immediately if something is wrong.
- *Setter-operating* the operator sets the machine and then operates it. This allows the operator to be more aware of the need for accurate setting and to make changes if the product goes out of tolerance.

It will be apparent from these examples that the required investment in training and development can repay the company in lower costs, higher throughput and improved quality.

PROBLEMS

1. Draw the product structure for a boxed pen- and pencil-set (use your imagination or, better still, take a set to pieces!).
2. Convert this into an indented bill of materials.
3. A bone china plate is made by forming a piece of clay on a platemaking machine. It is then dried, fired, decorated, glazed (by spraying), fired again, gold-lined, fired a third time, inspected and put into stock. Prepare a production routeing for this process complete with times for setting up and processing, with move and queue times. Note that firing is an operation that does not depend on quantity.
4. Use the times you have estimated in problem 3 to calculate the lead time for a batch of 100 plates.
5. Use these lead-time calculations to draw a Gantt chart similar to the one in Fig. 6.13 showing:
   (a) the gap type of scheduling—your original figures
   (b) the normal type of scheduling (no gaps)
   (c) the overlap type of scheduling assuming a 15 per cent overlap.
   How much time is saved over (a) by using (b)? How much time is saved over (a) by using (c)?
6. Use the data given in the example in Fig. 6.12 to calculate the current lead time, order completion percentage and priority value (CR), given that operations 10 and 20 are fully completed and operation 30 is 60 per cent completed. The current date is week 22/4. (See appendix A for solution).

FURTHER READING

Dilworth, J. B., *Production and Operations Management: Manufacturing and Non-manufacturing*, 4th Edn, McGraw-Hill, New York, 1989.

Mair, G., *Mastering Manufacturing*, Macmillan, London, 1993.

Mather, H., *Competitive Manufacturing*, Prentice-Hall, Englewood Cliffs, NJ, 1988.

Vollman, T. E., Berry, W. L. and Whybark, D. C., *Manufacturing Planning and Control Systems*, 3rd Edn, Irwin, Boston, MA, 1988.

CHAPTER

SEVEN

THE MANAGEMENT OF PRODUCTION SERVICES

THE LINK

The production function needs both direct operators, who work directly on the products, and indirects, who provide the supporting services. These services include the supply and movement of materials, the supply and maintenance of tools, the maintenance of plant and machinery, and the recruitment and training of personnel. In addition, the operations and processes are specified and planned by production engineers and planners, and the work is managed and controlled by supervisors.

The concept of team working is equally relevant in this area, where direct and indirect employees need to work together to optimize production throughput. This means that there must be no 'them and us'—everyone should be marching in step to the same tune!

This can be looked at in another way: the production process involves a sequence of steps, some being carried out by direct operators (fabrication, machining, assembly, etc.) and some by indirects (supplying materials and tools, moving materials between operations, maintaining machines, etc.), and this chain must not contain any weak links. In the past, management has often failed to recognize this fact, and has 'wasted' time in efforts to increase the output of machines and direct operators, often ignoring the fact that overall lead times can include up to 90 per cent of indirect time (waiting for tools or materials, change-overs, move and queue time, breakdowns, etc.). However, modern management recognizes the valuable contribution of the indirect members of the production team, and acts accordingly by tackling the problems in all areas.

7.1 INTRODUCTION

In this chapter the role of the various production services, and in particular the way in which they should work closely with Production, is examined. Each service is then considered in detail in order to highlight the functions that need to be performed and the manner in which these should be carried out. The chapter concludes with the communications with departments other than production which should be established, and the types of two-way information flow that should take place.

7.2 DEFINITION OF PRODUCTION SERVICES

Production services can be defined as a series of functions within the production areas which, if effectively applied, will ensure that production plans are achieved, production delays are minimized, standards of tidiness and cleanliness are maintained and statutory legislation is met. These services include:

1. production supervision
2. materials movement
3. production tooling
4. plant maintenance
5. personnel
6. work study/production engineering.

Production Planning and Quality Control should also be included but, for the purposes of this book, are dealt with separately (the former in chapter 6 and the latter in chapter 13).

7.3 PRODUCTION SUPERVISION

The role of the production supervisor has changed radically over the past few years, and the old title of foreman is now seldom used because it tended to be associated with duties that have become obsolete. These duties, which involved a high level of disciplinary activity and a low level of management (in the modern sense), were based on the concept that production employees were of low intelligence, unreliable and without initiative. Such employees, if they ever existed, had to be driven by fear of dismissal or promises of financial gain to achieve production targets.

This approach never really worked — it was a continuous conflict — and has been replaced by concepts such as team working and self-motivation which have totally changed the roles of both supervisor and operative. For example, the functions of the progess chaser, who was given a list of orders to follow through, have disappeared; instead it is now the supervisors who are responsible for ensuring that work flows through their section or department in the right sequence and at the right time based on work-to lists and priority codes.

The supervisor is now a production manager whose task is to lead a team that could include fabricators, machine setters, machinists, assemblers and materials movers, and this task will include the managerial functions discussed in detail in chapter 1. These functions will vary to some extent from one company to another according to company type, but in general terms are given in Table 7.1 in the form of a job specification. However, this type of specification, although necessary, does not fully define the supervisor's role. Supervisors should be two-way communicators who can ensure not only that their work teams know what has to be done, and how to do it, but also that information flows back to production planning and other relevant sections. They must also be 'enablers' who can, by example, motivate and help the teams of employees in their production areas to achieve agreed targets, and then, by a process of continuous improvement, set new targets for achievement (specified as objectives in the job specification).

This process, sometimes known by the Japanese term of '*kaizen*', requires a degree of cooperation and teamwork that the old style of 'foremanship' could never achieve. It requires of the supervisor a management style that can be developed only in a company where there is a high level of commitment to change, and where senior management accepts that all employees

Table 7.1 Typical job description for a production supervisor

Job title	Machine shop supervisor
Duties	1. To receive instructions from production planning in the form of works orders and work-to lists and to ensure that the operations specified are carried out in the required sequence, at the right time, using the tools, materials, machines, labour and other resources listed on these documents in accordance with the standard operating procedures. 2. To use personal initiative and experience to adapt and modify these instructions when circumstances make it impossible to carry out the work as specified and to inform production planning of any such changes made. 3 To ensure that any company employee or visitor within the machine shop work area complies with all company and statutory regulations (health and safety, security, hazardous substances, fire, etc.) 4. To ensure that all work done is recorded as specified in the standard operating procedures. 5. To use all possible means to ensure that non-productive and lost time is kept to the minimum and that all such time is recorded as specified in the standard operating procedures.
Objectives	To improve efficiency in their production area by: 1. reducing non-productive time by 10% from its current level 2. reducing wastage due to scrap by 15% from its current level 3. increasing throughput (in standard hours) by 5% from the current level.

have a contribution to make. In other words, a company that believes that it can compete with the best in the world and is prepared to develop the natural talents of all its employees and to use those talents to full advantage.

7.4 MATERIALS MOVEMENT

Materials movement does not always exist as a separate work group, since in some companies this function is carried out by the production operatives. However, in modern systems such as CIM (computer-integrated manufacturing), FMS (flexible manufacturing systems) and JIT (just-in-time) it is an essential ingredient. The role of materials movement can best be defined in terms of the objectives that should be achieved:

- that production operations are not delayed because materials, tools or other essentials are unavailable when required
- that delays between production operations are minimized.

It is therefore important to ensure that the materials movement team are given the information they need to achieve these objectives, and this can best be done by developing the production documentation to include the materials movement operations.

For example, operation cards can be printed defining the movements required between operations or at the start/end of a production route: examples of this type of card are shown in Fig. 7.1. Materials movement teams may also be given a form of work-to list which will show, in summary form, what is planned, taking account of priorities.

MATERIALS PROCUREMENT CARD

PRODUCT CODE: C/.6321 — WORKS ORDER: W.24653
DESCRIPTION: xxxxxxxxxxxxxxxxxxxxxxxxxxxxxx — ORDER QTY.: 10
ADDITIONAL DESCRIPTION: xxxxxxxxxxxxxxxxxxxxxxxxxxxxxx — DRG.NO. FR/86437

OPN. NO.	WORK CENTRE DESCRIPTION	MOVEMENT OPERATION DESCRIPTION	WEIGHT	START DATE	FINISH DATE
10	MATERIALS MOVEMENT	PROCURE RAW MATERIALS FROM STORE RM/1 PROCURE TOOLS FROM STORE TL/2 PROCURE DRAWINGS FROM PLANNING OFFICE DELIVER TO MACHINE 2346 (CR.LATHES)	5 Kg 30 Kg	DD/MM/YY	DD/MM/YY

STOREMAN		STOREMAN		STOREMAN	
ISSUE DETAILS		ISSUE DETAILS		ISSUE DETAILS	
COMPUTER		DATE STARTED		DATE COMPLETED	

INTER-OPERATION MOVEMENT CARDS

PRODUCT CODE: C/.6321 — WORKS ORDER: W.24653
DESCRIPTION: xxxxxxxxxxxxxxxxxxxxxxxxxxxxxx — ORDER QTY.: 10
ADDITIONAL DESCRIPTION: xxxxxxxxxxxxxxxxxxxxxxxxxxxxxx — DRG.NO. FR/86437

	OPN. NO.	WORK CENTRE DESCRIPTION	OPERATION DESCRIPTION	TOOLING	START DATE	FINISH DATE
FROM TO	20 30	LIGHT CR. LATHES VERTICAL MILLING	MOVE PART PROCESSED PRODUCT RETURN TOOLS TO STORE TL/3 PROCURE TOOLS FROM STORE TL/2	TK.305 TK.576	DD/MM/YY	DD/MM/YY

OPERATOR		COMPUTER		DATE STARTED	
QUANTITY GOOD		QUANTITY SCRAP		DATE COMPLETED	

COMMENTS:

Figure 7.1 Typical materials movement cards.

7.5 PRODUCTION TOOLING

Production tooling can be defined as any item (apart from the machine or work station) that is used to make a product but is not a part of the product, for example:

- *Cutting tools* these can be standard items (e.g. twist drills) that are readily obtainable from a stockist or specials (e.g. profile cutters) that are designed for a specific cutting operation
- *Work holders* these are designed to hold the work piece while an operation is being performed, e.g. welding jigs and fixtures that locate components during welding, or drilling jigs that hold the work and include guide bushes for positioning the drill bit
- *Press tools* these are specially made (and often complex) tooling sets which allow a component to be cut out and pressed to shape from metal sheet or continuous strip; forging tools can also be included in this category
- *Injection dies* these are also specially made single impression or multi-impression moulds which are used to make plastic components on an injection moulding machine

- *Measuring tools* these are usually standard items such as gauges, calipers and micrometers which are used to check dimensional accuracy. However, in some cases specials are required (e.g. to gauge a non-standard screw thread).

In most cases, standard tools are issued and held at a machine whereas specials are issued only when required and returned after use.

The functions involved in the control of production tooling can be split into three categories:

1. tool design and manufacture
2. tool usage management
3. tool stock management.

Tool design and manufacture

It is the function of the production engineering department, in conjunction with the design office, to determine how an item is to be made and thus to specify the tooling requirements, and if any special tools are required, to design those tools. These can then be made 'in-house' or by a subcontractor from the tooling drawings.

If the company employs its own toolmakers then they will also be responsible for the inspection and refurbishment of special tools when returned by the user. However, standard cutting tools are often cared for by the operator in terms of sharpening, grinding, tip replacement, etc.

Tool usage management

The cost of new tooling is often significant, and tool shortages can seriously disrupt production. It is therefore important to balance these two factors in terms of the policies for tooling availability. It is an essential part of good tool management to maintain accurate records of tool usage so that estimates of tool life can be obtained and new tools ordered in good time. Equally, records of tool breakages should be kept so that items subject to such failures can be backed up by a stock of replacements. However, regular checks should be made with the design and planning departments to ensure that the product or the tooling is not about to be altered or replaced!

Such records can be kept manually or on computer: an example of a typical manual tooling record is shown in Fig. 7.2. From these data it is possible to develop the required policies and to initiate investigations when a tool fails to meet its predicted life expectancy in terms of undue wear or breakage.

Tool stock management

Records of all tooling stocks must be kept either manually or on computer, with full details of production issues/returns and receipts from suppliers. In the case of standard tools that are issued to operators or setters as required, it is not generally possible to maintain usage records but, in order to maintain control, a replacement issue should be allowed only if the original tool is returned. This can then be examined for record purposes.

In the case of special tools, which are only issued against works orders, returns should be progressed to ensure that the tools are still in use and not being held in the production area (a

TOOLING RECORD CARD							Card 1 of 1
TOOL NO.	TP/3427	DESCRIPTION:	Profile Cutter for Horiz. Milling			AQUIRED DATE	15/2/93
SUPPLIER:	Apex Tools	COST:	35.50	USED ON:	Part SK.90756	DRAWING NO.	TP/5427.02

USAGE RECORD

DATE ISSUED	DATE RETURNED	ISSUED FOR WKS. ORDER	USED ON MACHINE	TO MAKE PART NO.	QTY. MADE/ CUMULATIVE	COMMENTS/INITIALS
21/4/93	27/4/93	W.24586	HM.3240	SK.90756	150/ 150	No apparent wear HJW
15/7/93	24/7/93	W.24754	HM.3240	SK.90756	200/ 350	Re-Sharpened TRS
	etc.		etc.		etc.	
	etc.		etc.		etc.	
	etc.		etc.		etc.	
	etc.		etc.		etc.	
	etc.		etc.		etc.	
	etc.		etc.		etc.	
	etc.		etc.		etc.	
18/5/94	23/5/94	W.30945	HM.3241	SK.90756	175/ 2350	Scrapped - Can't be Re-Ground Or Would Be Undersize JPR

Figure 7.2 Typical tooling record card.

common occurrence!). When such tools are returned they should not be booked back into stock in the tooling stores but into a holding store for inspection and refurbishment if necessary. They can then be transferred to the tool stores for issue when required.

Details of all special tools needed to make a product should be included in the bill of materials so that they can be included on the kitting list or the material requisition. This will enable the production planner to check tooling availability (using the trial kitting facility) before a new works order is issued. If the tooling is not available it can be located by computer screen enquiry as being at a production machine, in the holding stores, or on order from a supplier or the toolroom.

7.6 PLANT MAINTENANCE

In any manufacturing company the maintenance department provides an essential service to production, since its objective is to ensure that machines and ancillary plant not only are available when required but are in good working order (the latter includes not only volume and quality requirements but also safety, for example guards and emergency stop controls). If this objective is to be achieved and breakdowns are to be avoided, some form of planned maintenance is essential. Such a system allows machine servicing to be planned to minimize lost production time and to maximize the utilization of maintenance resources.

This is one of the main advantages of planned maintenance: it gives the production planners confidence that a plant item will not fail at a critical time and, conversely, allows plans to be made for an item to be taken out of service—hopefuly, at a time when it is not needed.

The basic principles

A planned maintenance system is based on the fact that any machine or plant item consists of various components, each of which has a definable working life. These can be classified into one of the following four item types.

- *Maintainable* A maintainable item is one that will last for many years if serviced regularly, e.g. an item that requires regular lubrication (such as a gearbox).
- *Replaceable* A replaceable item is one with a limited life which should be replaced at regular intervals if sudden and possibly disastrous failures are to be avoided, e.g. a drive belt. Inspection of such items is unlikely to identify a potential failure.
- *Inspectable* An inspectable item also has a limited life, but potential failure can be identified by some form of measurement, e.g. play in a bearing or wear on electrical contacts.
- *Long-life* A long-life item is one that is likely to last for many years without failure because it is not subject to any appreciable wear, vibration or stress, e.g. a machine baseplate.

This method of classification allows maintenance schedules to be prepared for any type of plant item in terms of a series of instructions listing what should be done to a specific item and at what frequency. A typical schedule of this type is shown in Fig. 7.3. However, it should be noted that in many cases an item of plant could have several schedules covering various tasks and frequencies, and these would then be combined to create planned work orders as shown in Fig. 7.4.

The databases

In order to set up a planned maintenance system it is therefore necessary to collect data for all the plant items being covered, and this can best be achieved by using one of the computer-based systems available commercially. Typically such a system would include the following database files.

PLANNED MAINTENANCE SCHEDULE

Plant Number: A.1752 Description: Rotary Engine Frequency: 6 monthly

Location: Power House #3

CODE	OPERATION	LABOUR	OTHER REQUIREMENTS
10	Disconnect all electrical connections	Electrician	
20	Drain off engine oil by releasing the sump plug located to the right of the Macpherson strut bracket.	Fitter	
30	Drain off the coolant (see operations manual for this engine)	Fitter	Operations Manual
40	Remove and wash air filter housing; fit new air filter	Fitter	Air Filter S/30492
50	Check all belts for wear and if any belt shows signs of such wear that a further 6 months of service is unlikely replace it.	Fitter	Belt Replacement Pack P/09743
60	Check all hoses for condition paying particular attention to the high pressure turbo unit connection. Replace any which display signs of hardening.	Fitter	Hose Replacement Pack H/10945
70	etc.	etc.	etc.

Figure 7.3 Typical planned maintenance schedule.

- *Plant register* This would include full details of all the plant items, e.g. plant number, description, supplier, date of purchase, original cost and current value, location. This file might also include historical data for each item, or this could be in a separate file.
- *Schedules* This would include full details of the schedules for each plant item as shown in the example in Fig. 7.3, including a defined start week for the annual cycle.
- *Stock register* This would carry details of all items likely to be stocked as spare/replacement parts or consumable items (e.g. lubricants or coolants) and, where appropriate, these would be referenced to a plant item and a supplier. Each item would also have a unique internal part number and a supplier's part number.
- *Suppliers* This would contain full details of all suppliers (i.e. code, name, address, contact, etc.) and might include other data such as payment terms. This file should also include details of subcontractors (e.g. painters or instrument engineers).
- *Service contracts* This would include details of all service contracts on plant items (including warranties) so that a user could check that specified work had been carried out and correctly charged.
- *Operatives* This would contain details of all labour employed by the department including special skills, wage rates, standard working hours, etc.

The programs

In addition to these files the system would contain a number of programs that would use these files to run the system and produce reports/enquiries. Some examples are given in Table 7.2.

Additional facilities

Some planned maintenance systems include additional facilities not mentioned previously. These can include the following.

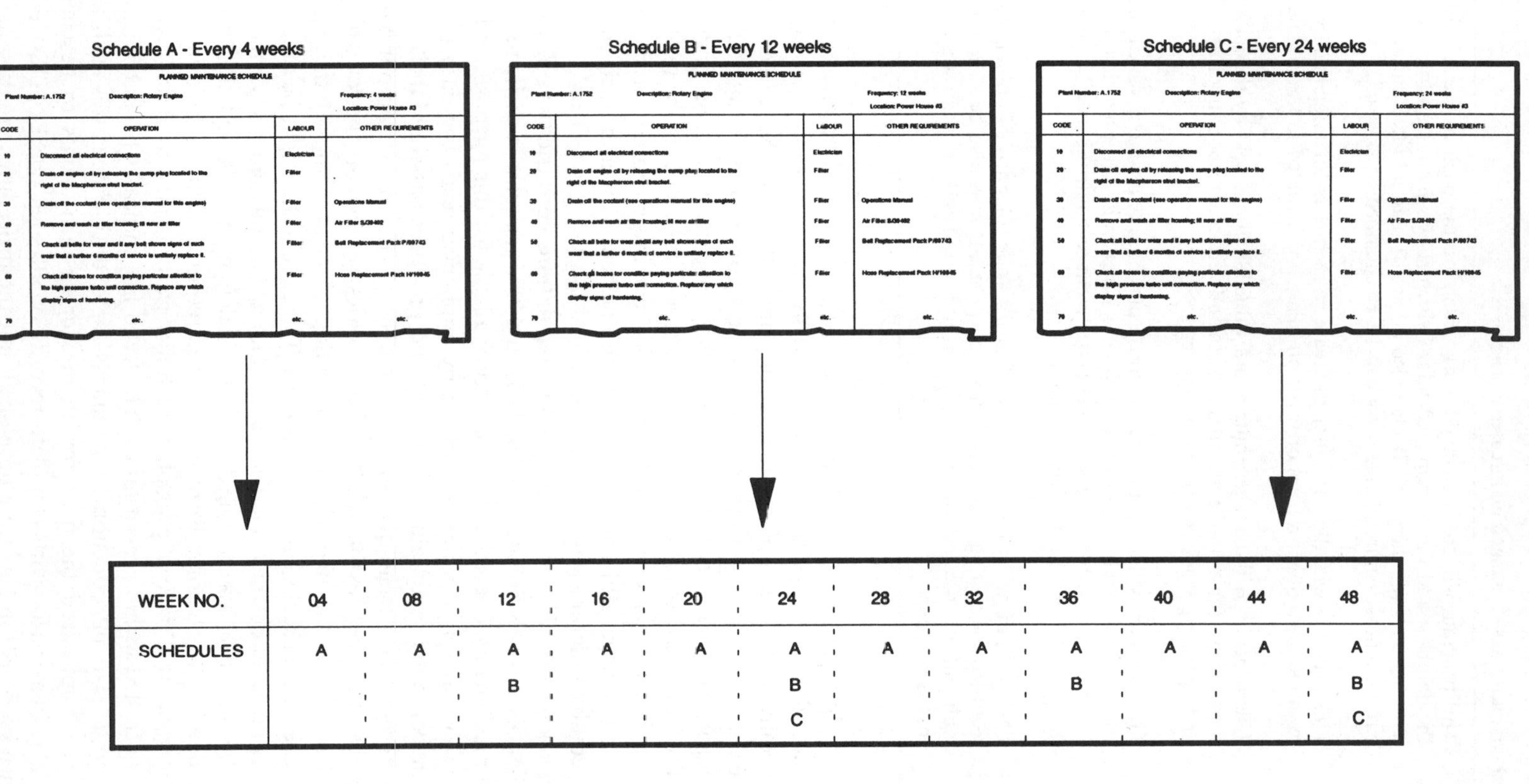

Figure 7.4 Combining planned maintenance schedules.

Table 7.2 Typical computer-based planned maintenance programs

Planned orders	This program will create planned work orders on the basis of the schedules held on the database and the working calendar (which shows plant shutdown, holiday and working days for the year). It may also allow for manual intervention where circumstances require planned work to be rescheduled or skipped.
Unplanned orders	This program will allow the user to create work orders for unplanned work, for example breakdown repairs or work on items not included in the planned system.
Stock file	This program will manage, control and value all stocks held by the maintenance department, for example recording issues and receipts, responding to stock enquiries and operating stock replenishment policies.
Purchase requisitions	This program will allow the user to create purchase requisitions which could then be passed to the buying department. (Note: it is generally accepted that all types of purchase should be handled by one department to control expenditure, goods inwards and purchase ledger posting.)
Work records	This program will enable details of all work done (planned or unplanned) to be recorded for costing and plant history purposes. It may also be able to calculate maintenance efficiency in terms of labour utilization and will show any planned work that is overdue.

- *Real-time input* Up to this point it has been assumed that planned maintenance schedules are based on calendar time, but this is not necessarily so. For example, it may be more effective to schedule on the basis of run-time or on some other unit. Schedules could be set on the distance travelled, on output tonnage or even on a combination value (5000 tonnes or 1000 hours, whichever occurs sooner). In a manufacturing plant running under computer control this information is often available on a continuous basis, and can be read at any time by the planned maintenance system so that work orders can be generated automatically when the required level (tonnage, run-time, etc.) has been reached. This approach should not be confused with condition monitoring, which is described later in this section.
- *Portable appliance testing* Under statutory safety regulations (Electricity at Work 1990), all electrical appliances have to be checked regularly and this includes such items as portable drills and heating elements. Special instruments are available for this purpose, and a software module may be available that schedules testing by appliance, reads the data from the test appliance to the computer, maintains test history and produces pass/fail labels.
- *Condition monitoring* This technique replaces the concept of time- or output-based scheduled maintenance and operates on the basis of a series of regular checks which seek to establish whether or not the plant item is operating efficiently and at the right quality level; in some cases these checks can be continuous. Typical checks might be on temperature, pressure, bearing noise, or chemical analysis. Some planned maintenance systems accept these readings and, by statistical forecasting techniques, calculate trends and thus forecast the probable time/date at which some form of action is required. This technique is illustrated in Fig. 7.5 showing how dosage of phosphate in boiler water is measured daily and the how the levels are controlled within pre-set limits. The trend line indicates probable future levels.
- *Calibration control* The introduction of standards for quality procedures such as BS5750 and ISO9000 has emphasized the need for all measuring instruments to be checked and calibrated on a regular basis, with complete records of all such checks. This requirement has existed since such instruments were developed, but the procedures are now more forma-

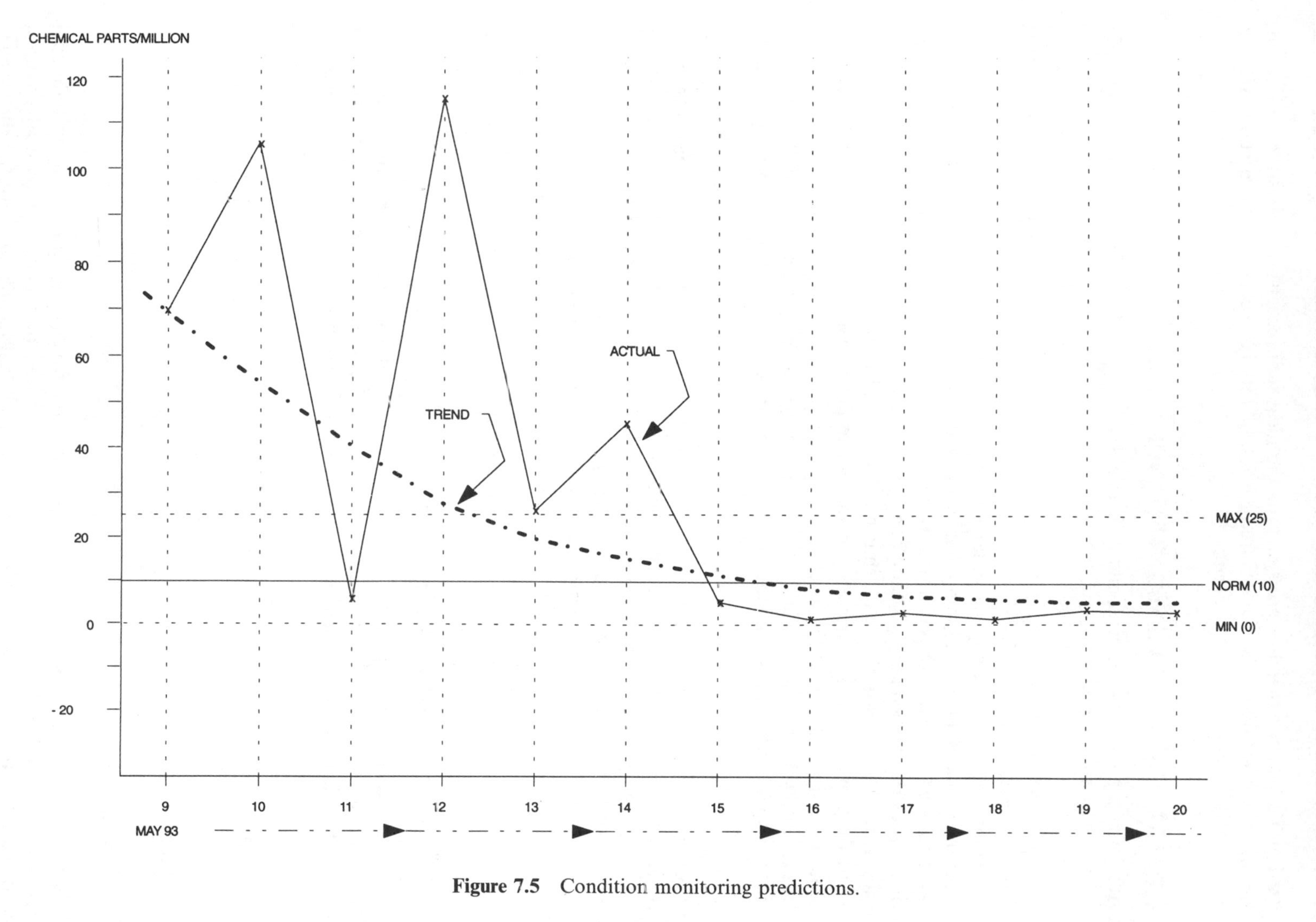

Figure 7.5 Condition monitoring predictions.

lized and a planned maintenance system is the ideal tool for ensuring that the appropriate checks are carried out at the required frequencies, that the correct check/test methods are used and that full records are maintained.

7.7 TOTAL PRODUCTIVE MAINTENANCE (TPM)

A full definition of TPM can be found in Kunio Shirose's *TPM for Workshop Leaders* (1992). TPM is based on five points:

1. it aims at getting the most efficient use of equipment (i.e. overall efficiency) by implementing improvement activities
2. it establishes a total (company-wide) PM system encompassing maintenance prevention, preventive maintenance and improvement-related maintenance, by establishing a system of autonomous maintenance performed by equipment operators
3. it requires the participation of equipment designers, equipment operators and maintenance department workers in the establishment of a planned maintenance system
4. it involves every employee from top management down, with the establishment of training courses designed to raise skill levels
5. it promotes and implements planned maintenance based on autonomous, small-group activities, with a system for MP (maintenance prevention) design and early equipment management.

In this context, MP design is defined as acting to prevent breakdowns and defects in newly designed equipment by applying preventive maintenance techniques during the design process. Early equipment management is defined as the process of minimizing the time to achieve stable operation during the installation, test-running and commissioning of a machine.

TPM is designed to bring the whole company together to enable goals such as zero breakdowns and zero defects to be achieved, leading to higher productivity and enhanced profitability. This is particularly important in the JIT (just-in-time) environment where — because no allowance is made for unplanned events — breakdowns and process failures must be eliminated.

7.8 PERSONNEL FUNCTIONS

The personnel department in the smaller manufacturing company may be responsible for functions that can be considered as production services, although in large companies these may be operated by separate departments. These functions — in respect of production — can be classified as follows:

- recruitment
- training
- company rules
- statutory rules.

Recruitment

The personnel department should endeavour to meet the needs of Production when additional staff are to be recruited. It should therefore assist production management to draw up a specification for the job (unless one already exists), and then use this to advertise the vacancy and as a basis for selection. It will be aware of statutory requirements and can therefore avoid problems such as sex or race discrimination, and will also be responsible for advising potential employees on such administrative matters as working conditions, pay rates, holidays and pension arrangements.

Training

The personnel department should assist production management to identify training needs for existing employees and also to arrange for training of new recruits. It should maintain records of all training courses attended by employees and keep a 'library' of all suitable training facilities (preferably on a computer database for easy search and access). It should also be aware of any statutory training requirements (e.g. fork-lift truck operation), safe working practices and the control of hazardous substances.

Company rules

The personnel department should ensure that all company rules are clearly understood by all employees. Preferably these should be set out in a handbook given to each employee as a part of the induction training course. These rules should cover all disciplinary matters and the procedures for dealing with breaches. They should also specify actions to be taken in case of fires and accidents. These rules should not be confused with operating procedures, which should be dealt with separately.

Statutory rules

There are now numerous regulations that can generally be classed as legislation designed to protect employees and others from damage to health, to prevent industrial accidents and to eliminate environmental damage. It is often the responsibility of the personnel department to ensure that such legislation is fully understood by all employees and that notices are posted as necessary. This department may also be responsible for maintaining records of any breaches of the regulations and reporting these to management and the authorities concerned.

7.9 WORK STUDY/PRODUCTION ENGINEERING

The functions of a company's work study section have changed radically in the past few years, and in many cases have been taken over by other sections. Historically, the term 'work study' meant work measurement, and this was often used as a basis for incentive schemes designed to increase productivity. Some of these schemes were heavily biased in favour of the unscrupulous employer and were used to reduce the labour force, causing industrial disputes and low morale.

It is therefore essential to recognize that work study techniques include a logical approach to methods improvements and also allow the users to determine standard time values from which standard costs can be calculated and standard production times set (for planning purposes). Thus, work study should no longer be considered as a stick with which to beat the employees, but rather as a management tool that can be used to provide valuable information and to improve productivity by developing best working practices.

The work study function is therefore now less concerned with work measurement (time study)-based incentive schemes, and more concerned with aspects of production engineering such as

- methods engineering
- workplace and factory layout
- work simplification
- time standards
- non-productive time analysis.

Methods engineering

Methods of performing an operation, or the sequence of elements used to perform an operation, are studied, recorded and analysed in order to determine the best method, which will then become the standard. The techniques that can be used all operate in basically the same way, using a series of steps. This 'sequence of analysis' is given in Table 7.3 (International Labour Organisation, 1960). Six stages have been identified as essential if methods are to be properly analysed. These are given in Table 7.4.

Once an operation has been selected there are basically two means of recording the sequence of events, namely manual and camera (video), but in either case in order to facilitate analysis it is necessary to use some system of classifying and quantifying each element. The standard procedure is to use a series of symbols as shown in the flow chart in Fig. 7.6, the symbols being defined thus.

- *Operation* An operation occurs when an object is intentionally changed in any of its physical or chemical characteristics, is assembled or dismantled from another object or is arranged or prepared for another operation, for transport, inspection or storage. An operation also takes place when information is given or received or when planning or calculating takes place.
- *Transport* A transport occurs when an object is moved from one place to another, except when such movements are part of an operation or are caused by the operator at the work station during an operation or inspection. This symbol is used whenever material is handled on or off trucks, benches, storage racks, etc.

Table 7.3 Methods engineering: the sequence of analysis

1. *Define* the problem
2. *Obtain* all the facts relevant to the problem.
3. *Examine* the facts critically but impartially.
4. *Consider* the courses open and decide which to follow.
5. *Act* on the decision.
6. *Follow up* the development.

Table 7.4 Methods engineering: the essential stages

1. *Select* the work to be studied.
2. *Record* all the relevant facts about the present method *by direct observation.*
3. *Examine* those facts critically and in ordered sequence, using the techniques best suited to the purpose.
4. *Develop* the most practical, economic and effective method with due regard to all contingent circumstances.
5. *Install* that method as standard practice.
6. *Maintain* that standard practice by regular routine checks.

FLOW PROCESS CHART

		SUMMARY			
PRODUCT	Used Bus Engines	ACTIVITY	PRESENT	PROPOSED	SAVING
ACTIVITY	Stripping. Cleaning and Degreasing prior to Inspection	Operation	4		
		Transport	21		
		Delay	3		
METHOD	Present/~~Proposed~~	Inspection	1		
LOCATION	Degreasing Shop	Storage	1		
OPERATOR(S)	A.B C.D.	Distance (m)	760		
		Time (Hrs)			

Description	Qty	Distance	Time	Symbol ○	⇨	D	□	▽	Remarks
Stored in old-engine shed	1	---	---					x	
Pick up engine					x				Electric Crane
Transport to next crane		25			x				Electric Crane
Unload to floor					x				
Pick up					x				Electric Crane
Transport to stripping bay		33			x				Electric Crane
Unload to floor					x				
Strip engine				x					
Clean & lay out main components				x					
Transport to engine inspection area		75		x					Trolley
Unload to ground				x					
Store (temporary) waiting inspection					x				
TOTAL		260		4	21	3	1	1	

SYMBOLS:

Figure 7.6 Methods engineering (flow process chart).

- *Delay* A delay or temporary storage occurs to an object when conditions (except those that intentionally change the physical or chemical characteristics of the object) do not permit or require the performance of the next planned action. Examples are work stacked on the floor of a work area between operations, cases awaiting unpacking, parts waiting to be put into storage bins or a document waiting to be signed.
- *Inspection* An inspection occurs when an object is examined for identification or is checked for quantity or quality in any of its characteristics.

- *Storage* A storage occurs when an object is kept and protected against unauthorized removal.

The difference between an operation and an inspection is that an operation takes the item a stage further towards completion and adds value, whereas an inspection is merely a verification process which does not add value. The difference between storage and temporary storage (delay) is that a document or other form of authorization is generally required to get an item out of storage. Alternatively, an item in storage is part of the stock whereas an item in temporary storage is usually a part of the work-in-progress.

The object of any method study is to cut out or minimize any wasteful elements (e.g. delays and transports) and to simplify or reduce the essential elements; this can usually be done by critically analysing and examining the record of the study. A new method can then be developed, tested and implemented after operator training.

Among the applications of these techniques are process charts (see Fig. 7.6), two-handed process charts (where the movements of each hand are plotted in an assembly operation) and flow diagrams (where the sequence of elements is plotted on a scale plan of the work area). Other techniques that can be used for method studies where very short cycle elements are involved (e.g. assembling a micro-switch) are known generically as PMT systems (predetermined motion time), for example the MTM (methods time measurement) family. These techniques use a series of standard movements, each of which has a code and a standard time. Thus, each work element is broken down into its basic 'components' which can then be analysed to eliminate redundant or awkward movements and delays, the standard times being used to compare the existing and proposed methods. A list of such movement 'components' is given in Table 7.5.

Table 7.5 Movement component classification/coding (ASME)

Basic elements	Operation	Code
Physical	Reach	R
	Move	M
	Change direction ¶	CD
	Grasp	G
	Hold ¶	H
	Release load	RL
	Pre-position ¶	PP
	Disengage	D
Semi-mental	Position ¶	P
	Search ¶	S
	Select ¶	SE
Mental	Plan ¶	P
	Examine	E
Objective	Do	DO
Delay	Avoidable delay ¶	AD
	Unavoidable delay ¶	UD
	Balancing delay ¶	DB
	Rest to overcome fatigue	F

¶ Ineffective movement element: where these occur an attempt should be made to eliminate them completely if possible.

Workplace and factory layout

A new layout is often developed on the basis of a method study in order to reduce wasteful movement and handling or to simplify the work. However, in addition to these factors, the new layout should take account of the ergonomic factors that are designed to improve productivity by providing better working conditions. An example of this approach can be seen in Fig. 7.7, showing normal and maximum working areas for a seated operator where ergonomic factors dictate that only *normal* movements should be required in order to limit fatigue.

Other ergonomic and methods based factors that should be considered include the following.

1. The hands should be relieved of all work that involves holding the workpiece where this can be done by a jig, fixture or foot-operated clamp.
2. Two or more tools should be combined wherever possible.
3. Where each finger performs some specific movement, as in a keyboard operation, the load should be distributed in accordance with the inherent capacities of the fingers.
4. Handles such as those on cranks and large screwdrivers should be designed so as to permit as much of the surface of the hand as possible to come in contact with the handle. This is especially necessary where considerable force has to be used.
5. Levers, crossbars and handwheels should be so placed that the operator can use them with the least change in body position and the greatest mechanical advantage.

Equally important is the layout of a working area or stores in order to eliminate unnecessary movements, to provide adequate space and to improve work flow between machines or benches. This is particularly important when a group of machines is to be set up for a production line or a manufacturing cell where a sequence of operations is to be carried out on a continuous basis.

Various techniques are available for this type of work, starting with the traditional string diagram, through modelling, to computerized simulation. However, in all cases the objectives are the same; to maximize efficiency by eliminating waste movements and redundant work elements.

Work simplification

This is really a development of methods engineering, the objective being to reduce work complexity such that mistakes are less likely to occur. The following examples of this technique are given to illustrate the principles.

1. If a round peg has to be inserted into a round hole, it is an easier task if the top of the hole is countersunk, especially if the hole is small.
2. If the peg has to be inserted in a particular way, (i.e. it must not be reversed), it is simpler if the peg is designed such that one end is different from the other so that it can only be inserted one way.
3. Where parts have to be assembled in a certain sequence, it is better if these can be laid out in that sequence on the work bench and it is easier still if they can be fed automatically and presented the right way round.
4. Where there are a number of parts in an assembly, every effort should be made to reduce the number to a minimum by redesign.

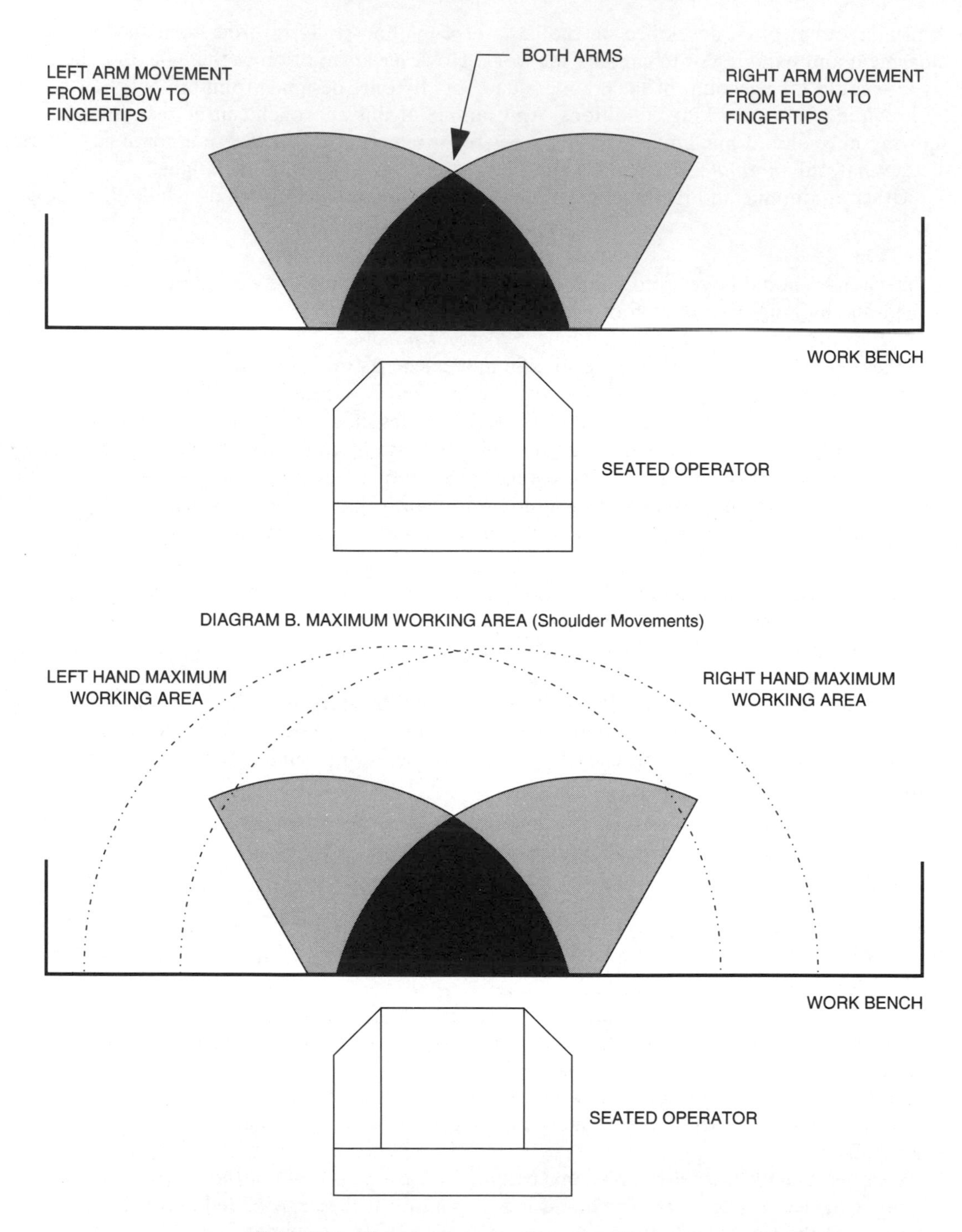

Figure 7.7 Hand movements for a seated operator.

Much of this simplification can be achieved by following the principles of DFM (design for manufacture) and an example quoted by Corbett, Doona and others (1991) gives details of a spindle housing redesign. In this case the assembly was originally designed in stainless steel and contained 10 separate parts (a spindle, a housing, two nylon bushes and six brass screws). This was redesigned with only two parts (the spindle and an injection-moulded nylon housing). This resulted in the assembly efficiency rating's increasing from 7 to 93 per cent.

The assembly efficiency rating is used to measure the effectiveness of a product simplification exercise using the DFA (design for assembly) approach. There are various means of measurement that take account of parts count reductions and assembly costs.

Time standards

It is necessary to establish time standards in manufacturing in order to plan effectively, to measure efficiencies, and to establish standard costs against which actual costs can be compared. A number of methods can be used to obtain these standard times, for example work measurement, synthetics and PMT systems.

- *Work measurement* This is the familiar stopwatch method whereby an operation is broken down into elements and then observed in order to establish a time for each element. This is the observed time and has to be corrected by a rating factor, which is the observer's estimate of the operator's performance while that element is being carried out. These ratings are based on a scale of values known as the 75/100 scale, where 100 is classed as standard performance and 75 as normal. This technique, which is sometimes known as 'levelling', is illustrated in an example in Table 7.6. This shows how the observed times for each element have been taken in decimal minutes over five cycles of an operation, each observed element being rated on the 75/100 scale as it takes place.

  On completion of the study, each observed elemental time is levelled by applying its rating (as a percentage). The average of the levelled times is then found for each element and thence, by adding all the levelled times, the total basic time at standard (100) performance is found for the operation. However, this is not yet the standard time since no allowance has been made for contingencies or fatigue (RA—rest allowance). It is therefore necessary to add percentages to cater for these factors. These percentages must be derived by observation and calculation, taking account of the type of work and the working conditions. In general terms, contingency allowances can be set by random observation or production studies, and rest allowances by using tables of values that take account of the factors that can cause fatigue. A typical example of such a table is shown in Fig. 7.8. It should, however, be noted that rest allowances should be calculated and applied to each element, whereas contingency is normally applied at the operation level. *If any form of work measurement is to be used to establish standard times it is essential to ensure that standard methods have first been developed using methods engineering techniques, and that these methods are being used.*

- *Synthetics* Synthetic values can be used to estimate the standard time for an operation if the method is known. For example, from data taken from a series of time studies it should be possible to establish that the times for various types of welding can be estimated on the basis of the number of millimetres of weld run. In general terms, synthetics are expressed in terms of 'time per unit', where the units can be length, volume, area, etc. but in some cases may be given as a formula or a table that combines various types of synthetic. For instance, a

Table 7.6 Example of levelling observed times

Times/ratings	Elements						
	A	B	C	D	E	F	G
Observed	1.25	0.25	0.72				
Rating	90	90	85	etc.	etc.	etc.	etc.
Levelled	1.13	0.23	0.61				
Observed	1.20	0.30	0.75				
Rating	95	85	90	etc.	etc.	etc.	etc.
Levelled	1.14	0.26	0.68				
Observed	1.16	0.28	0.80				
Rating	100	90	85	etc.	etc.	etc.	etc.
Levelled	1.16	0.25	0.68				
Observed	1.35	0.21	0.78				
Rating	85	105	90	etc.	etc.	etc.	etc.
Levelled	1.15	0.22	0.70				
Observed	1.30	0.22	0.67				
Rating	90	90	100	etc.	etc.	etc.	etc.
Levelled	1.17	0.20	0.67				
Average	1.15	0.23	0.67	1.35	0.15	0.63	0.92

Thus, the total levelled operation time is 5.10 minutes, to which contingency and rest allowance must be added to produce the standard time.

standard time for a turning operation on a lathe could be built up from the data given in Table 7.7.

Synthetic data are particularly useful if there is a very wide range of work/product types, for example in a jobbing machining or fabrication company where standard times are required for estimates and quotations for work never previously performed.

- *PMT systems* Many predetermined motion time systems such as MTM or MOST include standard time values for each element (see Table 7.5 for a list of typical basic elements), for example a 'reach' could be valued at 0.10 time units per 10 mm. However, over the years, combination time values have been developed which group such elements together to give values for more complex movements, for example inserting and driving home a screw or fitting a pulley to a spindle. These systems are commercially available for use for both method studies and work measurement, but special training is needed to use them correctly.

Non-productive time analysis

As part of the job-booking procedures, all lost and non-productive time, either within a job or between jobs, should be recorded and it should be a function of the work study/production engineering department to analyse these records with the objective of reducing such bookings to the minimum. These bookings should include a reason (or reason code) and it should be

REST ALLOWANCE TABLE (Values are percentages of Levelled Times)

	Male	Female
1. CONSTANT ALLOWANCES		
Personal Allowance	5	7
Basic Fatigue Allowance	4	4
2. VARIABLE ALLOWANCES		
(a) Standing Allowance	2	4
(b) Abnormal Position Allowance		
Slightly Awkward	0	1
Awkward (Bending)	2	3
Very Awkward	7	7
(c) Use of Force or Muscular Energy		
Weight Lifted (Kg)		
Up to 2	0	1
2 to 5	1	2
5 to 7	2	3
7 to 9	3	4
9 to 11	4	6
11 to 14	5	8
14 to 16	7	10
16 to 18	9	13
18 to 20	11	16
20 to 23	13	20
over 23	Modify method	
(d) Bad Light		
Slightly Below Recommended	0	0
Well Below	2	2
Quite Inadequate	5	5
(e) Atmospheric Conditions (Heat & Humidity)		
Cooling Power (Kata) Allowance		
16	0	0
14	0	0
12	0	0
10	3	3
8	10	10
6	21	21
5	31	31
4	45	45
3	64	64
2	100	100
(f) Close Attention		
Fairly Fine Work	0	0
Fine or Exacting	2	2
Very Fine/Very Exacting	5	5
(g) Noise Level		
Continuous	0	0
Intermittent, Loud	2	2
Intermittent, Very Loud / High-pitched, Loud	5	5
(h) Mental Strain		
Fairly Complex Process	1	1
Complex/Wide Attention Span	4	4
Very Complex	8	8
(i) Monotony		
Low	0	0
Medium	1	1
High	4	4
(j) Tediousness		
Rather Tedious	0	0
Tedious	2	1
Very Tedious	5	2

Figure 7.8 Typical rest allowance table.

possible, given this information, to determine the real cause and to ensure that corrective action is taken to avoid a repetition. Table 7.8 gives some examples of this approach.

7.10 COMMUNICATIONS WITH OTHER DEPARTMENTS

The people who have been classified in this chapter as Production Services include

- production supervisors
- materials movement operatives
- toolroom operatives
- maintenance engineers
- personnel staff
- work study/production engineers.

Table 7.7 Example of synthetic data

1. *Preparation*	10.0 SMs (standard minutes)
2. *Change chuck*	5.0 SMs
3. *Load work*	2.5 SMs (up to 5 kg)
	5.0 SMs (5 to 15 kg)
	7.5 SMs (over 15 kg)
4. *Face end*	0.5 SMs per 10 mm diameter
5. *Turn down (roughing)*	0.6 SMs per 100 mm length
(finishing)	0.3 SMs per 100 mm length
6. *Reverse work in chuck*	1.5 SMs (up to 5 kg)
	2.5 SMs (5 to 15 kg)
	4.0 SMs (over 15 kg)
7. *Remove work/aside*	2.0 SMs (up to 5 kg)

Table 7.8 Causes and actions: non-productive/lost time

Reason	*Cause*	*Action*
Machine breakdown	Broken drive belt	Alter planned maintenance schedule
Waiting for work	Supervisor not available	Supervisors to issue work before previous job completed
Waiting for setter	Setter on another job	Train operators to set up own jobs
Machine cleaning	Dust and oil on machine	Install dust extractors and operators to report oil leaks
Power failure	Machine overload	Check overload trip settings and limit operator feed/speed settings
Faulty material	Not to specification	Check out supplier or review specification

Obviously, these employees must maintain close communications with production operators, but, other links should also be established and maintained. Some examples are given in Table 7.9.

PROBLEMS

1. As a production supervisor, you are required to submit a written report on an accident that has happened in your department. The accident occurred when a machine operator slipped on an oil patch and broke his arm. Your report should give full details of the event and suggest causes. It should also include your recommendations for action to avoid a similar occurrence.
2. Design a format for a computerized work-to list that can be used by the materials movement team. (Note: it may be helpful to study the list shown in Fig. 6.8.) Your list should give all the necessary information to allow the materials movers to ensure that all material, tooling and other requirements are available at a work centre or machine when needed.

Table 7.9 Production service communications

Service group	Communication links with
Production supervisors	Production planning Plant maintenance Personnel and wages Toolroom and tool stores Materials stores
Plant maintenance	Production planning Purchasing
Toolroom	Design office Production engineering Purchasing
Production engineers	Design office Production supervisors Costing Sales estimating
Materials movement	Purchasing Materials stores Toolroom and tool stores
Personnel	Production supervisors Company secretary Outside training agencies

3. Prepare a draft documentation procedure that will cover the issue, return and inspection of tooling for production. This should include the format for a tooling stock record, specifying the data such a record should contain.
4. Prepare a job specification for a plant maintenance engineer, listing the duties and responsibilities together with limits of action (e.g. expenditure) and performance criteria.
5. As a personnel and training manager you are required to prepare a training course for production operators in advance of the implementation of a shop-floor data collection system which is to replace a daily manual work-booking sheet. Suggest the topics that should be included in such a course (i.e. an outline programme).
6. You, as a methods engineer, have to explain how flow process charting can improve efficiency. As an example that everyone can understand, you choose to demonstrate the technique by comparing three methods of domestic refuse collection. Prepare charts for each method showing the improvements (use a chart similar to that shown in Fig. 7.6). The three methods are:
   (a) using traditional dustbins
   (b) using dustbins with disposable bin-liners
   (c) using a specially adapted vehicle and 'wheelie' bins.

REFERENCES AND FURTHER READING

Corbett, J., Dooner, M., Meleka, J. and Pym, C. (Eds), *Design for Manufacture*, Addison Wesley, Wokingham, England, 1991.

International Labour Organisation, *Introduction to Work Study*, Geneva, 1960.
Lyonnet, P., *Maintenance Planning: Methods and Mathematics*, Chapman and Hall, London, 1991.
Mair, G., *Mastering Manufacturing*, Macmillan, London, 1993.
Mather, H., *Competitive Manufacturing*, Prentice Hall, Englewood Cliffs, NJ, 1988.
Nakajima, S. (Ed), *TPM Development Program*, Productivity Press, Cambridge, MA.
Shirose, K., TPM for Workshop Leaders, Productivity Press, Cambridge, MA, 1992.
Vollmann, T. E., Berry, W. L. and Whybark, D. C., *Manufacturing Planning and Control Systems*, 3rd Edn, Irwin, Boston, MA, 1988.

CHAPTER

EIGHT

THE MANAGEMENT OF PURCHASING

THE LINK

In the previous chapter it was stressed that production cannot operate effectively unless it is backed by a series of efficient service functions, one of which is materials movement, whose main objective is to ensure that work centres and machines are supplied with the right materials, components, etc. when required, to eliminate waiting time. To achieve this objective, the purchasing function has a vital contribution: to ensure that the raw materials and bought-in components/tools are delivered in the stores or production areas when they are needed.

However, Purchasing is also the link between the company and its suppliers and is therefore responsible for building up partnerships based on mutual trust and respect. This aspect of purchasing is often considered as secondary to factors such as price, quality and delivery, but this should not be the case since, if the right sort of relationships have been developed, these factors can be established at satisfactory levels by friendly discussion on the basis of mutual self-interest.

Purchasing is sometimes treated as the 'Cinderella' function by unenlightened management; this is perhaps because it is associated with buying, which is only one of the tasks involved. In fact, in some companies the term used is 'materials management', which is much more comprehensive and descriptively correct.

8.1 INTRODUCTION

In this chapter the purchase order processing functions are considered in detail, starting with the ways in which purchasing requirements can be identified. Next, the methods that should be used to process these requirements to ensure that production and other departments are effectively served are considered, together with the records that should be kept and the information that should be communicated to other departments and the suppliers.

The goods-receiving functions are then examined, including the procedures for document preparation and handling, together with suitable methods for dealing with returns to suppliers. Methods of managing the financial aspects of purchasing are also covered, i.e. is invoice control, invoice validation, purchase expenditure control and purchase ledger posting.

The use of EDI (electronic data interchange) in purchasing is discussed, and finally the various questions related to sourcing are considered, e.g. choosing a supplier, vendor rating and single or multiple suppliers.

Author's note: The other functions that may be carried out by Purchasing are covered in chapter 9, and involve the control of material stocks in relation to production needs.

8.2 THE PURCHASE ORDER PROCESSING FUNCTIONS

The functions that should be performed by Purchasing and other departments can be defined as follows.

Purchase price maintenance

Wherever possible, purchase orders should contain the prices that should be payable (in fact, in some companies no order can go out without prices). These can be obtained in a number of ways, for instance the following.

- *Price lists* These must be kept up to date and are often issued only by suppliers of the make-to-stock type.
- *Internal records* These should if possible be kept on computer for easy access, and could be based on general prices (e.g. £ per tonne) for different product types. These records should be updated on the basis of prices actually invoiced.
- *Quotations* A recording system for quotations should be set up so that these are readily available for use when an order is to be placed, or when an invoice does not agree with the order.

In some large companies the purchasing department includes an estimating section whose function is to calculate what a purchased product or service should cost, for comparison with the supplier's quoted price. This approach is usually applied only where a regular, high-volume supply is needed or where a high-cost and complex item is to be supplied.

Supplier/product data maintenance

The purchasing department should maintain, preferably on computer, full details of all suppliers and purchased goods or services. These two sets of records should be linked such that supply sources can be identified from a product, or products from a supplier. Typical examples of this type of record, using a computer enquiry screen, are shown in Figs 8.1 and 8.2.

If these records can be held on computer, then enquiries can be made that will prove extremely helpful to the users in the department, for example:

- who can supply a particular product or service?
- who is the lowest-cost supplier?
- who is the most reliable supplier in terms of delivery and quality?
- which supplier is likely to give the shortest lead time?

If possible, the records should also be linked to the current purchase orders file, so that an enquiry will show which orders are still open on a supplier for a given product/service, and to the stock file to show current stock levels.

SUPPLIER DATA ENQUIRY SCREEN

SUPPLIER CODE: AB.3682 SUPPLIER NAME: ABC BEARING CO. LTD.

SUPPLIER ADDRESS: xxxxxxxxxxxxxxxxxxxxxxxxxxxxxxxxxxxxxxxxxx
xxxxxxxxxxxxxxxxxxxxxxxxxxxxxxxxxxxxx
xxxxxxxxxxxxxxxxxxxxxxxxxxx
xxxxxxxxxxxxxxxxxxx POST CODE: XXXX XXX

TELEPHONE: (0642) 23569 FAX: (0642) 23796

CONTACT NAME: GEORGE BENSON

CONTACT TITLE: AREA SALES MANAGER

PRODUCT/SERVICE GROUPS: 12 16 18 24

TERMS: 30 DAYS FROM INVOICE DATE

DISCOUNT TYPE CODE: 23

CREDIT LIMIT: 30,000

ORDER NO.	ORDER QTY.	PART NO.	DUE DATE
P/95321	250	234/56	28/02/94

Figure 8.1 Supplier data record.

PRODUCT DATA ENQUIRY SCREEN

PRODUCT CODE: 234/56 PRODUCT DESCRIPTION: BALL BEARING 25mm RADIAL

SUPPLIER CODES:	PRICES	U.O.M.	LEAD TIME DAYS	QUALITY CODE	DISCOUNT CODE	SUPPLIER PART NO.
1. AB.3682	2.50	EACH	14	85	15	PT.18643
2. GE.4376	2.65	EACH	18	90	07	7465/90
3. DY.7425	2.53	EACH	15	90	13	12AN/432
4. WE.2798	2.85	EACH	12	80	05	DT.PR.4567

CONTROL METHOD: MRP

CURRENT STOCK: (FREE) 350 (ALLOCATED) 200

STD ORDER SIZE: 250

ORDER MULTIPLIER: 5

STOCK LOCATION: S.3/425

ORDER NO.	ORDER QTY.	SUPPLIER	DUE DATE
P/95321	250	# 1	28/02/94

Figure 8.2 Product data record.

Order creation

The purchasing department should be responsible for placing *all* orders within the company, regardless of origin or value. This will ensure that full control is maintained, and purchasing procedures are followed. Purchasing should ensure that:

- all orders are placed in response to a requirement, which should be identified on the order if possible, e.g. a purchase requisition or an MRP recommendation
- all orders clearly identify what is required, quoting the supplier's product code if possible, and including price, delivery date requirements, and delivery instructions (e.g. packing, delivery address, documentation)
- if a verbal order is placed, an order number is reserved and given at the time (it should then be confirmed in writing and entered to the computer, to ensure that there is a record when delivery is made or an invoice received; this order confirmation should be marked 'confirmation of verbal order' to ensure that the supplier does not view the verbal and the written as two separate orders)
- to make doubly sure, no verbal order is issued without an order number, which must be recorded, so that the hard-copy confirmation order can carry the same number
- all orders should be correctly specified and quote the correct unit quantity, for example an order for 100 ball-bearings may be interpreted by the supplier as 100 boxes each containing 25 ball-bearings!

The above examples indicate some of the functions that must be carried out by Purchasing when placing orders, to ensure that subsequent problems are avoided.

Order progressing

If a manual ordering system is used it is difficult to ensure that orders which are due for delivery are not overlooked. This is because—in most such systems—order copies are filed in order number sequence, or alphabetically by supplier, rather than in due date sequence.

Ideally, therefore, orders should be held on computer, using either a purchase order processing (POP) system or a standard database, so that enquiries can be made in various ways as shown in Fig. 8.3. This shows how a database can be used to present information in almost any way, either at a simple level (as shown) or at a compound level, for example '*list all orders for supplier X that are due or overdue by DD/MM/YY*'. This flexible approach makes it easy to select orders for progressing, supplier by supplier.

However, it may also be necessary to keep a record of progress 'calls', and many computerized systems do not have this facility. It is therefore useful for the purchasing department to keep hard copies of all orders, and to note on these the results of progressing activity.

Goods receipt

The maintenance of control over goods receipts is almost always the responsibility of Purchasing, and procedures for this activity should be included in the manuals that are a part of any quality system. These procedures should clearly define the records that should be kept, specify who is authorized to accept goods, and under what conditions. The objectives of such controls are to ensure that:

PURCHASE ORDERS FILE — Date: 21/02/94

Order No.	Supplier	Line No.	Part No.	Quantity	UOM	Due Date	Unit Price	Item Value
P/95321	AB.3682	001	234/56	250	Each	28/02/94	2.50	625.00
		002	478/57	100	Each	28/02/94	1.00	100.00
P/95322	HG.9086	001	355/92	20	Each	15/03/94	25.00	500.00
P/95323	FM.0973	001	208/09	750	Metre	12/02/94	0.90	675.00
P/95324	AB.3682	001	234/56	250	Each	14/03/94	2.50	625.00
		002	409/23	150	Each	14/03/94	1.50	225.00
P/95325	KK.7548	001	108/41	10	Each	18/03/94	120.00	1200.00
	etc.		etc.		etc.		etc.	

1. List of all orders due or overdue by 28/02/94

Due Date	Order No.	Supplier	Line No.	Part No.	Quantity	UOM	Unit Price	Item Value
12/02/94	P/95323	FM.0973	001	208/09	750	Metre	0.90	675.00
28/02/94	P/95321	AB.3682	001	234/56	250	Each	2.50	625.00
28/02/94			002	478/57	100	Each	1.00	100.00
	etc.		etc.		etc.		etc.	

2. List of all orders on Supplier AB.3682

Order No.	Line No.	Part No.	Quantity	UOM	Due Date	Unit Price	Item Value
P/95321	001	234/56	250	Each	28/02/94	2.50	625.00
	002	478/57	100	Each	28/02/94	1.00	100.00
P/95324	001	234/56	250	Each	14/03/94	2.50	625.00
	002	409/23	150	Each	14/03/94	1.50	225.00

3. List of all orders for Part 234/56

Order No.	Line No.	Supplier	Quantity	UOM	Due Date	Unit Price	Item Value
P/95321	001	AB.3682	250	Each	28/02/94	2.50	625.00
P/95324	001	AB.3682	250	Each	14/03/94	2.50	625.00

4. List of all orders with Value > £650.00

Order No.	Supplier	Line No.	Part No.	Quantity	UOM	Due Date	Unit Price	Item Value
P/95323	FM.0973	001	208/09	750	Metre	12/02/94	0.90	675.00
P/95325	KK.7548	001	108/41	10	Each	18/03/94	120.00	1200.00

Figure 8.3 Typical enquiry facilities on purchase orders file.

- a record is available to check which goods have been received, and on what date, in order to allow purchase invoices to be validated and cleared for payment
- goods are not signed for by an unauthorized person, who may not be aware of the procedures and therefore may fail to check what has been received, or to record the receipt correctly
- goods are accepted only if they have been ordered, that the order has not been cancelled or amended, that they are accompanied by a suitable document which quotes the correct order number, and that they are in good condition (this last may be difficult to check but the problem can be overcome if the recipient signs for them with a comment, e.g. 'subject to quality check').

The purchasing department is not normally responsible for the quality of goods received, but merely for the receipt itself, although there may be a requirement to check quantities (e.g. 10 cartons, 5 bags) or to record weights.

Invoice validation

The purchasing department is usually responsible for checking all purchase invoices against orders and receipts, but not for invoices for services such as electricity, rates, telephones and water, which should be dealt with by the accounts department.

Invoice validation is a relatively simple process if other procedures have been correctly applied (i.e. orders contain prices and clearly specified quantities, and receipts have been properly recorded). Some computerized purchasing systems include facilities to allow this type of validation, which is designed to ensure that payment is not authorized if prices are incorrect or goods have not been received as ordered. However, if there are any discrepancies or if the goods received are not up to the required standard, then the invoice should be 'held' pending an investigation. This may, at worst, involve return of the goods to the supplier and the creation of a debit note, which is designed to cancel all or part of the invoice, or, at best, cause a delay in payment.

8.3 IDENTIFYING PURCHASING REQUIREMENTS

The requirement to place a purchase order can be identified in a number of ways, depending on how the business is managed, and there is also a need to specify when a purchase order is not required, even if goods or services are to be supplied.

Purchase order required?

In manufacturing industry, most purchase orders are for the supply of the raw materials and components needed to make the company's products, and these are identified on the basis of MRP runs or stock replenishment policies. The actual methods of operation for these two policies are covered in detail in chapter 9, and are summarized in Table 8.1.

In addition, purchase orders should be created to cover 'one-off' goods or services such as capital expenditure (against a quotation and an authorization), and for indirect materials such as maintenance spares, production tooling, stationery and consumables (against a purchase requisition).

Table 8.1 Identifying purchase requirements

Material requirements planning	Regular computer-based MRP runs identify potential shortages of bought-in items against specified demand for products and sub-assemblies, and recommend that purchase orders be created to meet these shortages
Stock replenishment policies	Stock records identify when stocks of a bought-in item have fallen below a predetermined level and an order can then be placed to ensure that a supply of the item is received before a stock-out occurs

Purchase order not required?

To some extent this depends on the company concerned, although many service items cannot readily be covered by a purchase order and are dealt with when an invoice is received, e.g. electricity, gas, water, service agreements and local authority rates. However, other services such as auditing and consultancy should, where possible, be ordered on an annual or quarterly basis.

The most difficult area is probably subcontract work, e.g. metal finishing and transport. In these cases there may be some form of agreement, renewable from time to time, which stipulates the basis on which charges will be made, and these will be applied on invoices submitted monthly. The problem is that unless accurate records are kept (and this is not always the case!), such invoices are very difficut to check. However, most companies accept this, since they believe that the creation of a purchase order for every such item would be even more costly, and could lead to unacceptable delays.

Where small items of expenditure are involved, say up to £25, the petty cash system may be used to avoid the cost of purchase ordering (such orders can cost £20 each or even more!). In such cases (milk, window-cleaning, small stationery and maintenance items, etc.), control is exercised by preparing petty cash vouchers against cash payments so that such expenditure can be accounted for.

8.4 CREATING PURCHASE ORDERS

A purchase order can be created in three ways:

1. manual typing/word processing
2. computer-assisted
3. computer-generated.

In all cases pre-printed stationery should be used with conditions of purchase printed on the back.

Manual/word processing

This method must be used if there is no computer-based purchasing system, and is a somewhat laborious process, as can be seen from the typical purchase order layout in Fig. 8.4. However, if a word-processing package is available, time can be saved by setting up a 'skeleton' format specifically designed to suit the pre-printed stationery, and further time can be saved by pre-numbering the order forms and using made-up order sets, printed on NCR (no carbon required) paper.

SUPPLIER

xxxxxxxxxxxxxxxxxxxxxxxxxxxxxxxx
xxxxxxxxxxxxxxxxxxxxxxxxx
xxxxxxxxxxxxxxxxxx
xxxxxxxxxxxxxxxx
xxxxxxxxx

PURCHASE ORDER

REQ.NO. M/25890
NL CODE: 207.3842.2
SUPPLIER CODE AB.1369

XYZ Engineering Ltd.
Thetford, Norfolk

PURCHASE ORDER NO. P.17532
DATE: xx/xx/xx
DELIVER TO: xxxxxxxxxxxxxxxxxxxxx

PLEASE SUPPLY THE FOLLOWING ITEMS: PRICES ARE IN: Pounds Sterling

ITEM CODE (Drg.No.)	REV.	DESCRIPTION	SUPPLIER PART NO.	QTY.	UNITS	NET PRICE /UNIT	TOTAL NET PRICE	DUE DATE
234/56	2A	RADIAL BEARING 25mm	234/56	100	EACH	1.50	150.00	xx/xx/xx
346/50	O	RADIAL BEARING 40mm	234/65	150	EACH	1.50	225.00	xx/xx/xx
147/92	2B	RADIAL BEARING 50mm	234/80	200	EACH	1.20	240.00	xx/xx/xx
932/70	3C	THRUST BEARING 25mm	342/56	30	EACH	3.00	90.00	xx/xx/xx
502/07	1A	THRUST BEARING 50mm	342/80	250	EACH	1.00	250.00	xx/xx/xx
800/43	2A	DOUBLE RADIAL 35mm	507/60	300	EACH	1.25	375.00	xx/xx/xx

DELIVERY INSTRUCTIONS:
xxxxxxxxxxxxxxxxxxxxxxxxxxxxxxx
xxxxxxxxxxxxxxxxxxxxxxxxxxxx
xxxxxxxxxxxxxxxxxxxxxxxxxxxx
xxxxxxxxxxxxxxxxxxxxxxxxxxxx

SPECIAL INSTRUCTIONS:
xxxxxxxxxxxxxxxxxxxxxxxxxxxxxxx
xxxxxxxxxxxxxxxxxxxxxxxxxxxx
xxxxxxxxxxxxxxxxxxxxxxxxxxxx
xxxxxxxxxxxxxxxxxxxxxxxxxxxx

TOTAL 1530.00

See Reverse For Conditions

Figure 8.4 Typical purchase order layout.

Computer-assisted

This method requires a computerized POP (purchase order processing) module and allows the user to enter—to the screen—only the basic order details. This method is much faster: for example, using the order layout in Fig. 8.4, the number of key strokes is reduced from 580 to 160 (over 70 per cent fewer). These values are taken from the explanatory chart in Table 8.2.

As an example of automatic data entry, once the item code has been typed in, the computer can refer to the parts file for the description, supplier's part number, units, and unit price. The computer will then calculate the total prices and the grand total.

Table 8.2 Purchase order entry: manual v. computer assisted methods

Entry field	Comments		Key strokes	
	Manual	Computer	Manual	Computer
Supplier code	Type	Type	7	7
Address	Type	Auto	70	0
Req. No.	Type	Type	7	7
NL Code	Type	Type	10	10
Date	Type	Auto	8	0
Deliver to	Type	Auto	15	0
Item code (×8)	Type	Type	48	48
Description (×8)	Type	Auto	160	0
Supp. part no. (×8)	Type	Auto	48	0
Quantity (×8)	Type	Type	24	24
Units (×8)	Type	Auto	32	0
Unit price (×8)	Type	Auto	32	0
Total price (×8)	Type	Auto	48	0
Due date (×8)	Type	Type	64	64
Grand total	Type	Auto	7	0
Dely. instructions			See note 1	
Spec. instructions			See note 1	
		Totals	580	160

Note 1: In many cases these instructions will be standard text requiring only a two- or three-digit entry code for each on the computer system against perhaps over 80 each on the manual system.
Note 2: In the comments columns, 'type' means that the entry must be typed in full whereas 'auto' means that the data in this field are generated automatically by the computer using other files for reference.

Computer-generated

There are various levels of sophistication in MRP (materials requirements planning) systems. Those at the higher levels will generate suggested or recommended purchase orders, and hold them in a file.

These are not real orders, and will disappear at the next MRP run unless confirmed by the purchasing department (the methods used to confirm such orders are generally as shown in Fig. 8.5). This screen lists all the recommended orders in summary form and allows the user to view any order in full, to amend it if necessary, to return the amended order to the original screen, and then to confirm it. Alternatively, a recommended order can be confirmed without viewing it in full. On completion, the user can print all confirmed orders, which will then be removed from the screen.

This method is obviously the most efficient, since it does not require any typing apart from possible amendments. However, it should be noted that not all purchases can be controlled by the MRP system (e.g. capital expenditure items cannot), and in such cases one of the other two order-creation methods must be used.

If a computer-based system is being used, the confirmation of a purchase order will automatically update the stock file records so that subsequent stock enquiries will show not only what is in stock but also what is on order.

8.5 PURCHASE ORDER AMENDMENTS

In general terms it is not good practice to amend a purchase order once it has been sent to the supplier, unless this is very carefully controlled. This is because it is difficult to ensure that the

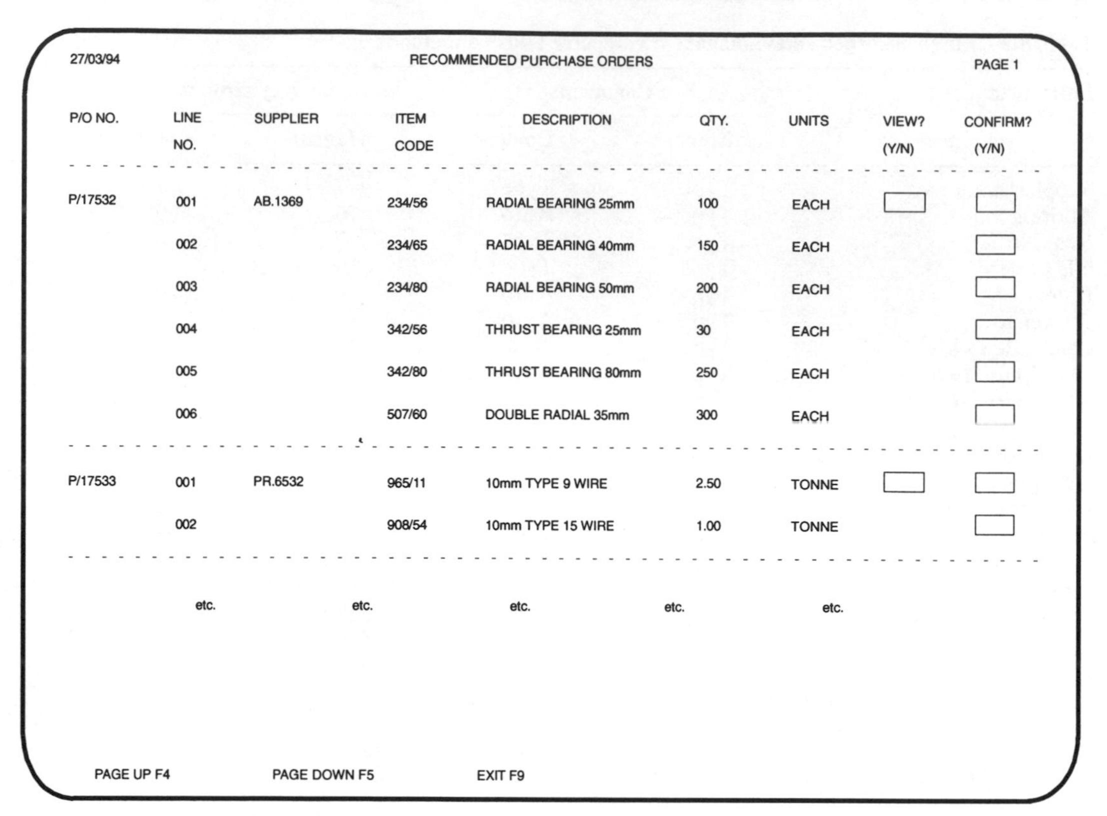

27/03/94 RECOMMENDED PURCHASE ORDERS PAGE 1

P/O NO.	LINE NO.	SUPPLIER	ITEM CODE	DESCRIPTION	QTY.	UNITS	VIEW? (Y/N)	CONFIRM? (Y/N)
P/17532	001	AB.1369	234/56	RADIAL BEARING 25mm	100	EACH	☐	☐
	002		234/65	RADIAL BEARING 40mm	150	EACH		☐
	003		234/80	RADIAL BEARING 50mm	200	EACH		☐
	004		342/56	THRUST BEARING 25mm	30	EACH		☐
	005		342/80	THRUST BEARING 80mm	250	EACH		☐
	006		507/60	DOUBLE RADIAL 35mm	300	EACH		☐
P/17533	001	PR.6532	965/11	10mm TYPE 9 WIRE	2.50	TONNE	☐	☐
	002		908/54	10mm TYPE 15 WIRE	1.00	TONNE		☐
	etc.	etc.	etc.	etc.	etc.			

PAGE UP F4 PAGE DOWN F5 EXIT F9

Figure 8.5 MRP recommended orders—view and confirm.

supplier has taken note of the amendment and corrected its records accordingly. The best compromise is therefore to limit the amendments to certain types, and to cancel orders for any other types accompanied by the issue of a new order. Many MRP systems will, in addition to recommending new orders, suggest amendments to existing orders, but these suggestions are usually limited to deliveries, as follows:

- *expedite* bring forward a delivery date
- *delay* put back a delivery date
- *cancel* eliminate an order.

Such amendments are generally acceptable. However, it is suggested that most others are not. This can be seen from the procedures in Table 8.3.

8.6 PURCHASE ORDER PROGRESSING

Earlier in this chapter (section 8.2) the question of identifying orders for progressing was considered. If a computer system is installed, there may be a facility to progress such orders and record the results. For instance, a screen can be called up which lists all the orders open on a particular supplier, that are due for delivery up to a given date, as shown in Fig. 8.6. This will then allow the user to telephone the supplier and record responses by entering 'flag' codes against each order line, the system recording the date of the call.

Table 8.3 Suggested purchase order amendment procedures

Amendment	Reason	Action
Delivery dates	To expedite or delay order	Send order amendment note
Products	To correct error/redesign	Cancel order and re-issue
Increase quantities	Demand change	Create new order for balance
Decrease quantities	Demand change	Send order amendment note
Units of measurement	To correct error	Cancel order and re-issue
Prices	To correct error	Cancel order and re-issue
Delivery instructions	To correct error	Send order amendment note
Special instructions	To correct error	Send order amendment note

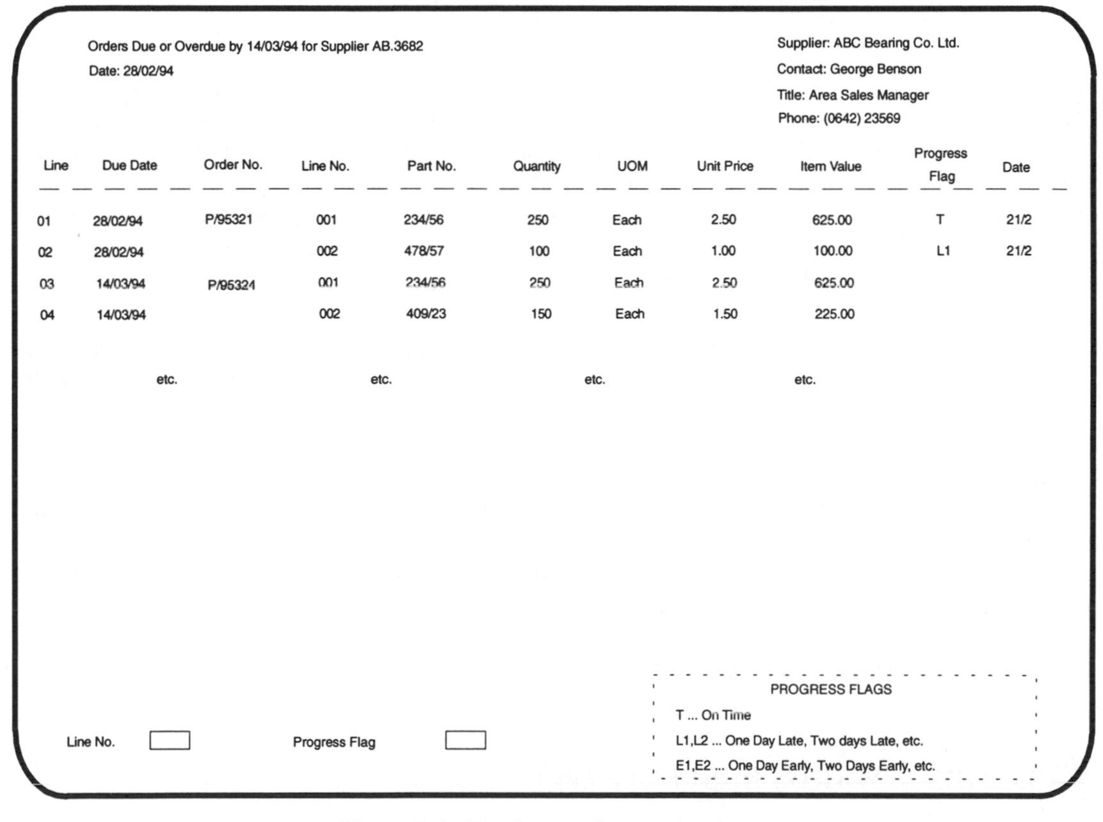

Orders Due or Overdue by 14/03/94 for Supplier AB.3682
Date: 28/02/94

Supplier: ABC Bearing Co. Ltd.
Contact: George Benson
Title: Area Sales Manager
Phone: (0642) 23569

Line	Due Date	Order No.	Line No.	Part No.	Quantity	UOM	Unit Price	Item Value	Progress Flag	Date
01	28/02/94	P/95321	001	234/56	250	Each	2.50	625.00	T	21/2
02	28/02/94		002	478/57	100	Each	1.00	100.00	L1	21/2
03	14/03/94	P/95324	001	234/56	250	Each	2.50	625.00		
04	14/03/94		002	409/23	150	Each	1.50	225.00		

etc. etc. etc. etc.

Line No. Progress Flag

PROGRESS FLAGS
T ... On Time
L1,L2 ... One Day Late, Two days Late, etc.
E1,E2 ... One Day Early, Two Days Early, etc.

Figure 8.6 Purchase order progressing.

It could, with some justification, be argued that it should not be necessary to progress orders—that unreliable suppliers should be replaced by others—and in some companies this is true. However, the facility to check on the progress of a purchase order is useful if properly used, since it allows the planning department to be warned if a delivery is likely to be late and thus perhaps to avoid costly delays by rescheduling.

8.7 GOODS INWARDS (RECEIVING)

Strict procedures should be enforced in the goods-receiving area to ensure that goods are accepted only if they have been ordered, and a clear sequence of steps should be followed as shown in Fig. 8.7. Obviously it is not always possible to reject a delivery if the goods are urgently required, and the normal practice is to telephone the supplier to query any discrepancy. However, the basic principle is correct.

Any goods received should be accompanied by the supplier's advice note, which should include:

- supplier's details (name, address, etc.)
- purchase order number
- advice note number
- details of the goods delivered
- details of packages (cartons, bags, etc.)
- total weight.

Goods inwards department can then check on the orders file (manual or computer) that the order exists, is still open, and is for the goods detailed. The quantity delivered should also be checked to ensure that it is not greater than the outstanding balance. At this stage there is unlikely to be any form of quality check.

Goods inwards can then sign the advice note copy which is to be retained by the driver as 'proof of delivery', and may add comments such as 'accepted subject to inspection'. The advice note copy to be retained by the purchaser can then be used

1. as a goods-received note if a manual system is in operation, for reference when the invoice has to be validated—it should therefore be passed to Purchasing, but the main details should be recorded so that the goods can be identified (e.g. on a label) and onto the purchase order copy
2. as the source document for data entry if a computer system is installed—This entry will record delivery details and reduce the quantity on the purchase order. The facility to print GRNs (goods-received notes) is available on most computer systems, and one GRN copy can be used to identify the goods, another being sent to Purchasing to advise it that a delivery has been made.

An example of a typical GRN is shown in Fig. 8.8.

8.8 GOODS INWARDS (INSPECTION)

The necessity for goods to be inspected on receipt from a supplier is questionable, and many companies accept goods uninspected as a part of their JIT (just-in-time) philosophy; this is because they believe that vendor rating and auditing (assessment of a supplier's ability to maintain agreed standards) is a less costly and more effective approach. This method of operation is dealt with later in this chapter (section 8.13).

There are three methods of dealing with goods after they have been booked in.

1. *Traditional* Goods are placed in a holding area and are not included in the available stock until they have been inspected. However, they are 'visible' since the holding area is classed as a non-available stock location. After they have passed inspection, a stock transfer transaction is used to move them into available stock. However, failures are held for return to the supplier.

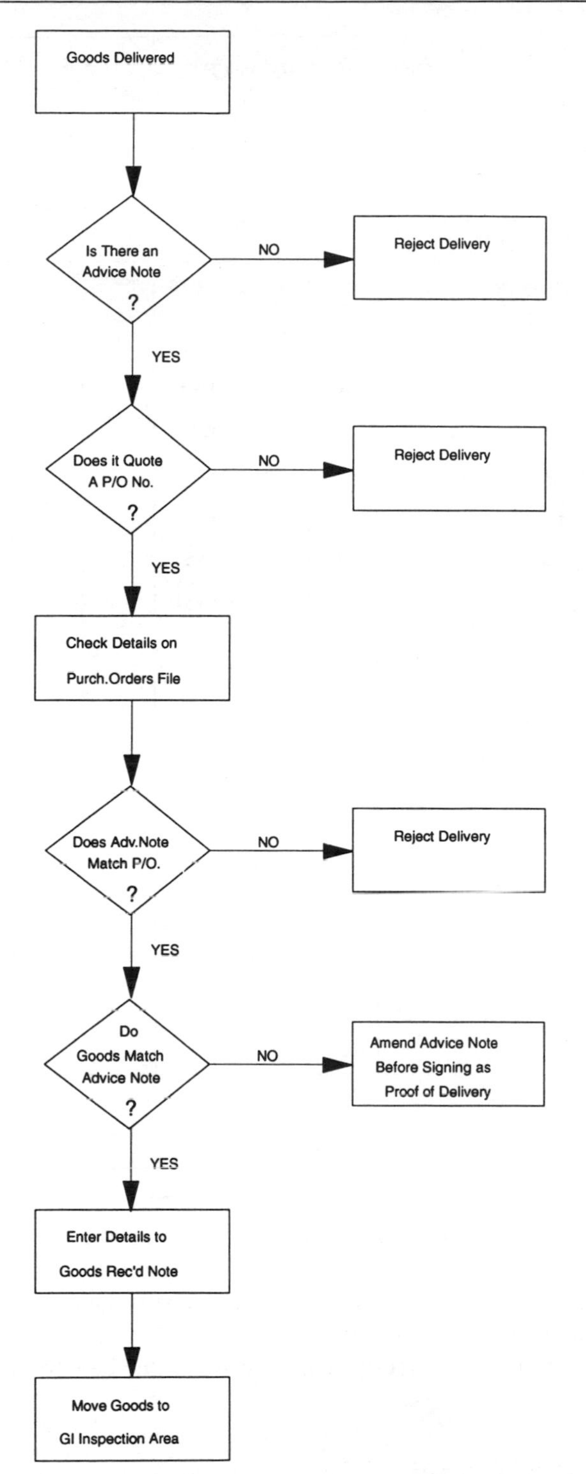

Figure 8.7 Flow diagram for goods inwards acceptance.

SUPPLIER

GOODS RECEIVED NOTE

G.R.N. NO. GR.96467

DATE: xx/xx/xx

xxxxxxxxxxxxxxxxxxxxxxxxxxxxxxx
xxxxxxxxxxxxxxxxxxxxxxxxx
xxxxxxxxxxxxxxxxx
xxxxxxxxxxxxxxx
xxxxxxxx

INSPECTOR NO. ....

SUPPLIER CODE AB.1369

ADVICE NOTE NO. xxxxxxxxxxx

THE FOLLOWING ITEMS HAVE BEEN RECEIVED:

ITEM CODE (Drg.No.)	REV.	DESCRIPTION	QTY. REC'D	QTY. STILL DUE	UNITS	INSPECTION COMMENTS			LOCATIONS	
						PASS QTY.	FAIL QTY.	REASON CODE	PASS	FAIL
234/56	2A	RADIAL BEARING 25mm	100	0	EACH					
234/65	O	RADIAL BEARING 40mm	50	100	EACH					
234/80	2B	RADIAL BEARING 50mm	0	200	EACH					
342/56	3C	THRUST BEARING 25mm	30	0	EACH					
342/80	1A	THRUST BEARING 80mm	250	0	EACH					
507/60	2A	DOUBLE RADIAL 35mm	300	0	EACH					

ADDITIONAL INSPECTION COMMENTS:

Figure 8.8 Typical goods-received note (GRN).

2. *Partial JIT* Goods are moved directly from the receiving bay to the stores and booked in immediately as available stock.
3. *Full JIT:* Goods are moved directly from the receiving bay to the production area for immediate use.

8.9 RETURNS TO SUPPLIERS

Goods can be rejected on receipt, after goods-inwards inspection, from stock, or from production, and in all cases procedures should be specified for this possibility. Normally, rejection is due to the goods being below the quality standard, but other reasons are 'wrong goods supplied', 'quantity greater than ordered', 'goods no longer required' or 'incorrect documentation'. However, in general terms the procedures are the same, thus:

- a return should not be made until the supplier has agreed to accept it
- all returns should be accompanied by a fully detailed delivery note
- a debit note (which is a request for credit) should be mailed to the supplier quoting, if known, the invoice number to be credited and the supplier's advice note number.

The debit note will normally be raised by the accounts department on receipt of a copy of the delivery note sent with the returned goods.

8.10 PURCHASING AND ACCOUNTING

The main function of Purchasing can be defined as *the provision of a service to the company which will ensure that all purchased items are supplied at the right time, in the right place, to the right standard and at the right price.* This requires close control of all aspects of the purchasing operation, including expenditure, and can be achieved only by accurately pricing all orders, and then checking that invoices reflect what has actually been delivered, at the correct prices and extensions.

Thus, all purchase invoices should be passed to Purchasing for validation, and this process should involve a check that

- the goods were actually ordered
- the goods have been received and accepted
- the prices charged agree with those on the order
- the price extensions (unit price × quantity) are correct
- the invoice goods total is correct
- sales tax has not been charged on non-taxable items
- the sales tax calculations are correct
- the goods have not been delivered too early (a potential cash-flow problem).

The invoice can then be confirmed as correct and passed to Accounts for entry to the accounting system. The purchasing department should also be in a position to value the outstanding purchase commitment at any time, although this is extremely difficult unless the orders are held on a computer. However, if a system is available, a report can be generated that compares expenditure—actual and projected—with budget, as shown in Fig. 8.9, where projected expenditure is based on a formula such as:

$$\text{projected expenditure for month } m = \text{total value of open orders due in month } (m - 1)$$

This assumes that orders delivered in a month have to be paid for, on average, a month later. This example also shows how data taken from a purchase orders system can be copied into a spreadsheet and then used to prepare a report.

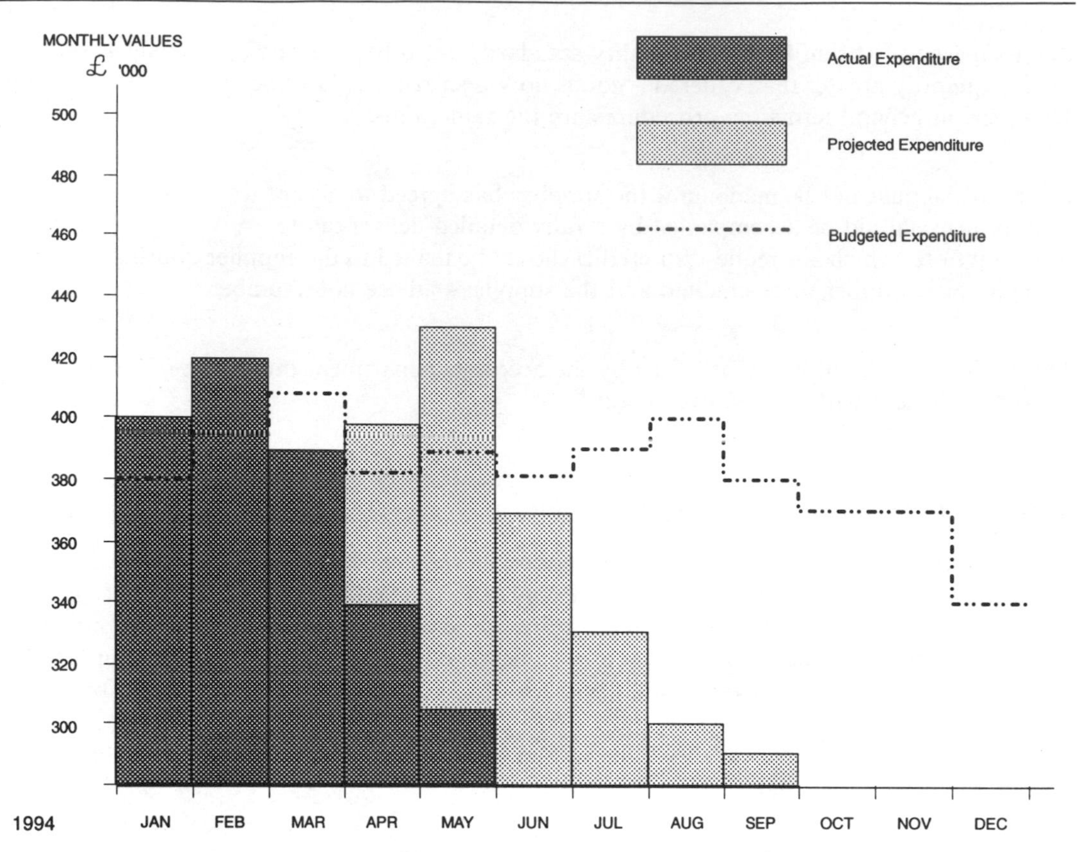

DATA TABLE £ '000

MONTH	JAN	FEB	MAR	APR	MAY	JUN	JUL	AUG	SEP	OCT	NOV	DEC
A	400	420	390	340	300							
B				60	130	370	330	300	290			
C	380	390	410	385	390	385	390	400	380	375	375	340
D	400	820	1210	1630	2060	2430	2760	3060	3350			
E	380	770	1180	1565	1955	2340	2730	3130	3510	3885	4260	4600
VARIANCE	−20	−50	−30	−65	−105	−90	−30	+70	+160			

NOTES:
A. Actual Expenditure per Month
B. Projected Expenditure per Month
C. Budget Expenditure per Month
D: Cumulative Expenditure for Year ([A] + [B])
E. Cumulative Budget for Year
VARIANCE Row[E] −Row[D]

Figure 8.9 Purchase expenditure (budget v. actual).

8.11 SOURCING AND SUPPLIER SELECTION

A great deal has been written about the question of sourcing, i.e. is it better to obtain supplies of a particular type from a single source or to use several sources? The advantages and disadvantages of both these policies are set out in Table 8.4.

Companies moving over to single sourcing believe that the advantages easily outweigh the disadvantages, because the latter can be eliminated by the partnership concept. This replaces the traditional approach, where a supplier was mistrusted and threatened—which never really worked! The modern approach is to work together for the benefit of both.

Traditionalists may not yet be convinced, but there are many examples of the financial benefits of single sourcing if all aspects are taken into account. For example, although prices may not be the lowest in the market, the savings in terms of better service, elimination of goods inwards inspection, improved quality and joint development far outweigh the possibly higher prices. It is also possible for a 'no price rise' agreement to be arranged with the supplier, based on the continuous improvement principle.

Most 'world-class' companies accept and use the single sourcing approach, but have not reached this point without a great deal of hard work in the development of their relationships with the suppliers concerned. After all, if you are going to put all your eggs in one basket, it is advisable to make sure that the basket is 100 per cent sound! If single sourcing is a company objective (and it should be in most cases), then Purchasing will have a major role to play in developing the right environment, which will turn suppliers into associates and trusted partners.

Table 8.4 Single v. multiple sourcing

Single sourcing	Multiple sourcing
Advantages	*Advantages*
Can develop partnership to build loyalty	Can ensure supply if one supplier fails
Can obtain best volume discounts	Can negotiate best terms through competition
Can work on joint developments	Can use several small suppliers
Can more easily audit for quality/service	
Can develop workable purchase schedules	
Disadvantages	*Disadvantages*
Can be problems if supplier fails	Cannot get best volume discounts
May not get best prices	Can have problems with delivery schedules
May not get best deliveries	Difficult to build loyalty
May not get best quality	Costly in quality audit terms
Choice of goods restricted	Problems with joint developments

8.12 PURCHASING AND EDI

EDI (electronic data interchange) can be used to link a computer terminal in the purchasing department to computer systems at supplier companies. This allows the purchaser to enter orders directly to its supplier's system or to amend orders within agreed limits. This facility saves time and effort, and is particularly useful if the company being supplied is working on a JIT system, with supplies being delivered against schedules, since the schedule quantities and delivery times/dates can be updated at very short notice.

EDI can thus be used as a two-stage process: first, an order or amended schedule is received from a customer, which can then be processed through the MRP (materials requirements planning) system to determine the effects on the suppliers. Then, as a second stage, these amended purchasing schedules can be sent via EDI to the suppliers. This saves a great deal of time, which is essential in a JIT environment.

However, this type of operation requires a high level of partnership and excellent 'understanding' between customers and suppliers, to ensure that the customer's requirements are reasonable and acceptable to the supplier. This is usually achieved only after a series of joint seminars or training sessions.

8.13 VENDOR RATING AND QUALITY AUDITS

To some extent these areas are the responsibility of the quality assurance department, and are covered in chapter 13. However, Purchasing has a role to play in terms of monitoring performance and supplier selection.

Vendor rating

Six factors can be used to measure the 'performance' of a supplier:

1. *price stability* this can be measured in terms of the frequency with which prices are increased
2. *price level* this can be measured in terms of a factor (e.g. the cost of living index)
3. *delivery promise* this can be measured in terms of the variation in promises against some agreed standard
4. *delivery performance* this can be measured in terms of the variation against promise
5. *quality* this can be measured in terms of the ratio of rejects against total supplies
6. *accuracy* this can be measured in terms of the number of inaccuracies found on advice notes and invoices, and also in terms of total quantities delivered against quantities ordered.

Records for these factors can be maintained for major suppliers on a computerized spreadsheet, and this information can then be used to select the best supplier for a group of items, as a basis for the selection or retention of a single source supplier or to revise the list of multiple sources. An example of such a spreadsheet is shown in Fig. 8.10.

Quality audits

The operation of an effective vendor rating system can be time-consuming and costly, and a number of companies have moved to quality auditing as an alternative (although in theory both methods could be used).

Quality auditing approaches the problem in a different way, by looking at the internal procedures and control systems employed by the supplier, to ensure that the vendor rating factors are met. Thus, an investigation is carried out at the supplier's premises by an auditor, and points are awarded on the basis of this, to give an assessment. This is then used to advise the supplier of his status, to suggest possible areas of improvement and, where appropriate, to award a certificate as an approved supplier.

For example, the auditor could question the methods and procedures used to deal with the control of goods inwards to the supplier, or sales order management, and the answers to questions on these topics could then be evaluated on an agreed scale. The purchasing department

COLUMNS

ROWS	A	B	C	D	E	F	G	H	I	J	K	L
1	**VENDOR RATING REPORT**			SUPPLIER CODE: AB1369			SUPPLIER: ABC BEARING CO. LTD.					
2	PRICE STABILITY		PRIOE LEVEL		DELIVERY PROMISE		DELIVERY PERFORMANCE		QUALITY		ACCURACY	
3	DATE	VALUE	DATE	VALUE	DATE	VALUE	DATE	VALUE	DATE	VALUE	DATE	VALUE
4												
5	DD/MM/YY	2	DD/MM/YY	3	DD/MM/YY	1	DD/MM/YY	4	DD/MM/YY	2	DD/MM/YY	3
6	DD/MM/YY	2	DD/MM/YY	3	DD/MM/YY	2	DD/MM/YY	4	DD/MM/YY	2	DD/MM/YY	3
7	DD/MM/YY	3	DD/MM/YY	3	DD/MM/YY	2	DD/MM/YY	4	DD/MM/YY	3	DD/MM/YY	4
8												
9	etc.		etc.		etc.		etc.		etc.		etc.	
65	AVGE. (*)	3.2		3.5		2.2		4.0		2.9		3.7
66	TOTAL	19.5	(Max = 30)									
67	RATING	65 %										

VALUES:

Excellent..........5
Good.................4
Average............3
Poor..................2
Unacceptable...1

(*) 12 month moving average

Figure 8.10 Vendor rating using a spreadsheet.

should therefore liaise with quality assurance, to ensure that any failure by the supplier to maintain the required standards is taken into account at the next audit.

PROBLEMS

1. Prepare a job specification for the manager of the purchasing department, listing duties and responsibilities together with limits of action (e.g. expenditure) and performance criteria.
2. As purchasing manager you have been asked to prepare a report on the possibility of your company's changing to a single sourcing policy. Draft such a report for the board of directors setting out your views with estimates of financial benefits and costs.
3. As purchasing manager, prepare a draft annual budget for your department, setting out not only the costs of purchases in relation to various levels of activity but also the administrative and other costs of running the department.
4. As a part of the development of a computerized purchasing system, the computer department has agreed to allocate a programmer to your department in order to allow you to obtain information in the form of screen enquiries and printed reports. Specify, in outline, the enquiry/report facilities that you feel would be of most value to you and other members of the purchasing team.
5. An auditing system is to be set up by your company in order to determine which of its major suppliers should be registered as 'approved'. Suggest which aspects of their procedures and control systems should be examined, together with criteria against which performance could be measured.

FURTHER READING

Anderson, E. J., *The Management of Manufacturing*, Addison-Wesley, Wokingham, 1994.

Mair, G., *Mastering Manufacturing*, Macmillan, London, 1993.

Smith, S. B., *Computer-based Production and Inventory Control*, Prentice-Hall, Englewood Cliffs, NJ, 1989.

Vollmann, T. E., Berry, W. L. and Whybark, D. C., *Manufacturing Planning and Control Systems*, 3rd Edn, Irwin, Boston, MA, 1988.

CHAPTER

NINE

THE MANAGEMENT OF MATERIALS

THE LINK

In the previous chapter, the importance of the purchasing functions was shown in terms of supplying production with the right goods, at the right price and time. However, for this to be possible, purchasing must be provided with information that specifies what should be ordered and when it is required.

The provision of this information is the task of materials management and is based on customer requirements. Thus, in a make-to-stock company, when stocks fall below pre-set levels a demand to replenish would generally lead to a need to purchase raw materials for production. Alternatively, in a make-to-order company, customers' orders will create a production demand followed by a need to purchase raw materials.

The main objective of materials management is to balance the need for minimum stocks and maximum service; in other words, the target is zero stock with no stock-outs — a very difficult task! In fact, as with many other management functions, the answer is optimization (the best possible) rather than perfection (the impossible), and this is achieved by very tight control on high-cost items scaling down to minimal (low-cost) control on low-cost items.

This tight control can be achieved only if suppliers can be relied on to keep delivery promises — another good reason for single sourcing and partnership.

9.1 INTRODUCTION

In this chapter the need to control the materials used in production is considered in terms of the conflicting demands of low stocks (to minimize costs) and high availability (to allow production flexibility). This can best be achieved by matching stocks with demand, and therefore methods of defining and measuring demand are considered; this is followed by the alternative methods available for maintaining stocks at the correct levels. The various types of stock that may need to be held are then defined in terms of the type of business, and the policies and procedures needed to manage such stocks are examined.

Reference was made in chapter 6 (section 6.7) to methods of issuing materials to production (e.g. kitting): this process is now considered in much greater detail, with various alternatives suggested. Finally, the overall function of materials management is examined and defined, with particular reference to the financial aspects of the operation.

9.2 MATERIALS MANAGEMENT—THE BASIC PRINCIPLES

The materials management function is usually considered to include not only the control of materials within the company, but also the purchasing of such materials. However, for convenience, these two aspects are separated into two chapters in this book, the purchasing functions having already been covered in chapter 8. Therefore, the basic principles of sound materials management can be defined in overall terms as follows:

- to maintain stocks at the lowest possible level consistent with demand
- to purchase all bought-in goods and materials in the most economic manner
- to work with other departments to eliminate all possible sources of material waste
- to maintain accurate stock records
- to ensure that production is not hampered by material shortages.

As can be seen from these principles, the management of materials is a delicate balance. On one side is the need to ensure that stocks are adequate to meet the requirements of production and the customers; and on the other is the need to minimize the costs of holding stocks.

This task would be difficult enough if a static situation prevailed (constant demand and reliable supplies), but of course this is never the case. In fact, the only constant factor in materials management is that nothing is ever static or reliable! For example, customer demands are often highly variable, suppliers cannot be relied on to meet delivery promises, and production plans frequently have to be changed.

In the materials management context, all types of production materials must be considered as stocks, including bought-in components and raw materials, process materials/consumables, inter-process stocks, work-in-progress, and finished goods. It is therefore essential, if the principles of sound materials management are to be applied, that materials management work in close cooperation with Production Planning, Production Management, Design and Quality Assurance.

9.3 DEFINING AND MEASURING DEMAND

The main reason for holding stocks is to satisfy demand (although an alternative reason might be to stockpile against predicted price increases by suppliers), and it is therefore essential to understand what demand really is, and how it can be measured. Therefore, demand should not be considered merely as a series of orders held in the sales office, or a manufacturing programme of products that the company expects to sell. Instead, in order to have a clear understanding of demand, it should classified, broken down into its component parts, and measured.

Demand can be classified into two types, independent and dependent, and these are defined in Table 9.1. This may appear to be a confusing concept, but the example in Table 9.2 may help to clarify the position. This example shows that there is an *independent* weekly demand for 1200 cars, all of which require a common windscreen, plus a further *independent* demand for 50 windscreens as replacements. There is therefore a total *dependent* demand for windscreens of 1250 per week. Thus, any item that is not a top-level item can be subject to both independent and dependent demand, and in this example the only top-level items are the complete cars. It can therefore be seen that the relevance of the demand type to an item is not really the quantity required (this is dealt with in another way) but the means by which that item is controlled and managed.

Table 9.1 Independent and dependent demand

Independent demand	A demand from customers for items over which the supplier has no direct control, for example complete machines and spare parts.
Dependent demand	A demand for items that is related to the independent demand and can be managed and controlled by the supplier.

Table 9.2 Example of demand types

A car-making company supplies its customers with both cars and spare parts and on this basis is able to classify its weekly demand as follows.

Independent demand

Item	Quantity
Cars—model A	500
Cars—model B	300
Cars—model C	200
Cars—model D	200
Spare gearboxes for A and B	60
Spare gearboxes for C and D	30
Spare windscreens	50
Spare engines for A	25
Spare engines for B and C	25
Spare engines for D	10
etc.	

Dependent demand

Item	Quantity
Wheels for A and B (5 per car)	4000
Wheels for C and D (5 per car)	2000
Gearboxes for A and B	860
Gearboxes for C and D	430
Windscreens—all models	1250
Engines for A	525
Engines for B and C	500
Engines for D	210
etc.	

Independent demand is identified and measured by some form of sales forecasting (as described in chapter 3) and, on the basis of this information, dependent demand is determined using one of the techniques described in the following sections. Such techniques can be used to ensure that—as far as possible—stocks of materials match the pattern of demand. They are:

1. MRP (material requirements planning)
2. stock replenishment policies
3. service level setting
4. inventory management

and in some cases a combination of these techniques may be used.

9.4 MAINTAINING STOCKS IN LINE WITH DEMAND—MRP

MRP, which should not be confused with MRP II (manufacturing resource planning) is a technique that has become available in its present form only since the development of powerful, low-cost computers. This is because an MRP run requires a very large number of transactions to be carried out in a reasonably short time—a task that is beyond the capabilities of the human brain.

The way in which an MRP system operates is shown in diagrammatic form in Fig. 9.1. Essentially, MRP looks at the independent demand that has been set up in a file called MPS

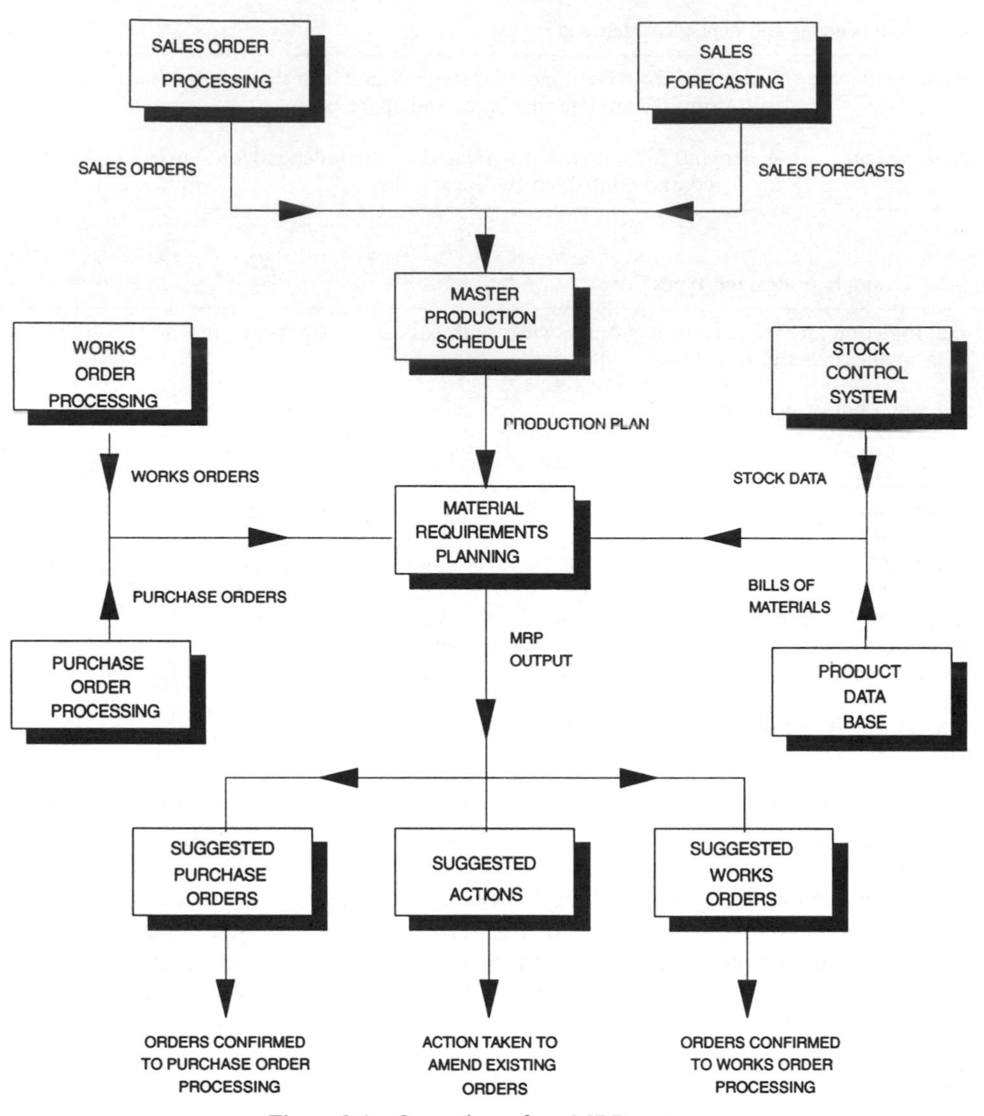

Figure 9.1 Operation of an MRP system.

(master production schedule) which is often a combination of orders and forecasts. MRP then breaks this demand down into a series of 'gross requirements', on the basis of the product structures in the bills of materials. These gross requirements are all the materials, components and sub-assemblies, that would have to be purchased or manufactured to meet the MPS demand if there were no stocks or outstanding purchase/works orders. However, this is seldom the case, and therefore MRP compares the gross requirements with existing free stocks and open orders, and calculates the potential shortages. These shortages are known as the 'net requirements'.

In order to overcome these shortages MRP will then make various recommendations, which fall into three basic types:

- create purchase orders for items where there are no existing orders or where existing orders are not sufficiently large
- create works orders for items where there are no existing orders or where existing orders are not sufficiently large
- suggest that an attempt be made to reduce the quantities on specific open purchase and works orders, or to cancel them, if the quantities on order are greater than the requirements.

The recommendation to cancel an order, or reduce the order quantities, may not be acceptable if the order is already part-processed, in which case it can be ignored. However, if the orders concerned are part of an order schedule then this recommendation can be used as a basis for amending the schedule.

But MRP is not just a material requirements calculator. It is also a scheduling tool which, having looked at the requirements and recommended the actions, suggests when these actions should take place. In other words, MRP will suggest the dates when works or purchase orders should be placed such that completion or delivery will be at the time when the items are required.

This can be done if the bill of materials for the items concerned includes purchase or production lead times for use during an MRP run. This process is illustrated in Fig. 9.2, which shows how each item in the BOM contains a lead time and a code to indicate whether the item is bought-in (B) or manufactured (M). it also shows how, by scheduling backwards from the due date for the top-level assembly, the start dates and due dates for all the orders for the lower level items can be determined.

Assembly A is an independent demand item and this demand for 15 units has been taken from the master production schedule (MPS) on the basis of the requirements for the next period only (overdue plus period 4), as shown in Fig. 9.3.

In the example in Fig. 9.2, certain assumptions have been made to avoid undue complexity, although in real life they would probably not apply. For instance:

1. order quantities have been assumed to be equal to net requirements although they might be modified by factors such as a scrap allowance, a minimum order/batch size or an order multiplier
2. net requirements have been assumed to cover one period only, although they might be grouped together for several periods ahead in order to make quantities required more economic
3. all items in the BOM have been assumed to be controlled by MRP although, in fact, some items might be controlled by some other form of stock replenishment policy
4. it has been assumed that there are no outstanding open order balances, although in fact this may not be true, and MRP would first try to meet the requirements from these orders.

Order quantities

MRP systems allow the users to set up a series of 'MRP policies' which modify the order quantities, for example:

- *scrap allowance* where predictable levels of scrap apply to an item and these cannot be eliminated (e.g. bar ends), an allowance, usually in the form of a percentage factor, can be set up in the MRP system (e.g. a factor of 1.04 allows for 4 per cent scrap)

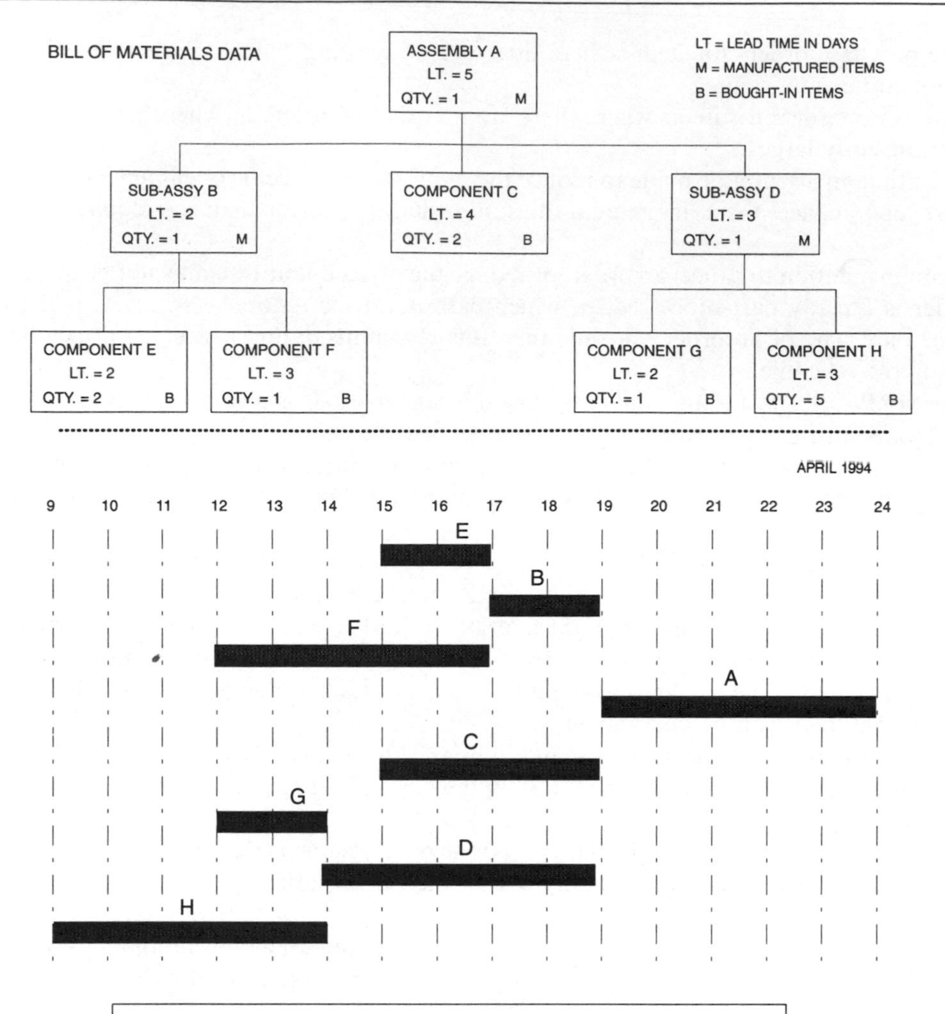

There is a requirement for 15 Assemblies 'A' due on 24th April 1994 and the stock situation is as shown in Table I. There are no open orders for any of the items B to H.

TABLE I (Requirements)

Item	Gross Reqmt.	Free Stock	Nett Reqmt.
A	15	0	15
B	15	0	15
C	30	5	25
D	15	0	15
E	30	8	22
F	15	0	15
G	15	2	13
H	75	40	35

TABLE II (Recommended Orders)

Item	Order Type	Nett Reqmt.	Start Date	Due Date
A	M	15	19/4/94	24/4/94
B	M	15	17/4/94	19/4/94
D	M	15	14/4/94	19/4/94
C	B	25	15/4/94	19/4/94
E	B	22	15/4/94	17/4/94
F	B	15	12/4/94	17/4/94
G	B	13	12/4/94	14/4/94
H	B	35	9/4/94	14/4/94

Figure 9.2 Scheduling orders in an MRP system.

PRODUCTION PERIODS - 1994

PRODUCT		O'DUE	4	5	6	7	8	9	10	11	12	1	2
ASSEMBLY A	ORDERS	3	10	8	4	2	1	0	0	0	0	0	0
	F'CASTS	0	2	5	8	11	14	14	12	11	12	14	13
	TOTALS	3	12	13	12	13	15	14	12	11	12	14	13
ASSEMBLY B	ORDERS	6	18	21	12	9	4	1	0	0	0	0	0
	F'CASTS	0	3	0	6	10	16	21	22	21	19	18	20
	TOTALS	6	21	21	18	19	20	22	22	21	19	18	20
ASSEMBLY C	ORDERS												
	F'CASTS	etc.	etc.	etc.	etc.	etc.	etc.	etc.	etc.	etc.	etc.	etc.	etc.
	TOTALS												

Figure 9.3 Typical format of a master production schedule (MPS).

- *order multiplier* where an item is sold in multiple units, or where several are made from a fixed-size material (e.g. boxes of 50 or 12 units cut from a standard size steel sheet), an order multiplier can be set up in the MRP system
- *minimum quantity* where, for economic or other reasons, a batch minimum has been set, this can be set up in the MRP system.

The operation of this process is shown in Table 9.3, based on the net requirement figures in Fig. 9.2, the method of operation being based on what is known as an algorithm, which sets the rules to be used by the system to determine the correct order quantity.

In general terms, this operates by calculating the order quantity against each of the three policies based on the net requirement for an item, and selects the highest. It then tests this value against the other two policies and adjusts it upwards if necessary. For example, in Table 9.3, all three policies for item G are greater than 1 (the normal value) and, if these are all applied, the order quantities are found to be 15.5 (scrap factor), 16 (order multiplier) and 17 (minimum quantity), the largest being 17. However, this does not meet the order multiplier policy and therefore the order quantity must be raised to 20 to satisfy all three policies.

Table 9.3 Order quantity modification for policies

Item	Net requirements	Scrap allowance	Order multiplier	Minimum quantity	Order quantity
A	15	1.00	1	12	15
B	15	1.02	1	12	16
C	30	1.00	8	24	32
D	15	1.05	3	12	18
E	30	1.00	1	24	30
F	15	1.00	2	12	16
G	15	1.05	4	17	20
H	75	1.00	8	60	64

Net requirements

In the example in Fig. 9.2 it was assumed that net requirements are to cover one period only, i.e. period 4 of 1994 (see Fig. 9.3—the master production schedule), the demand, including overdue orders, being 15 for assembly A. This is known as 'lot-for lot' and ignores the economics of batch size. However, it is sometimes considered more practical to allow MRP to look further ahead, in order to justify placing orders for batches of a more economic size. To achieve this, many MRP systems include the facility to specify the number of time periods (known as time buckets) that should be visible for processing. The way in which this facility would operate is shown in Table 9.4 based on the master production schedule previously used, but to avoid complexity it has been assumed that these sub-assemblies and components are not used on any other assembly. In fact it is highly probable that this assumption is not valid, and the effect of this on requirements is illustrated in Fig. 9.4 (Table II), which shows on which products a sub-assembly is used; this is known as a 'used on' list and, on this basis, the overall requirements can be determined.

Non-MRP items

Up to this point it has been assumed that, if an MPS/MRP system has been installed, all independent demand items are on the master schedule (MPS) and both independent and depen-

Table 9.4 Order quantity modification for time-buckets

Item	Time periods	Net requirements	Scrap allowance	Order multiplier	Minimum quantity	Order quantity
A	2	28	1.00	1	12	28
B	2	28	1.02	1	12	29
C	2	56	1.00	8	24	56
D	2	28	1.05	3	12	30
E	2	56	1.00	1	24	56
F	2	28	1.00	2	12	28
G	2	28	1.05	4	17	32
H	2	140	1.00	8	60	144

dent items are managed by MRP; however, this is not necessarily so. In certain types of manufacturing company it is more convenient to control some low-cost, high-usage items by using stock replenishment policies. For example, where a 'bulk issue' system is in operation, items in this category, which are always of the dependent demand type, are excluded from the MRP system.

Bulk issue items can be defined as components or materials that are issued in bulk, on request, rather than in defined quantities related to a works order. Such items are held at a work bench or alongside a production line and can be used on any job, as required. Typical of such items could be small screws, bolts, nuts and washers or consumable items such as welding rods, tubes of adhesive or rolls of protective tape.

Because they are a part of the product, bulk-issue items should appear on the bill of materials, but because they are flagged as 'bulk-issue', they will be ignored in an MRP run and excluded from kitting lists.

Outstanding open order balances

Having determined the requirements, MRP looks for the most economical means of meeting each requirement. Thus, if possible, existing free (unallocated) stock will be allocated, but if this is insufficient then MRP will look at open order balances. In some cases, these orders will be 'pegged', i.e. based on a particular demand in the form of a sales order. In this case, MRP may be set up to compare the actual demand from that sales order with the actual quantity on the works or purchase order, to see if there is a surplus balance for use against another requirement.

However, if lot traceability is in operation, or for some other reason, this approach may not be acceptable, and therefore MRP will look only at works or purchase orders that are not pegged, either because pegging is not being used or because the order is for stock (or a cancelled sales order).

This is a complex area, and can best be specified in flow-chart form as shown in Fig. 9.5.

9.5 TYPES OF MRP

Most computer-based manufacturing systems include an MRP module, but these can vary considerably in the degree of complexity and the facilities available, as follows.

MASTER PRODUCTION SCHEDULE

PRODUCT		O'DUE	4	5	6	7	8	9	10	11	12	1	2
		PRODUCTION PERIODS - 1994											
ASSEMBLY A	ORDERS	3	10	8	4	2	1	0	0	0	0	0	0
	F'CASTS	0	2	5	8	11	14	14	12	11	12	14	13
	TOTALS	3	12	13	12	13	15	14	12	11	12	14	13
ASSEMBLY B	ORDERS	6	18	21	12	9	4	1	0	0	0	0	0
	F'CASTS	0	3	0	6	10	16	21	22	21	19	18	20
	TOTALS	6	21	21	18	19	20	22	22	21	19	18	20
ASSEMBLY C	ORDERS												
	F'CASTS	etc.	etc.	etc.	etc.	etc.	etc.	etc.	etc.	etc.	etc.	etc.	etc.
	TOTALS												

NOTE: For the purposes of this example all MRP policies are assumed to be 'normal', that is, will not affect order quantities.

TABLE I (Top Level Net Requirements)

Item	Time Periods	Net Reqmts. (From MPS)
ASSEMBLY A	1	15
ASSEMBLY B	2	48
ASSEMBLY C	2	30
ASSEMBLY D	1	18
ASSEMBLY E	3	64
ASSEMBLY F	2	32
ASSEMBLY G	1	16
etc.	etc.	etc.

TABLE II (Components/Sub-Assys used on Assemblies)

Item	Order Type	Qty Used on Assembly						
		A	B	C	D	E	F	G
SUB-ASSY B	M	1	0	O	1	0	O	1
SUB-ASSY D	M	1	1	0	0	0	0	0
COMPONENT C	B	2	0	2	0	0	2	0
COMPONENT E	B	2	0	0	2	2	0	2
COMPONENT F	B	1	0	0	1	0	0	1
COMPONENT G	B	1	1	0	0	0	0	2
COMPONENT H	B	5	5	0	0	0	8	0
etc.	etc.			etc.			etc.	

TABLE III (Components/Sub-Assys Required for Assemblies)

Item	Qty Required by Assembly							Total Reqd.
	A	B	C	D	E	F	G	
SUB-ASSY B	15	0	O	18	0	O	16	49
SUB-ASSY D	15	48	0	0	0	0	0	63
COMPONENT C	30	0	60	0	0	64	0	154
COMPONENT E	30	0	0	36	64	0	32	162
COMPONENT F	15	0	0	18	0	0	16	49
COMPONENT G	15	48	0	0	0	0	32	32
COMPONENT H	75	240	0	0	0	8	0	315
etc.		etc.			etc.			etc.

Figure 9.4 Modification of net requirements due to variable time periods.

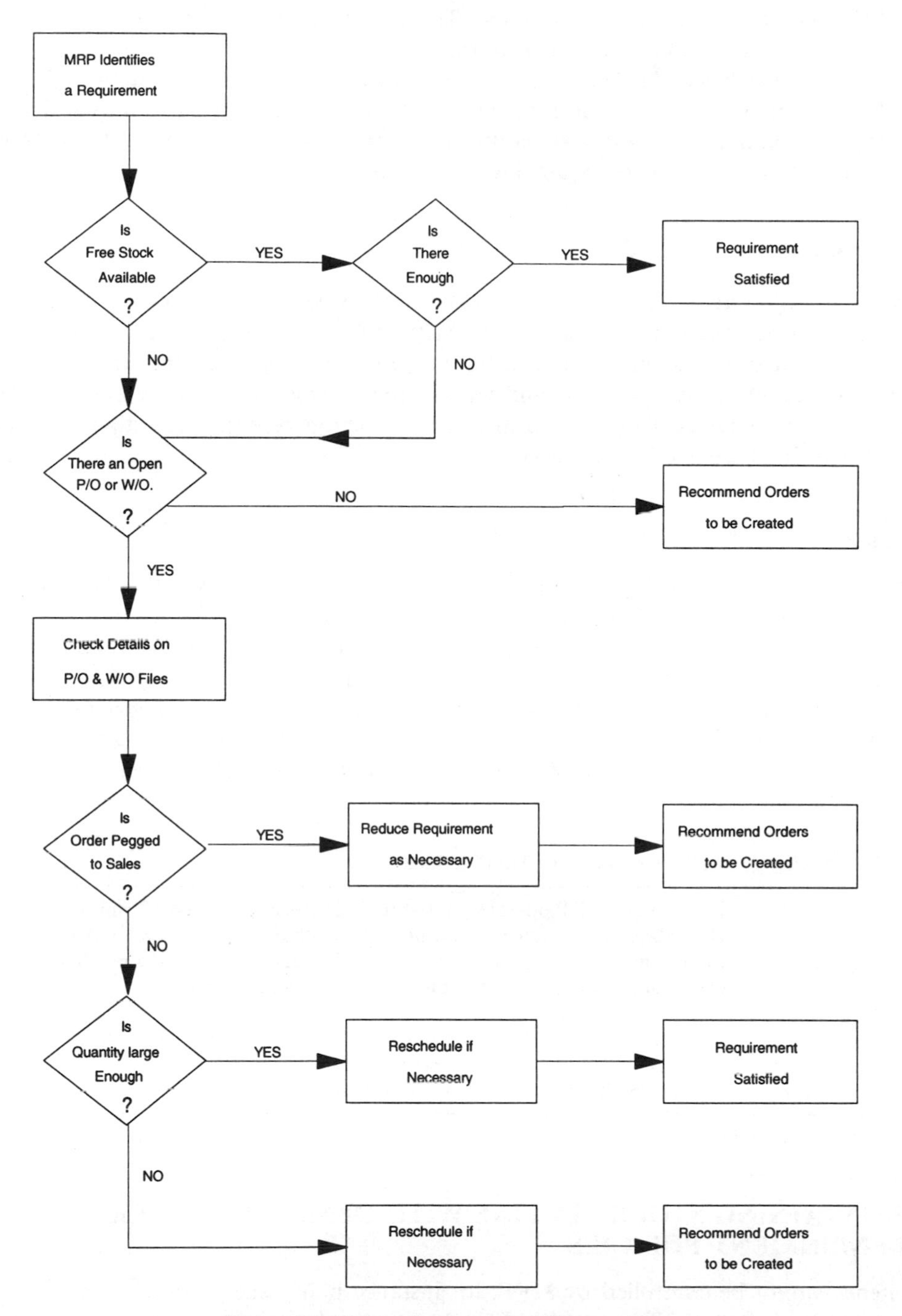

Figure 9.5 Flow diagram of MRP processes.

Basic MRP

A basic MRP system is not linked to the sales order processing module for automatic generation of the master production schedule (sales orders plus forecasts). Instead, it requires the user to set up the MPS data for each production period. This may be ideal for a company that is making machines to sell from stock, rather than to order.

Apart from this, basic MRP operates as previously described, by breaking the MPS demand into net requirements and then checking against stock and purchase or works orders before suggesting the creation of new orders. However, it probably does not create these orders but merely lists details of the orders required as a tabulation.

Standard MRP

A standard MRP system will probably be linked to the sales order processing module so that sales orders can be copied automatically into MPS, but sales forecasts may need to be set up manually. Such a system should include a full range of batching rules and MRP policies, and will generate recommended purchase and works orders, which can readily be converted into confirmed orders if required. It should also include action reports suggesting that specified orders be expedited, delayed or cancelled.

Full MRP

A full MRP system will include all the features available in standard MRP, but in addition should allow user-specified MRP runs, for example by selection of a group of products in the MPS. It will also allow much greater flexibility in the choice of time bucket sizes so that, over a given period, buckets can be of mixed size as shown in Fig. 9.6.

Finally, full MRP will allow the user to run either 'net change' or 'full regenerative' MRP as necessary. These are defined in Table 9.5. Generally, if both systems are available, the net change option may be run two or three times weekly, and the full option weekly or fortnightly.

Table 9.5 Net change and full regenerative MRP defined

Net change	Net change MRP operates on the basis of the previous MRP run by examining all the data used for that run and noting any changes. It then re-runs taking account of the changes and suggests actions and orders that need to be changed or confirmed. This is considerably faster than a full run.
Regenerative	Full regenerative MRP ignores any previous MRP runs and the data associated with such runs and creates a completely new series of recommended actions and orders which replace those previously suggested.

9.6 MAINTAINING STOCKS IN LINE WITH DEMAND—STOCK REPLENISHMENT POLICIES

Some items cannot be controlled by MRP, for instance bulk issue items or items with a long procurement lead time. Such items should be classified as 'non-MRP' and controlled by some form of stock replenishment policy, such as one of the following:

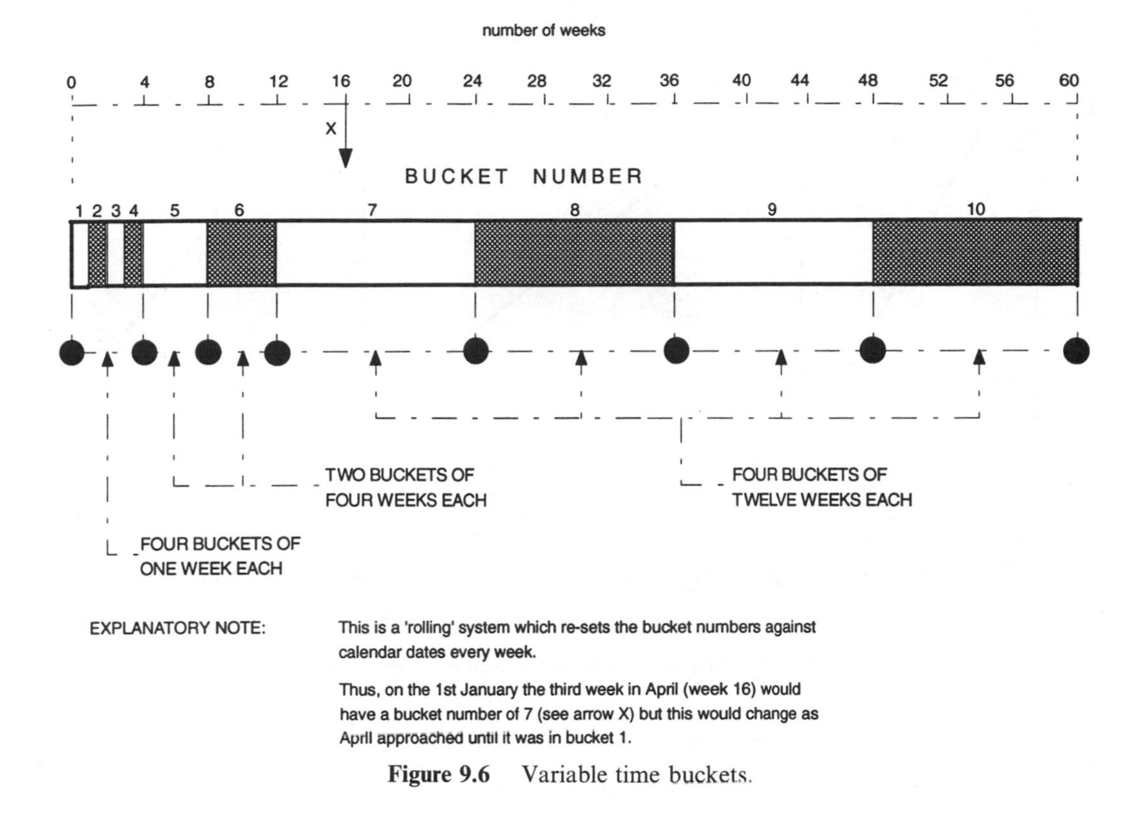

Figure 9.6 Variable time buckets.

1. reorder point
2. reorder cycle
3. combined reorder point and cycle.

Reorder point

This is somewhat similar to the so-called 'max–min' system where the objective is to keep the stock quantity between maximum and minimum values, but in this case the two values are known as *reorder level* and *reorder quantity*. They are calculated for each stock item using a formula based on average demand and replenishment lead time with, in some cases, a safety factor (safety stock).

The way in which such a system works is shown in graphical format in Fig. 9.7, with stock level plotted against time. As time passes, demand eats into the stock until it has fallen to the reorder level (RoL). At this point, an order is placed for the quantity set up as the reorder quantity (RoQ) and, as the lead time passes, demand continues to reduce the stock until the goods are received. Stock then returns to an acceptable level.

Figure 9.7 also shows the safety stock level, which is designed to prevent a stock-out if the actual lead time is greater than planned, or if demand during the lead time exceeds expectations. However, most companies nowadays try to avoid holding safety stocks, by a careful choice of a supplier that can give a short lead-time promise and keep it.

The methods of calculating the reorder level and the reorder quantity are shown in Table 9.6. If no safety stock is required the safety factor is set to zero. The reorder point system is most

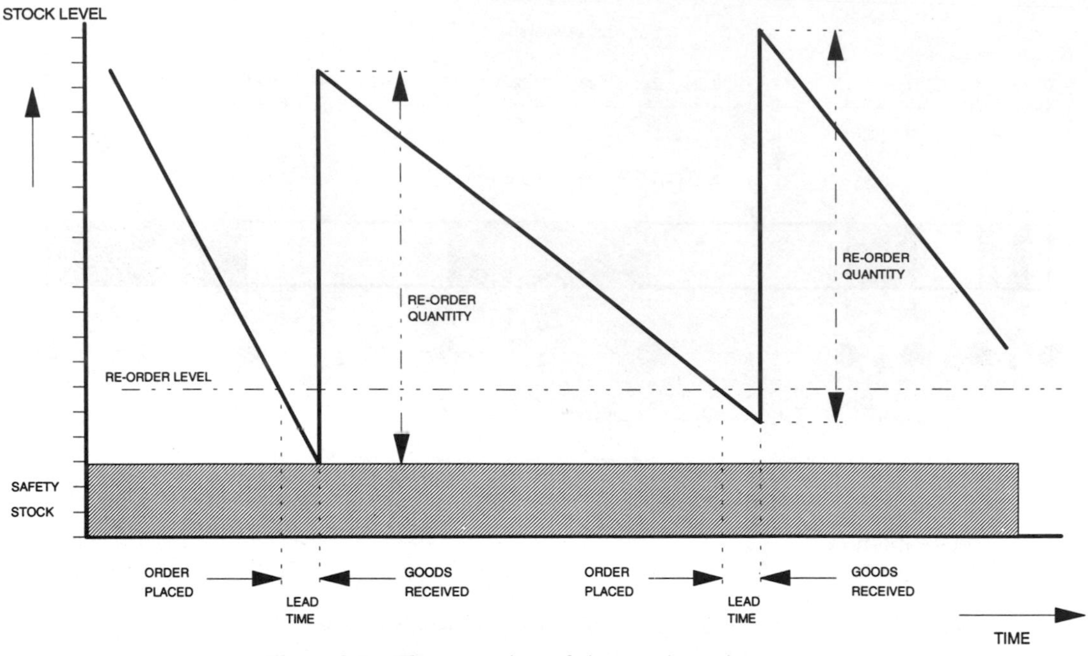

Figure 9.7 The operation of the reorder point system.

Table 9.6 Formulae for reorder level and quantity

RoL	Determine: A = average demand per week (e.g. 200 units/week) L = lead time for supply in weeks (e.g. 1.3 weeks) S = safety factor (e.g. 0.5) Then, RoL $= A\ (L+S)$ (e.g. $200[1.3+0.5] = 360$ or 1.8 weeks' demand)
RoQ	This can be set either (a) in terms of a number of weeks average demand N, on the basis of an order frequency policy, e.g. for 6 orders per year $N = 8$ for a 48 week year, giving RoQ $= N \times A$, or (b) in terms of an economic order quantity (EOQ) using the formula given in Fig. 8.9

likely to be used by a company that wishes to minimize stock levels and is prepared to order at irregular intervals.

Reorder cycle

This system is based on the setting of a regular ordering frequency, such that an order will be placed every *n* days or weeks, the size of the order depending on the stock level at that time. The way in which such a system works is shown in graphical format in Fig. 9.8, with stock level plotted against time. As time passes, demand eats into the stock until it is the end of a cycle and another order is to be placed. At this point, the system calculates the size that this order should be, based on the difference between a predetermined *maximum stock* and the *current stock*. However, in some cases this can lead to orders being placed for small, uneconomic quantities if demand has not been up to expectations, or to a stock-out if demand has been higher than planned.

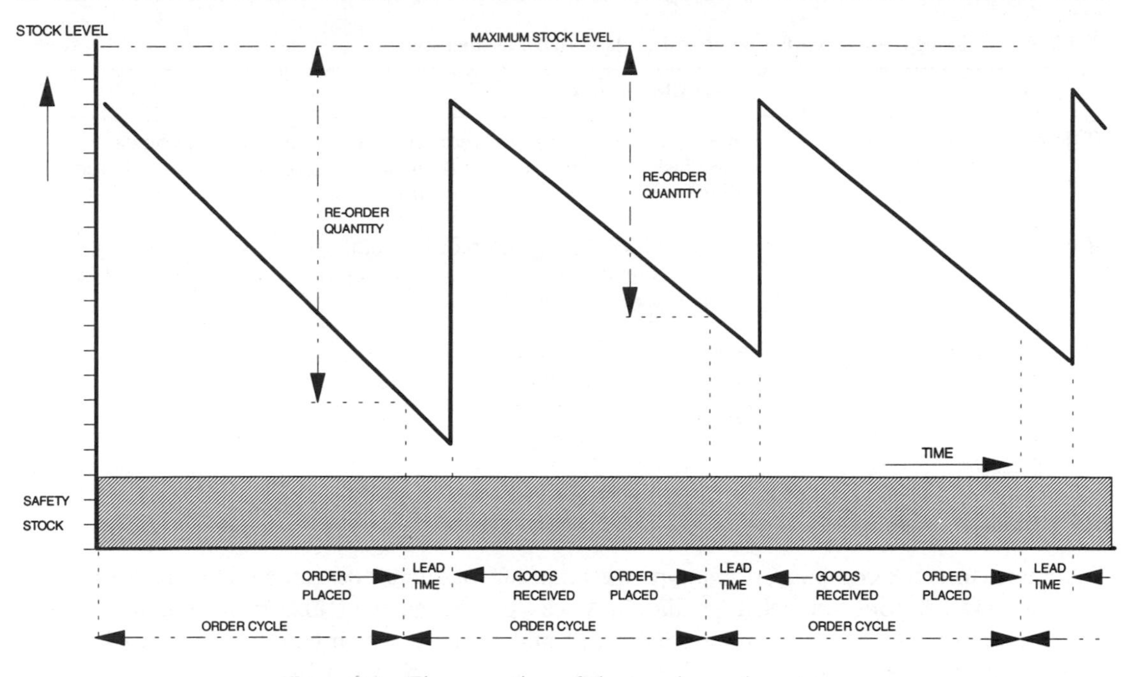

Figure 9.8 The operation of the reorder cycle system.

This system is therefore most likely to be used by a company that has fairly consistent demand levels, and wishes to keep the number of purchase and works orders to a minimum. The system can also incorporate safety stocks if required, and should be set up for an item by specifying the following three parameters.

- *Cycle time* Usually in weeks on the basis of n orders per year. Thus if $n = 8$ in a 48 week year, the cycle time = 6 weeks.
- *Maximum stock* Usually based on predicted weekly average demand and the cycle time, although the lead time could be added. For example, if demand = 120/week, cycle time = 6 weeks and lead time = 1 week then maximum stock would be $120(6 + 1) = 840$.
- *Safety factor* This could be included if suppliers are unreliable and would be set in terms of a number of weeks of demand. For example, if factor = 0.5 and demand = 150/week then safety stock level = 75. However, if the lead time is included in the maximum stock calculation the safety factor could probably be set to zero.

Combined reorder point and cycle

This system is designed to use the strengths of the other two systems and to eliminate the weaknesses. It achieves this by allowing both sets of reordering rules to operate and then using whichever gives the better result.

This is illustrated in Table 9.7 on the basis of the demand being above or below expectation. In general terms, the reorder point system is easier to operate and understand, but the reorder cycle system can be of advantage if scheduled orders have been placed on suppliers with the schedule dates matching the cycle frequencies. Thus, each time the next schedule quantity is due to be advised to the supplier, that quantity can be adjusted to match the reorder quantity from the reorder cycle system.

Table 9.7 The combined reorder point and cycle systems in operation

Variable	Operating method
If demand above expected	Reorder level setting (on reorder point system) would trigger a replenishment order earlier than would have been the case if the reorder cycle system had been used.
If demand below expected	Reorder level setting (on reorder point system) would trigger a replenishment order later than would have been the case if the reorder cycle system had been used.
If demand as expected	Reorder cycle frequency setting would trigger replenishment orders at regular intervals and of economic size.

9.7 MAINTAINING STOCKS IN LINE WITH DEMAND—SERVICE LEVEL SETTING

In a make-to-stock company, of the type where an out-of-stock position could mean a lost sale, it is not always good enough to maintain stocks on the basis of historical demand. This is because an average demand will conceal peaks and troughs, which can lead to a stock-out at a peak. It is therefore necessary to try to overcome this problem by holding safety stocks. However, the method of determining these should not be by 'guesstimate' but by a more scientific statistical method, such as service level setting.

The service level can be defined in terms of the statistical probability that an item will be in stock when required. For example, a VSL (vendor service level) of 100 per cent means that there will always be stock, whatever the demand, and a VSL of 50 per cent that there is a 50:50 chance of stock being available. In fact, a VSL of 100 is statistically impossible (since this would require a stock level of infinity!), although 99.99 could be achieved.

The calculations involved are based on a series of demand quantities for an item over a period of time. Then, by means of a statistical technique known as standard deviation, and by using a statistical table, a stock reorder level can be found to meet a required service level.

The method of calculation is illustrated in Fig. 9.9 and the statistical table is shown in Fig. 9.10. Figure 9.9 shows how a standard deviation σ is calculated from the demand series. In this case, the demand is fairly level with a σ of 15 based on actual deviations above and below the average (333) of about ± 7 per cent. Then, by using the statistical table in Fig. 9.10, the k factor can be found for any specified service level which, when multiplied by one standard deviation, gives the extra stock required to match the required VSL (i.e. there is a 0.95 probability that stock will be available if the stock is not allowed to fall below 358 units (see Fig. 9.9), assuming that stock can be replenished in less than one month.

Author's note: The basic principles of normal distribution are described in greater detail and more general terms in appendix B.

9.8 MAINTAINING STOCKS IN LINE WITH DEMAND—INVENTORY MANAGEMENT

Any method of stock replenishment (whether computerized or manual) relies totally for its effectiveness on the accuracy of the stock records, and it is therefore essential to ensure that

PERIOD	MONTHLY DEMAND (X)	DEVIATION FROM AVGE. $(X-\bar{X})$	SQUARES OF DEVIATIONS $(X-\bar{X})^2$
1	340	+ 7	49
2	330	− 3	9
3	350	+ 17	289
4	330	− 3	9
5	325	− 8	64
6	335	+ 2	4
7	345	+ 12	144
8	310	− 23	529
9	355	+ 22	484
10	310	− 23	529
Total	3330	Sum of Squares	2110
Avge $\bar{X}$	333		

USING MONTHLY DEMAND FIGURES FOR THE ITEM PERFORM THE FOLLOWING STEPS:-

1. ADD ALL DEMAND FIGURES [sum of (X)s]
2. CALCULATE AVERAGE DEMAND = TOTAL (From Step 1)/NO. OF PERIODS
3. CALCULATE MONTHLY DEVIATIONS FROM AVERAGE (eg. 340 − 333 for period 1)
4. SQUARE ALL THESE MONTHLY DEVIATIONS
5. ADD SUM OF DEVIATIONS SQUARED
6. DIVIDE THIS TOTAL BY (NO. OF PERIODS − 1) $\frac{2110}{(10-1)}$ = 234 (Variance)
7. TAKE SQUARE ROOT OF VARIANCE 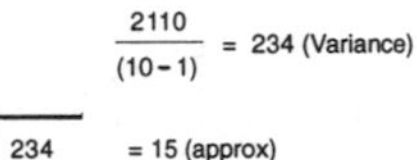 $\sqrt{234}$ = 15 (approx)

THIS IS KNOWN AS THE STANDARD DEVIATION (Greek Letter σ)

THUS, IF ONE STD.DEVIATION (1 σ) = 15

TWO STD.DEVIATIONS (2 σ) = 30

THREE STD.DEVIATIONS (3 σ) = 45

FROM STATISTICAL TABLE (Figure 8.10) IT IS POSSIBLE TO DETERMINE WHAT VALUE OF 'k' (VSL FACTOR) IS REQUIRED TO GIVE A SPECIFIED VENDOR SERVICE LEVEL (VSL), FOR EXAMPLE:

SERVICE LEVEL %	VALUE OF k	EXTRA STOCK* k VALUE x 15	RE-ORDER** LEVEL
80	0.85	13	346
85	1.04	16	349
90	1.28	19	352
95	1.65	25	358
99	2.33	35	368

*BASED ON EXAMPLE, (k x 1 σ), eg. 90% VSL = 1.28 x 15 = 13

**BASED ON EXAMPLE, ($\bar{X}$ + Extra Stock), eg. 90% = 333 + 19 = 352

Figure 9.9 Calculation of reorder level for a given service level.

inventory management practices and procedures are such that mistakes do not occur. However, if they do (and they always do!) they must be found and corrected as quickly as possible. A number of techniques can be used for this purpose, including:

1. perpetual inventory
2. ABC classification
3. back-flushing
4. effective documentation
5. effective procedures.

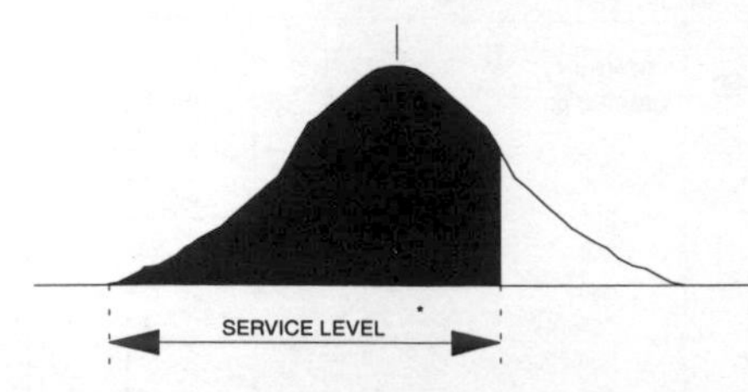

* THE SERVICE LEVEL IS DEFINED AS THE PERCENTAGE RATIO OF THE SHADED AREA UNDER THE GRAPH TO THE TOTAL AREA UNDER THE GRAPH.

VALUES OF k

SERVICE LEVEL	0.0	0.5
75	0.68	0.69
76	0.70	0.72
77	0.74	0.76
78	0.77	0.79
79	0.81	0.82
80	0.84	0.86
81	0.88	0.90
82	0.92	0.94
83	0.95	0.97
84	0.99	1.02
85	1.04	1.06
86	1.08	1.10
87	1.12	1.15
88	1.17	1.20
89	1.23	1.25
90	1.28	1.31
91	1.34	1.37
92	1.41	1.44
93	1.48	1.51
94	1.55	1.60
95	1.65	1.70
96	1.75	1.81
97	1.88	1.96
98	2.05	2.17
99	2.33	2.58

TO USE THIS TABLE:

READ OFF VALUE OF k TO MATCH SERVICE LEVEL REQUIRED, FOR EXAMPLE, FOR 85.0% k = 1.04 AND FOR 85.5% K = 1.06.

THEN R.O.L. = $\bar{X} + (k \times \sigma)$

WHERE $\bar{X}$ = AVERAGE MONTHLY DEMAND AND σ = THE STD. DEVIATION

Figure 9.10 Statistical table—normal distribution areas.

Perpetual inventory

The traditional method of stock-taking, whereby all stocks (including work-in-progress) were checked annually, was based on the need for stock valuation as a part of the requirement for a company to produce annual accounts. This method is totally unsatisfactory from a number of points of view, i.e. it is time-consuming, disruptive to production, possibly inaccurate, unselective, and far too infrequent. For example, an annual stock-take can take several days to complete, during which time production is virtually halted; it may also involve staff who are unfamiliar with the items concerned, and therefore cannot relate part numbers to parts. Such staff may also be confused by the various units of measurement (UOMs).

For these reasons, many companies have now changed to the perpetual inventory (PI) system, which can eliminate such problems and if carried out correctly will allow the auditors to accept the figures on the stock records (with some random physical checks) for the annual accounts. Perpetual inventory is generally used only on stores stock (not on work-in-progress) and operates on the principle of 'divide and rule', i.e. by dividing the stock into a number of small groups it is possible to check each group, in rotation, on a regular basis. For example, if a company has 4000 stock items and wishes to check stock on a three-monthly cycle (say 60 working days) then, by dividing the stock into groups of between 65 and 70 items (4000/60 = 66.7) and checking one group per day, all stock will have been checked by the end of the three-month period.

The PI system can offer a number of advantages over the traditional annual stock-take: firstly, stock is checked more frequently and therefore errors can be corrected more quickly; secondly, the daily check need involve only experienced staff (fewer errors); thirdly, the daily check does not take long and therefore production is not held up.

However, the most important advantage is that there is an opportunity to investigate major discrepancies (hopefully only a few!) between the physical stock and the stock records, on a daily basis, and thus to eliminate the causes of errors which will generally fall into one of the following categories:

- inaccurate counting of the physical stock (quantity or UOM)
- transaction document mislaid or not created (record not updated)
- incorrect entry from document to record
- incorrect physical issue against document
- incorrect count at goods inwards (document incorrect).

Tracing such errors can be a time-consuming process, but is well worth while, because it can not only lead to the elimination of the causes, but also ensure that a further error does not occur. For example, if a stock issue document has been mislaid and not entered to the computer, a discrepancy will be found. If this is then corrected without an investigation, and the document is later found and entered, a double adjustment will have taken place!

However, if a computerized system of stock recording is in use this task is greatly simplified, since the system will include an audit trail facility. This will take the form of a report (on screen or printed) which lists all the transactions that have taken place against a stock item since the last stock check. Each of these transactions should be authorized by a numbered document which can be found and checked against its computer entry. A typical example of this is shown in Fig. 9.11.

In addition, if a computer is being used, a daily stock check-list can be printed, which can then be used to enter physical counts, and subsequently to identify and correct discrepancies. This method is described in more detail below ('Effective procedures').

ABC classification

The Pareto, or 80:20, rule is now an accepted management tool, and can be used in numerous business areas, including inventory management. In this example, all stock items are to be placed into one of three categories, based on their annual usage value (AUV), although another basis could be chosen (e.g. unit cost):

A. high usage-value items
B. medium usage-value items
C. low usage-value items.

AUDIT TRAIL ENQUIRY

PART NO. 275/8742 — Preformed Radiator Inlet Hose - 40mm Dia. — UOM: Each — DATE: DD/MM/YY

DATE	TIME	TRANSACTION TYPE	QUANTITY	LOCATION	PREVIOUS BALANCE	NEW BALANCE	DOCUMENT NO.
dd/mm/yy	hh-mm	Stock Check	1755	S4/A231	1750		ST.2310
dd/mm/yy	hh-mm	Stock Adjust.	+ 5	S4/A231	1750	1755	SA.7654
dd/mm/yy	hh-mm	Issue	- 100	S4/A231	1755	1745	WO.2367
dd/mm/yy	hh-mm	Issue	- 250	S4/A231	1745	1495	WO.5643
dd/mm/yy	hh-mm	Return	+ 10	S4/A231	1495	1505	RT.9805
dd/mm/yy	hh-mm	Issue	- 200	S4/A231	1505	1305	WO.5743
dd/mm/yy	hh-mm	Receipt	+ 500	S4/A231	1505	1805	GR.0277
	etc.	etc.	etc.		etc.	etc.	
dd/mm/yy	hh-mm	Stock Check	1490	S4/A231	1490		ST.2317

F1 Exit — F2 Page Down — F3 Page Up — F6 Another Part — F9 View Summary

Figure 9.11 Typical audit trail enquiry screen.

These principles can be seen in graphical form in Fig. 9.12 which shows

class A	20 per cent of items	80 per cent of AUV
class B	30 per cent of items	10 per cent of AUV
class C	50 per cent of items	10 per cent of AUV.

Then, with every stock item carrying an A, B or C code, it is possible to use these codes to make the perpetual inventory system more cost-effective and efficient.

This is achieved in two ways. The first is by setting stock check tolerances, which allows minor discrepancies to be adjusted without costly investigations into causes on low-value items, and by increasing the stock-check frequency on high-value items. Table 9.8 shows how the tolerance system works. The second is the selective approach to perpetual inventory: this operates by allocating different stock-check frequencies to each class, such that high usage-value (class A) items are checked much more frequently than the low usage-value (class C) items. Table 9.9 shows how this works.

In fact, the number of items to be checked daily is only 10 more than in the earlier example for unselective perpetual inventory (65 to 70 per day), and this can easily be justified by the improved control on the A class items. The Pareto-based ABC system can be applied in many

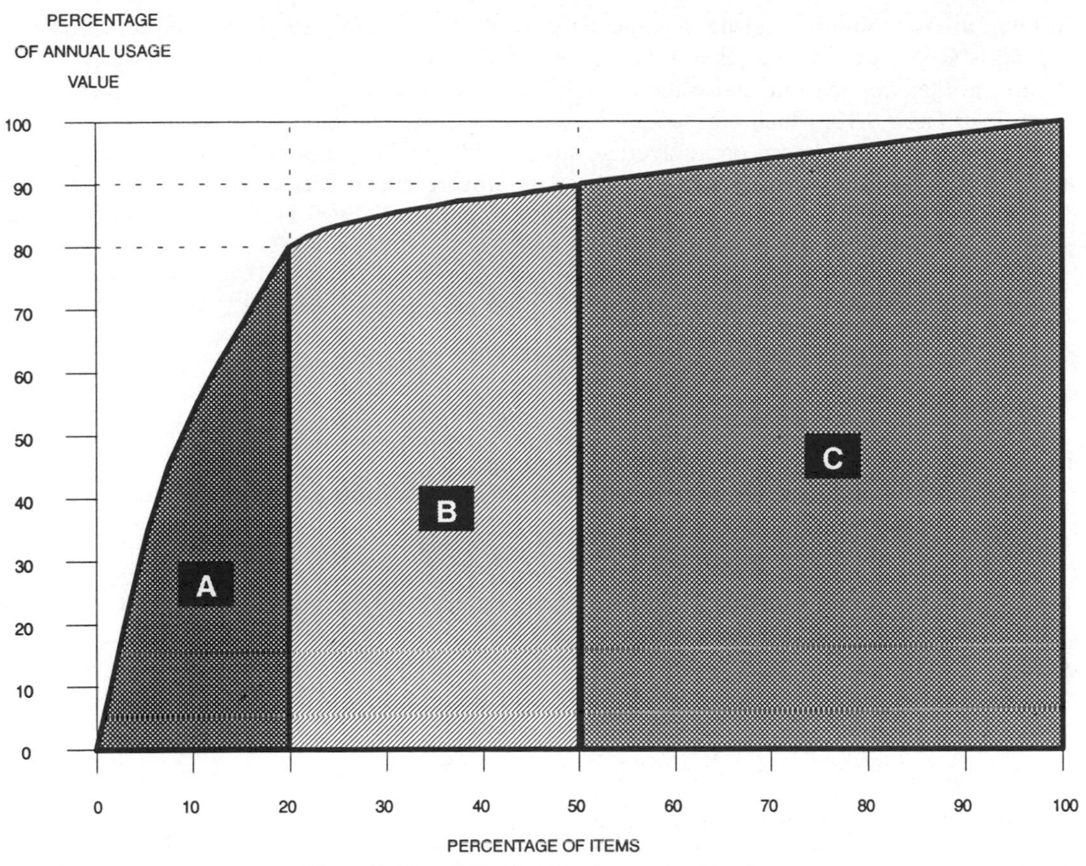

Figure 9.12 ABC classification using Pareto.

Table 9.8 Stock check tolerances set by ABC class

Class	Tolerance	Item	Physical	Theoretical	Variance	OK?
A	±1%	Part 231	40	39	–1 (2.5)%	No
B	$\pm 2\frac{1}{2}\%$	Part 747	120	122	+2 (1.7)%	Yes
C	±5%	Part 289	900	925	+25 (2.8)%	Yes

Table 9.9 The use of ABC classes in a PI system

Class	Item quantity	Frequency (weeks)	Working days	Items/day
A	800	4	20	40
B	1200	12	60	20
C	2000	24	120	17
Total	4000			77

different areas of manufacturing management; for example, safety stocks can be set selectively (e.g. class C = 2 weeks; class B = 1 week; class A = zero).

In another application, the value of ABC classification to a company can be seen in the example in Table 9.10, which is based on the following data. Total annual usage value (AUV) of stock held is £360 000 based on an average stock-holding of three months (this means that the stock is sufficient, on average, to cover three months of sales); it is calculated as stock value divided by material cost of sales, as shown in the example in Table 9.11. The distribution of this stock is as follows:

class A 80 per cent of AUV = £288 000 or £72 000 for 3 months
class B 10 per cent of AUV = £36 000 or £9 000 for 3 months
class C 10 per cent of AUV = £36 000 or £9 000 for 3 months.

This gives a total average usage of £90 000 for 3 months.

However, by changing the methods of control in the following ways (see Table 9.10), the stock-holding can be reduced to £51 000:

class A stock-holding reduced to one month by tighter control (e.g. lower reorder levels and quantities and improved supply times)
class B no change
class C stock-holding increased to 6 months by easier control (e.g. higher reorder levels and quantities).

Table 9.10 The value of selective stock control

By changing methods of monitoring and control selectively, considerable savings were achieved in the form of a stock reduction of £90 000 – £51 000 = £39 000

Class	Old method	Old value	New method	New value
A	3 months	72 000	1 month	24 000
B	3 months	9000	3 months	9000
C	3 months	9000	6 months	18 000
Total		£90 000		£51 000

Table 9.11 Average stock value and usage

If the value of the stocks of materials issued is found on a monthly basis as shown below, then the average value for the year can be calculated.

Month	1	2	3	4	5	6	7	8	9	10	11	12	Total
Value (£000)	30	32	27	34	31	32	29	30	28	31	30	26	360

Thus annual usage value = £360 000. If the policy is to cover for three months' usage, then stocks of £90 000 must be maintanied.

There may also be a saving in clerical work since, although there is more work on class A items, these are only about 20 per cent of the total number of items, with less clerical work on the class C items which are about 50 per cent of this total.

Back-flushing

Where items are issued in bulk to a production line or work area on a 'bulk-issue' basis, they are normally booked out of the main stores by means of a stock transfer to another store, e.g. a work centre location. In stock accounting terms these items are still in stores, rather than in work-in-progress. The back-flushing technique can be used to deal with this situation, if a suitable computer system is installed, by using the records of work completed on the line or in the work area, as follows.

1. The program scans the bill of materials for each of the work-completed products and identifies any component that is 'flagged' as bulk issue for back-flushing, noting the quantity required for one unit of the completed product.
2. The program then multiplies each of these values by the product-completed quantity and stores the results in a temporary file against the works orders concerned.
3. Finally, this file is used to book these quantities out of their work area stock locations and into production, such that this information is available for costing and stock-check auditing.

This technique is not available on all computer systems but can save a great deal of clerical time. However, it is not entirely accurate unless a control is set up for items rejected from the bulk-issue stocks.

Effective documentation

The term 'paperless system' has been used to describe the factory of the future, but it is certainly not yet practical as far as materials management is concerned. This is because of the need to ensure that stocks are properly controlled and secured, in order to maintain accurate stock records. If a computerized system is in operation the task of creating documents is greatly reduced, but the need to use such documents is still paramount, both as an authority for stock transactions and as a record for stock accounting purposes. It is therefore suggested that some, if not all, of the following documents should be used.

- *Goods received note* This should be created in order to record details of incoming goods. It can then be used to book these goods into stock, to record any other activity (e.g. goods returned to supplier) or to deal with invoice queries.
- *Materials requisition* This often forms part of the works order documentation set; it is used to authorize an issue from stock and to enter details of that issue to the stock-recording system. However, an MR may also need to be created if additional material is required, or if a stock issue is to be made for an unplanned purpose (e.g. a development project, a customer sample or a bulk issue).
- *Stock transfer note* Where stock is to be transferred from one location to another, it is advisable that the details be recorded on a transfer note for entry to the records and as an auditing aid. For example, a transfer could be needed to take some items from the main store and relocate them as a part of the stock held at a service centre or as stock in a service

engineer's vehicle. Alternatively, stock could be transfered to a quarantine area from main store or vice versa.

- *Return-to-stores note* Where stock has been issued to production or any other department and is then no longer required, it should be returned to stock, or a quarantine stores location, on an RTS note so that a record is available for future reference.

It is accepted that, if the company is fully computerized with terminals at all points where stock can be issued or received, it may be possible to eliminate some or all of these documents, but there is then no easy way of auditing stock movements and stock accuracy may be compromised.

Effective procedures

As well as effective documentation, it is necessary to lay down clear procedures for all operations and transactions that may affect stock accuracy. These would normally form a part of the quality manuals which are an essential part of approval for BS 5750 or ISO 9000.

As an example of this requirement, the method of stock checking and stock adjustment in a perpetual inventory system can be defined as a series of steps, as follows.

- *Step 1* Issue a daily listing showing which items should be checked (see Fig. 9.13). The quantity of each listed item is to be entered by the stock checker.
- *Step 2* Enter these quantities to the computer using the list number to display the entry screen.
- *Step 3* Call up and view the discrepancy list (Fig. 9.14); flag (Y) any item where a discrepancy is to be accepted.

DAILY STOCK CHECK LIST

LIST No. 124/91
DATE: XX/XX/XX
PAGE: 1 of 6

LINE No.	ITEM CODE	DESCRIPTION	U.O.M.	STORES	BIN/ RACK	PHYSICAL STOCK	COMMENTS
01	12/345	xxxxxxxxxxxxxxxxxxxxxxxxxxx	EACH	STORE A	23/1		
02	13/444	xxxxxxxxxxxxxxxxxxxxxxxxxxx	EACH	STORE A	24/2		
03	11/907	xxxxxxxxxxxxxxxxxxxxxxxxxxx	LTRS	STORE A	30/5		
04	275608	xxxxxxxxxxxxxxxxxxxxxxxxxxx	KGS	STORE A	33/3		
05	P.43356	xxxxxxxxxxxxxxxxxxxxxxxxxxx	EACH	STORE A	34/1		
06	17/0987	xxxxxxxxxxxxxxxxxxxxxxxxxxx	EACH	STORE B	62/2		
07	12/0978	xxxxxxxxxxxxxxxxxxxxxxxxxxx	EACH	STORE B	73/2		
08	23/4689	xxxxxxxxxxxxxxxxxxxxxxxxxxx	MTRS	STORE C	92/9		
09	32486/1	xxxxxxxxxxxxxxxxxxxxxxxxxxx	EACH	STORE C	97/2		
10	W.90865	xxxxxxxxxxxxxxxxxxxxxxxxxxx	EACH	STORE D	34/1		
11	15/35278	xxxxxxxxxxxxxxxxxxxxxxxxxxx	EACH	STORE D	62/2		
12	11/85732	xxxxxxxxxxxxxxxxxxxxxxxxxxx	EACH	STORE D	73/2		
13	30/79409	xxxxxxxxxxxxxxxxxxxxxxxxxxx	MTRS	STORE D	92/9		

Continued

CHECKED BY: ....

Figure 9.13 Typical stock check-list.

DAILY DISCREPANCY LIST

LIST No. 124/91
DATE: XX/XX/XX
PAGE: 1 of 6

LINE No.	ITEM CODE	DESCRIPTION	U.O.M.	STORES	BIN/ RACK	PHYSICAL STOCK	DISCREPANCY QTY.	DISCREPANCY VALUE	ACCEPT ? [Y]
01	12/345	xxxxxxxxxxxxxxxxxxxxxxxxxxxx	EACH	STORE A	23/1	150	+5	0.75	[Y]
02	13/444	xxxxxxxxxxxxxxxxxxxxxxxxxxxx	EACH	STORE A	24/2	600	+12	2.75	[ ]
03	11/907	xxxxxxxxxxxxxxxxxxxxxxxxxxxx	LTRS	STORE A	30/5	25	0	0	[Y]
04	275608	xxxxxxxxxxxxxxxxxxxxxxxxxxxx	KGS	STORE A	33/3	355	−22	7.65	[ ]
05	P.43356	xxxxxxxxxxxxxxxxxxxxxxxxxxxx	EACH	STORE A	34/1	25	−1	1.23	[Y]
06	17/0987	xxxxxxxxxxxxxxxxxxxxxxxxxxxx	EACH	STORE B	62/2	216	−5	4.25	[ ]
07	12/0978	xxxxxxxxxxxxxxxxxxxxxxxxxxxx	EACH	STORE B	73/2	43	+1	8.90	[ ]
08	23/4689	xxxxxxxxxxxxxxxxxxxxxxxxxxxx	MTRS	STORE C	92/9	134	+3	1.80	[Y]
09	32486/1	xxxxxxxxxxxxxxxxxxxxxxxxxxxx	EACH	STORE C	97/2	1205	−45	0.90	[Y]
10	W.90865	xxxxxxxxxxxxxxxxxxxxxxxxxxxx	EACH	STORE D	34/1	48	−2	18.90	[ ]
11	15/35278	xxxxxxxxxxxxxxxxxxxxxxxxxxxx	EACH	STORE D	62/2	23	0	0	[Y]
12	11/85732	xxxxxxxxxxxxxxxxxxxxxxxxxxxx	EACH	STORE D	73/2	NIL	0	0	[Y]
13	30/79409	xxxxxxxxxxxxxxxxxxxxxxxxxxxx	MTRS	STORE D	92/9	250	−10	1.40	[Y]

Continued

CHECKED BY: ....

Figure 9.14 Typical stock discrepancy list.

- *Step 4* Print out a list of all other discrepancies and investigate reasons.
- *Step 5* Correct theoretical or physical stock as necessary.
- *Step 6* Correct causes of these other discrepancies.

Any discrepancies found between theoretical and actual stock with a value of less than an agreed amount (e.g. £5) can be accepted at step 3, but those of higher value should be thoroughly investigated to determine the cause before the error is corrected.

9.9 BATCH SIZES

Various methods can be used to determine the size of batch that should be ordered from a supplier, or through production, but all of these are based on the need for economy. The questions that must therefore be asked and answered, assuming a given level of demand, are as follows.

1. The smaller the batch, the more frequently an order must be placed, so how much does it cost to place an order?
2. The larger the batch, the higher the average stock level, so how much does it cost to hold and manage stock?
3. If a production order is too large, will it occupy a limited production facility for so long that other work is delayed? What could this cost?
4. If a production order is too small, the set-up cost may be excessive, so what is the cost of setting up?
5. If a purchase order is too small, will this result in extra cost due to lost quantity discounts? What could this cost?

This is a complex problem which cannot easily be resolved on a day-to-day basis. Companies must therefore decide on a policy and review it at regular intervals. There are two basic types of policy: the MRP approach and the economic approach.

The MRP approach

This was considered earlier in this chapter (section 9.4) and involves the setting of a batching policy for every MRP item. This could be 'lot-for-lot', whereby only the requirement for the period concerned is ordered, or some form of batching where an order is placed to cover the demand for a specified period ahead, measured in time buckets. The setting up of such policies can be simplified if the ABC classification system is used, e.g.

class A lot-for-lot
class B batching demand for *n* time buckets ahead
class C economic batch (see below).

However, it may be decided to operate a combination of policies such that, for example, the order size is set on the basis of lot-for-lot or economic batch, whichever is the larger.

The economic approach

This method is based on optimizing the costs of ordering and stock-holding, to determine the EOQ (economic order quantity) for purchase ordering, or EBQ (economic batch quantity) for works ordering. It is shown in graphical format in Fig. 9.15.

The general principle of this method is that costs can be established for placing an order, and for holding stock, and that these costs can then be used to calculate a variety of scenarios, such that the lowest cost option can be selected. However, it should be noted that these calculations do not always reflect the true costs, e.g. loss of volume discounts or production delays due to overloading of limited resources.

The data used to prepare the graph (Fig. 9.15) are given in Table 9.12, the symbols and values used to find Q (order quantity) being as follows:

V (item unit value) = £2.50
S (stock-holding cost—percentage of stock value) = 20 per cent
A (annual usage of item) = 5040
C (cost of placing an order) = £35.00.

Table 9.12 shows that the total cost falls as the number of orders placed increases to eight per year, but then starts to rise as the number of orders rises beyond eight. Thus, from the table the number of orders that should be placed is between four and eight, giving EOQ/EBQ values of between 1260 and 630.

In fact it is not necessary to prepare tables and graphs in this way, since the value of Q can be calculated from the formula

$$Q = \sqrt{(2 * C * A * 100)/(V * S)}$$

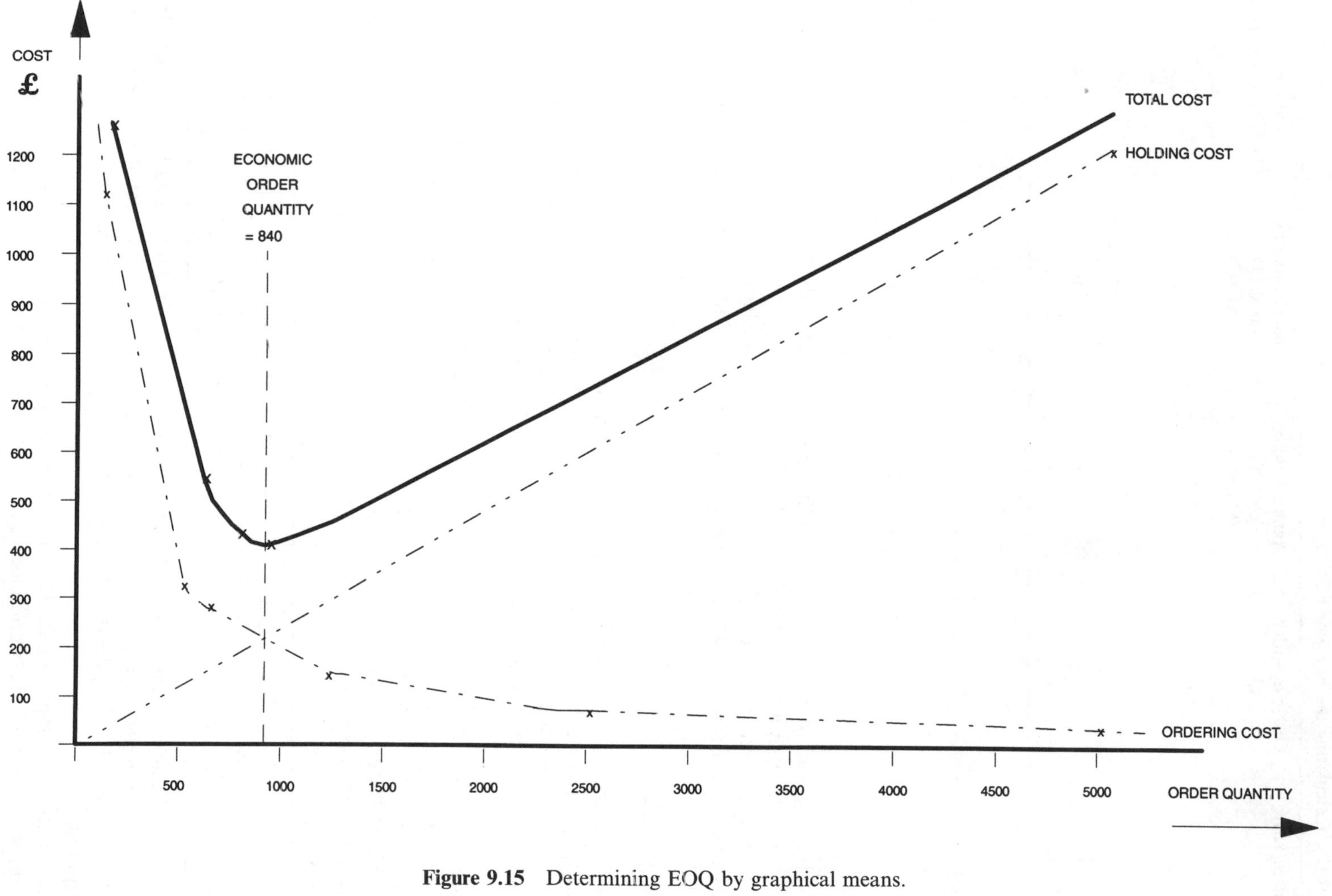

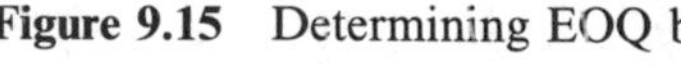

Figure 9.15 Determining EOQ by graphical means.

Table 9.12 Calculating the EOQ or EBQ

Orders/year	Order quantity Q	Annual holding cost (£) (QVS/2)	Annual ordering cost (£) (AC/Q)	Total cost (£) [QVS/2] + [AC/Q]
1	5040	1260	35	1295
2	2520	630	70	700
4	1260	315	140	455
8	630	158	280	438
16	315	79	560	639
32	158	39	1120	1159
64	79	20	2240	2260

If this is used on the values from the example:

$$Q = \sqrt{}(2 * 35 * 5040 * 100)/(2.50 * 20) = \sqrt{}(35\,280\,000)/50$$
$$= \sqrt{}705\,600 = 840$$

This value of Q can then be substituted in the formula to calculate that this is equivalent to six orders per year, with a total annual cost of £420.

It will be apparent that the difficulty with the economic ordering method is to determine the two cost factors, C (order cost) and S (stock-holding cost). This is dealt with in chapter 14.

PROBLEMS

1. As the materials management manager of a company that manufactures a range of machine tools, to prepare a brief report in support of your recommendation that an MRP system be installed on the computer. The company makes a range standard of machine tools which are sold from stock and some specials which are made to order. There is also a steady demand for spares.
2. You work for a company that supplies automotive components to a vehicle manufacturer against weekly schedules which are quite variable in terms of quantities. The sales department wishes to be able to improve its service level and has requested that this be increased to a VSL of 98 per cent. Based on the figures given below for one of the components, calculate the minimum stock level that should be held to meet that level of service (see appendix A for solution).

Week no.	1	2	3	4	5	6	7	8	9	10	11	12
No. of items called	125	140	164	138	121	96	135	157	148	168	179	165

3. You have been given the task of specifying a spreadsheet that can be used to set up an ABC classification system based on the annual usage values of a range of stock items. Prepare this specification, including details of the columns to be used and examples of the calculations required.
4. The company wishes to calculate the costs of ordering and holding stock as a basis for preparing economic order and batch quantities. Suggest the factors that should be included in those costs for

(a) the cost of holding stock
(b) the cost of placing a purchase order
(c) the cost of placing a works order.

5. The company for which you work wishes to reduce its stocks of purchased items, and has asked you to prepare a report suggesting how this can best be done without adversely affecting production. Suggest possible methods that should be considered and the safeguards that should be applied to achieve this objective

FURTHER READING

Anderson, E. J., *The Management of Manufacturing*, Addison-Wesley, Wokingham, 1994.
Mair, G., *Mastering Manufacturing*, Macmillan, London, 1993.
Plossl, G. W., *Orlicky's Materials Requirements Planning*, McGraw-Hill, New York, 1993.
Smith, S. B., *Computer-based Production and Inventory Contorl*, Prentice-Hall, Englewood Cliffs, NJ, 1989.
Vollman, T. E., Berry, W. L. and Whybark, D. C., *Manufacturing Planning and Control Systems*, 3rd Edn, Irwin, Boston, MA, 1988.
Waters, C. D. J., *Inventory Control and Management*, Wiley, Chichester, 1992.

CHAPTER

TEN

THE MANAGEMENT OF RESOURCES

THE LINK

In the previous chapter, the supply of materials for production was considered in detail. In this chapter, the way in which resources (including materials) should be provided to meet a company's plans are examined in terms of those plans.

One of the major tasks of management is to look ahead in order to foresee potential threats and opportunities and, having identified these, to prepare plans that will minimize the former and maximize the latter. To achieve this, management must be able to identify the strengths and weaknesses of the organization. This form of forward planning is known as the 'SWOT' technique (strengths, weaknesses, opportunities, threats) and is designed to ensure that the resources of the company are used to the best effect in particular circumstances.

The SWOT technique can be applied in almost any situation where change is taking place, and allows management to decide which, of a series of possible options, is best for the company. For example, if an opportunity arose as a result of political changes in a particular market, should the company set up a factory in that market, develop a network of agents, or set up its own sales organization? The answers to such questions should be based on what the company does best — its strengths.

However, for this technique to be used effectively, management must be able to identify the company's resources and its ability to optimize their use. This chapter therefore defines the resources that are available and the methods that can be used to compare a resource with a requirement.

10.1 INTRODUCTION

In this chapter the various resources needed to operate a manufacturing business are identified, defined and examined in detail. The need to ensure that these resources are managed effectively is considered, and the effects of under-utilization are examined in relation to business efficiency. The 'closed loop' MRP II system (manufacturing resource planning) is then defined and considered in terms of the various elements of business and resource planning and how these plans can be applied.

Finally, the chapter examines alternative means of resource management such as JIT (just-in-time) and OPT (optimized production technology), and their relationship with the two MRP systems (materials and resources planning).

10.2 DEFINING THE RESOURCES

In order to operate a manufacturing company, the following resources must be available to management:

1. finance
2. materials
3. plant, machinery and tools
4. land and buildings
5. people.

Finance

This is the most vital resource. The money needed to operate a business can be obtained in two ways, by borrowing in the financial market or by selling at a profit. Most companies need to employ both of these methods.

Initially, when a company is first established, it is necessary to find investors who are willing to put money into the business on the understanding that they will share in its success by receiving a return on their investment. These investors may be individuals or institutions (such as banks or pension funds), and a number of alternative methods can be used.

- *Debentures or loans* In this case, money is invested on the basis of an agreed annual percentage return at a specific rate of interest over a fixed period. These dividends nust be paid even if the business is making a loss and at the end of the term the original investment must be repaid.
- *Preferred shares* Shareholders in a company are part owners of that company and, as such, share in its success or failure. Dividends are calculated and paid yearly or half-yearly on the basis of the level of profit, and if the company makes a loss it is possible that no dividend will be paid. If the company fails then the shareholders have the right to a share in whatever is left when the assets of the company are realized, subject to certain other priorities. Shareholders may also have voting rights in terms of how the company is run, although this usually means that they can vote to decide who runs the company, i.e. the board of directors. Thus shares can be 'voting' or, in a few cases, 'non-voting'.
- *Ordinary shares* Ordinary shareholders have similar rights to the preferred shareholders, the only real difference being in their level of 'priority'.

Priority really applies only if a company is to be wound up, and it is a statutory requirement that the assets of the company be paid out in priority sequence. This sequence lays down that the 'debts' of a company that is being wound up must be settled first by payment to the tax authorites, followed by debenture and loan holders, preferred shareholders, ordinary shareholders and finally other creditors.

Once a company has started to manufacture and sell it should be able to rely on the income from these sales to finance its operations, and should need additional investment only for expansion. The prudent company sets aside money in a good year to finance the bad years.

Materials

In chapter 9 the requirements for an effective system of materials management were considered in some detail. However, in that chapter the emphasis was on the short-term needs for materials to be available when needed by production — but not before — so that the level of stocks could be minimized.

When considering materials as a resource it is necessary to look further ahead, as is the case with all resources. This means that plans must be drawn up so that the materials requirements can be built in to the business plan in terms of financial commitment.

For example, in certain industries the major raw materials are not manufactured and sold to a customer in the normal way by a supplier, because they are available only in limited quantities and possibly at particular times. These materials are usually classified as commodities and, as such, are bought and sold through commodity markets. This group of items includes natural fibres for the textile industry, some non-ferrous metals, skins and hides, timber, rubber and certain agricultural products used in the food and drinks industries.

Where an industry depends on such 'natural' materials it is essential to ensure that the forward requirements are known so that contracts can be agreed in the 'futures' market. This is a specialized area of expertise which is outside the scope of this book, but management in industries that need to purchase such materials must be aware of market trends so that purchases can be made at economic prices and in adequate quantities. It is worth noting that the same principles apply to foreign exchange if a company needs to purchase goods or services from abroard.

Plant, machinery and tools

These resources require careful planning to ensure the optimum value for money. The starting point should be a forward production plan that sets out, over a period of at least 12 months, what is to be made. This must then be converted into requirements for machines with the associated tooling and services. This can be extremely difficult due to the wide variety of machines available and the wide band of prices. Other decisions also have to be made, for example the following.

- Is it better to have one high-capacity machine or several smaller machines that may be more costly but can provide greater flexibility?
- Is it better to set up an in-house toolmaking unit or to purchase from an outside supplier? Should tools be refurbished in-house?
- Is it better to buy machines and associated plant or to finance them in some other way such as leasing or taking out a loan?

It is therefore necessary to be able to evaluate the plant and machinery requirements in both production and financial terms, in order to obtain the best solution, and this can best be done by setting up a series of computerized spreadsheets so that the various factors (output volumes, sales forecasts, etc.) can be applied.

Land and buildings

In this case the company is faced with a different set of problems. In most cases the costs of these resources will be marginally reduced only if the requirements can be cut by better utilization; these savings could come from a reduction in local taxation if some buildings were unoccupied and there could also be small savings in the costs of heating, lighting and insurance.

These resources can, in fact, be considered to be in the fixed costs category since the need for them will alter only when there is a very marked change in the level of demand. It is therefore pointless to spend too much management time in planning for the optimum use of such resources unless the production plan shows that there is likely to be a very large change in the level of output requirements.

However, it is important to recognize that the term 'fixed costs' is somewhat misleading since, although such costs cannot be reduced per production unit (i.e. by better utilization) they can perhaps be reduced on a global basis. For example, it may be possible to reduce heating costs by better insulation or to offset other fixed costs by subletting some areas to other companies, on a short-term basis, if better utilization leads to reduced space requirements.

People

Unlike land and buildings, people are very much a variable cost and it is therefore important for management to match the 'people requirement' to output demand levels. This requirement for people is related to other factors such as the plan for plant and machinery, and therefore it is essential that resource management combine and integrate all plans into one overall plan.

Two other factors must be considered when looking at the 'people plan': the costs involved in recruitment and training when increasing staffing levels, and the costs of termination when reducing these levels. Ideally, from a production management point of view, a stable work force should cause minimum problems, but unfortunately this is not the case as can be seen from Table 10.1, which shows that no single resource management area can be totally satisfied and that the usual rules of compromise must apply.

Table 10.1 The conflicting demands of resource management

Management preference	Attitudes	Effects
Production	Long production runs	High stock levels
	Minimum variety of products	Limited product range
	Minimal set-ups	Limited product range
Marketing	High product availability	High stock levels
	Maximum variety of products	Wide range
Finance	Minimum capital employed	Low stocks
Personnel	Stable manning levels	Fluctuating stocks

10.3 THE MANAGEMENT OF MATERIALS

The relationship between material costs and levels of output is almost totally 'direct' in that the ratio is virtually constant regardless of the output level. Therefore, if this resource is to be properly controlled and managed, it is essential to employ one of the established techniques such as MRP I (materials requirements planning). This is described in detail in chapter 9 (section 9.4) but, as a reminder, can be defined as shown in Table 10.2.

Table 10.2 MRP I defined

MRP I is a method of determining the materials, components and sub-assemblies required to complete a production plan for a specific period. It breaks down each independent demand item into its component parts and compares these with what is available in stock or on order. Finally, it suggests what orders should be placed (or what actions should be taken on existing orders) to cater for any discrepancies.

MRP I allows management to ensure that all items (such as raw materials and tools) are available when required, while at the same time permitting management to keep stock levels at a minimum. It is therefore a good means of matching the production plan to materials supply, thus ensuring that this resource is properly managed and controlled.

JIT (just-in-time), which is more fully described in chapter 12, is not really an alternative to MRP I in this instance, but can be used to refine and develop the way it operates. This is because JIT is not a planning technique but a method of minimizing waste of all types and at all levels, for example by limiting production to what is actually required at the time rather than what is likely to be required if the production plan is to stay 'on target'.

10.4 THE MANAGEMENT OF PEOPLE

The people employed in a business can be classified as:

1. direct production employees
2. indirect production employees
3. administration and management employees.

Direct production employees

These can be defined as employees whose costs of employment can be directly related to levels of production output. It is thus fairly easy to define how many 'directs' will be required for a given level of output.

Thus, if some form of standard time is used to measure work content for a particular product or product group, in terms of the various skill categories, then it is possible to determine the number of standard hours required to make one unit of the product or group. Figure 10.1 shows how the requirements for direct production employees could be set up. In Table A the unit values for each product group have been calculated. Table B shows how these have been extended to give the total required standard hours for each skill category. These standard hours have then been divided by an annual average output figure (also in standard hours) for each 'skill', to give the levels of manning required for the given outputs to meet the sales quantity requirements.

Indirect production employees

Indirect employees can be defined as all works employees who are not 'direct' and could include works management staff, supervisors, machine setters, quality controllers, materials handlers, storekeepers and others.

In the case of these employees the costs of employment are not directly related to levels of production, although there may be some form of semi-direct relationship that can be applied. For instance, it may be possible to set 'bands' of production levels between which particular

TABLE A

PRODUCT CODE	STANDARD HOURS REQUIRED FOR ONE UNIT OF PRODUCT					
	WELDING	TURNING	MILLING	DRILLING	ASSEMBLY	PAINTING
A	10	15	11	4	12	--
B	21	5	--	16	10	5
C	14	9	3	2	13	6
D	7	17	8	1	15	4
E	17	4	--	3	9	3
F	3	--	7	12	14	6
G	--	10	5	3	7	5
H	19	7	3	--	13	4

TABLE B

PROD. CODE	ANNUAL SALES	TOTAL STD.HOURS REQUIRED FOR GIVEN ANNUAL SALES					
		WELDING	TURNING	MILLING	DRILLING	ASSEMBLY	PAINTING
A	100	1000	1500	1100	400	1200	--
B	220	4620	1100	--	3520	2200	1100
C	150	2100	1350	450	300	1950	900
D	400	2800	6800	3200	400	6000	1600
E	180	3060	720	--	540	1620	540
F	650	1950	--	4550	7800	9100	3900
G	320	--	3200	1600	960	2240	1600
H	525	9975	3675	1575	--	6825	2100
TOTAL STD.HRS.		25,505	18,345	12,475	13,920	31,135	11,740
NO.OF DIRECT EMPLOYEES		13.3	9.6	6.5	7.2	16.2	6.1

BASED ON AN AVERAGE OF 1920 STD.HRS PER YEAR PER DIRECT EMPLOYEE

Figure 10.1 Direct production employee requirements.

manning levels will apply. An example of this approach is shown in Fig. 10.2. This shows that different types of indirect labour change their manning requirements at different output levels, e.g. at up to 50 per cent of standard capacity two supervisors are required but this rises to three between 50 and 100 per cent and to four if 100 per cent is exceeded.

Administrative employees

These can be defined as all non-works employees, and would include most clerical staff, office supervisors and managers covering such departments as Accounts, Sales, Purchasing and Design/drawing. They would also include outside employees such as sales representatives and drivers.

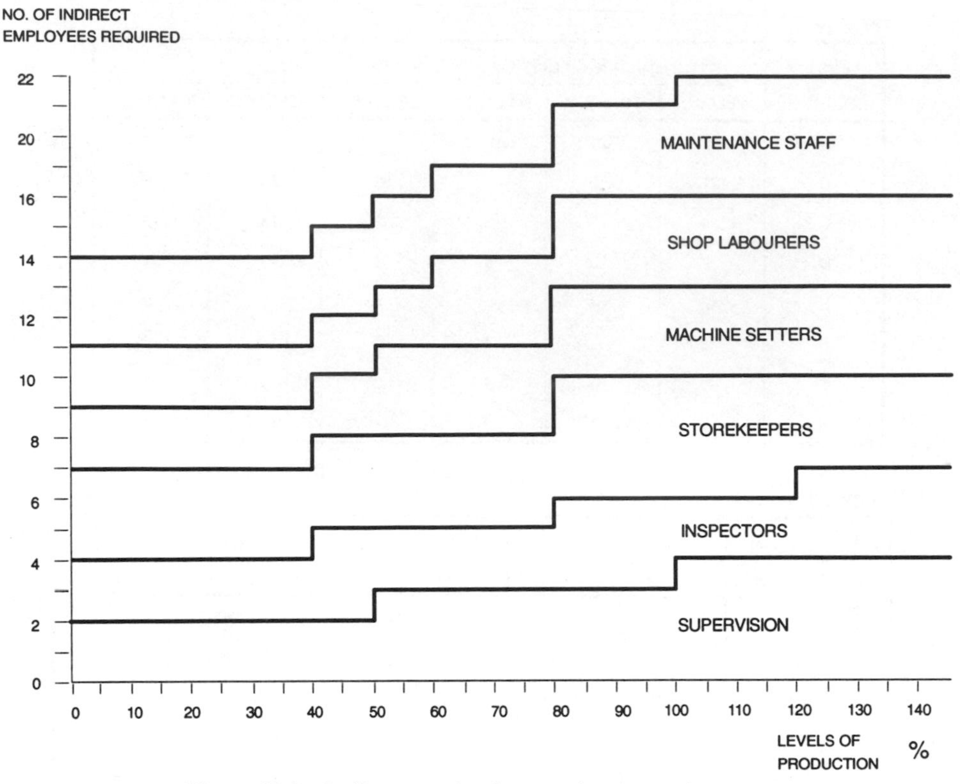

Figure 10.2 Indirect production employee requirements.

In the case of these employees there is no obvious relationship between production levels and manning. In some areas it may be possible to derive some method of calculation, but this often bears no relationship to the work involved. It is therefore better to consider an alternative based on levels of departmental activity in relation to departmental manning. This system, which is similar to the ABC (activity based costing) approach described in chapter 14, allows management to treat such employees as 'directs' (or semi-directs) against their own departmental levels of output.

For example, the levels of manning in the accounts department could be related to the following factors and by observation, a manning formula derived:

- number of sales invoices S
- number of purchase invoices P
- number of people on payroll W.

Thus, for example, the accounts department manning could be found from the formula

$$\text{Manning} = S/60 + P/40 + W/50$$

10.5 THE RESOURCES BREAK-EVEN CHART

The break-even chart as a costing tool is discussed later (chapter 14, section 14.10), but it can also be used in resource planning as Fig. 10.3 shows. This chart shows that, at zero output, there

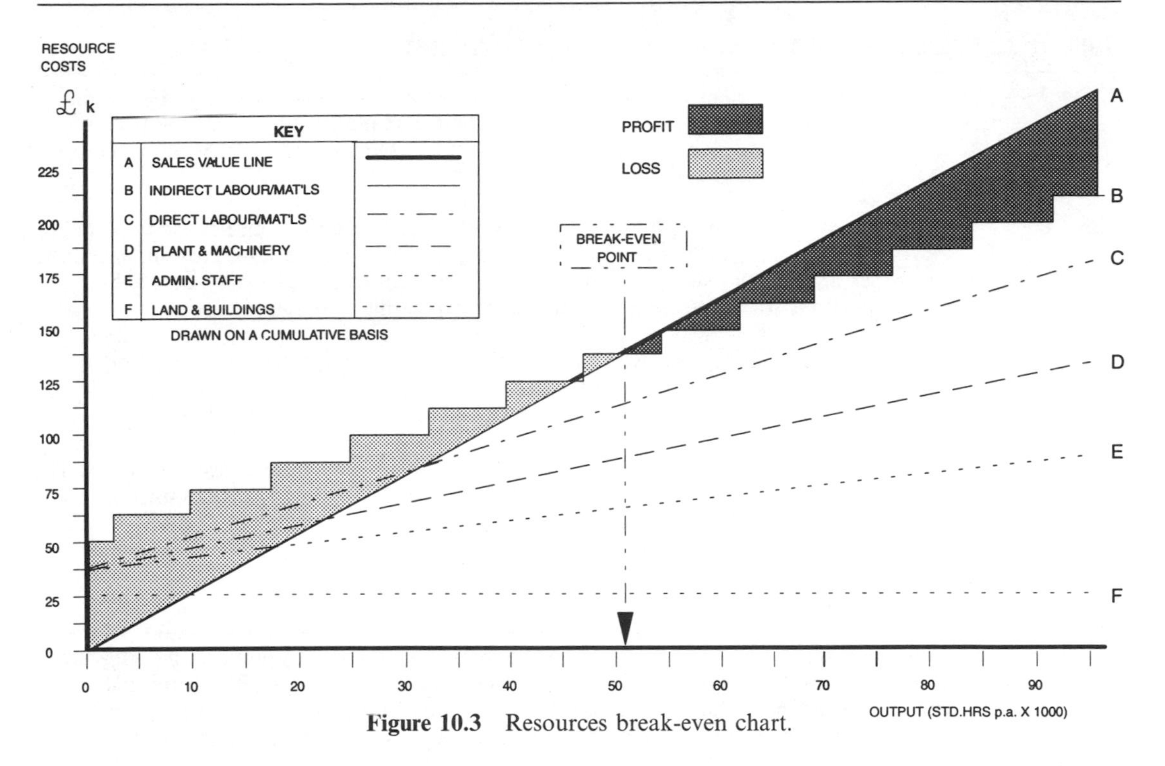

Figure 10.3 Resources break-even chart.

is a resource cost of about £50 000 per year and this rises to about £190 000 when output reaches 90 per cent of planned capacity. At this level, the profit is about £50 000 per year on a sales turnover of about £240 000 (20.8 per cent gross). However, if the turnover falls to about £140 000 (51 per cent of capacity) then there is no profit, and turnover lower than this will mean that sales are being made at a loss.

Figure 10.3 illustrates the need for management to maintain tight control over resource costs and to cut these wherever possible, particularly in a period of low activity.

10.6 THE CLOSED-LOOP MRP SYSTEM (MRP II)

Closed-loop MRP, or MRP II (manufacturing resources planning), is a computer-based system designed to assist management with the preparation of the business and associated plans, and then to monitor these plans so that, if necessary, they can be adjusted in line with the current situation.

The 'father' of MRP II was Oliver Wight, who classified manufacturing companies in terms of their ability to plan their operations and match resources to those plans. These classifications are given in Table 10.3. Wight also specified certain basic criteria by which a company could be measured in terms of MRP achievement; for example, stock accuracy could be measured in terms of the percentage of 'hits' at a stock check. In this context, a hit is where the theoretical stock of an item is the same as (or within an agreed tolerance of) of the actual stock. Wight suggested that a score of 95 per cent hits should be the minimum for effective MRP II operation.

The concept of MRP II can be defined as shown in Table 10.4; it operates on the principle of hierarchical scheduling, whereby a series of interrelated schedules is prepared using a top-down, level-by-level process. This means that plans are first prepared in broad general terms and then, if these check out satisfactorily, in greater detail, for example:

Table 10.3 Classifications of MRP users (source: Oliver Wight, 1981)

Class A	Closed-loop system used for priority planning and capacity planning. The master production schedule is levelled and used by top management to run the business. Most deliveries are on time, inventory is under control, and little or no expediting (progress chasing) is done.
Class B	Closed-loop system with capability for both priority planning and capacity planning. In this case, the master production schedule is somewhat inflated, top management does not give full support, some inventory reductions have been obtained, but capacity is sometimes exceeded and some expediting is required.
Class C	Order-launching system with priority planning only. Capacity planning is done informally with a problably inflated master production schedule. Expediting is used to control the flow of work; a modest reduction in inventory is achieved.
Class D	The MRP system exists mainly in data processing. Many records are inaccurate. The informal system is largely used to run the company. Little benefit is obtained from the MRP system.

Table 10.4 MRP II defined

MRP II closed-loop MRP is a method for determining the resources required to prepare the integrated business, marketing and production plans for a manufacturing company for a specific period. It then allows management to convert these plans into detailed schedules and to check performance against these schedules. Finally, it allows management to react quickly to deviations from the plans in terms of adjustments to the schedules.

- plans based on broad product groupings can be refined down to smaller groupings and then perhaps to individual products
- plans based on long-term planning horizons can be refined down to shorter horizons
- time buckets spanning long periods can be refined down to shorter time buckets
- top management responsibility can be moved down to operational management responsibility.

Thus, the first step in operating an MRP II system is for top management to prepare a *business plan* that sets out the long-term future of the company as a series of financial statements and targets. These should include projected profit and loss, cash flow and investment plans. From this plan a *marketing plan* can be prepared that gives—in fairly general terms—what is to be sold, and what income will therefore be derived to meet the sales turnover figures in the business plan. From this plan a *production plan* can be prepared that sets out—again in fairly general terms—what is to be produced, what resources will be needed to produce it, and what this mix of products and output volumes will cost.

At each stage in this process the plans can be checked against resource availability, since the details of the various resources and what levels of output they can support will already have been set up in the system. This phase of the MRP II process is illustrated in the 'top management planning box' in Fig. 10.4.

10.7 DEFINITIONS IN THE MRP II SYSTEM

In order to operate a satisfactory MRP II system it is necessary to define the data that should be available in terms of resources and plans, as follows.

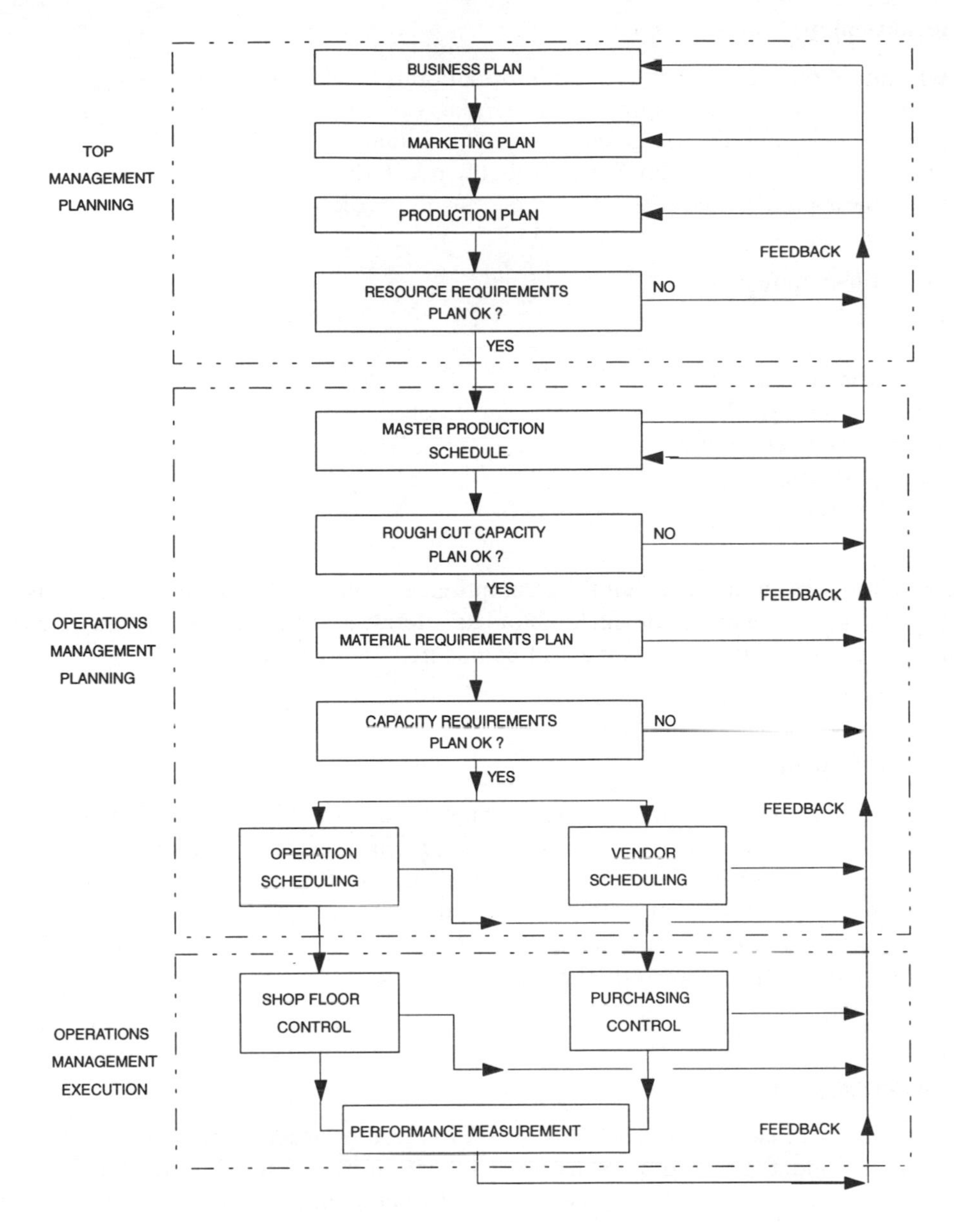

Figure 10.4 MRP (the closed-loop MRP system).

Resource data

Data should be available in the system that can be used to measure the resource requirements against various possible levels and combinations of production and administrative activity. For instance, this information should be so set up that, given that a particular volume of a product group has to be processed in a specified time, the system will be able to calculate the percentage load against capacity for each production process. This is explained in greater detail later in this chapter.

The business plan

This plan should include all the financial information that will set the parameters for the other plans. In many companies the initial business plan can cover a period of up to five years, or possibly even longer in industries such as aircraft manufacture where development time is high, and will be reviewed on a regular basis, probably quarterly.

The business plan should include such data as the projected

- profit and loss statement
- balance sheet
- capital expenditure
- revenue expenditure
- cash-flow statement
- return on capital employed
- dividend payments
- loans and repayments.

In addition, the plan should show all the assumptions that have been made and the basis used to make them. These should include such factors as interest and taxation rates, changes in wage and material costs and the likely impact of competitors' activities.

The marketing plan

The marketing plan should show what the company expects to sell during the life of the business plan, probably split into a series of time periods (time buckets) and markets. It would not be expected to show sales by individual products unless the industry was of the type where each sale has a very high value, for example, large power station turbines and generators. Instead the plan would show projected product sales by groups or families for each time bucket and market. A typical plan of this type is shown in Fig. 10.5.

The production plan

The production plan should show what is to be produced in order to meet the marketing plan, how it is to be made, the rates of production and the resources required. It should also show the projected investment in new plant, machinery and tooling and the likely expenditure on sub-contract work.

The objectives of the production plan are:

- to provide a basis that, if agreed, can be used to break down the product group figures into specific end items (saleable products) which can then be set up in the master production schedule (MPS)
- to provide input for a resource requirements planning (RRP) run so that decisions can be made if plant expansion or acquisitions are needed
- to try to stabilize production by stock-building if the marketing plan shows marked seasonal fluctuations in sales demand.

TABLE A - 1992 SALES PLAN - PERIOD 1

PROD. GROUP	UNIT PRICE		MARKET					TOTALS
			UK	EUROPE	N.AMERICA	S.AMERICA	AFRICA	
A	5.50	QTY.	1500	3000	2500	1000	700	8,700
		£ k	8.25	16.50	13.75	5.50	3.85	47.85
B	3.20	QTY.	3600	2100	800	250	450	7,200
		£ k	11.52	6.72	2.56	0.80	1.44	23.04
C	8.40	QTY.	1800	1400	900	350	500	4,950
		£ k	15.12	11.76	7.56	2.94	4.20	41.58
D	1.25	QTY.	5500	1200	3500	400	1100	11,700
		£ k	6.87	1.50	4.37	O.50	1.38	14.62
E	12.50	QTY.	1200	400	750	200	350	2,900
		£ k	15.00	5.00	9.37	2.50	4.38	36.25
TOTALS		£ k	56.76	41.48	37.61	12.24	15.25	163.34
PERCENTAGES			34.75	25.40	23.00	7.50	9.35	100.00

TABLE B - PLANNED ANNUAL SALES EXPENDITURE - 1992

ITEM	DETAILS			COST £ k
ADMINISTRATION	MANAGER x 1 GRADE A x 2 GRADE B x 5	24.00k 12.00k 8.00k	24.00k 24.00k 40.00k	88.00
ADVERTISING	CAMPAIGN A x 3 CAMPAIGN B x 6 CAMPAIGN C x 4	4.00k 2.00k 3.00k	12.00k 12.00k 12.00k	36.00
SALES PROMOTION	PRODUCT B BROCHURE x 3000 PRODUCT B LEAFLET x 5000 PRODUCT F BROCHURE x 2000	1.80k 1.40k 1.30k	1.80k 1.40k 1.30k	4.50
DISTRIBUTION	AIR FREIGHT ROAD SERVICES A ROAD SERVICES B	12.50k 32.50k 22.50k	12.50k 32.50k 22.50k	67.50
TRAVELLING	SALES REPS. x 3 EXPENSES x 3 CAR LEASING ETC x 3	18.00k 15.00k 4.00k	54.00k 45.00k 12.00k	111.00
			TOTAL ANNUAL COST £ k	307.00

Figure 10.5 Part of a typical marketing plan.

An example of part of a typical production plan is shown in Fig. 10.6. This shows, for one department only and a limited number of product groups, how many units of each group would be made in each period and the number of standard hours that such quantities would represent.

The bar chart shows the work loads that these standard hours would impose on the department: it can be seen that unacceptable overloads would occur in periods 1 and 5. However, the problem in period 5 could be overcome by increasing output in periods 2 and 3 such that stock can be built up to cater for this situation. As far as period 1 is concerned, if stocks have not been built up sufficiently in previous periods it may be necessary to subcontract some of the work.

TABLE A - 1992 PRODUCTION PLAN - FABRICATION DEPT.

PROD. GRP / STD. HRS.	UNITS AND SHs.k	PERIOD 1	2	3	4	5	6	7	TOTALS
A	QTY.	8,700	3,500	5,200	7,500	9,000	6,400	3,100	62,000
3.50	S.HRS	30.45	12.25	18.20	26.25	31.50	22.40	10.85	217.00
B	QTY.	7,200	3,000	4,600	5,900	7,800	5,600	1,800	51,600
4.00	S.HRS	28.80	12.00	18.40	23.60	31.20	22.40	7.20	206.40
C	QTY.	4,950	2,100	3,700	6,400	7,300	4,200	2,600	41,650
5.50	S.HRS	27.22	11.55	20.35	35.20	40.15	23.10	14.30	229.08
D	QTY.	11,700	4,200	7,400	8,300	12,200	5,300	4,800	82,600
2.00	S.HRS.	23.40	8.40	14.80	16.60	24.40	10.60	9.60	165.20
E	QTY.	2,900	1,100	700	1,750	2,800	2,150	1,050	19,150
7.50	S.HRS.	21.75	8.25	5.25	13.12	21.00	16.13	7.87	143.63
TOTALS	STD.HRS.k	131.62	52.45	77.00	114.77	148.25	94.63	49.82	961.31
PERCENTAGES		13.69	5.46	8.01	11.94	15.42	9.84	5.18	100.00

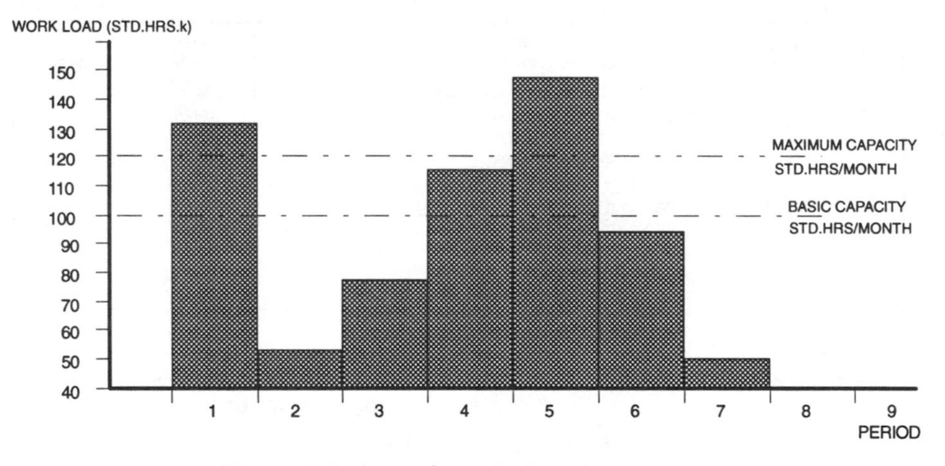

Figure 10.6 Part of a typical production plan.

10.8 RESOURCE REQUIREMENTS PLANNING

If an MRP II system is in operation in a company, it is possible to use the RRP facility to determine the loads on the resources imposed by the production plan, and to compare these with the theoretical capacity. In order to do this it is necessary to set up a series of conversion factors in the computer system such as:

- the number of standard man-hours per unit of a product group
- the number of standard machine-hours per unit of a product group

- the area required for assembly per unit of a product group
- the quantities of the various materials required per unit of a product group
- the storage requirement volume per unit of a product group.

However, if an RRP facility is not available on the main system it is possible to set up the necessary data and calculations on a PC-based spreadsheet. Figure 10.7 shows such a spreadsheet as table A, with the graphics facility being used to present the results in bar chart format as table B. For ease of presentation, this spreadsheet is much simplified. For instance, the data should be split into production periods, and the required production quantities shown, with extensions, for the total values used in the graph (as shown in Table 10.5).

TABLE A: RESOURCES REQUIRED PER PRODUCTION UNIT

	A	B	C	D	E	F	G	H	I	J
1		PER UNIT OF PRODUCT IN GROUP								
2	PROD. GRP.	FABRICATION					MACHINE SHOP			
3–4		MAN HRS.	M/C. HRS.	AREA REQD.	MATL. REQD.	STORE VOL.	MAN HRS.	M/C. HRS.	AREA REQD.	MATL. REQD.
5		S/HRS.	S/HRS.	SQ.MTRS.	TONNES	CU.MTRS.	S/HRS.	S/HRS.	SQ.MTRS.	TONNES
6										
7	A	0.25	0.15	2.00	0.03	0.11	etc.			
8	B	0.36	0.13	1.65	0.05	0.16	etc.			
9	C	0.21	0.20	1.55	0.13	0.21	etc.			
10	D	0.27	0.19	2.05	0.09	0.23	etc.			
11	E	0.32	0.08	2.50	0.14	0.15	etc.			
12	F	0.42	0.17	1.95	0.15	0.21	etc.			
13	G	0,37	0.13	2.09	0.19	0.17	etc.			
14–15	etc.		etc.		etc.					

TABLE B: GRAPHICAL PRESENTATION OF TABLE A DATA

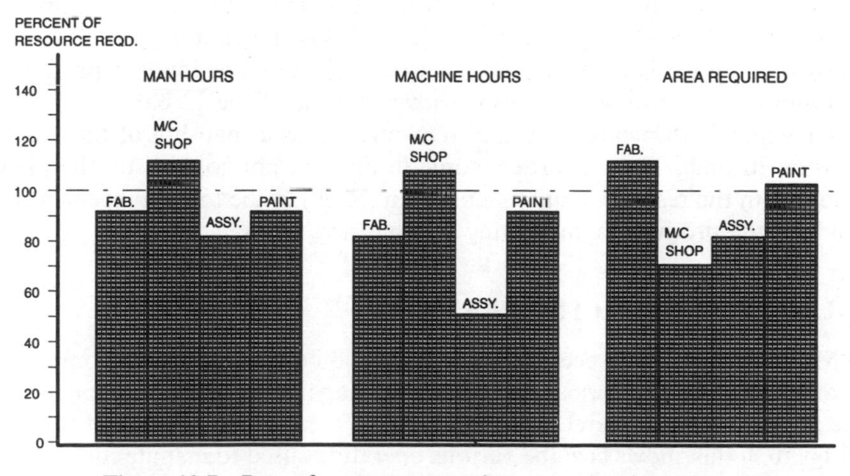

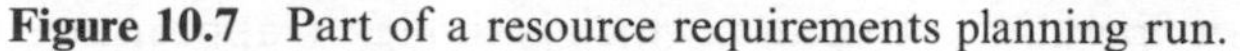
Figure 10.7 Part of a resource requirements planning run.

Table 10.5 Spreadsheet columns for an RRP exercise (period 1—fabrication)

Product group	Quantity required	Manual hours/unit	Total man-hours	Machine-hours/unit	Total machine-hours/unit	Area required /unit	Total area required
A	500	0.25	125.00	0.15	75.00	2.00	1000
B	350	0.36	126.00	0.13	45.50	1.65	577
	etc.		etc.		etc.		etc.

An RRP run determines whether a production plan is feasible, and if the plan is too demanding then it must be amended, either by providing additional resources or by reducing the volumes of production. Equally, if the RRP run shows a considerable under-use of resources then consideration should be given to increasing the planned output levels or reducing the resources available.

If either of these situations occurs, it will probably be necessary to amend the marketing and business plans in line with the changes made to the production plan. As so often happens in life, the answer lies in a compromise since the conflicting demands of finance, marketing and production are unlikely to be met in full.

This iteration process is time-consuming if a full MRP II system is not available, but it is essential within MRP II to ensure that all three plans are in balance at the 'top management' level before moving to the next level down (operations management).

10.9 THE MASTER PRODUCTION SCHEDULE IN MRP II

The MPS is a statement of what a company plans to manufacture in detail, i.e. at individual product level. It is therefore different from the production plan, which was at product group level. In MRP II, the MPS should be based on the agreed production plan by breaking down the product groups into separate products on an agreed ratio or percentage basis: this is shown in operation in Fig. 10.8.

The MPS provides the basic data from which MRP I (materials requirements planning) operates and initially will contain only the information obtained by breaking down the production plan. However, once it has been agreed, firm orders from the sales order processing (SOP) system can be fed in to replace the planned orders. This is shown in Fig. 10.8. It should also be noted that each bar in the bar chart is in fact a series of blocks, one for each product for planned orders, and another for firm orders: this is shown for one of the 12 bars.

A master schedule should be viewed as dynamic, i.e. as a snapshot of the order position at any given point in time. Thus, as orders come in that are not covered by the plan, it may be necessary to re-run the resource requirements plan (RRP) in the top management part of MRP II and to adjust the production, marketing and business plans.

10.10 ROUGH CUT CAPACITY PLANNING

When the Master Schedule has been prepared it should be run against the computerized RCCP facility in order to identify overloads or underloads against production capacity. This is normally done at product group level. In many systems, these capacities are set up as 'profiles' as shown in Fig. 10.9; this shows how the various operations used to manufacture a product group have been set out in an operations structure, which is similar in appearance to a bill of materials

PROD. GRP.	RATIO %	PRODUCT	QTY.FOR P.1		QTY.FOR P.2		QTY.FOR P.3		QTY.FOR P.4			ANNUAL TOTALS
			GRP.	PROD.	GRP.	PROD.	GRP.	PROD.	GRP.	PROD.		
	10	A.10		48		40		52		60		615
	15	A.20		72		60		78		90		850
A	25	A.30	480	120	400	100	520	130	600	150		1550
	30	A.40		144		120		156		180		1800
	20	A.50		96		80		104		120		1180
	15	B.10		111		105		128		138		1300
	35	B.20		259		245		297		322		3070
B	30	B.30	740	222	700	210	850	255	920	276		2875
		etc.		etc.		etc.		etc.		etc.		etc.

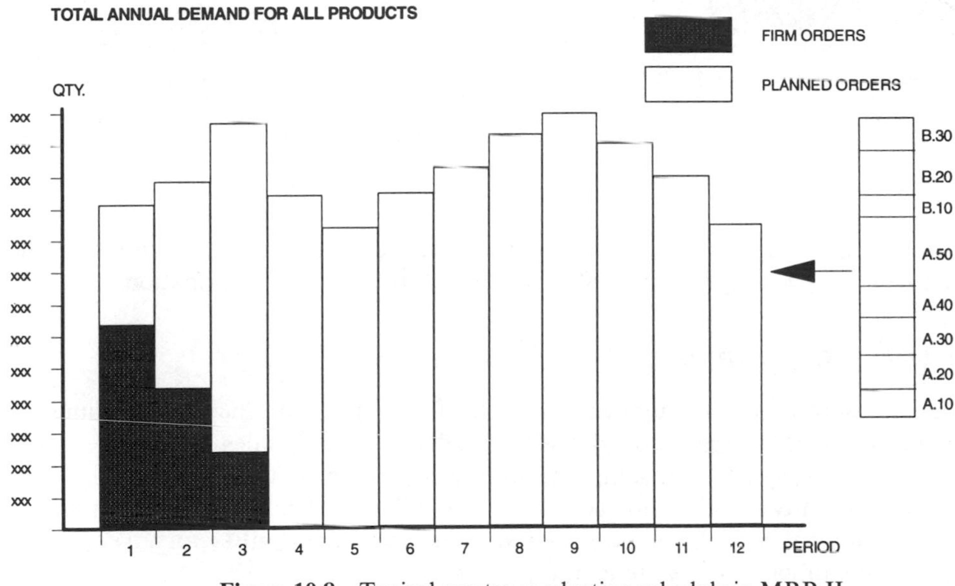

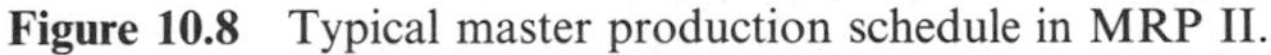

Figure 10.8 Typical master production schedule in MRP II.

structure. This has then been converted into the profiles which show the load one unit of the product group would impose on the various production departments.

From this information, the system can be used to prepare a projected labour hours chart for each department: a typical example is shown in Fig. 10.10. At this stage the loadings on individual work centres or operators are not known, since the RCCP only works at department and product group level. However, as soon as a material requirements planning (MRP I) run

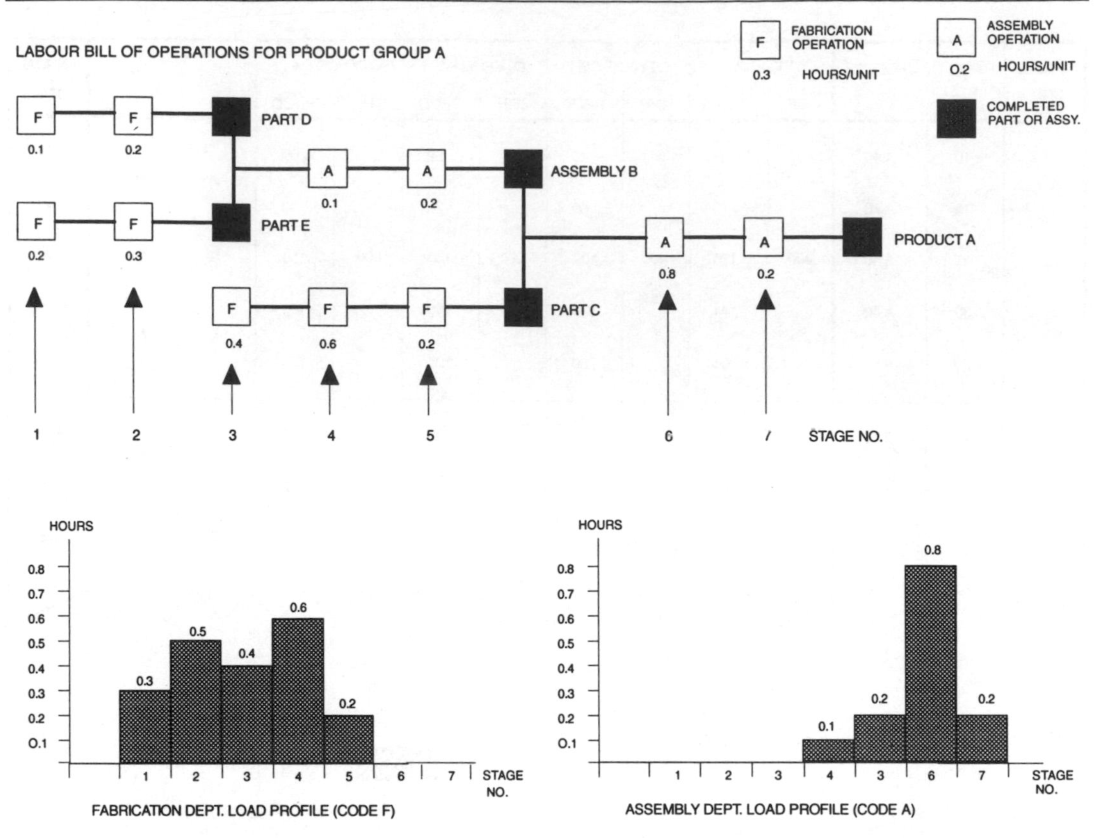

Figure 10.9 Rough cut capacity planning.

has taken place, and planned works orders have been created, it is possible to run the detailed capacity requirements planning (CRP) facility, which will provide this information.

10.11 CAPACITY REQUIREMENTS PLANNING

The CRP facility allows the planning department to 'fine tune' a production programme by highlighting any work centre that is overloaded or under-utilized. It does this by taking each open works order in the system (including unconfirmed orders from an MRP I run if desired) and applying a series of calculations to each, as shown in Table 10.6. This table gives a rather simplistic version of what really happens, since it ignores three important factors, namely work centre capacity, type of capacity planning and priority.

The capacity of a work centre in standard hours can be used in one of two ways, depending on the type of capacity planning being used, since there are two types of capacity planning, as follows.

- *Finite* In this case the system will not allow a work centre capacity to be exceeded. Thus, work is loaded into the work centre for the various time periods, in priority sequence, until the capacity is reached. The system will then load any work left over into the next time period, and so on.

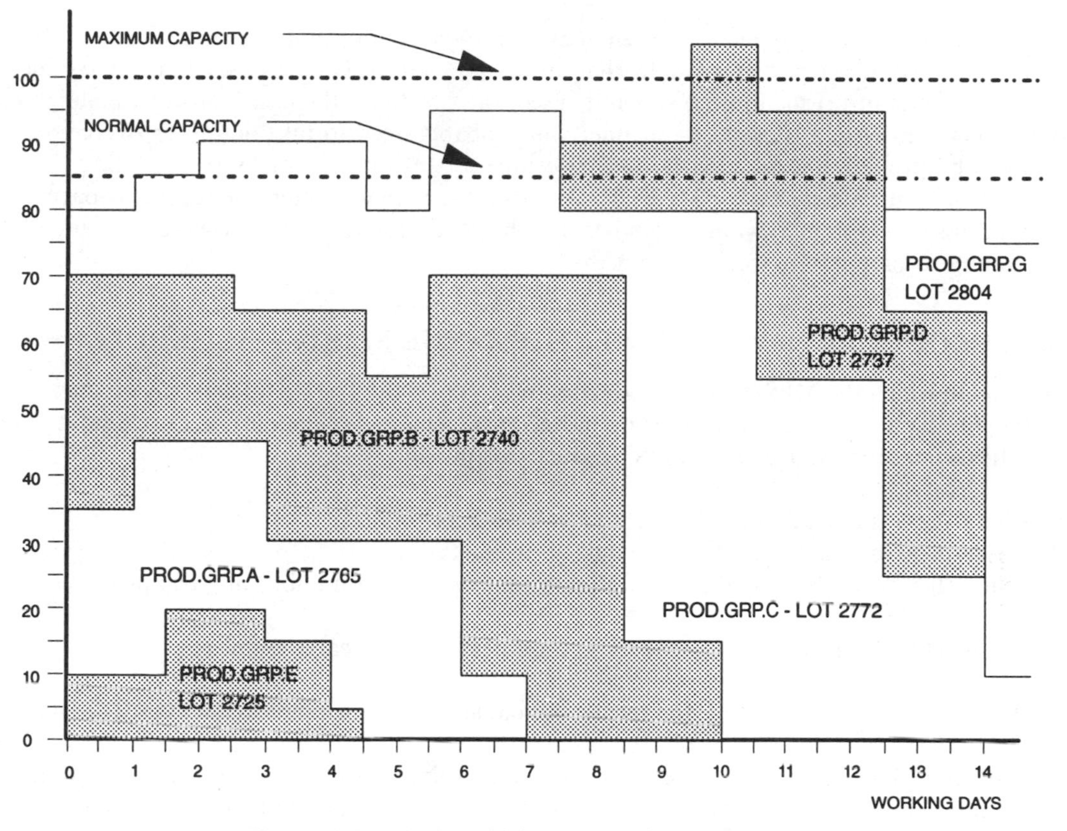

Figure 10.10 Projected labour hours from RCCP run.

Table 10.6 Capacity requirements planning sequence (for each open works order)

Stage	Operation
1	From product routeing and works order data, calculate a start date for each operation based on back-scheduling from the order-due date
2	Use the same data to calculate the number of standard hours that will be needed at each work centre for each operation
3	Combine the information found in stages 1 and 2 to set up a file that contains a record for each operation on each works order: each of these records will contain a works order number, a work centre, an operation start date and an operation time in standard hours
4	Sort this file by works order within start date within work centre such that for each work centre there is a load in standard hours in total and by works order for each time period (day or week).
5	Print out a report showing this information such that a user can see that on any given work centre there is a load of *h* hours due to specified works orders in any given time period

- *Infinite* In this case the system ignores work centre capacity such that an operation is always loaded into its scheduled start period. This means that a work centre could be loaded to 150 per cent or more of its capacity.

Both capacity planning types have their uses in providing information to the planner. If the finite system is used, reports may show that certain jobs will be late because of late starts due to potential overloads; if the infinite system is used, the overloads themselves will be highlighted. But, whichever system is used, the planner will probably need to take action to overcome the problem by providing extra capacity (e.g. overtime or subcontract) or by rescheduling.

In companies where both systems are available, the planners often use infinite capacity for their initial planning and 'what-ifs' and then, when the problems have been sorted out, finite scheduling to prepare the working plan.

10.12 PRODUCT CHARACTERISTICS AND THE MASTER SCHEDULE

The way in which the MPS (master production schedule) should be set up depends on the nature of the business and the characteristics of the products. It is generally accepted that three classifications can define a company's products for this purpose:

A. where there is a small number of standard products
B. where there is a large number of standard products
C. where there is a large number of configurations on a small number of basic products.

Figure 10.11 illustrates and describes these classifications in greater detail.

A. With this class of products, the master schedule should include all the standard saleable items.
B. With this class of products, the master schedule should include the product groups and a 'planning bill' would then be used to break these down into individual items. This bill would be based on market research or historical records (or both), and would define the proportion of the group that would be expected to be taken by each product.
C. With this class of products, the master schedule should include the basic products, plus a series of modules covering the most popular configurations (a Pareto approach may be helpful in choosing these). The production plan would aim to put into stock sufficient basic products and optional 'add-ons', and then a futher plan would be issued at the final assembly stage.

10.13 PLANNING HORIZONS AND TIME BUCKETS

In order to maximize the utilization of resources it is necessary to plan sufficiently far ahead not only to cater for the immediate production requirements but also to allow management to make decisions at the strategic level. Thus, the planning horizon within a master scheduling system should be at least three months longer than the manufacturing/purchasing lead times to allow decisions to be taken on such issues as

- make or buy?
- economic batch sizes?
- best supplier?

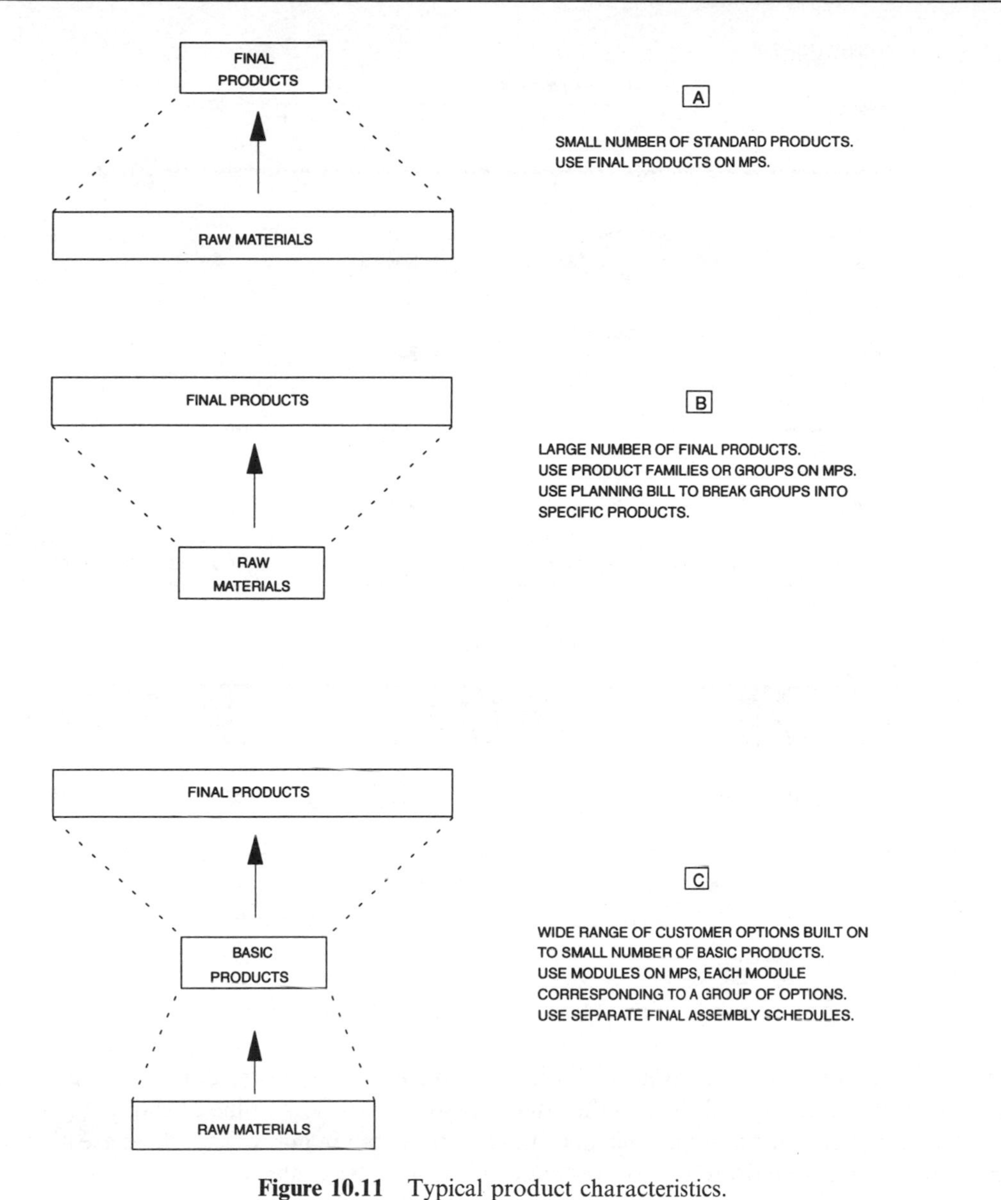

Figure 10.11 Typical product characteristics.

- employment policies?
- capital equipment policies?

Figure 10.12 shows a typical planning horizon and an arrangement for splitting it down into time buckets.

The planning horizon is a device that can be set up in a computerized production planning system to enable the users to restrict forward vision, and does not necessarily have to be applied at all levels. For example, it should not be used at the top level in MRP II where resource planning could involve a new development project with a time-to-market of two or three years. However, at the operational level it may be necessary to ignore certain long-term sales orders or

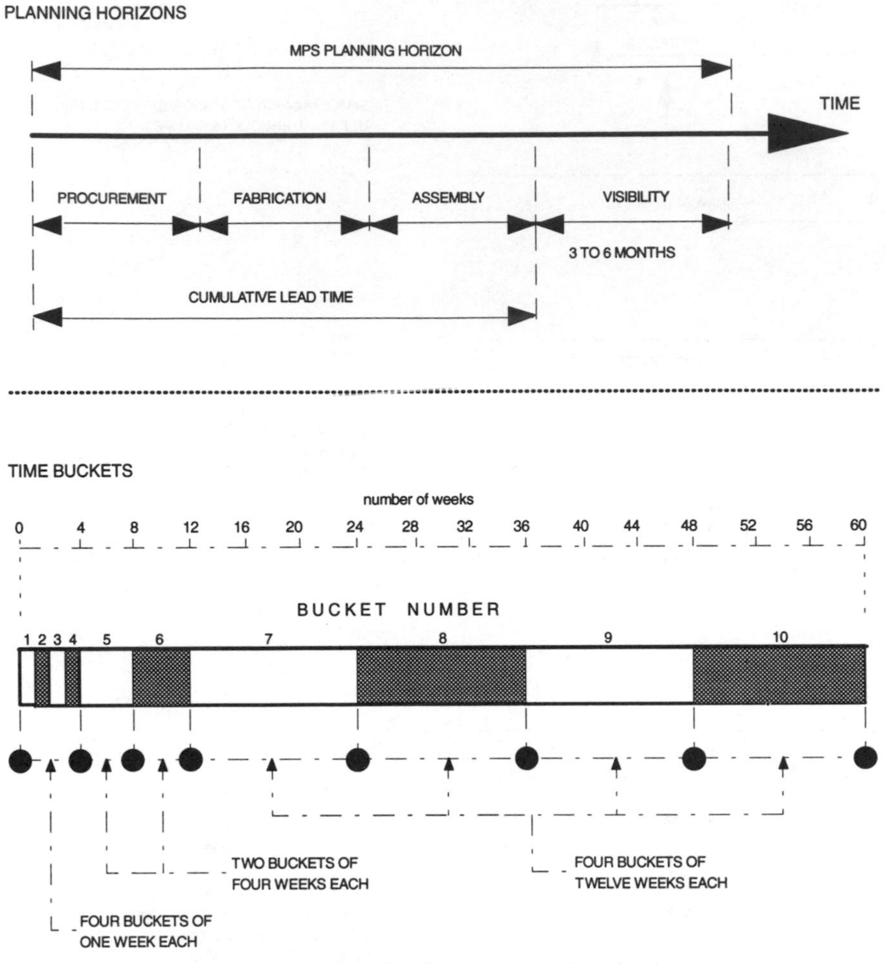

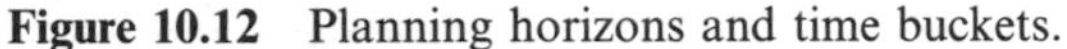

Figure 10.12 Planning horizons and time buckets.

forecasts in the master production schedule when planning, to ensure that efforts are concentrated on definite confirmed work rather than possible work some time ahead.

All companies work on some system of time buckets, but in many cases these are all of equal size, often one week. However, variable-length time buckets, as shown in Fig. 10.12, should be used by a company that has a long planning horizon. This allows such a company to follow the sound management principle of taking a broad view at the strategic stage of planning and then narrowing that view when tactical decisions have to be made.

10.14 RESOURCE MANAGEMENT IN A JIT ENVIRONMENT

Up to this point in this chapter, the 'tools' that have been examined for planning and managing resources have been based on MRP concepts, both MRP I and MRP II. However, a number of companies have now decided to follow a different approach using the JIT (just-in-time) or OPT (optimized production technology) philosophy, and do not use conventional MRP II for resource planning. The full JIT approach is considered in detail in chapter 14, but it is necessary to look at it now from the point of view of resource management.

There are many misconceptions about JIT, and it is therefore necessary to define what it really is, and what it is not; Table 10.7 gives this definition in outline. In the industries where it is applicable, JIT can be used to manage resources because it is a waste elimination system, as the following examples show.

1. *Finance* JIT ensures that the costs of holding stocks are minimized, and that production work is carried out only when there is a definite demand. It aims to reduce set-up times (which cost money and produce nothing) and this allows smaller batch sizes to be economic. JIT also cuts the costs of indirect work in the production areas (fewer storekeepers and setters), but may increase these costs in other ways, by demanding more maintenance and tooling manufacture. However, these extra costs are normally offset by the savings in other areas.
2. *Materials* JIT can help to maximize the effectiveness of materials management in several ways. Firstly, it can reduce the quantities purchased to the minimum, although this may involve some additional cost due to greater ordering frequency. Secondly, it can build partnerships with selected suppliers, so that the quality of goods and services is greatly improved. Thirdly, it aims to minimize the wastage caused by defective materials and products by improving production methods.
3. *Plant, machinery and tools:* JIT is not obsessed with plant utilization to the extent that machines are run to produce items for which there is no immediate need. However, high levels of utilization are achieved by minimizing setting times, by improving plant and machine layout, and by ensuring that breakdowns are minimized by planned preventive or total productive maintenance. JIT recognizes that 'there is no gain without pain' and that to ensure fast changeovers it may be necessary to duplicate some tooling, in order to be able to 'pre-set' for the next job while the current job is running.

Table 10.7 An outline of the JIT philosphy

JIT is concerned with producing and/or stocking only the necessary items in the necessary quantities at the necessary times, and the JIT concept requires

- deliberate lowering of inventory
- streamlining of material flow
- reduced manufacturing lead times
- exposure of manufacturing problems to resolve them permanently.

JIT comprises the following elements:

- management commitment and employee involvement
- lowering of inventory to expose problems
- reductions in batch sizes and set-up times
- reduction of the number of suppliers
- reversal of traditional customer/supplier relationships
- forward visibility
- effective manufacturing layout
- total preventive maintenance
- timely and effective planning information.

JIT is not a panacea for all manufacturing problems, it is not applicable in all types of industry and it is certainly not a 'stockless' system.

4. *Land and buildings* JIT is not a long-term planning tool and cannot therefore be used to determine the land and buildings requirements for the future. However, by minimizing stocks and cutting throughput times it can contribute to the effective use of these resources, thus allowing management to plan for the future needs on the basis of minimum requirements.
5. *People* The JIT concept requires involvement at all employee levels if it is to succeed. It cannot be imposed from above and can only be implemented with the full cooperation of a well-trained and motivated work force. Thus, the JIT approach requires that the people resource is used to the maximum effect, and encourages participation in improving individual, group and company performances. These improvements are made not by working harder, but by working more effectively — by cutting out waste.

It is thus apparent that JIT has much to contribute in terms of making the most of existing resources, but it is not a suitable tool for planning what resources should be made available in the future. It will therefore be necessary for a JIT-based company to use other tools for this purpose, such as RRP (resource requirements planning), which is described in section 10.8 of this chapter. RRP can be set up without the need to operate a full MRP II system, as shown in the example in section 10.8, by using a specially designed spreadsheet.

10.15 RESOURCE MANAGEMENT IN AN OPT ENVIRONMENT

Unlike JIT, which is a set of tools and techniques for exposing problems, OPT (optimized production technology) is a production system which recognizes that there can be process problems at production bottlenecks. However, both OPT and JIT are 'pull' systems based on the need to meet demand (from the despatch end of the production line for JIT and from the bottleneck operation for OPT).

The main concept of OPT is that manufacturing should be a money-making system with a fair return on the investment, as shown in outline in Table 10.8. There are therefore three simultaneous aims which can be used to measure effectiveness, and these are related to key resources as follows.

- *Throughput* This is the rate at which money is generated by selling the goods produced. The first OPT objective is to maximize throughput.

Table 10.8 An outline of the OPT philosophy

1. OPT is a manufacturing planning and control system based on the work of a team of four Israelis led by Dr Eliyahu Goldratt.
2. It is based on the assumption that the goal of any manufacturing organization is to make money.
3. It recognizes the existence of two manufacturing phenomena—dependent events and statistical fluctuations.
4. This means that production processes rely on the completion of proceding processes and that process times fluctuate around an average.
5. It is therefore argued that the capacity of a plant is unbalanced and that bottlenecks must exist. Thus material should be put into production only at the rate at which it can be consumed by the bottleneck process.
6. Furthermore, the production on the bottleneck operation should be protected by a buffer stock to ensure that production is never stopped by shortages.

- *Inventory* Money is invested in purchasing items for resale through the system. Inventory is that which has been paid for but not yet sold; the second OPT objective is to minimize inventory.
- *Operation expenses* These are the costs of turning inventory into throughput; the third OPT objective is to minimize these costs.

This approach runs counter to the traditional concept of machine utilization, which aims to keep machines running on the basis that an idle machine is costing but not earning. The third OPT objective makes this clear by recognizing that running a machine to produce goods that are not required is likely to be more costly than letting it stand idle. In fact, OPT principles state that only at the bottleneck process is high utilization a virtue; at all other processes it may well be a vice!

Unlike JIT, OPT can be computerized, and it can be more clearly defined in terms of a set of rules, as shown in Table 10.9. Conventional scheduling generally assumes the opposite of these 10 rules. The OPT approach attacks these assumptions; this can sometimes cause problems in an MRP system. OPT accepts the JIT philosophy of manufacturing, but formalizes the planning and scheduling functions; in other words, the materials requirements parts of MRP are still required, but are combined with many of the shopfloor practices and attitudes of JIT.

Table 10.9 The 10 rules of OPT

1. The utilization of a *non-bottleneck* resource is not determined by its own capacity; it is decided by some other constraint in the system.
2. The *activation* of a resource is not the same as its utilization.
3. Time lost at a *bottleneck* operation is time lost for the total production system.
4. Time saved at a *non-bottleneck* operation is an illusion.
5. *Transfer batches* between operations may not, and often should not, equal the process batches.
6. The *process batch* size should be variable, not fixed.
7. *Capacity* and *priority* should be considered simultaneously, not sequentially.
8. The unexpected is not unknown. Its damage can be isolated and minimized.
9. Unlike in many other systems, *plant capacity* should not be balanced. It is the *flow* that should be balanced.
10. The sum of the local optimization factors is not equal to the optimization of the whole system.

There is often some confusion between JIT and OPT, but there should not be. JIT is a comprehensive company-wide system which embraces many techniques, whereas OPT is basically a production planning and scheduling system. These differences can best be illustrated in diagrammatic form as shown in Fig. 10.13. It is therefore apparent that, although OPT is a system that aims to optimize the use of resources, it is not a medium- to long-term planning tool and it cannot be used to prepare the sort of plans that are fundamental to MRP II.

PROBLEMS

1. You have been asked to suggest methods for linking the requirements for certain indirect production employees to production output (which is measured in standard hours) so that these employees can be included in the resource requirements plan. Suggest how you might approach this task. A number of suggestions have already been made (see below). You are asked to give your comments on these.

THE JIT CONCEPT

In a JIT System throughput is dictated by the demand for finished goods.

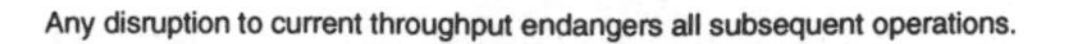
Any disruption to current throughput endangers all subsequent operations.

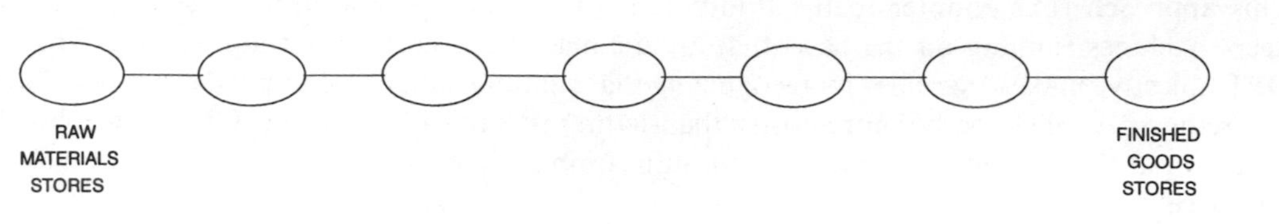

THE OPT CONCEPT

In the OPT System throughput is dictated by the major capacity constraint.

All resources prior to the constraint work to feed the buffer.

The rate of flow from the buffer dictates the flow of raw material into the system.

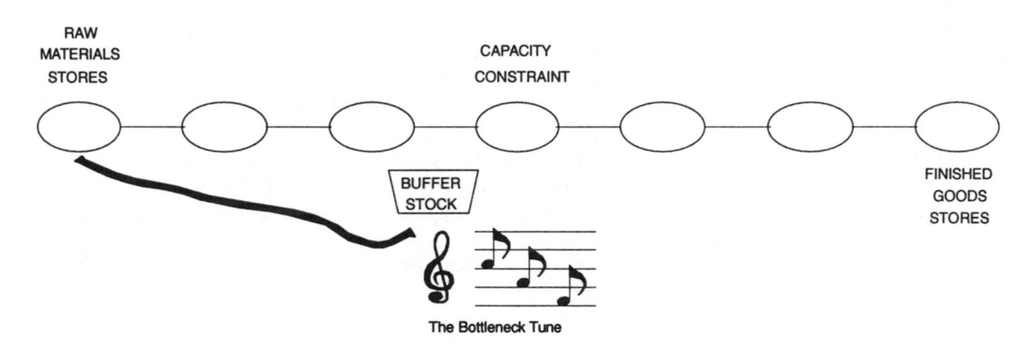

Figure 10.13 The JIT and OPT concepts compared.

(a) Toolmakers
- (i) based on the number of tool changes per year
- (ii) based on the number of tool sets in use
- (iii) based on the number of works orders per year
- (iv) based on the machines using tool sets.

(b) Supervisors
- (i) based on the number of standard hours per year
- (ii) based on the number of work centres
- (iii) based on the number of direct employees

2. It has been suggested that the sales department is overstaffed in relation to the activities involved. You are asked to investigate this and to suggest some means of relating the number of staff (including outside sales reps and delivery drivers) to the various activities that take place. Prepare a brief report showing which activities you consider to be relevant (and which not), and relate these to the various departmental functions.

3. The company for which you work is installing an MRP II system and as a part of this project you, as purchasing manager, have been asked to prepare a purchasing plan which will form a part of the production plan. Suggest what should be included in this purchasing plan in terms of activities and costs. (You may wish to use the marketing plan in Fig. 10.5 as a guide.)
4. Your company manufactures a wide range of saleable items and therefore sets up its master production schedule on the basis of product groups rather than individual products. Suggest a suitable basis for grouping the products. Also, suggest how the ratios for individual products within a group could be determined for use in the planning process.
5. In Fig. 10.11, three types of product characteristic are shown. Suggest not more than three industries and their products for each of the three types that would match the characteristics described.

FURTHER READING

Browne, J., Harhen, J. and Shivnan, J., *Production Management Systems: CIM Perspective*, Addison-Wesley, Wokingham, 1988.

Goldratt, E. and Cox, J., *The Goal* (revised edition), Gower, London, 1989.

Johnson, K., *Implementing OPT*, IFS Publishing, Bedford, 1990.

Smith, S. B., *Computer-based Production and Inventory Control*, Prentice-Hall, Englewood Cliffs, NJ, 1989.

CHAPTER

ELEVEN

THE MANAGEMENT OF AUTOMATION

THE LINK

In order to remain competitive in a highly competitive world, a manufacturer must not only make the best use of the resources available, but also reduce costs, increase productivity, and improve quality.

Investment in new plant, in people and in methods is the best means of achieving these objectives, and this means some form of automation.

The logic behind automation is based on the strengths and weaknesses concept. The strengths of a machine are that it does not suffer from the boredom of repetition, does not get tired (and thus make mistakes), does not need holidays, and does not have 'bad days'. Its weaknesses are that it cannot think for itself, cannot make decisions (unless pre-programmed) and has limited flexibility.

The strengths and weaknesses of people are the exact opposite to those of a machine. It is therefore logical to give the machines the tasks that are boring, repetitive and tiring and thus to release people to do what they can do best, i.e. to apply themselves to problems that require a reasoned approach. Thus, automation should be seen not as a means of getting rid of people, but rather as a common-sense approach that maximizes the strengths of machines and people and minimizes the weaknesses.

11.1 INTRODUCTION

The trend towards automation in manufacturing has led to a number of misconceptions. The most common of these is that it will be possible within the next few years to eliminate people totally by using robots.

This chapter tries to define automation and its place in the real world (rather than the world of the science fiction writer). This is not to deride the latter (after all, Arthur C Clarke defined how communication satellite systems could work over 50 years ago!), but merely to consider the practical applications that must be justified in financial terms.

The chapter therefore starts by examining the various levels of automation, and continues by looking at the available technology and its potential applications. The roles of the various types of automated equipment (robots, CNC machines, vision systems, etc.) are considered separately, together with the computer-based equipment is needed to control them.

The chapter concludes by considering the practicality of a fully automated production unit in financial terms, and the need for people involvement in such a unit.

11.2 A DEFINITION OF AUTOMATION

Automation is defined as 'The automatic control of the manufacture of a product through successive stages; the use of automatic equipment to save mental and manual labour' (Concise Oxford Dictionary). Thus, any piece of equipment that does not require, for some part of its operation, any human input can be defined as having some automation characteristics. However, the range is wide, from a simple limit switch on a machine that stops it automatically when it reaches a certain point to a fully automated manufacturing cell that requires only occasional attention.

It is therefore necessary to try to define these levels, and this can be done by use of a spectrum (Fig. 11.1), a form of which was used earlier to define company types. This Fig. includes a graph indicating for each level a percentage value, and this value can be said to be equivalent to the percentage of 'direct' operator time that has to be spent in attendance at a machine.

In other words, it is possible to define the degree of automation in terms of direct attendance time. However, this ignores the time that must be spent in 'indirect' work, e.g. on change-overs or computer programming, and this type of indirect work increases as the direct work decreases. This aspect will be considered later when the justification for automation is being examined.

Automation equipment can best be categorized in terms of the functions that can be performed and the areas where it can best be employed. The next two sections look at these in detail.

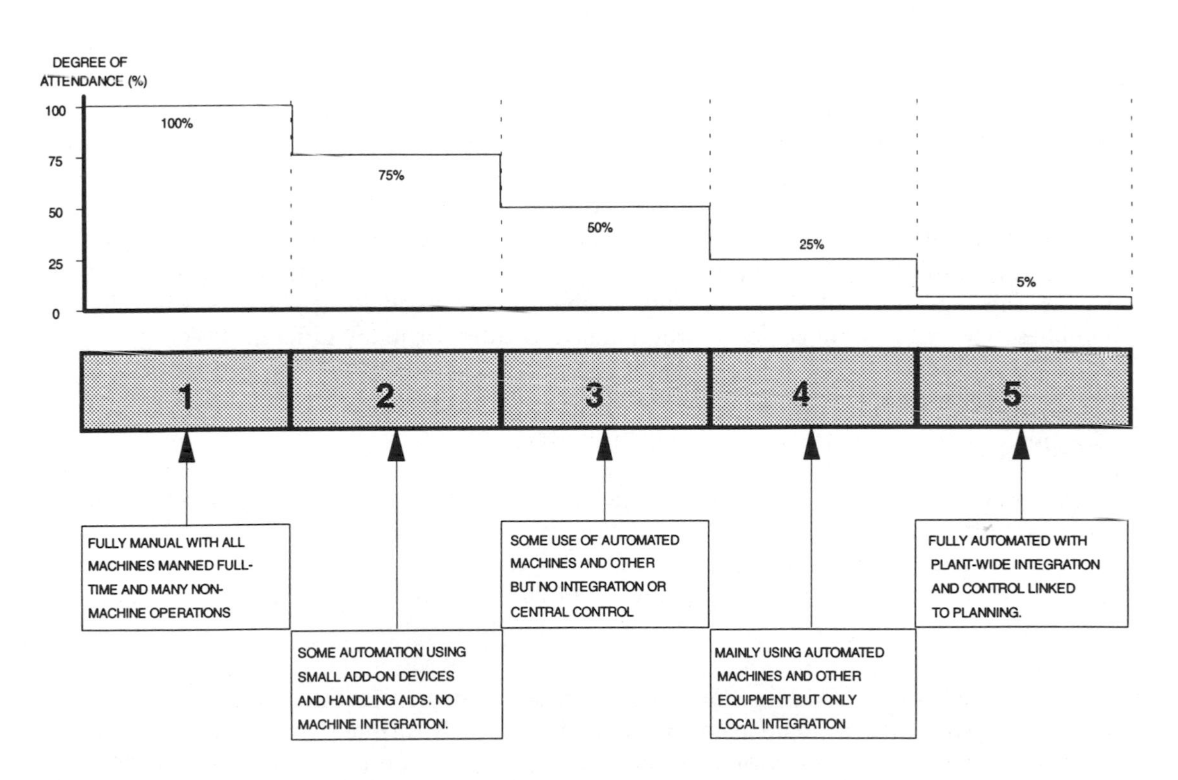

Figure 11.1 The spectrum of automation levels.

11.3 AUTOMATION EQUIPMENT—FUNCTIONS

Automation is, by definition, the transfer of work from people to machines and equipment that can imitate human behaviour, and in some cases improve on it. For instance, machines can handle heavier loads, can measure more accurately, do not take days off, and need no sleep! Four basic functions can be performed by automation machines and equipment, which are based on human capabilities:

1. manipulating
2. controlling
3. measuring
4. moving.

Manipulating

The manipulation function is often combined with the moving function, but for the purposes of this book it is easier if the two are dealt with separately, so that manipulation covers such functions as turning, lifting, lowering, gripping, releasing and positioning. Manipulation equipment imitates some or all of the movements that can be performed by a human, using shoulder, arm, elbow and wrist movements, with a device at the end of the 'wrist' that can grip a tool or component and therefore imitates a hand.

It is not true that all manipulation devices are robots, although a good many of them are. In his book *Dynamic Factory Organisation* (1992), Alastair Ross defines a robot as 'A reprogrammable manipulator, capable of moving various payloads, consisting of parts or tools, in a controlled fashion, to perform a variety of tasks.' This shows that the one factor that differentiates a robot from any other type of mechanical manipulator is the flexibility that comes with the programming facility. This means that a robot can be used for many different tasks, whereas a mechanical manipulator is designed for one specific task.

For example, a welding manipulator could be designed to hold a number of sheet steel components in the correct locations, and to manipulate this 'assembly' so that it is presented to the operator, or machine, for the necessary welds to be made. However, such a unit could not be used for any other type of work, although it could be designed to be used for a variety of assemblies, whereas a robot could be programmed to cater for many different tasks.

Controlling

This category probably covers the widest range of items, from small control switches to factory-wide computer networks. However, they all have in common their ability to control part of a process, a complete process or a sequence of processes. In the context of automation, control can mean to manage, to restrain, to regulate, to authorize or to identify. Thus, a control device or system can

- *manage* a process by ensuring that the correct sequence of operations is carried out; for example, a sequence of component insertions into a PCB (printed circuit board) whereby each component will be drawn from the appropriate location, positioned correctly and pressed into the base board.

- *restrain* a process by ensuring that a machine stops when certain pre-set conditions are not present; for example, it will stop a thread-tapping tool from proceeding if the hole to be tapped is missing or is blocked by a broken drill
- *regulate* a process by making adjustments if measurements show that the process is going out of control; for example, if a machine tool is linked to an automatic measuring machine, then the control system will adjust the machine tool settings in accordance with the readings from the measuring machine
- *authorize* a process if the correct conditions are present; for example, a user can have access to a machine or area only if the correct identity card is presented to the control unit (in this case a card reader): this is the reverse of the 'restrain' function
- *identify* an item or condition in order to ensure that the correct process takes place; for example, an identity marker (e.g. a barcode or electronic signalling device) can be attached to a car body such that, when it arrives at the spray painting process, it is identified as a particular model to be painted in a particular colour.

In other words, control devices or systems are designed to automate the human ability to make decisions based on observation and experience, and to initiate the necessary actions to allow those decisions to be implemented.

Measuring

This category covers all the devices that measure dimensions or conditions and, from these measurements, signals can be sent to control devices to take the appropriate action and/or to data storage units (e.g. computers) to record the values and other data. Such devices are now available in many forms, and measurements can be in terms of size, temperature, pressure, composition (e.g. gas analysis), weight, surface finish, hardness, quantity, etc. In fact, it is difficult to think of any feature that cannot now be measured!

Measuring devices are used in automation systems to replace the human ability to compare a reading with a standard scale, e.g. a linear scale or a thermometer, but in some cases they can carry out measurements that the human brain could not easily handle. For example, a measuring device can determine the exact location of the tip of a tool in a three-dimensional space relative to a datum point. Such an ability is extremely useful when a CNC (computer numerical control) machine is being programmed or is operating.

These devices can also be used for other compound measurements (measurements that combine more than one factor). For example, in the leather processing industry it is necessary to measure the area of an irregular shape, since this information is used to invoice the customer; such a measuring machine can also mark the area (in defined units) on the back of the skin or hide, for customer checking and work planning. This 'measuring device' category should include equipment that imitates the human senses of vision, hearing, smell, taste and touch, as the following examples show.

- *Vision* Vision devices can 'see' an object, compare it with a standard, and pass on/record the variations. They can also 'read' data and act on the information thus obtained (e.g. a barcode on a product which can then be used to sort by product type or destination).
- *Hearing* In this case the device can locate a sound and compare it with sounds in its memory. This attribute is useful for identifying faults on a machine (e.g. noise from a worn bearing or gearbox), for fault diagnosis or as an alarm system.

- *Smell/taste* These are been combined since such sensing devices can identify an unspecified gas or similar emission and sound an alarm (or trip a switch) using methods that imitate both smell and taste. These methods are based on chemical analysis techniques or on an examination of physical characteristics, e.g. the pH (hydrogen ion concentration) of a liquid or the spectral analysis of a material.
- *Touch* Once again sensors can be used, in this case to feel for the presence or absence of an item or, by use of a series of sensors, the shape or orientation of an item. Such sensors can also 'feel' excessive levels of temperature.

Moving

The 'moving' function in this context could also be described as transporting, where the main objective is to carry an item from A to B, rather than the moving that is associated with the handling function. Almost any type of equipment that can be used for moving can be automated to some degree, and thus the range is very varied, but in general terms a number of conditions should be observed, including the following.

- *Safety* Since the move is likely to be unsupervised, the equipment involved must be so designed that accidents cannot occur. For example, an AGV (automated guidance vehicle) should carry an audible/visual warning and be fitted with sensors so that it stops if contact is made with any obstruction.
- *Security* Similarly, the design should ensure that any load being carried is secure, for example by means of automatic clamps. This is particularly important if the load is carried overhead (a good place for goods movement, since there are not many obstructions).
- *Economy* Distances to be moved should be made as short as possible by ensuring that the most direct route is taken and that the factory or warehouse layout is compact. The speed of movement is also a factor: if higher speeds can be obtained, the number of move units can be reduced: However, this must defer to safety and security.
- *Convenience* The route chosen for the movement should cause minimal inconvenience to other operations, for example the routeing of a powered conveyor should allow for through access at various points by having overhead sections at regular intervals.

The moving function is ideal for the application of automation techniques, since it is often possible to save space by having narrower gangways. For example, a typical palletized warehouse serviced by manned fork-lift trucks requires gangways two to three metres wide (depending on truck size and type), whereas in an automated warehouse these need only be just over a metre wide (to cater for the pallet width). This type of automation has another advantage, since storage heights can be up to 12 metres; this is not possible if manned fork-lift trucks are used, for safety reasons. Figure 11.2 illustrates these points.

11.4 AUTOMATION EQUIPMENT—APPLICATION AREAS

Automation is defined above as '. . . equipment to save mental and manual labour' and can therefore be applied in any area where this takes place, shop floor or office. This can be demonstrated by considering a number of such areas and suggesting how automation could be applied in each case.

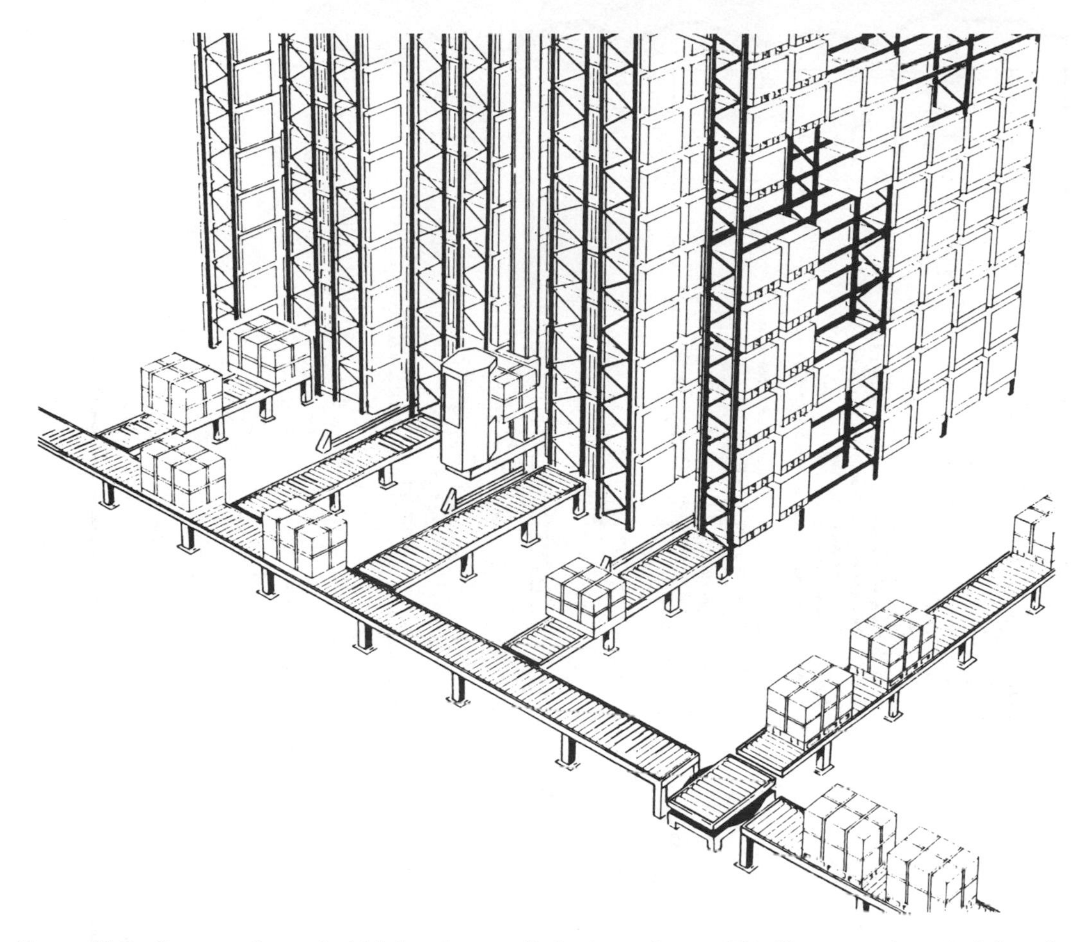

Figure 11.2 Layout of a typical high-volume palletized warehouse. This illustrates the principles of a narrow-aisle, high-bay, warehouse, with the goods stored on pallets. The storage and retrieval unit can be clearly seen, supported on a central column, with rails to guide horizontal movement. In this example the unit is manned but it could equally be unmanned with control by computer (drawing by courtesy of Egemin UK of Huntingdon, Cambridge).

Fabrication

This is a good area for automation, since the work is often heavy, dirty and hazardous. The preparation area (metal cutting and forming) can benefit from the use of numerical control (NC) on equipment such a punches, guillotines and bending machines. These units can usually be fitted with automatic handling equipment, to extract sheets or components from a storage area and feed them to the machine, and then from the machine to the next operation.

A typical application can be found in structural steel fabrication, and Fig. 11.3 shows a possible layout. In this example, the raw materials are rolled steel sections (joists, channels, angles, etc.) and steel plates. Both of these have to be cut to size/shape by sawing (rolled sections) or plasma cutting (plates) before drilling; they are then passed to the welding area for assembly.

In this example, these processes have been fully automated apart from the selection and presentation of the rolled sections by the overhead crane. All the machines are numerically

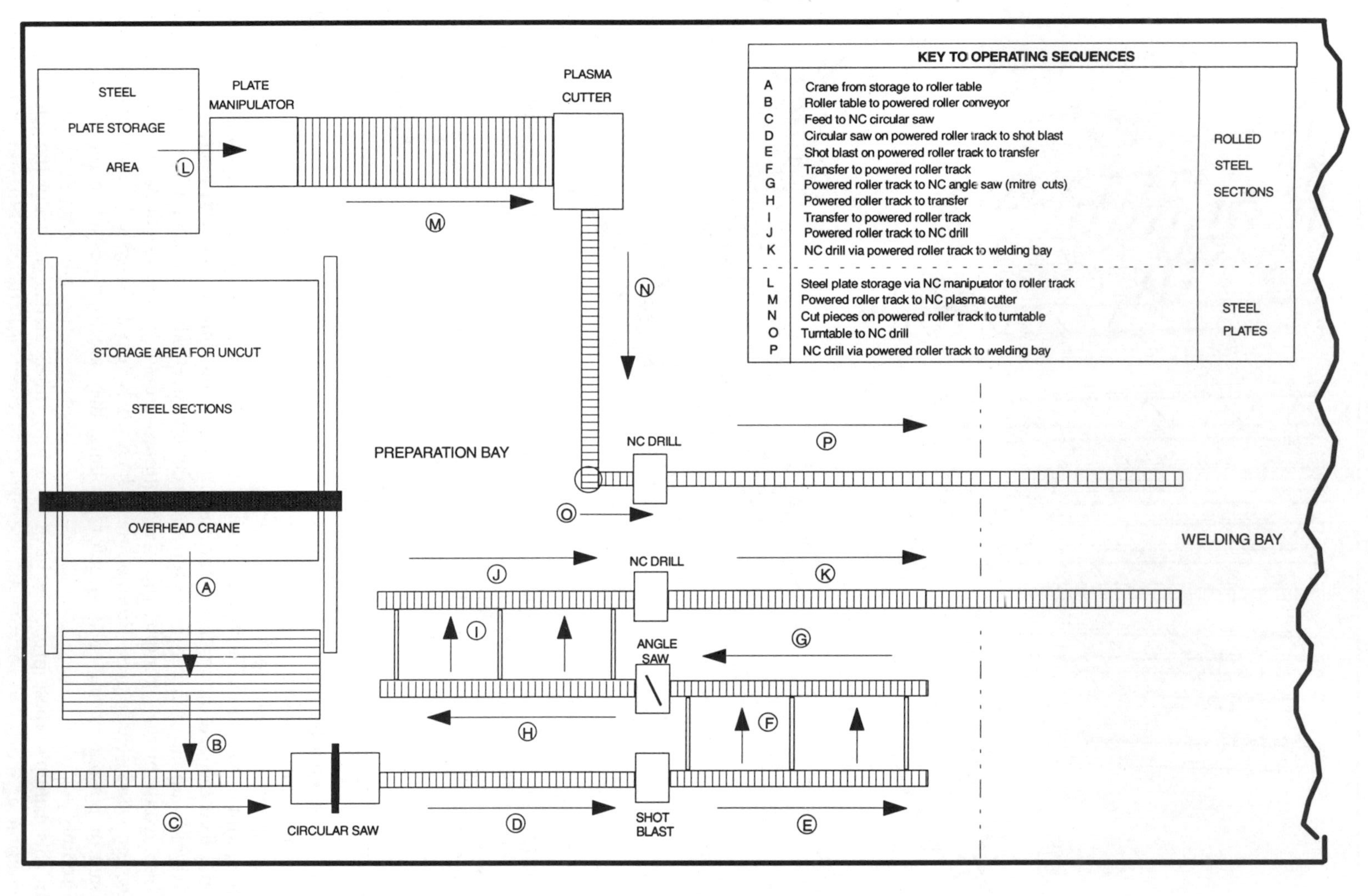

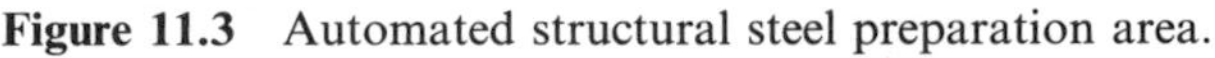

Figure 11.3 Automated structural steel preparation area.

controlled, and all handling is by powered conveyors, which operate when sensors are triggered by the presence of an item; the whole system is integrated by a master computer system with local control by PLCs (programmable logic controllers).

Press shop

Where multiple pressing operations are required (i.e. where the components cannot be formed in a single stroke, but require a number of presses, each with a different set of tools), the ideal arrangement is a manufacturing cell, as described in chapter 12 (section 12.10) and shown in Fig. 12.9. Such a set-up can often be automated by means of suitable devices to transfer components from one press to the next, with sensors to ensure that accidental collisions do not occur, these sensors being connected to electrical cut-outs and brakes. However, this type of automation may not be justifiable if frequent changeovers are required, since the cost (in time taken to reset the devices) may cancel out the savings from lower manning levels.

Figure 11.4 shows a CNC mechanical press which has been designed for automated operation, with a built-in computer to store tooling and feeder configurations; in this example, raw material (eg. steel strip or coil) is fed from the front, and pressed parts are discharged at the rear, on a conveyor that is linked to the press.

It should be noted that there are two approaches to the design of a press line. The first option is as described above, where a series of presses is linked together, each standing alone with its own power source. The second is based on the 'progression' concept, where a large, single, multi-stage press carries out the operation sequence. In this case, each station has tooling designed to carry out a specific operation.

Machine shop

The principles that apply in the press shop also apply in the machining area, where cellular manufacturing and automated transfer can reduce manning to a minimum and eliminate inter-process stocks. However, a further refinement can apply in this case if measuring devices are interposed between machining units. These devices can check whether the machined components are moving towards an 'out of tolerance' situation and, if they are, switch off the machine or adjust it to reverse the trend. Such a system could also, in certain cases, change the settings on the succeeding machine to correct the error.

Finished goods warehouse

Fully automated warehouses have been installed in a number of companies, and many more have some form of partial automation. Figure 11.5 illustrates the principles involved in the form of a flow chart. In this example all goods are stocked in stillages and the complete system is controlled by a computer, which keeps a record of all locations and their contents.

Thus, when a stillage (with a barcoded identity label) arrives in the marshalling area, this label is read automatically into the computer which then finds a free location in the racking, and records the stillage number/contents against that location. The marshalling system directs the stillage, via a conveyor, to the correct racking bay; it is then put into the correct rack by the narrow aisle handling unit.

When a customer's order is to be despatched, the computer locates suitable stillages from its records (using a specified policy, (e.g. 'first in, first out'), and instructs the narrow aisle handling unit(s) to retrieve the stillages in a sequence based on minimum travel. The required goods can

Figure 11.4 Example of a CNC press to suit JIT operation. This illustrates the design features of a CNC single operation press which can be used in an automated JIT-type manufacturing cell. It has its own built-in computer which can be used to store programs and control the process (e.g. speed and press weight). This computer can also control peripheral devices such as input and output units (photograph by courtesy of Joseph Rhodes Ltd of Wakefield, Yorkshire).

then be picked manually from the stillages for packing and despatch, and all stillages returned by conveyor to the marshalling area. Empty stillages can be returned to production, but those containing surplus items will be re-coded and put back into stock in the racks.

There are obviously many possible variations on this theme, depending on the type/size/weight of the products. For example, a semi-automated system would rely on manual order picking from the racks, where the operator rides on the narrow aisle unit. The computer will still

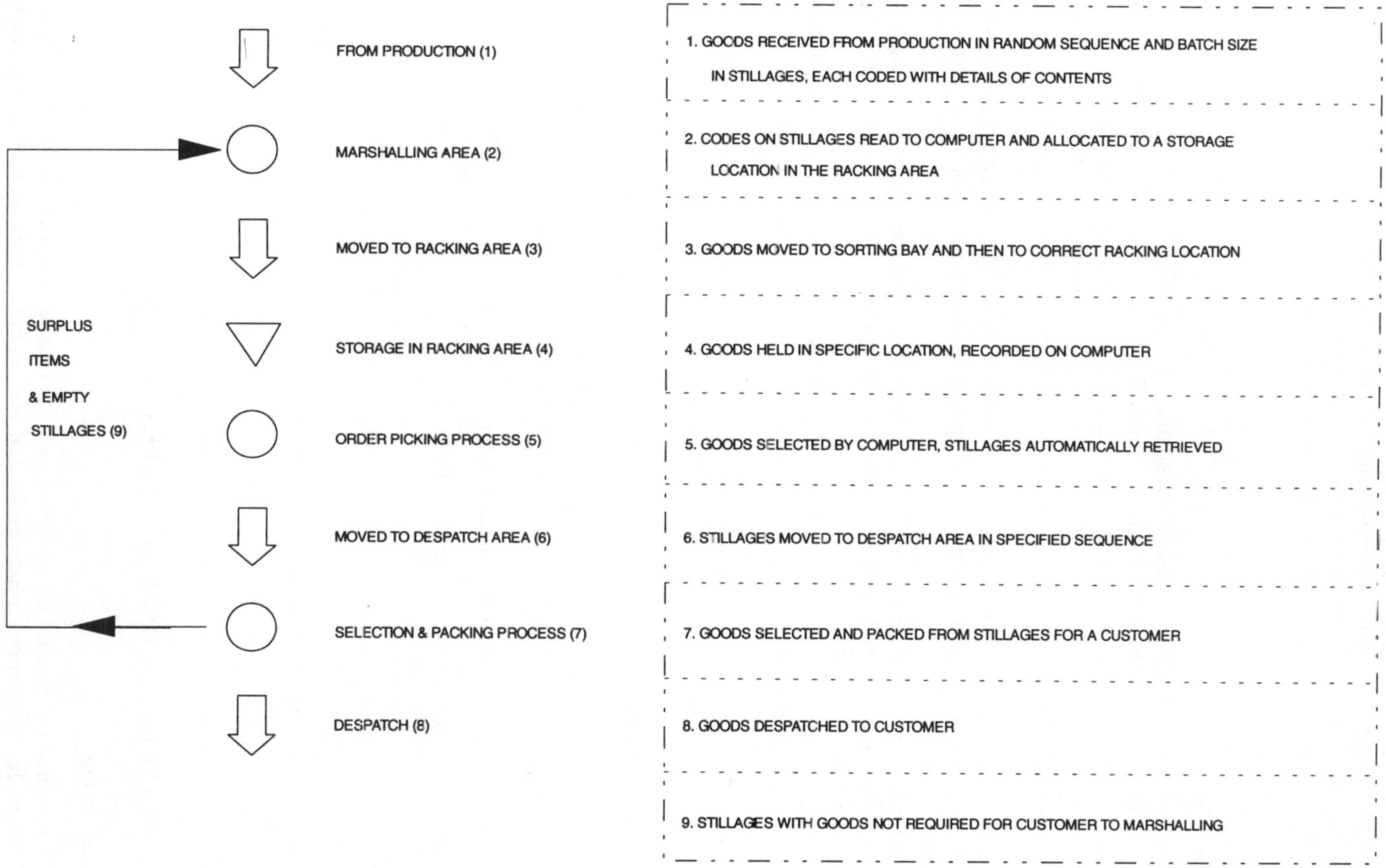

FROM PRODUCTION (1)

MARSHALLING AREA (2)

MOVED TO RACKING AREA (3)

STORAGE IN RACKING AREA (4)

ORDER PICKING PROCESS (5)

MOVED TO DESPATCH AREA (6)

SELECTION & PACKING PROCESS (7)

DESPATCH (8)

SURPLUS ITEMS & EMPTY STILLAGES (9)

1. GOODS RECEIVED FROM PRODUCTION IN RANDOM SEQUENCE AND BATCH SIZE IN STILLAGES, EACH CODED WITH DETAILS OF CONTENTS

2. CODES ON STILLAGES READ TO COMPUTER AND ALLOCATED TO A STORAGE LOCATION IN THE RACKING AREA

3. GOODS MOVED TO SORTING BAY AND THEN TO CORRECT RACKING LOCATION

4. GOODS HELD IN SPECIFIC LOCATION, RECORDED ON COMPUTER

5. GOODS SELECTED BY COMPUTER, STILLAGES AUTOMATICALLY RETRIEVED

6. STILLAGES MOVED TO DESPATCH AREA IN SPECIFIED SEQUENCE

7. GOODS SELECTED AND PACKED FROM STILLAGES FOR A CUSTOMER

8. GOODS DESPATCHED TO CUSTOMER

9. STILLAGES WITH GOODS NOT REQUIRED FOR CUSTOMER TO MARSHALLING

Figure 11.5 Flow chart for an automated warehouse.

move the unit to the correct location and then show what should be picked on the operator's computer screen. The operator can key in what has been picked to the computer, which will then move to the next location. A similar set-up can be used to operate a raw materials and components store, but current thinking (on JIT lines) is for deliveries to be made directly to production areas.

Food processing

An example of near-total automation can be found in the production of bread, where a sequence of operations must be carried out, at specific times from start-up, if the process is to be successful. A typical sequence for this process is given in Table 11.1. This sequence can be automated by controlling the complete process from a central computer, which is linked to a series of small computers at each process, as a network. It would operate as follows.

- An order for a given quantity of a certain type of bread is fed to the central computer, which uses data from its memory to set up the programs in the network of small computers (controllers).
- These controllers set up the sensors and control devices to meet the specification, e.g. oven temperature, throughput speeds and slicer settings. The system also ensures that the correct wrappers and labels are in place.
- The central computer calculates (for the batch size and product type) the required weights/volumes of raw materials, which are then be fed into the mixer in the right sequence by the local control computer.
- At each process, a series of sensors ensures that any process fault is quickly identified and reported to the central computer via the controller. Such faults are then compared with a 'library' of information containing details of any actions that should be taken in a given situation, and the appropriate action is implemented.

It is apparent that such a plant is almost completely self-reliant, and will require human intervention only when a problem arises or when plans need to be changed.

Table 11.1 Sequence of operations in an automated bakery

1. Raw materials weighed in specific proportions for batch
2. Batch fed to mixer and blended
3. Batch broken down into baking tin lots
4. Lots fed to prover to 'grow' yeast
5. Lots fed to oven for baking
6. Lots fed to cooler
7. Loaves emptied from tins (tins returned to stage 3)
8. Loaves fed to slicing machine
9. Loaves fed to wrapping machine
10. Despatch

11.5 AUTOMATION EQUIPMENT—TYPES

Sensors and actuators

Sensors are designed to measure an environmental factor and to convert it into an electrical signal. For example, a sensor can measure a level of heat, light or sound and detect any change;

this change will then be reflected in the proportional strength of the signal. This signal can then be transmitted to an actuator which will adjust the factor in order to bring it back to the original, pre-set level. There are various types of sensor with differing applications, as Table 11.2 shows.

Actuators can also take many forms, but are generally designed to make an adjustment to a control unit. For example, in its simplest form, an actuator can switch power on or off, but it is more generally used to make incremental changes such that the control unit mirrors the responses from the sensor.

Actuators can be pneumatic, hydraulic or electric, although the availability of low-cost electric servomotors has tended to favour that type over the others in recent years.

Programmable logic controllers (PLCs)

These are computer-like devices designed to operate in tough industrial environments and to perform high-speed control of processes in real time. For example, a PLC can receive electrical signals from sensors and use this input, together with a stored program, to monitor and control a sequence of operations by sending signals to control devices; it is therefore a form of low-cost micro-computer without the screen and keyboard.

Table 11.2 Types of sensor and their applications (From *Dynamic Factory Automation* by Alastair Ross)

Type	Means of operation	Application
Microswitch	Target moves mechanical lever	Part present—must touch sensor lever arm
Reed switch	Magnet in target causes two ferromagnetic reeds to make contact	Position of piston on pneumatic actuators
Strain gauge	Change in electrical resistance due to distortion of thin foil sensor	Force measurement in intelligent grippers
Hall effect	Detection of magnetic field from target plate mounted on object	Position detection
Microphone	Vibrations converted to electrical current	Machine cutting tool breakage
Optical retro-reactive	Infra-red beam relected off shiny surface	Parts present with reflective surface
Optical through-beam	Infra-red beam blocked from receiver by object	Parts present—access needed both sides
Capacitive	Object makes change in emitted electrical field	Presence of non-metallic parts or liquids
Inductive/proximity	Induced magnetic field affected by presence of ferromagnetic body	Presence of ferromagnetic item(s)
Ultrasonic	Sound pulses emitted and phase displacement of reflective beam measured	Distance measured
Machine vision	Digitization and image processing to provide data on key features of an item	Parts measurement and orientation; picking of random location parts
Geiger counter	Electron emision via photoelectric effect	Determination of radiation levels
Pressure switch	Fluid pressure moves diaphragm to actuate a microswitch	Adequate supply pressure to pneumatic actuators
LVDT*	Movement of target creates an e.m.f. due to induced voltages	High-accuracy positional movement
Thermocouple	E.M.F. generated by temperature difference between two dissimilar metals	Temperature measurement

* Linear variable differential transformer.

A PLC normally consists of two major elements, the CPU (central processing unit) and I/O (input/output unit); this can be seen in Fig. 11.6 in the form of a block diagram, and Fig. 11.7 shows a pair of typical PLCs.

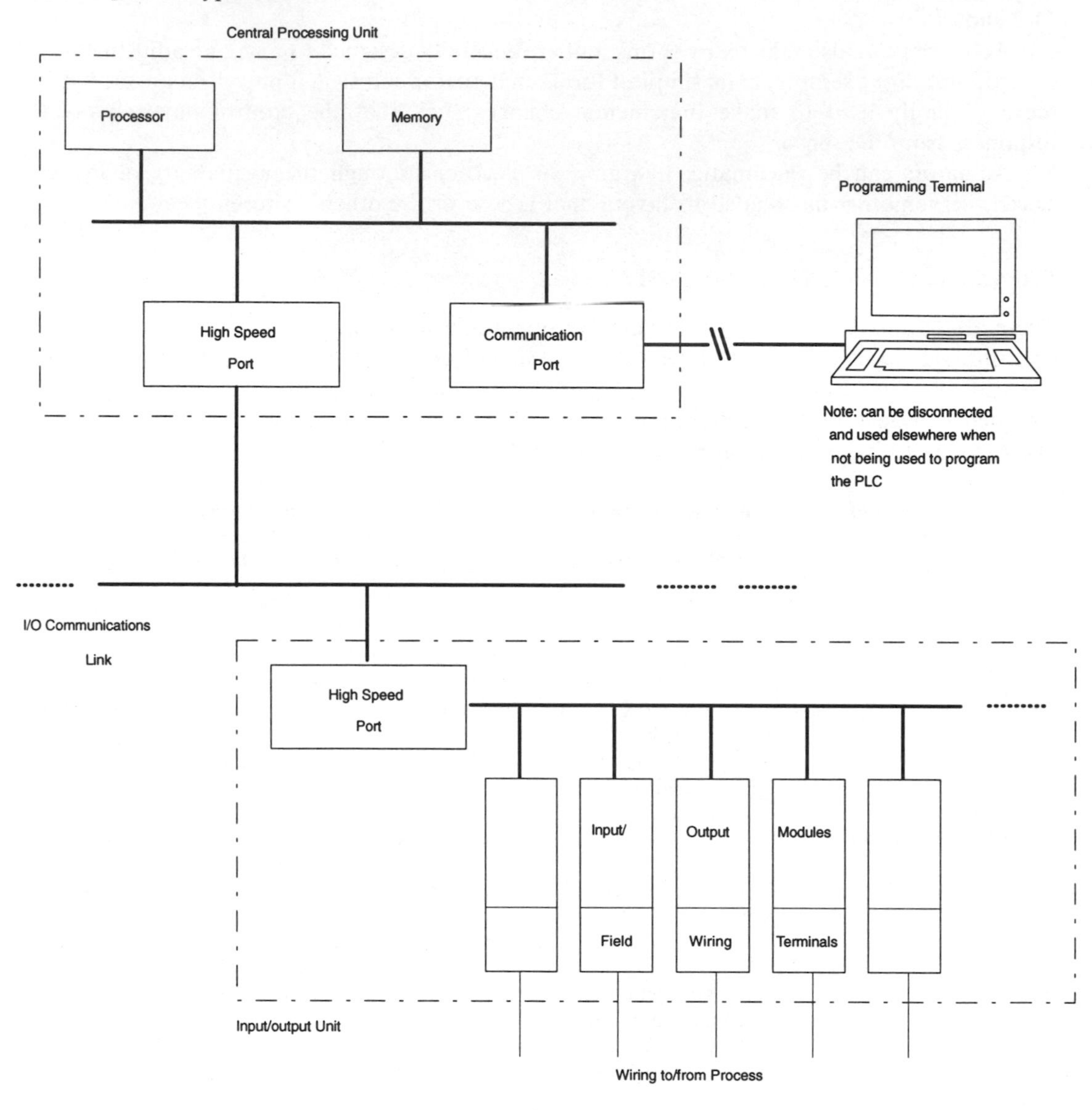

Figure 11.6 Block diagram of a typical PLC.

A PLC can differ from a microcomputer in four significant ways.

- PLCs are designed to operate in a harsh industrial environment; they can tolerate high temperature, humidity, vibration, shock, and electromagnetic interference at levels that would cause a general-purpose micro to fail.
- PLCs are designed for a specific purpose, i.e. to receive and send signals. They are thus able to handle a large number of individual I/O (inlet/outlet) points.

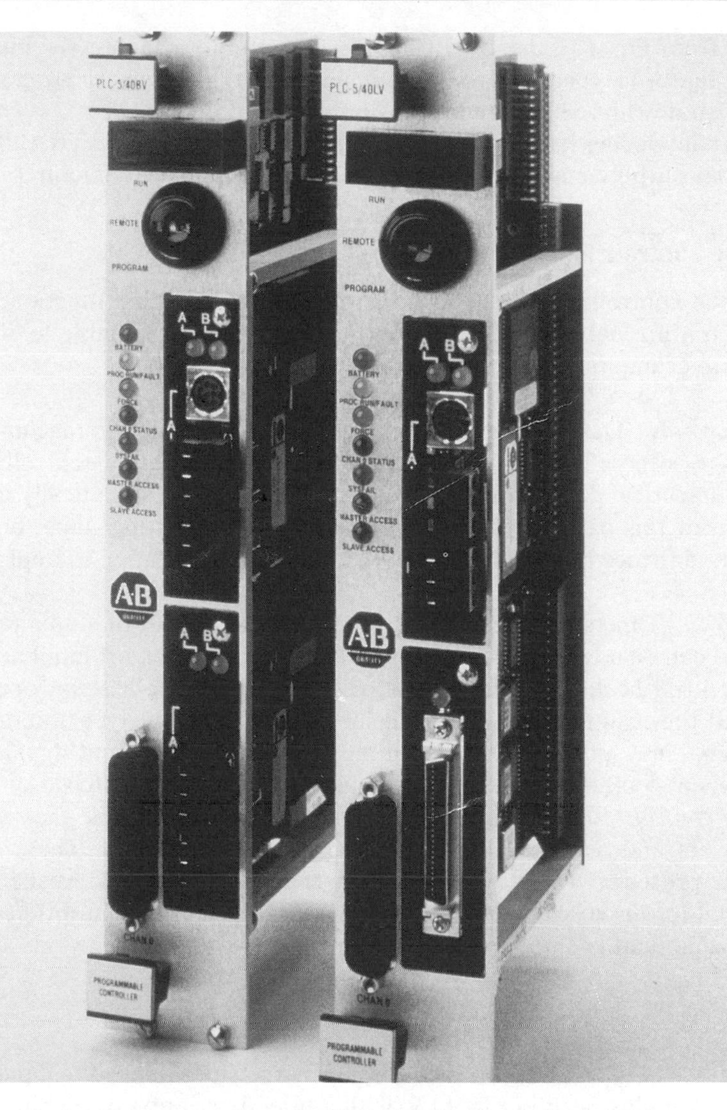

Figure 11.7 Example showing two versions of a PLC. This illustrates the design features of a typical PLC, this model being available in two versions, for remote and local connection. It can be seen that these units are designed to be mounted in a standard frame (or chassis). They can be directly connected to a host computer via an RS232C or RS422 port (photograph by courtesy of Allen-Bradley of Milton Keynes).

- PLCs are designed to operate in real time, at very high speeds, because of the need to respond immediately when controlling high-speed processes. Some PLCs can execute as many as a 1000 logical control instructions in less than one millisecond. This is much faster than a conventional microcomputer.
- PLCs and microcomputers use different software, for both the application and operating systems. This is because PLCs are designed to service one dedicated task very efficiently at high speed, whereas microcomputers are designed to deal with a range of tasks, possibly in a multi-user environment.

It will be noted from Fig. 11.6 that a PLC cannot be programmed directly; instead a portable computer (e.g. a laptop) is connected to the communications port and the program—previously prepared—is then downloaded to the PLC. A number of different languages can be used from low-level RLL (relay ladder language), through high-level such as FORTRAN, Pascal, C, and BASIC, to special-purpose application programs from the PLC manufacturer.

Special purpose controllers

There are various controller types designed to carry out a specific function. These may be programmable, but are not strictly PLCs because they are not adaptable to a range of tasks. The following are examples.

1. *Numerical controls* Designed to control the functions of a specific machine or robot. NC equipment has been defined as '. . . a system in which actions are controlled by the direct insertion of numerical data at some point. The system must automatically interpret at least some portion of this data.' Originally the programs ran on paper tape, but they are now computer-based (hence NC has become CNC). A simple application example is illustrated in Fig. 11.8.
2. *Timing controls* Timers and timer delay relays are used to start or stop a process at pre-set intervals, and can be used in a wide variety of applications, the most familiar being to switch a domestic central heating system on and off as required. The basis of operation for such devices is that they can receive an input signal and, based on the type and time setting, send an output signal in response. In the case of the heating timer the input signal comes from the clock, and when clock time matches pre-set time the heating is switched on or off.
3. *Temperature controls* Temperature/process controllers are available that combine the features of sensors and actuators in one microprocessor-based unit. These are particularly applicable in processes where not only must changes in temperature be monitored and controlled, but also a 'schedule' applies, e.g. given periods of time at different temperatures to build up, soak and cool the product.

Robots

There are basically five types of robot, each having characteristics suited to a particular application. These types are illustrated in Fig. 11.9, with a brief description of each in Table 11.3. The following terms used in the field of robot technology should also be defined.

- *Control systems* Most robots are sold with a controller which is programmable either directly through the console or by loading through a communications port from a computer. This unit often includes facilities to program items associated with the robot, e.g. a conveyor, a welder or a paint spray gun, together with a 'teach' module which helps the user to operate to the best advantage.
- *Degrees of freedom* The flexibility of a robot is governed to some extent by the number of degrees of freedom (DOF), i.e. the number of axes it can move through. The normal minimum is three DOF, but this only caters for movement of the tip of the robot arm in three planes. However, if the work holder is required to rotate in any plane then additional DOFs are required as shown in Fig. 11.10.
- *Point-to-point* This type of robot will move between points on a path determined by the robot configuration and its software system. This will normally be the optimum path (for

TRADITIONAL METHOD

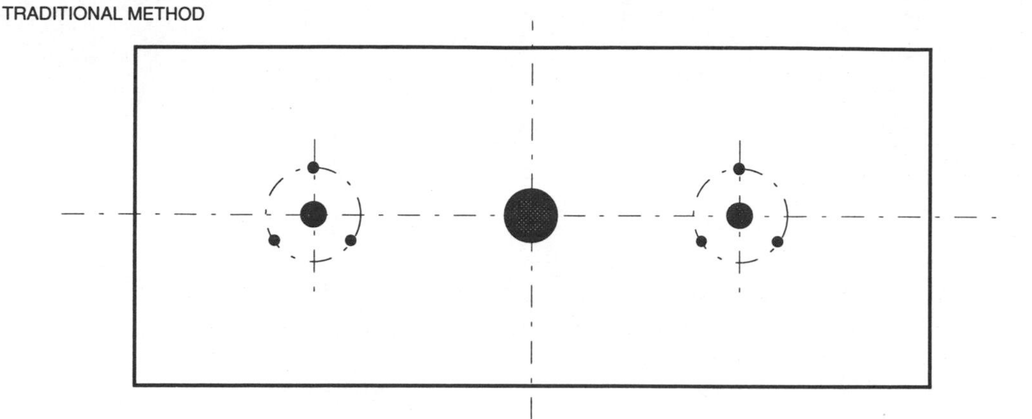

Nine holes (three sizes) are to be drilled in a steel plate. Using traditional methods, these would be marked out and then centre punched. A radial drill would then be used and this would involve three tool changes. Accuracy is hard to achieve and if repeatability is a requirement then a template or a drill jig would have to be designed and made.

The marking out process is complex, involving the use of geometric instruments to draw circles and mark centre lines. Total operation time (one off) about 15 minutes.

NUMERICAL CONTROL METHOD

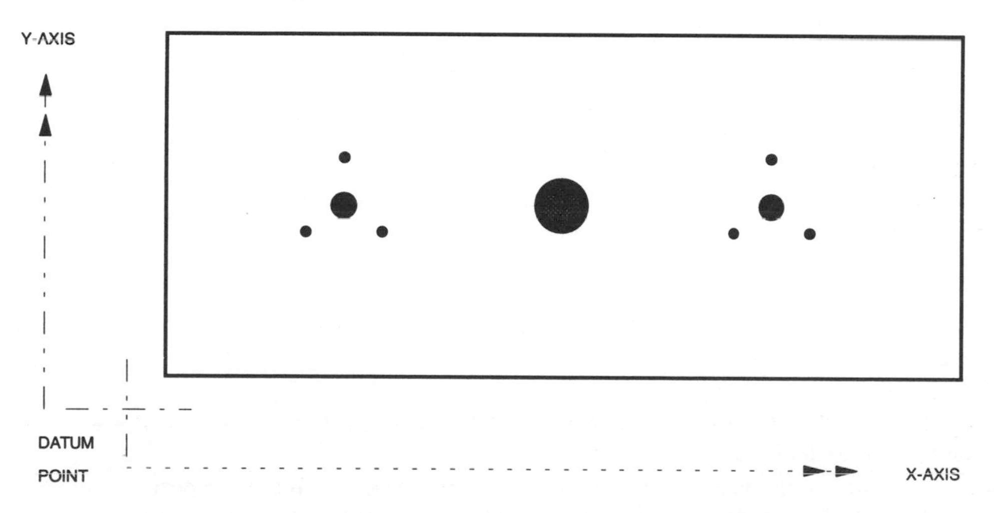

To drill the same nine holes using NC or CNC methods does not require any marking out. Instead, a program is prepared 'off-line' by specifying the position of each hole in terms of its co-ordinates from the datum point (distances along the X and Y axes). The size of each hole would also be specified. This would be very easy if a CAD system was available.

In this case, the program would be written by the software supplied with the NC/CNC drilling machine. This would work out the optimum tool paths for minimum process time, allowing for the drill head to return to the magazine for drill changes. Total operation time about 3 minutes. Accuracy and Repeatability 100%.

Figure 11.8 Example of the application of numerical control.

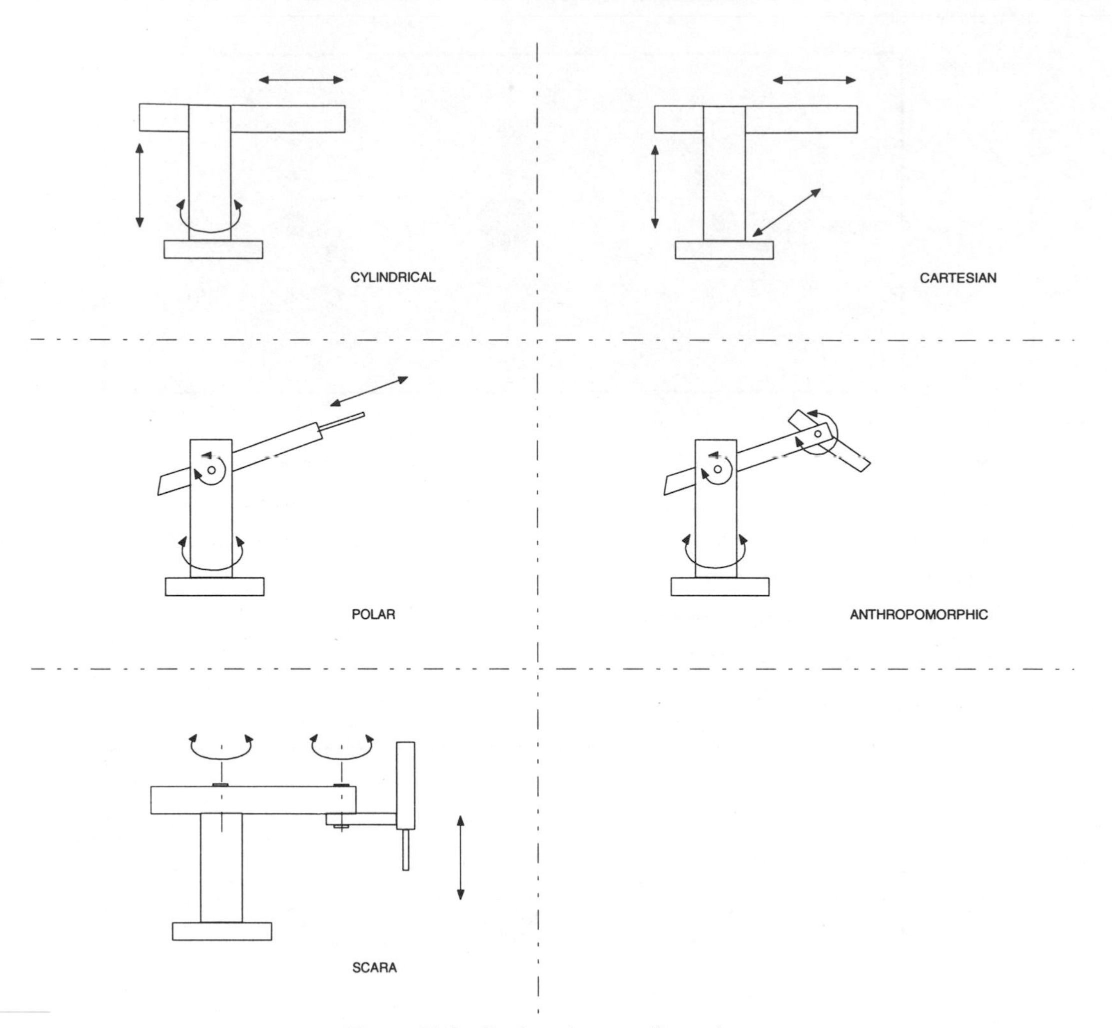

Figure 11.9 Basic robot configurations.

minimum time). The shortest path (straight line) may not always be the quickest since possible obstructions have to be allowed for!

- *Continuous path* This type of robot will move between points in a way specified by the user or by a sensing device. For example, if the robot is being used for seam welding or paint spraying, the path will be determined by the work. Sensing devices can sometimes be incorporated at the end of the robot arm to ensure that the path is followed (e.g. a fixed distance from the work).
- *Accuracy and repeatability* The difference between accuracy and repeatability is shown in the four diagrams in Fig. 11.11. These show that, when aiming at a target, a close 'cluster' indicates good repeatability, and if this cluster is in the 'bull' then the accuracy is also good.
- *Attachments* Many of the attachments fitted to the end of the robot arm can also be controlled by the programs. For example, 'beads' or 'runs' of adhesive or sealant can be dispensed along a predetermined path, with control over the cross-section to suit the application. Other possible attachments are a movable knife (for trimming plastic extrusions or

Table 11.3 Basic robot types

Cylindrical	Can sustain long reach on radial arm, but requires space behind the arm for retraction; fairly simple control system and usually reliable. Applications: pallet loading and press feeding.
Cartesian	Can be gantry or floor mounted; high repeatability due to rigid structure; easy, point-to-point programming using X-Y-Z coordinates; occupies minimal space. Applications: warehouse storage and machine tool changeovers.
Polar	An early configuration, not now commonly used; easy to control and drive. Applications: spot welding
Anthropomorphic	Best configuration for workng in a confined space or through a small opening; load capability may be limited due to rotational joints; more complex and difficult to control and therefore more expensive; may have lower reliability and repeatability. Applications: paint spraying, seam welding, 3-D assembly.
Scara	(Selective compliance assembly robot arm). Usually used for assembly work; easy to control; fast travel; usually only four degrees of freedom; high reliability. Applications: 2-D assembly.

mouldings), a gun for inserting self-piercing rivets, and a series of pneumatic suction pads for inserting a windsceen into a car body.

Automated guidance vehicle systems (AGVs)

An AGV is a driverless vehicle used for transporting goods or services (e.g. tooling) throughout a production area. It can be guided by wires embedded in the floor, by radio signals, or by a combination of the two. An AGV includes an 'in-board' computer, which allows it to be programmed to carry out specified tasks when the correctly coded signal is received.

In some cases an AGV carries some form of robot which allows it to act as a self-contained 'shuttle' with self-loading and unloading capabilities — virtually a mobile robot. An AGV system comprises two elements, transportation and information, and these are defined in Table 11.4. Typical applications where AGV systems could be used can be summarized as four types.

Table 11.4 Definition of AGV system elements

1. *Transportation system* This consists of three components:
   - *vehicles* these can be trains (a truck towing several trailers), pallet trucks (moving one or more pallets or stillages) or special-purpose transporters
   - *load handling* the links between the vehicle and the load/unload points in the traffic network. These can be a part of the vehicle (a robot arm) or static (conveyors, manipulators or humans)
   - *control* this system must coordinate the functions of vehicle control, load-handling control and traffic control
2. *Information system* This consists of two components:
   - *transmitters* these are located on the vehicle, at various points on the network, and in the central control room—they transmit information about a vehicle location and its load
   - *receivers* these are often located with the transmitters and receive similar information, but they can also receive information from sensing devices about obstructions or other problems and trigger appropriate action.

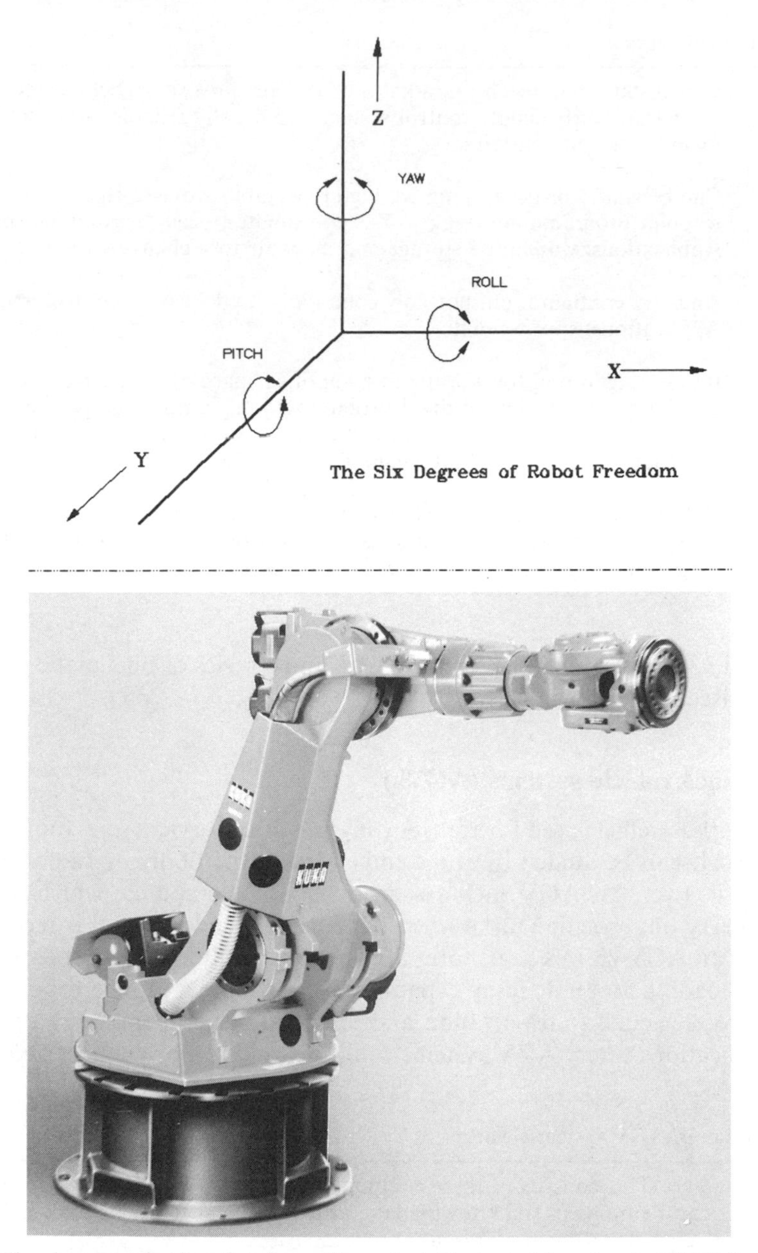

Figure 11.10 The degrees of robot freedom. The robot illustrated is a 6-axis IR760 from Kuka with a payload capacity of 120 kg. It has a repeatability of ± 0.02 mm inside a 38.6 cubic metric envelope. A seventh axis can be arranged by mounting the robot on a linear traversing unit (photograph by courtesy of KUKA Welding System + Robot of Reading, Berkshire).

1. *Manufacturing* To control and manage material flow between manufacturing cells and also to move tooling from the stores to the pre-setting area.
2. *Warehousing* To put goods into stock and to retrieve from stock. A remote control, narrow aisle racking truck is an example.

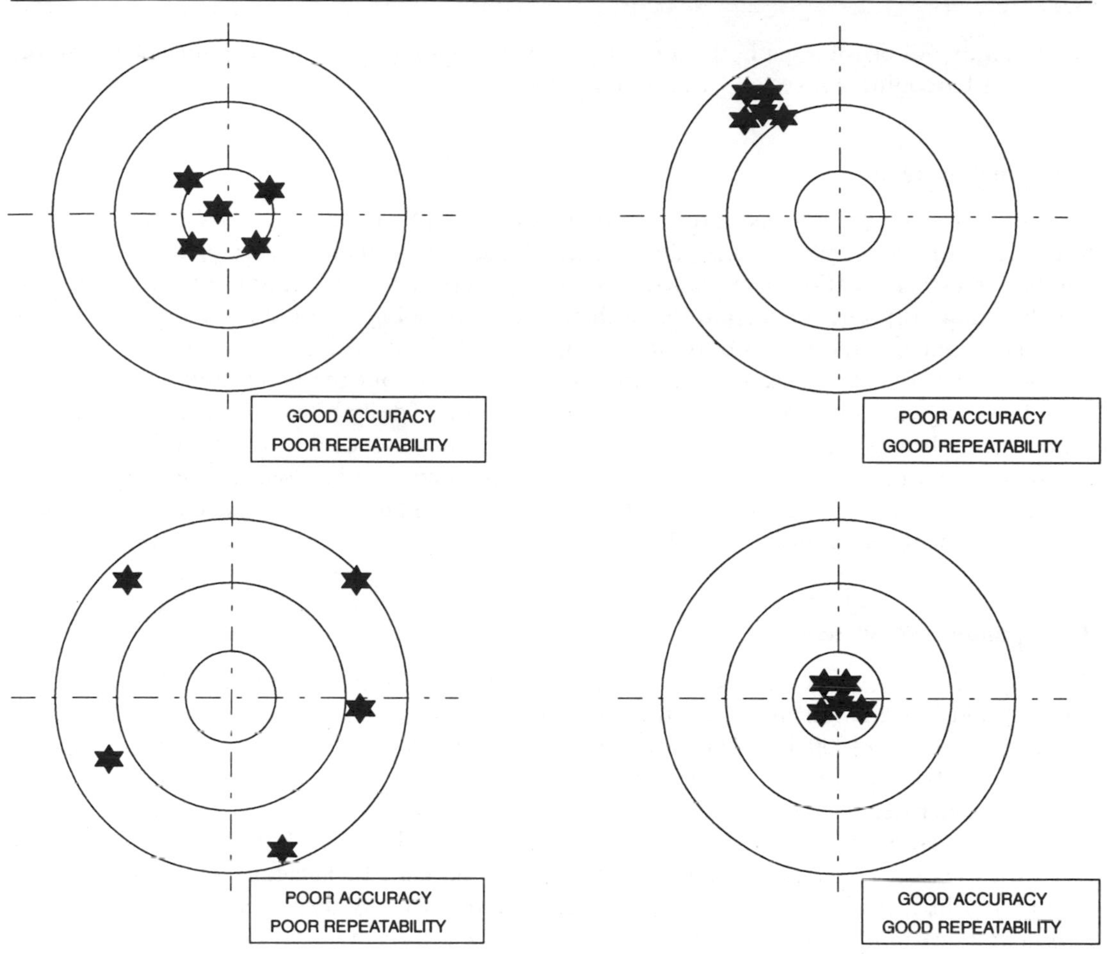

Figure 11.11 Accuracy and repeatability defined (based on figure in *Dynamic Factory Automation* by Alastair Ross).

3. *Transporting* To move goods from one location to another, for example in airports or rail terminals.
4. *Hostile* To move items or to carry a manipulator in a hostile environment, for example where there is danger from radiation or fumes.

11.6 AUTOMATION EQUIPMENT—IDENTIFICATION AND MEASUREMENT

In any development involving automation, it is necessary to identify items arriving at a process in order to ensure that the correct process is carried out. Devices are available for this purpose. It may also be necessary to measure what is to be processed in quantitative terms. For example, if a spray painting robot is dealing with a succession of items that arrive randomly on the carriers of an overhead conveyor, the robot needs to identify each item, and then to check its memory, in order to load the correct spray pattern and paint colour. Alternatively, in a batch process such as textile dyeing it will be necessary to know the weight and type of a batch in order to ensure that the right proportions of the recipe are used, and the right process time is applied.

A number of possible methods can be used. Two that are now widely available are based on the use of barcoding devices and transponder tags.

Barcoding systems

Barcodes, in the form of labels or overprinting, can be read by hand-held or fixed devices. They can contain such information as product codes, product details (colour, size, price, etc), a batch number or a stock location. They can also be used for personal identification for security or job/time booking purposes. Some 14 different barcoding 'symbologies' are used in different industries and countries, and it is important to ensure that the appropriate system is used.

Many benefits can be gained from the use of barcoding but the main disadvantage in the manufacturing environment is that it is inflexible: once a code has been printed on an item, it cannot easily be changed. However, this is not a disadvantage once the goods leave the factory.

In the field of automation in manufacturing, barcoding can be used not only for process control but also for sorting if an automated warehousing and distribution system is being used. Such a system is shown in Fig. 11.12.

Transponder tag systems

A transponder tag is a data storage device that can be attached to a component or product. It is used to hold information which can then be read for identification or tracking purposes. Such tags can allow up to 1000 'bits' of data to be read or written selectively from remote units up to 3 metres from the tag. Tags can carry different data such as manufacturing or process instructions, customer details and delivery instructions.

An unusual application of this technology can be found in a milking parlour where each cow wears a collar with a transponder tag (or it can be implanted). This tag contains data on the animal and its nutritional requirements and operates the feeding device, such that the correct weight and mix of food is automatically fed when a cow arrives at a stall. The data contained in the tag can be easily changed by the read/write unit at this time.

This technique is not yet widely used because of the cost, but has the advantage of flexibility. For instance, the original data can be amended, or extra data added, as the product passes through its operation sequence—in other words, the chip is re-programmed. Thus, information obtained from a quality check after a process can be added so that, when the product reaches the next operation, the appropriate action can be taken (e.g. adjust machine setting or divert to a rework process).

The disadvantage of such a system is that the equipment is more expensive than barcoding and can only really be justified when the number and value of the products to which such a device can be attached, is reasonable; in other words, it is ideal for car or white goods production but not for decorative tiles or tins of soup! Care is also needed to ensure that the devices are removed from the product before despatch, and the data deleted. It should also be noted that such devices may be affected by certain processes (e.g. excessive heat or high magnetic fields). An example of such a system is shown in Fig. 11.13.

Measurement systems

In the context of automation, these can apply at two levels, measurement of an individual component or measurement of a batch.

Figure 11.12 Warehouse automation—order sorting at Littlewood's distribution centre. In this example, the sorting system is linked to the order picking system and each package carries a barcoded label. These are scanned at points along the conveyor and when a 'code match' occurs the package is diverted, via a chute, to a despatch point.

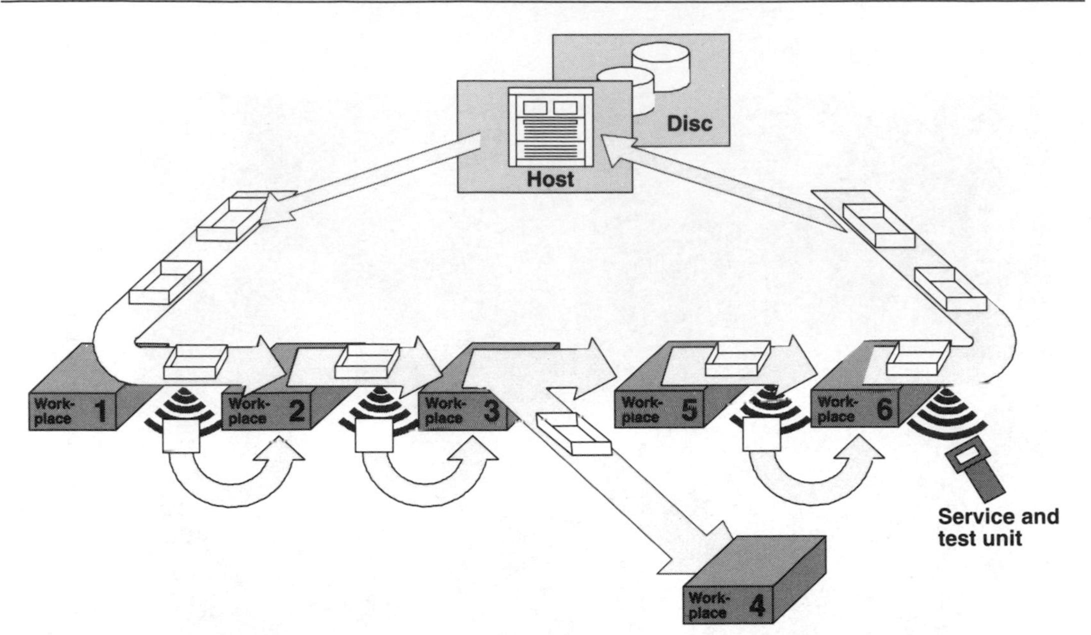

Figure 11.13 An identification system for production control. This shows a radio frequency identification system which can be used for almost any type of production process including car assembly lines and electrical appliance production. Mobile data carriers (MDS) are attached to the product and are read as the item passes over a read/write unit (SLG). The system is connected to controlling PLCs through an interface module and can store large amounts of data (e.g. 32 kB) which can be read by the SLG (photograph by courtesy of Siemens plc of Manchester).

1. *Component* This level is covered in section 11.3 of this chapter, which described various methods to determine whether an item meets the required quality standards. However, it should be noted that measurement can also refer to the location and orientation of a component, and this is vital information if it is to be picked up by a robot or fed through a machine.
2. *Batch* This level usually concerns counting or weighing, and devices are available for both functions. For instance, if the items are moving on a conveyor they can be counted by means of a photoelectric cell, and if this is connected to a PLC, the information can be recorded and used to adjust the process. Alternatively, if weight is a factor the same principles can apply, using a weighing device and a PLC.

Figure 11.14 shows examples of both types.

11.7 COMPUTERS IN AUTOMATION

Automation is totally dependent on computers, and full automation requires a sophisticated computer network. At the lowest level, PLCs (programmable logic controllers) are needed to 'manage' an operation or process, and all automated machines and robots have their own controllers which include a special type of computer. At a higher level, PCs (personal computers) are used to program PLCs and, in some cases, machine controllers; they may also be used to run a section or cell. At the top level, a large mini, or mainframe, may be used to coordinate all lower level activities into a single production plan and for costing purposes.

EXAMPLE 1:

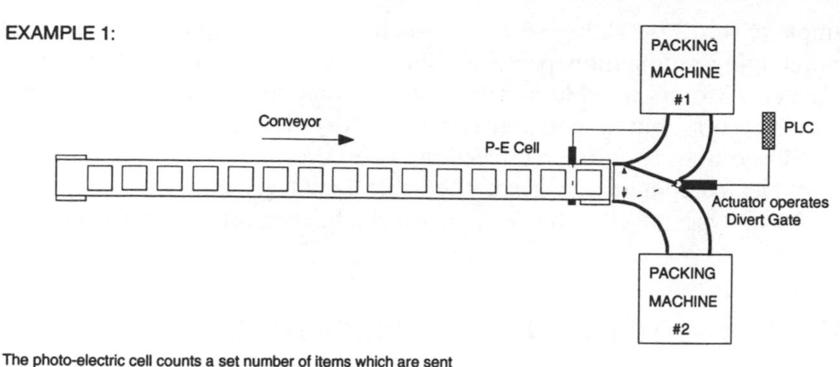

The photo-electric cell counts a set number of items which are sent to machine #1. The PLC which has monitored this process then actuates the divert gate and the next batch is sent to machine #2. This allows each machine time to complete the packing operation and to discharge the package before the next batch of products arrives.

EXAMPLE 2:

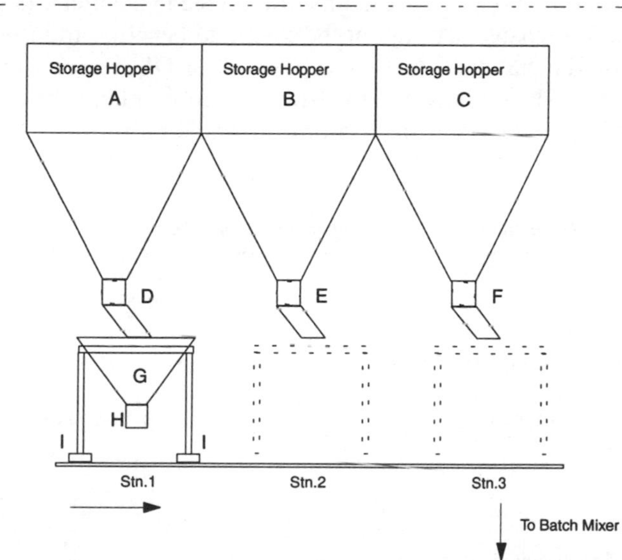

Three materials are to be blended in specific proportions depending on the product and the material weights in the blend depend on the size of the production batch. The three materials are stored in hoppers A to C, each having a rotary valve to regulate discharge. The whole system is controlled by a host computer and a series of PLCs. When a batch of product is required, the computer checks the recipe and batch quantity and calculates the three required material weights.

The weigh hopper (G) is moved to Station 1 and valve D is actuated. When the correct weight of the material (A) (measured by load cells I) is reached, valve D is closed and hopper G is moved to Station 2 where the process is repeated and then to Station 3. Hopper G then discharges (through rotary valve H) to the batch mixer.

Figure 11.14 Examples of batch control based on measurement.

As an example of this, Fig. 11.15 shows how such a network might be set up, and the use of MAP (manufacturing automation protocol) should be noted. This protocol was initially developed by General Motors in 1980 and has now become an industrial standard. MAP is not just another network, but a comprehensive open-system communication architecture based on the OSI (open systems interconnection) reference model. OSI is an international standard that enables computers to 'speak the same language' and thus to communicate, regardless of make or type; MAP is the way in which this communication is controlled and managed.

11.8 AUTOMATION—THE FINANCIAL IMPLICATIONS

In chapter 12, section 12.10, the costs of changeovers are considered and a graph (Fig. 12.7) is used to illustrate how the costs are related to the degree of changeover reduction. Exactly the same situation applies in automation, where the costs escalate as automation (the reduction in manning) approaches 100 per cent. This can be seen in Fig. 11.16. It is therefore essential to evaluate every automation project in the manner described in chapter 2 (section 2.6). This means that a report should be prepared setting out the costs and benefits, and that these should then be evaluated by one of the standard methods, payback, ROI (return on investment) or DCF (discounted cash flow). The savings due to reductions in manning should be given in detail in the justification report, as shown in the example in Table 11.5.

Table 11.5 Savings due to reductions in manning from automation

Item	Quantity	£/year
Annual savings:		
Reduction in number of machine operators	4	36 000
Reduction in number of materials movers	1	7500
Reduction in costs of employment	10% of wages	4250
	Total	46 750
Annual costs		
Increase in number of technicians	1	12 000
Increase in costs of employment	10% of wages	1200
	Total	13 200
One-time costs		
Redundancy payments for machine operators	3	9000
Redundancy payments for materials movers	1	2500
Training for m/c operator to technician	1	1500
	Total	13 000

The major benefits of automation can usually be expressed in financial terms, as can the costs of implementation. Some typical examples of possible benefits are given in Table 11.6.

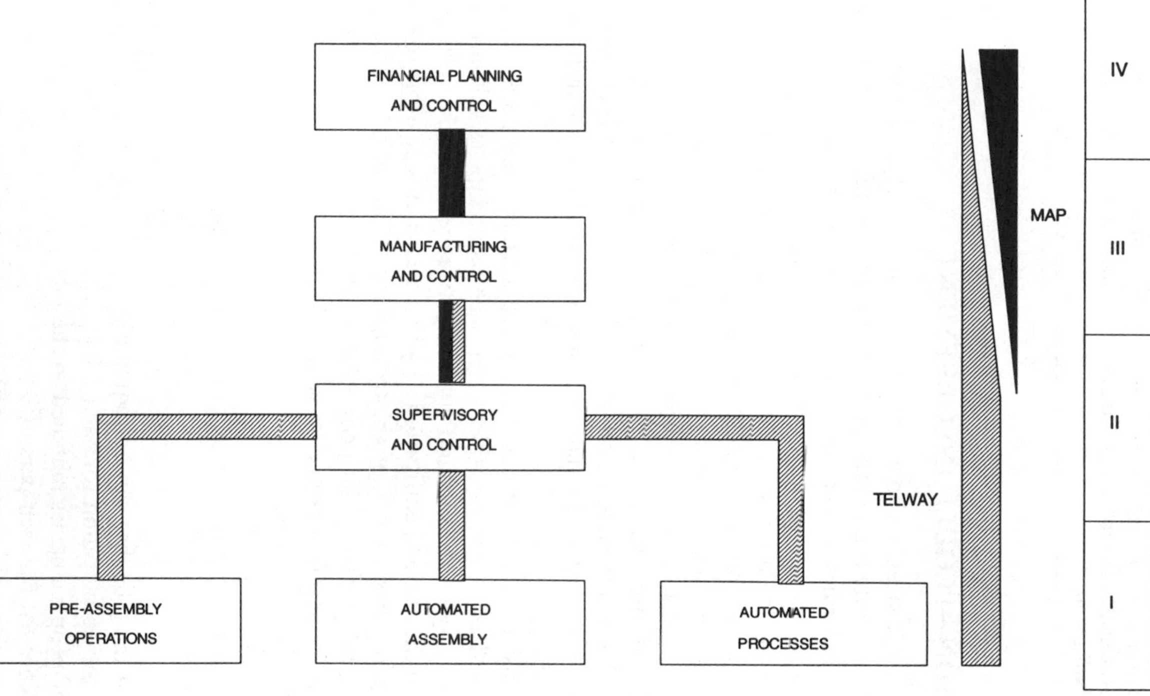

Figure 11.15 Fully automated factory communications.

Table 11.6 Typical benefits of automation

Quality	Reduced losses due to improved reliability/accuracy
Materials	Less material waste (e.g. paint overspray)
Labour (1)	Reduction in number employed
Labour (2)	Elimination of special skills not available locally
Monotony	Monotony of menial tasks leads to high absenteeism/labour turnover
Hazards	Reduction in work of a hazardous nature
Overtime	Reduction in overtime and shift premiums
Handling	Reduction in handling losses
Utilization	Plant can run 24 hours per day if required
Floor space	Automation generally requires less space
Lighting	Automated equipment can work in darkness
Heating	Automated equipment is less affected by heat or cold than humans
Fatigue	Automated equipment, properly maintained, does not require rests
Holidays	Not required!

11.9 AUTOMATION—PEOPLE INVOLVEMENT

Automation needs people: people who are commited to the concept, people who understand the problems, and people who are trained to operate it. In fact, without the right people automation will not work. From the earliest days of the industrial revolution people have fought against the idea of mechanization—the use of machines to replace them—and automation is just a continuation of this conflict. The problem can be summed up in one word—fear. People are afraid of change because they are afraid of the unknown, they fear for their jobs and for their future. It is vital that this fact be recognized and understood by management.

An automation project, if it is to succeed, must therefore start by involving everyone concerned in a series of meetings and discussions, at which management should explain what is planned, how it will be implemented, and what will result. It must dispel the rumours (many of which will be wildly exaggerated) by telling people what is really going to happen, and why automation is needed.

Management must be honest: if compulsory redundancies are necessary then the people concerned must be told as soon as possible. Often such redundancies will not be required if the plan has allowed for early retirements and voluntary arrangements (the costs of which must be taken into account in the project justification). By this means, it is possible to eliminate many of the people problems, but this is not enough. Throughout the life of the project it is necessary to continue the 'education' process, to let people know what is happening, and where possible to involve them, so that the new factory and its systems become everybody's property. Many of the best ideas have come from the least expected sources, but they will not come if management is not open, accessible and receptive.

It is therefore essential that the project plan includes a personnel plan, which should cover the following elements:

- *manning* a list of the jobs that will be required
- *specifications* a job specification for each job
- *nominal* a list of people's names matched to the jobs
- *abilities* an abilities list for each person
- *deficiencies* a list of 'gaps' between specification and ability/person
- *training* a list of training needs to eliminate the deficiencies.

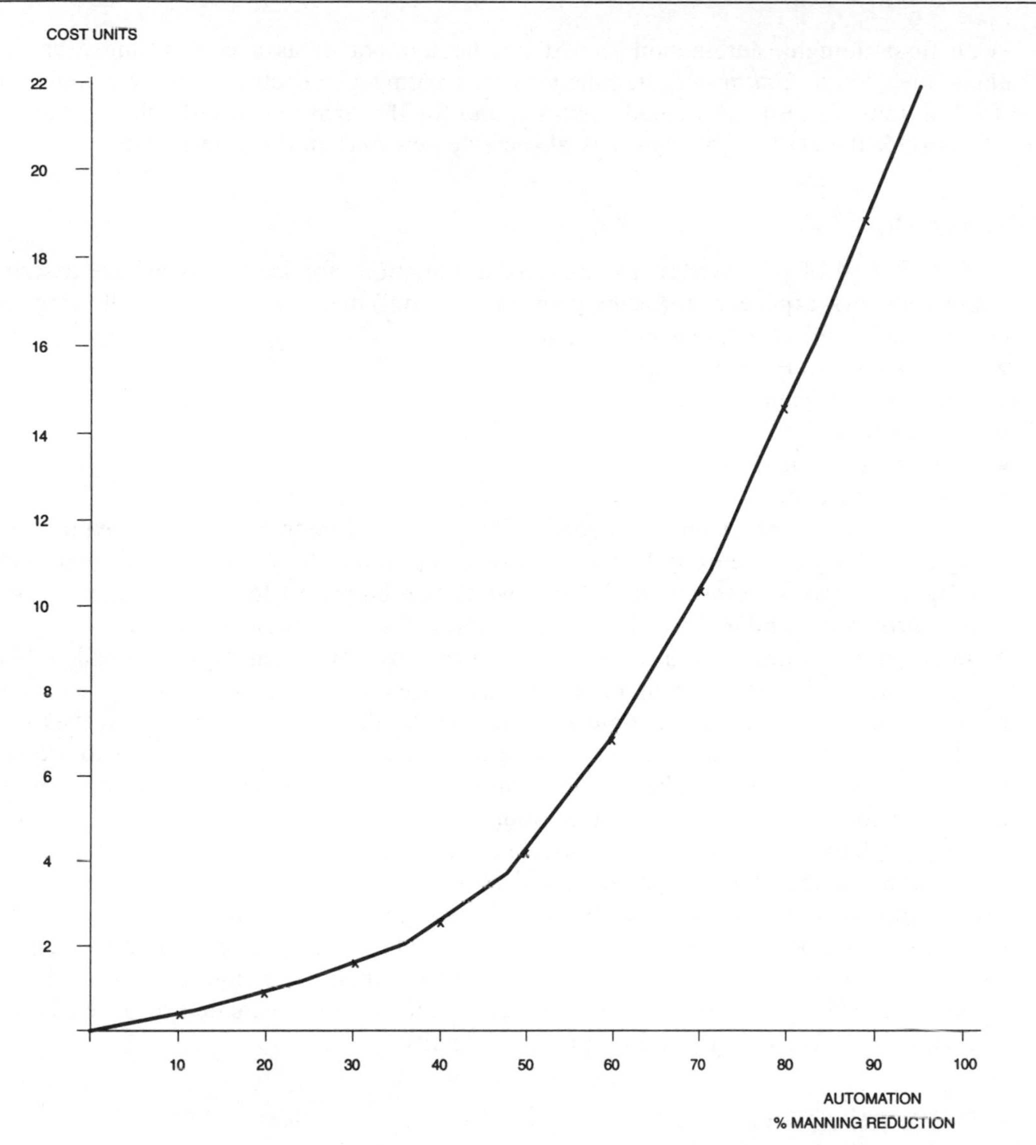

Figure 11.16 Comparative costs of automation levels.

It should be borne in mind during this exercise that jobs and their specifications need to take account of the importance of developing multi-skilling, so that all employees are given the capability of taking on an alternative job if necessary. This is not only to avoid shortages due to illness or holidays, but also to allow people to change jobs from time to time, to reduce monotony and to increase their awareness of the work being carried out and the problems in other areas.

Automation requires skills that are not always readily available in the locality, and the costs of 'imports' are often high (they are also often unsuccessful due to resentment if compulsory redundancies have been necessary). It is therefore almost always better to recruit internally, and train. This can be done by running internal courses or, perhaps better, by using local facilities such as the 'Investment in People' scheme.

One final thought: automation should not be thought of as a cost-cutting exercise to eliminate jobs, but as a means of retaining jobs, by creating a competitive edge in a competitive world. If automation is implemented correctly, and for the right reasons, it will not only safeguard many of the existing jobs but may also create new ones in the longer term.

PROBLEMS

1. In section 11.4 of this chapter a number of automation 'application areas' are described. From your own experience (or using your imagination!) describe one of the following 'non-engineering' application areas in the same way:
   - a dairy farm milking parlour
   - a sewage treatment works
   - a supermarket
   - an airport terminal (baggage handling)
   - a market garden.
2. Describe, in non-technical terms, a specification for a program that is to be written for a PLC (programmable logic controller) which is to be set up to manage a domestic oil-fired, central heating system with sensors in each room, which can be pre-set for a given room temperature, control valves on each radiator, and a timing device on the heating unit.
3. Your company wishes to use a robot to transfer cartons from a conveyor onto pallets. There are three types of carton arriving in random sequence and three pallet positions; each pallet has a different stacking pattern. Your specification should include details of how the cartons will be identified and include a sketch of the proposed layout. Each pallet will be located on a turntable which, when the pallet is filled, will revolve 180° to present an empty pallet. This movement should be controlled by the robot.
4. (a) Suggest how barcoding could be used to control stock in a finished goods warehouse based on portable terminals and computerized stock records.
   (b) Suggest how barcoding could be used in a supermarket or catalogue showroom to eliminate shelf stocking and pilferage (clue: why display more than one of any item?).
5. Your company is considering the installation of an automated warehouse for finished goods. Suggest the benefits that could accrue and could therefore be evaluated to offset the costs (see Table 11.5 for some ideas, but give more details).

FURTHER READING

Groover, M. P., *Automation, Production Systems and Computer Integrated Manufacturing*, Prentice-Hall, Englewood Cliffs, NJ, 1987.
Hartley, J., *FMS at Work*, IFS Publishing, Bedford, 1984.
Mair, G., *Mastering Manufacturing*, Macmillan, London, 1993.
Moir, P. W., *Efficient Beyond Imagining*, Ellis Horwood, Chichester, 1989.
O'Grady, P. J., *Controlling Automated Manufacturing Systems*, Kogan Page, London, 1986.
Owen, T., *Assembly with Robots*, Kogan Page, London, 1985.
Parrish, D., *Flexible Manufacturing*, Butterworth Heinemann, Oxford, 1990.
Rhodes, E. and Wield, D., *Implementing New Technologies: Innovation and the Management of Technology*, Blackwell NCC, Oxford, 1994.
Ross, A., *Dynamic Factory Automation*, McGraw-Hill, London, 1992.

CHAPTER

TWELVE

THE MANAGEMENT OF JUST-IN-TIME

THE LINK

Automation, which is described in the previous chapter, aims to make a company more effective and efficient by using the talents of people and machines to the best advantage. Although just-in-time (JIT) is, to some extent, designed to achieve similar objectives, the approach is different. Whereas automation is used to increase the efficiency of people and machines by maximizing strengths, JIT is applied in order to reduce waste, in all its forms.

In the JIT approach, everything that adds to cost without adding to value is waste. It is therefore wasteful to store goods, to produce more goods than are immediately needed, to move goods around, to produce scrap, and to carry out unnecessary work (e.g. rectification).

In JIT, changeover times must be cut to the minimum, not just because this can increase machine utilization but because it allows batch sizes to be reduced to the minimum economically. In fact, machine utilization is not always a critical factor. For example, it is less wasteful to have a machine standing idle than to keep it running, producing products that are not immediately required.

The application of JIT requires a massive change in attitudes within a company—from top to bottom—and this change in culture destroys many of the old traditional values, for example the idea that any increase in output or utilization is beneficial, regardless of the level of demand.

12.1 INTRODUCTION

In this chapter, the just-in-time (JIT) philosophy is examined and its application in various types of manufacturing industry is considered. JIT consists of a series of methodologies designed to work together to eliminate waste, and waste is defined as any activity that adds cost to a product without adding value. Such waste can occur in any area of a business, therefore the implementation of JIT requires a total systems approach, which will involve every employee and every area.

JIT covers a wide range of integrated management tools and techniques, and any company wishing to become JIT-based should be prepared to sacrifice many of its traditional values—nothing is sacred, and there is no place for the faint-hearted to hide!

JIT is not a panacea, and the chapter therefore opens by defining the suitability of the approach in terms of the various manufacturing company types. It continues by examining the techniques that can be employed and the benefits that can be obtained, and concludes by suggesting what changes need to be made in the organization if the implementation is to succeed.

12.2 THE SUITABILITY OF THE JIT APPROACH

In chapter 1 (section 1.4), a method of classifying companies on the basis of 'make-to-stock' or 'make-to-order' is described and illustrated in the form of a spectrum. This diagram has been modified to illustrate the suitability of the JIT approach, as shown in Fig. 12.1. This gives levels of suitability for each of the five types. Table 12.1 breaks these down in greater detail.

The techniques listed in Table 12.1 are described later in this chapter. All are all designed to allow a company to follow the JIT philosophy, which is set out in Table 12.2.

Table 12.1 Matrix of JIT techniques and company types

JIT techniques	Company type				
	1	2	3	4	5
Total quality management	√	√	√	√	√
Total preventive maintenance	√	√	√	√	√
Supplier development	√	√	√	√	
Change-over reductions	√	√	√	√	
Continuous improvement groups	√	√	√	√	
Flow-chart simplification	√	√	√		
Cellular manufacture	√	√			
Kanban system	√	√			
Levelled schedules	√	√			

Table 12.2 The basis of the JIT philosophy

JIT is concerned with producing and/or stocking only the necessary items in the necessary quantities at the necessary times. The concept requires:

- deliberate lowering of inventory
- streamlining of material flow
- reduction of manufacturing lead times
- exposure of manufacturing problems to resolve them permanently.

12.3 DEFINING JIT IN DETAIL

JIT is designed to eliminate waste, and waste is created by manufacturing problems. Thus, if these problems can be identified and attacked, waste can be reduced to the bare minimum. This can be achieved by deliberately lowering levels of inventory, since excess stocks (in stores or work-in-progress) are held to allow a company to cope when manufacturing problems occur (for instance, safety stock may be held to cater for late deliveries by suppppliers (raw materials) or production (finished goods)).

Fig. 12.2 shows this process in pictorial form. When the water level (inventory) is high the ship can proceed without difficulty, and the fact that manufacturing problems exist is concealed. However, when the water level is lowered these problems are exposed, and unless they are tackled urgently the ship comes to a halt. This must be a multi-stage, iterative process, whereby

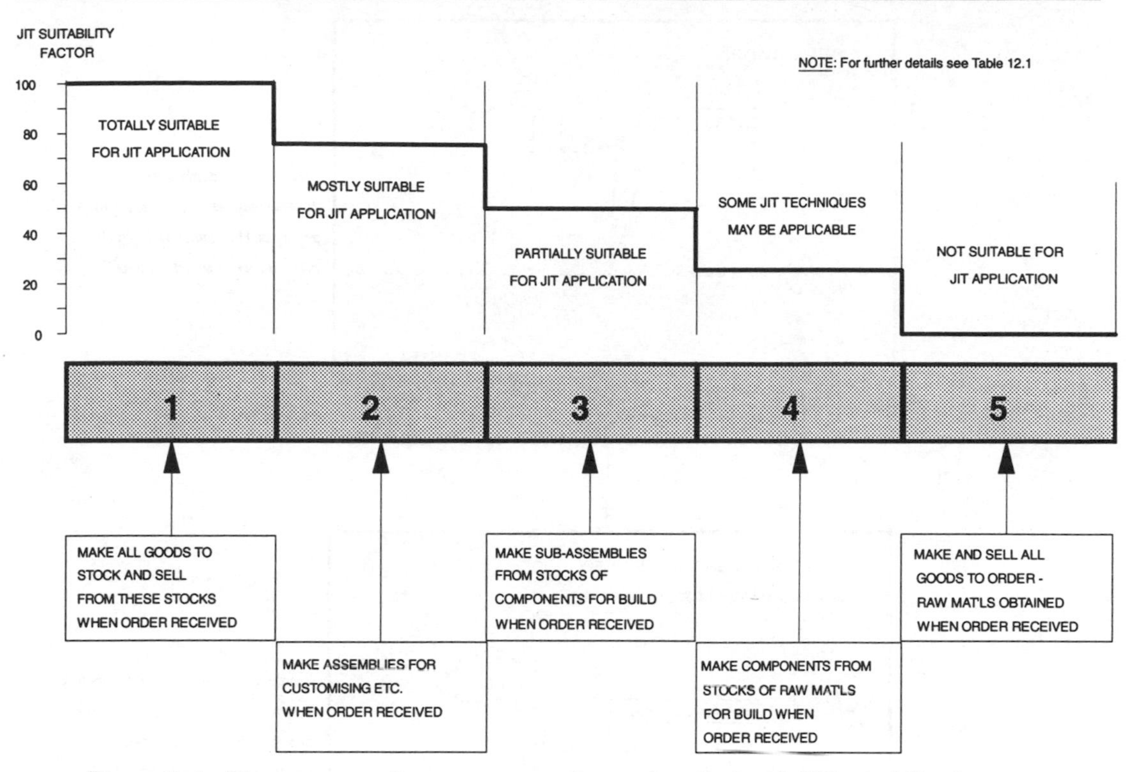

Figure 12.1 The spectrum of company types (by stock policy) with JIT suitability factors.

each lowering of the level is followed by an attack on the exposed problems, after which the level can be lowered further, and so on.

JIT can therefore be defined as a set of tools that can be used to identify and resolve problem areas in manufacturing, requiring:

- management commitment and employee involvement
- deliberate lowering of inventory
- reduction in batch sizes and set-up times
- reduction in number of suppliers
- development of customer/supplier partnership
- better forward visibility of demand
- improved manufacturing layout
- total productive maintenance (TPM)
- total quality management (TQM)
- improved planning information.

By using these tools and meeting these requirements, a company can greatly improve its manufacturing environment and thus become more efficient and competitive. This can be seen in the data given in Table 12.3, which are based on surveys of companies that have applied the JIT methodologies. However, it should be noted that these 'payback' figures have not been achieved by short-term, intensive measures, but by a steady and persistent campaign over several years, using the continuous improvement process.

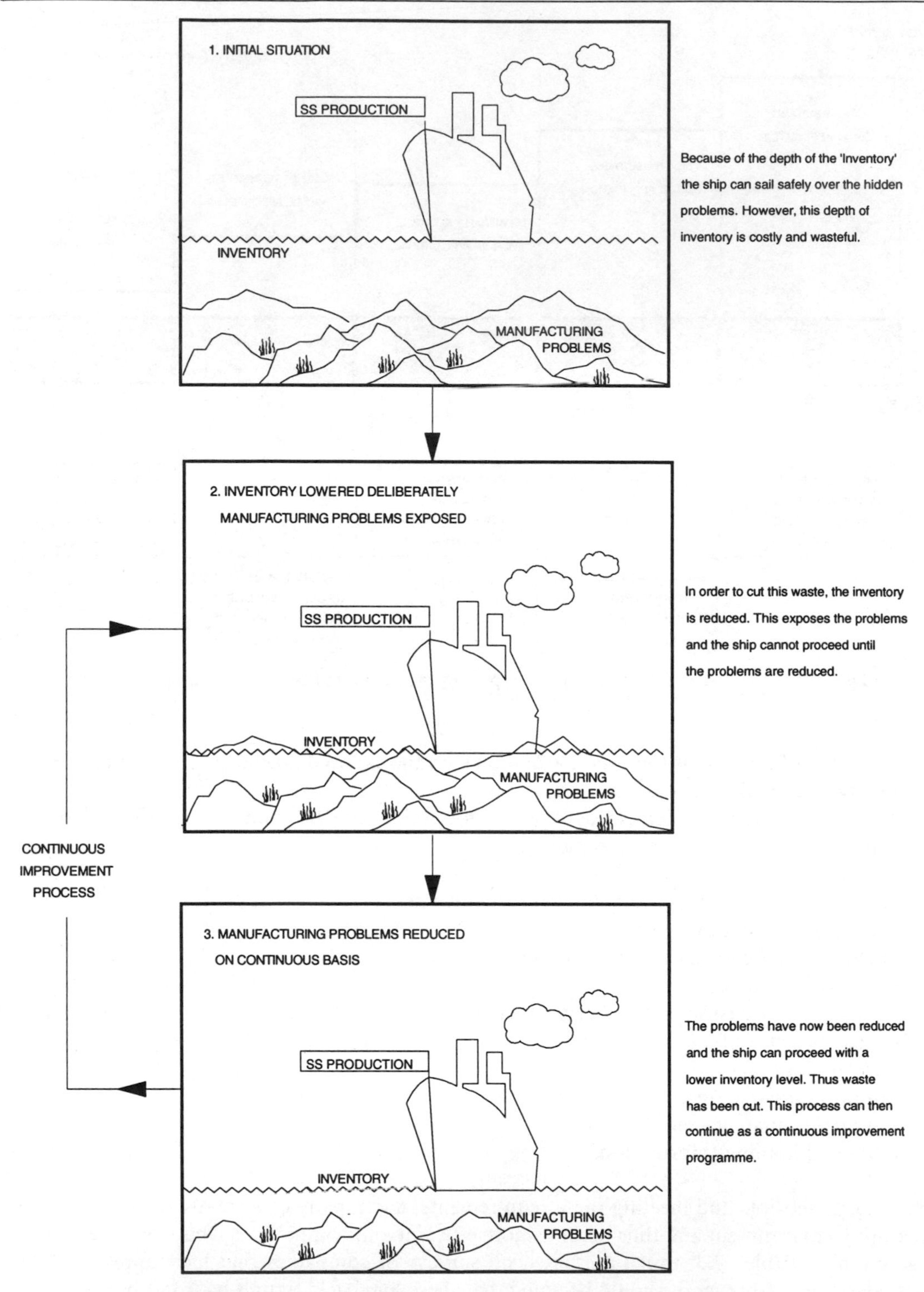

Figure 12.2 Exposing and reducing manufacturing problems.

Table 12.3 JIT potential paybacks

Manufacturing lead times reduced	80–90%
Productivity increases (direct)	5–50%
Productivity increases (indirect)	20–60%
Purchase price reduction	5–10%
Inventory reductions:	
raw materials	35–75%
work-in-progress	30–90%
finished goods	50–90%
Set-up reduction	75–95%
Space reduction	40–80%
Quality improvements	50–55%
Material stock-outs reduction	50–95%
Scrap reductions	20–30%

12.4 THE JIT PRODUCTION SYSTEM

In a JIT system, throughput is dictated by the demand for finished goods; thus, if there is no demand there is no production. This is known as a 'pull' system, where work is pulled through the operation sequences, when required, for packing and despatch. The inherent risks of such a system are obvious; any disruption to production throughput will endanger all subsequent operations in the sequence, leading to late delivery, but these risks can be minimized by the other JIT techniques such as TPM (total planned maintenance), and soundly based supplier partnerships to eliminate machine breakdowns and late deliveries from suppliers.

This underlines the fact the JIT is a total system, and although some of the 'tools' can be used in isolation, the real benefits are achieved only by a complete implementation.

12.5 THE KANBAN SYSTEM

The kanban system is designed to ensure that work-in-progress quantities are minimized, and develops the concept of JIT production described above. Kanban means a *sign* in Japanese, and in this context it is really a signal—an authorization to manufacture a specified quantity. The kanban does not need to be a written card, for example it could be a tote pan or a space on a shelf, but in every case the operational basis is the same. This can be seen in Fig. 12.3, which shows the original concept, whereby each machine operates only when the next machine has used some (or all) of the parts in the kanban location, and then makes only enough to meet the kanban quantity.

A further system has been developed using two kanbans, and this is shown in Fig. 12.4. This second system was designed to eliminate stocks between work centres, and to reduce the need for work centres to be close together for visibility of the shared kanban. This is known as the 'two kanban' system and shows how a kanban chain must exist throughout the production process, in order to realize JIT principles and to give line balance (i.e. a steady flow) for each process and between processes.

In this example, the system (which includes the supplier—see section 12.6) shows that each production area (line) has a *withdrawal kanban* (W) to signal to the previous area that it has a demand, and each area also has a *production ordering kanban* (PO). The sequence of events between any two production areas (say C and D) is therefore as follows.

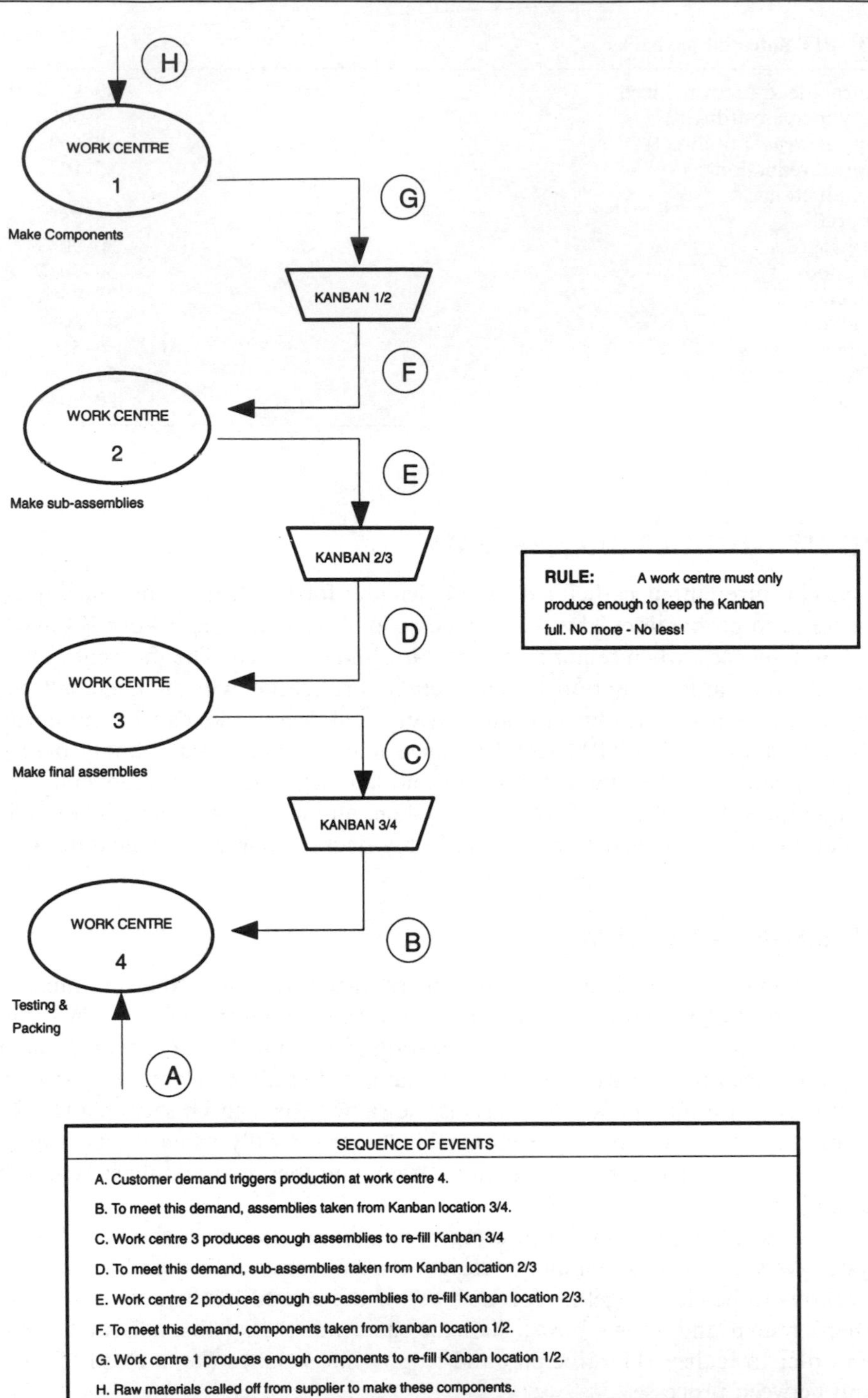

Figure 12.3. The kanban concept (the 'pull' system).

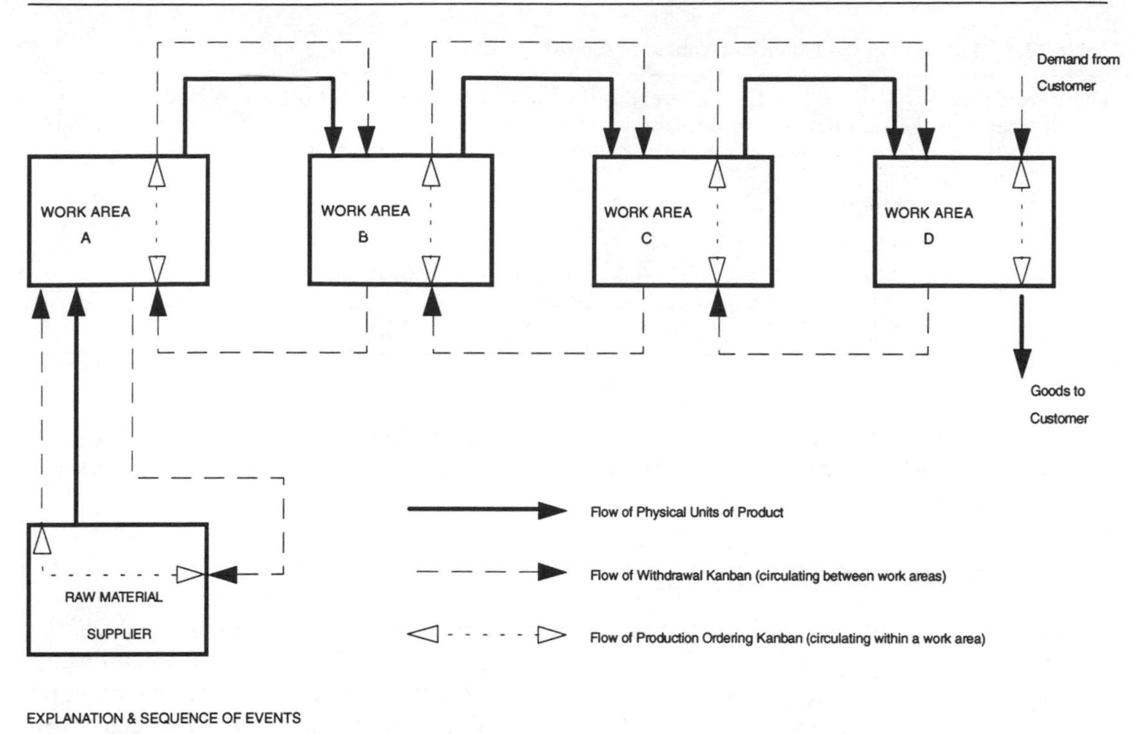

EXPLANATION & SEQUENCE OF EVENTS

1. There are two types of Kanban - 'Withdrawal' and 'Production Ordering'
2. There is one 'Withdrawal' Kanban between each pair of production areas (eg. Areas C and D)
3. There is one 'Production Ordering' Kanban for each production area.
4. The Withdrawal Kanban is sent to the previous work area (C) when a demand occurs in the final work area (D)
5. In the work area (C) it joins the Production Ordering Kanban for that area
6. The two Kanbans then pass through the area (C) as production takes place.
7. On completion of production the PO Kanban is returned to its start point.
8. The W Kanban accompanies the products to the final area (D) and is then returned to its original location.

Figure 12.4 The two-card kanban system.

1. When a demand arises at area D, The W-kanban from this area is passed back to area C.
2. This releases the PO-kanban, which is waiting at the start of area C, and the two kanbans then travel through area C with the parts being processed.
3. On completion of the process, the W-kanban is passed back to area D with the completed parts, and the PO-kanban is returned to the start of area C to await the next demand.

The same sequences are used between lines B and C, between lines A and B, and between the supplier and line A. In this example it has been assumed that only one kanban of each type is required for each production line, but this may not be the case.

A formula developed by Toyota (which originated the kanban system) allows the user to calculate how many kanbans are required, as shown in Table 12.4. At Toyota, the policy variable is expected to approach zero, and it can be seen that other adjustments (such as a change in container size (a)) can be used to reduce the number of kanbans required.

As stated above, it is not necessary to use cards as kanbans. Where production machines or processes are fully automated the use of cards, or even tote pans, is unpractical since there will

Table 12.4 Calculating the number of kanbans required

The formula used is $Y = (DL + W)/a$, where Y is the number of kanbans, D is the expected demand per unit of time, L is the lead time (i.e. processing time plus waiting and queue times), W is a policy variable (not more than 10% of DL), and a is container capacity (not more than 10% of daily demand).

Examples

D	L	W	a	Y
40/h	2 h	10	100	0.90 (say 1)
80/h	2 h	16	100	1.76 (say 2)
120/h	2 h	15	80	3.19 (say 3)

be no human operator. In such cases, the problem can be solved by what is called a *'full work system'*, which links the machines electronically so that when a machine is full, a signal is passed to the preceding machine to stop its operation.

12.6 KANBANS AND SUPPLIERS

Figure 12.4 shows the two-card kanban system in operation, including the supplier, which has an important role to play. Japanese companies tend to foster the parent–child relationship between vendor and manufacturer, a relationship based not on competition but on life-long association. This interdependence ensures that both quality standards and delivery schedules are strictly adhered to by the supplier — a vital part of any JIT system. A successful kanban system requires very short runs and frequent deliveries directly to the manufacturer's production line, so that the supplier tends to be located nearby.

Goods inwards inspection is eliminated and batches are expected to be totally free of defects. In many cases the supplier company is owned by the manufacturer (or is a joint venture) and is classed as *'first-tier'*, which means that the prime responsibility is to the owner, although a degree of independence is accepted.

12.7 JIT AND SUPPLIER DEVELOPMENT

Supplier development (sometimes called supplier integration) is a key element in any JIT system (in fact JIT cannot work without it), the basic principle being to build partnerships with a limited number of suppliers based on mutual trust and esteem. This is quite contrary to the traditional approach which adopted the master–servant relationship — the master gave the orders and the servant carried them out without question — or else! This just will not work in a JIT environment, and it must be appreciated that the two parties have common aims; both should be able to see the benefits of growing together and solving mutual problems.

Table 12.5 suggests how such a partnership policy might be implemented by a manufacturer, by choosing the right suppliers and helping them to develop into totally dependable associates. The manufacturer and the supplier should be capable of drawing up an agreement that sets out the conditions under which the partnership should operate, and in most cases this will include a 'single-sourcing' clause (see section 12.8). The sort of conditions that could be included are:

- means of transmitting purchase order schedules to supplier
- schedule amendments — acceptable arrangements

Table 12.5 Implementing a JIT-based supplies management system

The manufacturer should:

- be a significant customer
- offer extra volume to the supplier
- negotiate a 'no price rise' agreement
- arrange deliveries on a reliable daily basis
- expect the supplier to attain the highest standards in terms of quality and labour relations.

- tender and quotation procedures
- order contents (e.g. supplier part numbers, delivery instructions)
- arrangements for QA rejection and replacement
- vendor auditing.

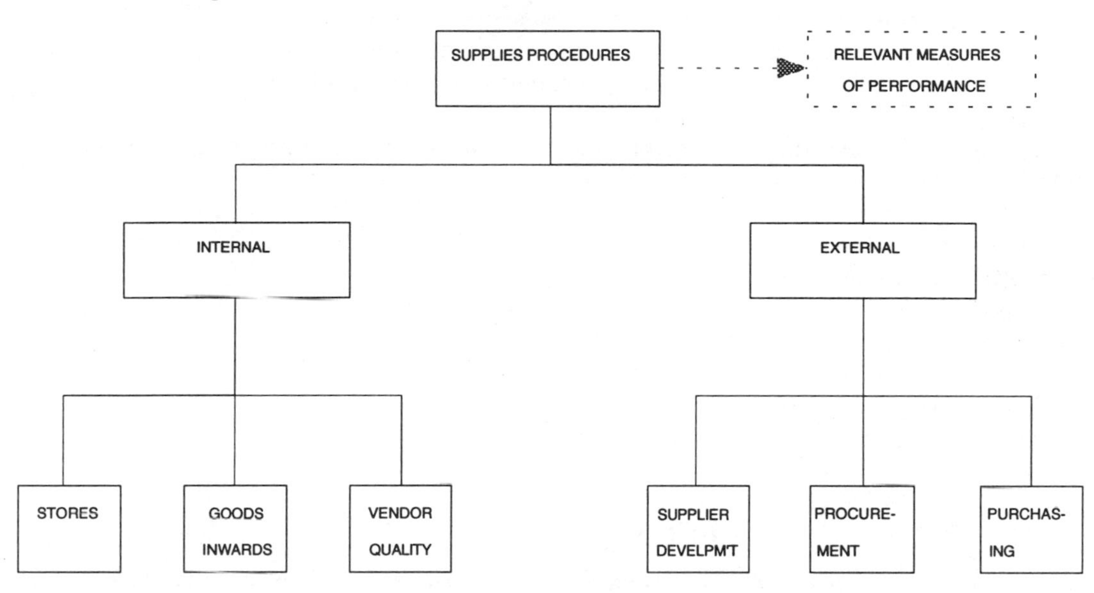

Figure 12.5 Supplier integration procedures.

The areas that need to be addressed during the supplier development phase are shown in Fig. 12.5, and can be detailed as follows.

- *Stores* —incoming material holdings to be minimized and monitored
  —minimum stocks (e.g. 1 per cent of sales)—ship direct to shop floor
  —stock only strategic items
  —packaging to suit usage by production (e.g. kanban units)
  —ABC classification system of control.
- *Goods inwards* —parts received on a JIT basis
  —frequent deliveries (a few hours of production)
  —kanbans used as signals for more supplies
  —majority of items delivered directly to production
  —ABC classification for control.

- *Vendor quality* —minimum inspection (by exception)
  —vendor conformance recording and analysis
  —total performance feedback
  —vendors responsible for corrective action
  —evidence of continuous improvement schemes.
- *Purchasing* —negotiated contracts for schedules etc.
  —reduced number of suppliers
  —regular planning meetings with suppliers
  —blanket orders based on levelled schedules (see section 12.9)
  —cost reduction exercises to agreed targets
  —quality/delivery standards clearly defined.
- *Procurement* —logistics (e.g. call-offs and means of delivery)
  —material movement authorization from supplier (e.g. kanban).
- *Supplier development* —strengths and weaknesses of suppliers recognized
  —common rules for planning, control, tooling, etc.
  —train supplier development engineers to help suppliers.

Measures of performance should be established to record continuous improvements and provide control, such as:

- stock turnover ratio
- turnover per employee
- cost index
- lead times
- supplier conformance Fig.s
- delivery performance ratios
- number of suppliers.

12.8 JIT AND SINGLE SOURCING

The pros and cons of single sourcing have already been covered (see chapter 8, section 8.11), but it should be stressed that, in a JIT environment, single sourcing is highly desirable if not essential. The reason for this is that a company operating on a JIT basis depends on total reliability from its suppliers, in terms of on-time deliveries of the required defect-free quantities. This is difficult when dealing with a few, single-source vendors, and impossible when many sources are involved. For example, how on earth could a kanban system operate?

The problems of single sourcing are more than offset by the benefits when JIT is in operation. For instance, the premiums that would probably have to be paid for small, frequent, deliveries would be more than counterbalanced by the savings from reduced stocks and lower quality auditing costs.

12.9 JIT AND LEVELLED SCHEDULES

Levelled scheduling is the process of producing schedules where both the production volume and the changes due to product mix have been smoothed out, over short time intervals, by varying the mix manufactured; this should ensure that capacity utilization is maximized, while small batch flow is catered for. The ideal levelled schedule would be so arranged that batches of goods were made in the same time periods in which they were sold, for example a product sold monthly should be made monthly; and a product sold daily, made daily.

Figure 12.6 shows examples of schedules based on both the 'traditional' and 'levelled' methods. In the traditional method, three large batches have been scheduled, whereas in the levelled method there are 10 small batches. From this it will be apparent that levelling can be successful only if small batches can be processed at reasonable cost. This is a JIT requirement that can be achieved only if set-up and change-over times can be cut to the minimum.

In the examples in Fig. 12.6 products have been categorized into three types, thus:

A. *runners* items that are in virtually continuous demand and are therefore produced daily
B. *repeaters* items that are required regularly but not continuously
C. *strangers* items that are required only occasionally.

The general rule for levelled scheduling is that *runners* should be put into the production program first, and *strangers* put in last, in order to achieve a levelled load (i.e. in this example to make high-volume items every day and low-volume items every week).

TRADITIONAL SCHEDULE

ITEM	WEEKLY VOLUME	BATCHES	DAY 1	DAY 2	DAY 3	DAY 4	DAY 5
A	100	1 BATCH OF 100	27	27	27	19	
B	27	1 BATCH OF 27				8	19
C	5	1 BATCH OF 5					5
TOTAL LOAD	132	3 BATCHES	27	27	27	27	24
CAPACITY	135		27	27	27	27	27

LEVELLED SCHEDULE

ITEM	WEEKLY VOLUME	BATCHES	DAY 1	DAY 2	DAY 3	DAY 4	DAY 5
A	100	5 BATCHES OF 20	20	20	20	20	20
B	27	4 BATCHES		6	7	7	7
C	5	1 BATCH OF 5	5				
TOTAL LOAD	132	10 BATCHES	25	26	27	27	27
CAPACITY	135		27	27	27	27	27

Figure 12.6 Examples of traditional and levelled schedules.

12.10 JIT AND CHANGE-OVER REDUCTION

JIT requires flexibility—the ability to run a variety of small batches—and to achieve this it is necessary to be able to change quickly from one product to another. In the past, change-over time, which is a non-value adding (wasteful) operation, was often ignored by management, who concentrated on reducing process time. There is no obvious explanation for this, apart from the possibility that change-overs were carried out by skilled engineers (who could be trusted not to waste time) whereas processes were carried out by semi-skilled or unskilled operators (who could not be trusted)—a totally misguided view! However, it is now accepted that reductions in change-over can be extremely cost-effective, in terms of both increased capacity and lower inventory (smaller batches).

Reductions in change-over times do not necessarily mean very high expenditure, as the graph in Fig. 12.7 shows since many of the time savings can be obtained by simple changes in methods rather than large capital projects. One of the easiest ways of cutting the change-over

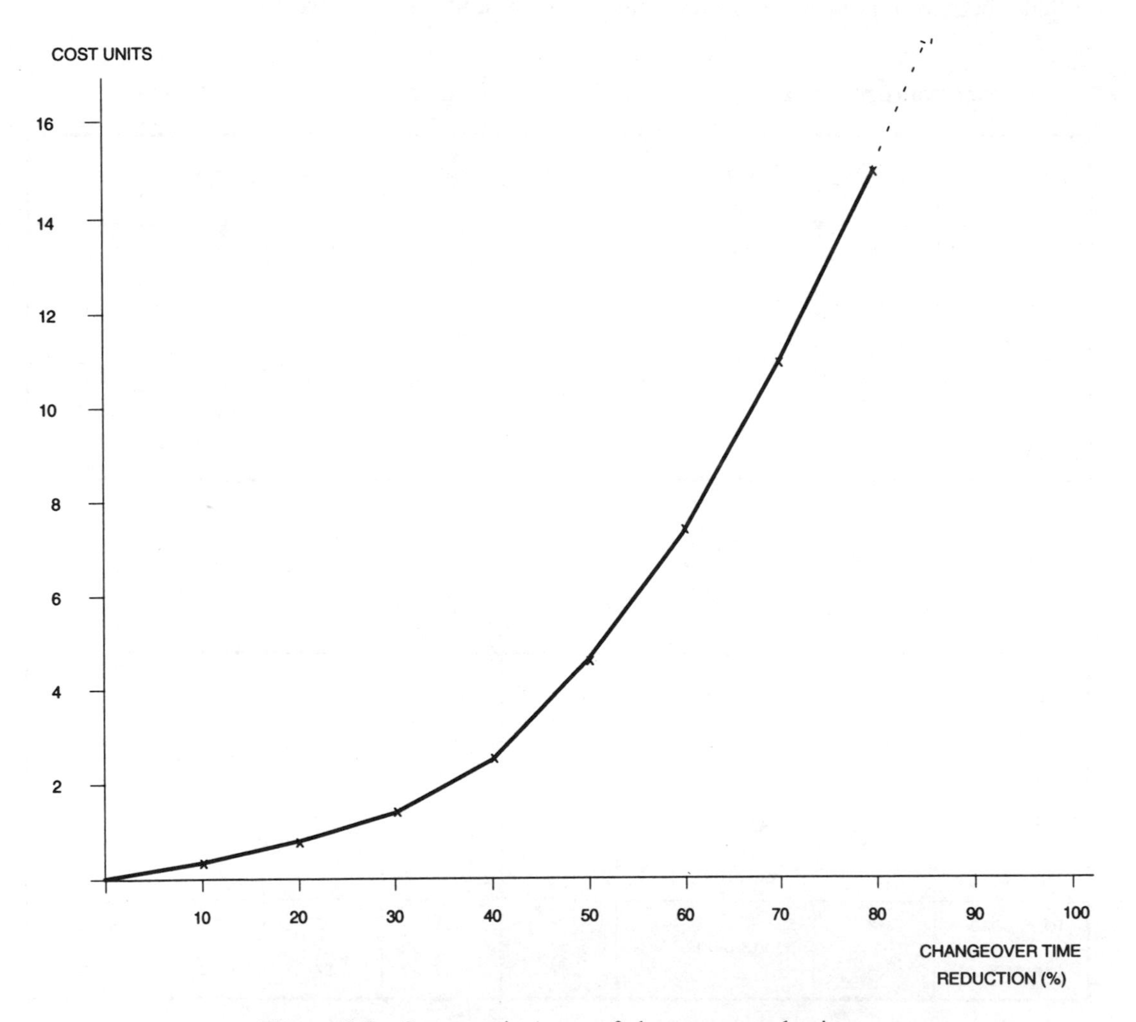

Figure 12.7 Comparative costs of change-over reduction.

time is illustrated in Fig. 12.8. This shows that all work involved in the change-over can be classified as either an internal or an external operation; these can be defined as follows:

- *internal* internal operations can be carried only out when the machine has stopped for the change-over
- *external* external operations can be carried out while the machine is still running (i.e. before it has stopped for the change-over).

Only the time spent on internal operations should be classed as change-over time, thus any operation that can be converted from internal to external will reduce total change-over time (e.g. the use of tooling that can be pre-set before it is fitted to the machine, the pre-setting of gauges, the pre-loading of NC programs and the pre-loading of the process material).

It is then possible to study the remaining internal operations and elements, using one of the work study techniques described in chapter 7 (section 7.9) in order to eliminate wasteful elements. For instance, quick-action clamps can be used instead of screw types, and specially designed quick-change tooling in place of conventional tooling. However, in many cases the elimination of lost time (i.e. waiting time) will yield the biggest savings of all. As an example of

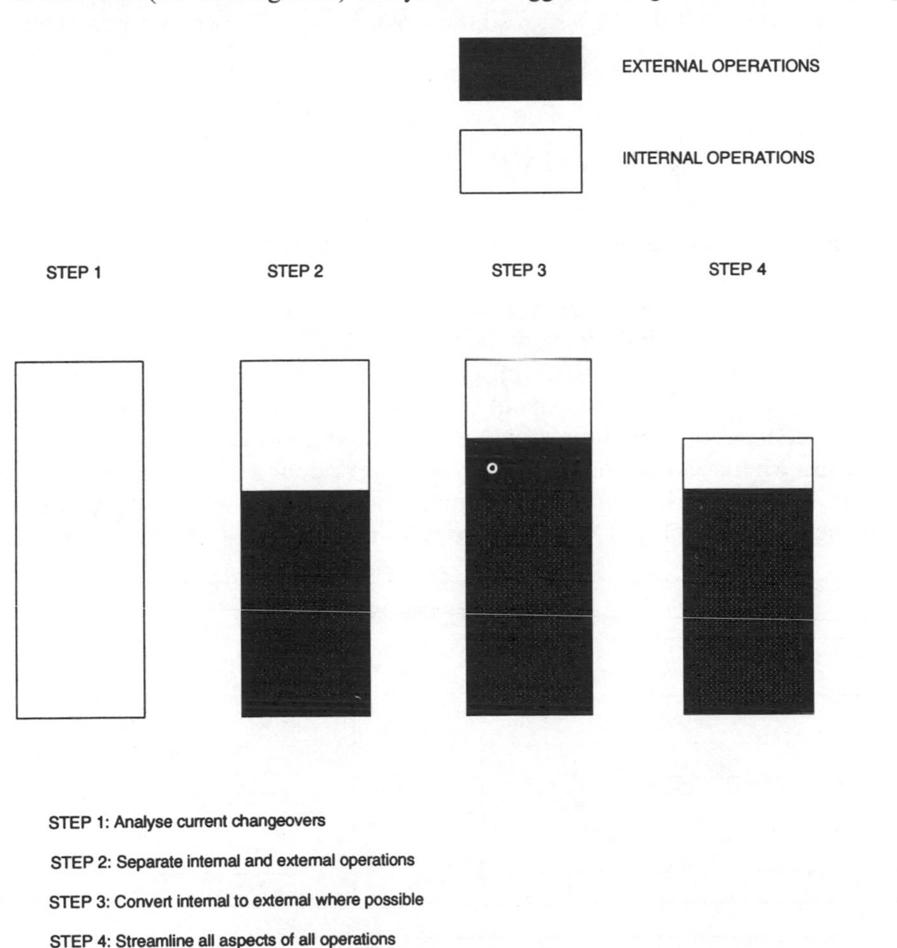

Figure 12.8 Change-over operations (internal and external).

Table 12.6 Reduction in set-up time—3000 tonne press (times in hours)

Element	Original	Revised	Comments
A. Remove/replace stillages	0.11		Made external operation
B. Remove/replace guards	0.18	0.07	Quick-release
C. Wait for tools	0.04		Eliminated
D. Clamping	0.92	0.04	Quick-action
E. Remove/replace tools	0.55	0.07	Pre-set
F. Cleaning	0.29	0.07	Mostly external operation
G. Release/replace air lines	0.15	0.15	No change
H. Adjust weight	0.07		Made external operation
I. KO bars/shut height adjustment	0.22		M/C modified
J. Paperwork	0.07		Made external operation
K. First-off pressing	0.04	0.04	No change
L. Inspection (first-off)	1.03	0.04	By operator
Total time	3.67	0.48	Reduction = 87%

what can be achieved, Table 12.6 shows how a reduction of 87 per cent was made in the set-up time for a 300 tonne press.

12.11 JIT AND CELLULAR MANUFACTURE

Cellular manufacture is based on the concept of 'devolution' whereby, within a *business unit*, there can be one or more *product units*, each of which can contain a number of *cells*, as shown in Fig. 12.9.

The business unit is self-contained, with its own management team and support services, and each product unit within it manufactures a single, logical group of products. The cells within a product unit contain a natural group of machines and people (the cell team), being laid out to meet the JIT requirements of efficient process flow, job flexibility, and multi-skills. The cellular concept does not apply only in production, but is equally effective in the administration area, where cells can be based on type of work done and information given or received.

A manufacturing cell should allow a number of machines to be operated by a very small team: in the example in Fig. 12.10 there are eight machines and two operators. Each operator

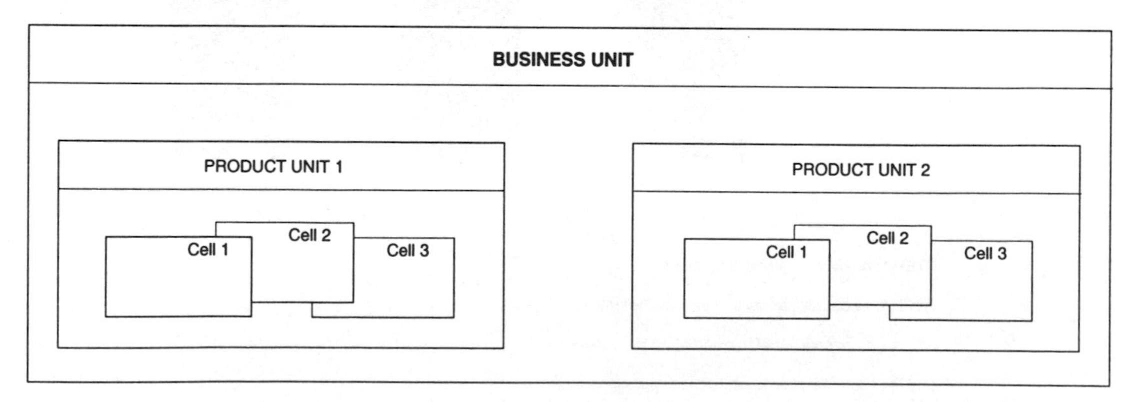

Figure 12.9 Organization for cellular manufacture.

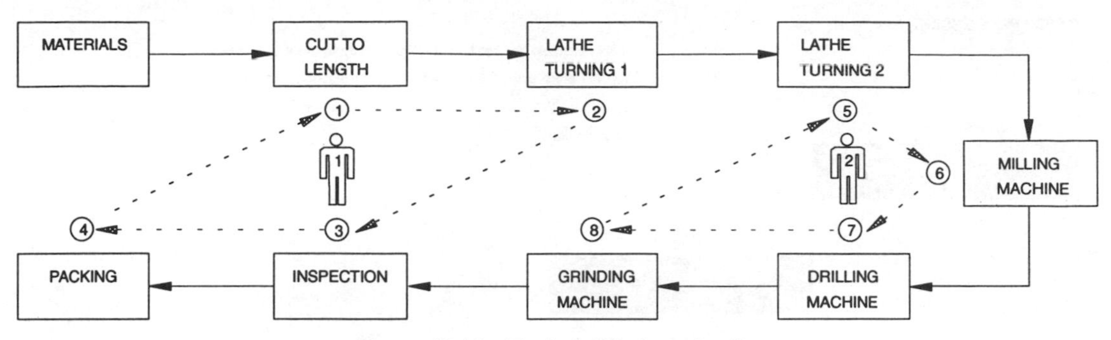

Figure 12.10 Typical 'U' shaped cell.

ensures that production flow is maintained by unloading a work station, reloading it with another piece, starting the process and taking the unloaded piece to the next work station.

It will be noted that this example shows two turning stations; this is to try to maintain line balance by splitting the various turning operations in half. It will also be apparent that such a system can work to the maximum effect only if the machines operate automatically, such that, once loaded with a workpiece, they will carry out a pre-defined sequence of operations and then stop.

The benefits of such a system can be seen in Fig. 12.11 which shows how, by replacing batch working with continuous flow, considerable reductions in time taken can be achieved. The delays in batch working occur when the first of the batch has to wait until the last of the batch has completed an operation before moving to the next operation. However, when using the flow process method, care must be taken to ensure that batch traceability is not lost.

12.12 JIT AND CONTINUOUS IMPROVEMENT

If a company is to be competitive and to retain a competitive edge over its rivals, it needs to follow a programme of continuous improvement, and this involves two fundamental changes in approach.

1. A complete reconstruction and simplification of the manufacturing system should be carried out to eliminate the complexities that have almost certainly crept in over the years. It may not be possible to install a full JIT system, but often many of the concepts can be applied.
2. A systematic process of continuous improvement should be set up, with employees trained and motivated to apply the required techniques. This is often achieved by setting up a series of 'local' improvement teams, their work being coordinated by a company-wide team.

Figure 12.12 shows how the process works, and the way in which a *step change* is used to break free from the inhibitions of the past ('we have always done it this way'). This example also shows that the competitors may be improving their performances at the same time, and without the step change would never be caught, let alone surpassed.

To achieve the first objective (step change) it is necessary for a company to apply a coordinated approach by a trained project team, with the full cooperation of all other employees; no area of the business should be ignored. The main tasks of this team are to involve everyone in determining what should be done, with clearly defined targets (the plan), and then to monitor the implementation.

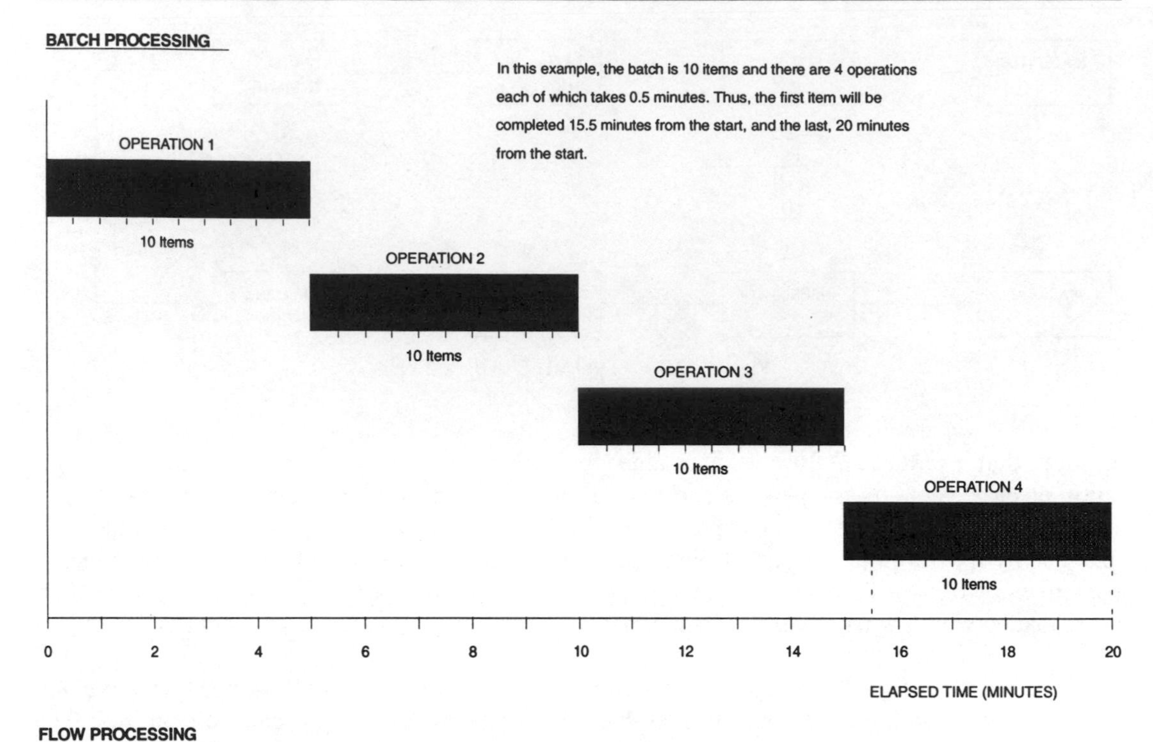

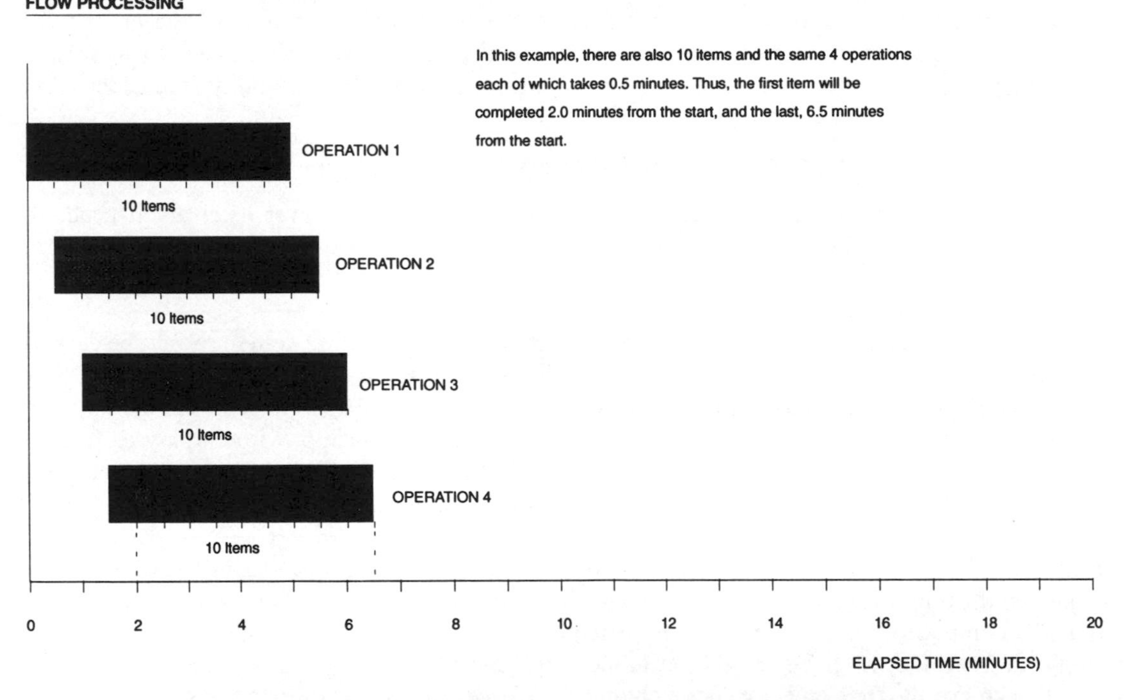

Figure 12.11 Batch processing v. flow processing.

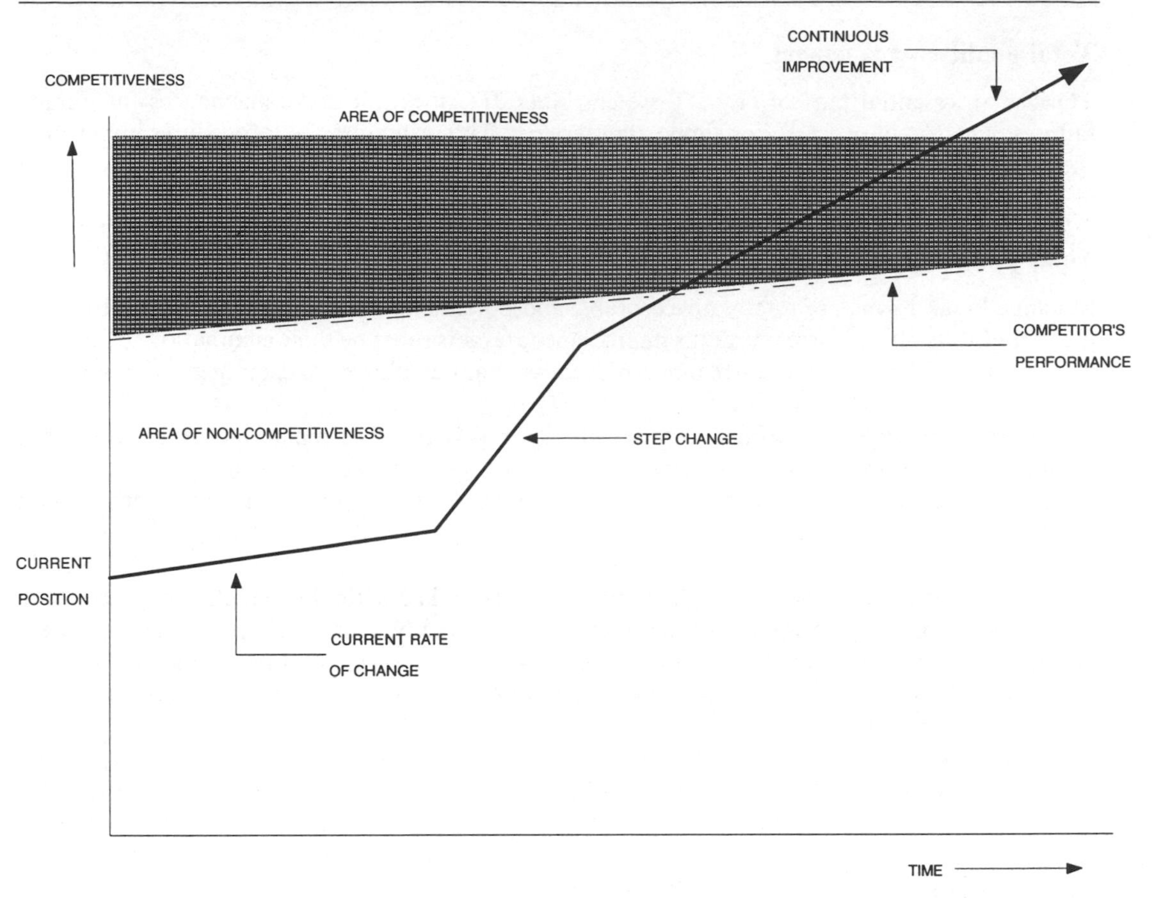

Figure 12.12 A continuous improvement policy.

The second objective (continuous improvement) can be achieved in the same way, but in this case the targets are improvements rather than system changes, and the problem is often one of motivation. This is because, once a target has been achieved, a new target must be set, and the process can never be said to have been completed.

12.13 OTHER JIT TECHNIQUES

Various other techniques can be applied to achieve the JIT ideal, which requires the elimination of all work that is non-value adding (NVA).

Flowchart simplification

This technique has already been examined to some extent in the section on cellular manufacture (section 12.11 above), but consideration should also be given to the overall flow through the factory, and between cells. In the section on work study (chapter 7, section 7.9) various methods of flow charting were described, and these should be used to eliminate all unnecessary movements, to reduce distances travelled, and to cut out such non-productive elements as inter-process (temporary) storage, inspections and delays.

Total quality management

TQM is an essential part of any JIT system, since JIT does not make allowances for quality failures, whether in processes or supporting services. The principles and procedures involved in the operation of a TQM system are described in detail in chapter 13 (sections 13.14–13.16).

Total productive maintenance

Machine breakdowns are totally unacceptable and potentially disastrous within a JIT environment, and thus one of the continuous improvement targets must be their elimination. In chapter 7 (section 7.6) the principles of planned preventive maintenance were examined, whereby:

- *preventive* measures are defined in terms of inspections and part replacements to ensure that all possible causes of breakdown are dealt with before a breakdown occurs
- *planned* inspections take place at the required intervals subject to minor adjustments to suit production requirements.

Chapter 7 (section 7.7) also covers the various aspects of TPM (total productive maintenance), which goes further by considering other factors such as the way in which plant and machinery are being used, and how all users can be involved, e.g. by observing and reporting any abnormalities (noise, vibration, overheating) or variations from standard in terms of process reliability or instrument readings.

In addition, all plant and machinery suppliers should be included in the TPM process, and be prepared to submit their equipment to reliability and risk analysis checks. They should also be able to recommend preventive maintenance schedules for their products, and be able to supply full maintenance manuals with diagnostic, fault-tracing charts (preferably in the form of 'decision trees'—see example in Fig. 12.13).

TPM is a part of the total quality concept, due to the need for the process capability of a machine to be maintained to match the design requirements. Thus, if the process capability falls, due to inadequate maintenance, the machine must be halted for servicing. This is a form of breakdown and is therefore unacceptable.

12.14 ORGANIZATIONAL CHANGES FOR JIT

The most suitable organization for a JIT-based company has been described in section 12.11 of this chapter, where the company consists of one or more *business units* within which there are *product units* and *cells*. The business unit should be a self-contained business with a general manager, an executive team and a number of support service units. These last should be organized into 'natural groups', i.e. multi-skilled groups or teams who 'own' a complete flow chart, and work together in one office or area.

This concept can best be defined by means of an example, as given in Table 12.7. In this case, the purchase ordering group should include all these functions, and therefore the team should include not only a purchasing clerk but also an accounts clerk and a goods inwards clerk.

Each product unit, within a business unit, should have a unit manager and be responsible for a clearly defined series of products. These products should all have similar characteristics, so that the various cells within the unit can be designed to produce any type of product mix without difficulty. The product unit should be responsible for its own production planning,

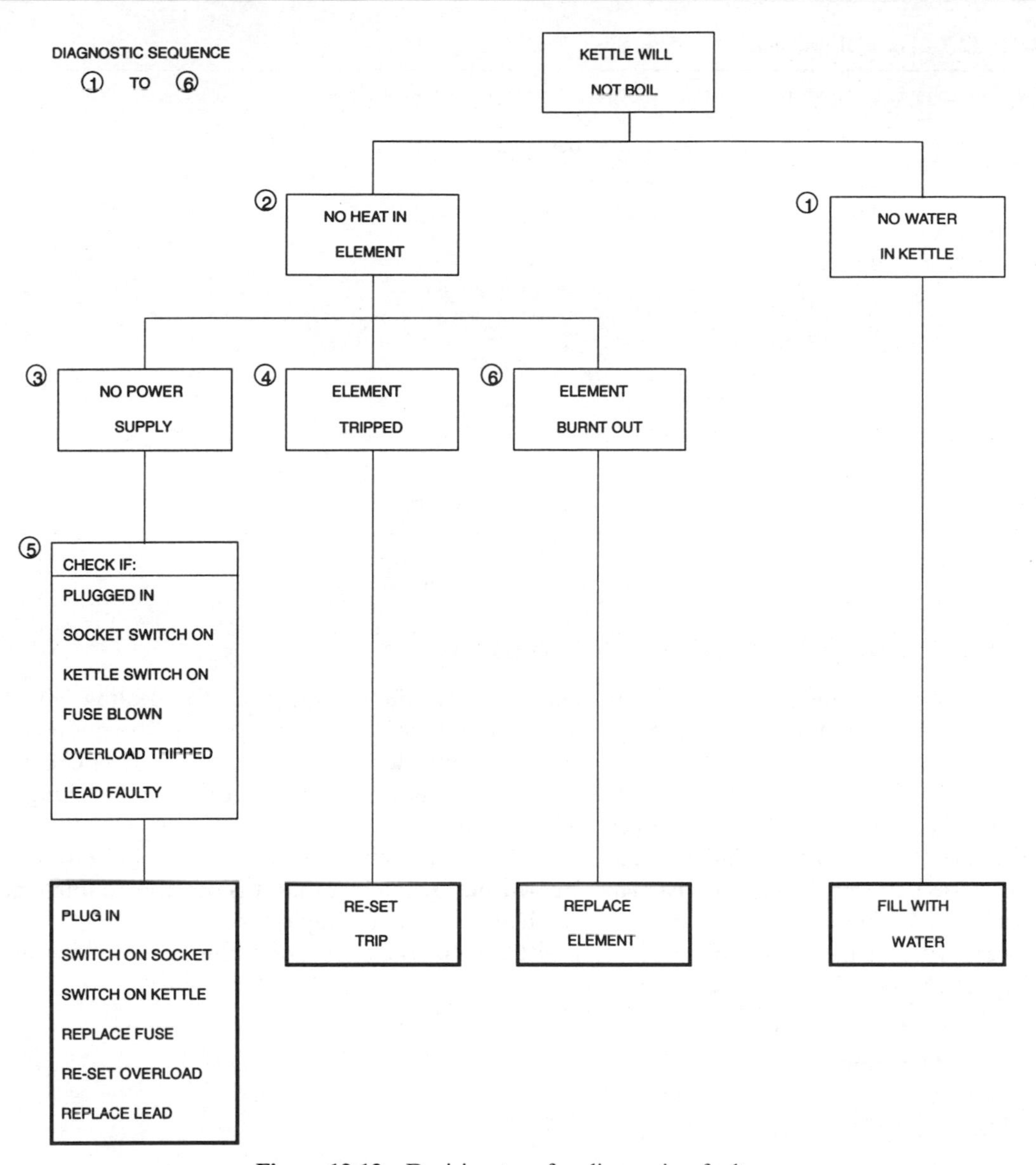

Figure 12.13 Decision tree for diagnosing faults.

scheduling, tooling, and quality control but should use the services of the business unit for other functions such as purchasing, accounts, maintenance and payroll.

Each manufacturing cell within a product unit should contain a natural group of machines/people (a complete flow chart) and their internal customers and suppliers should be clearly defined. They should also be given clear targets, and understand how achievement of these will be measured.

This type of organization (with both administrative and manufacturing cells) provides an ideal basis for the successful implementation of JIT.

Table 12.7 Natural groupings

If a flow chart is drawn for a purchase order, the elements will include:

order preparation
↓
order to supplier
↓
order progressing
↓
goods receiving
↓
invoice validation
↓
purchase ledger posting

12.15 COMPUTERS WITHIN A JIT SYSTEM

It has been said that JIT is a way of manufacturing life, an approach for dealing with the problems and demands of the contemporary world of manufacturing; it is therefore not possible to set up a computer system that can be used to implement JIT. In other words, a computer system cannot be used as a substitute for the work that needs to be done to make the changes in culture that are essential to a successful JIT implementation.

Thus, the role of the computer in a JIT system can be defined as one of the helpful tools that can be used, rather than a panacea. This is borne out by the fact that it is possible to implement many of the concepts of JIT without a computer in a small company.

The role of the computer in JIT can best be illustrated by examining some of the areas where it can be of most value, for example:

1. Manufacturing cells
2. Customer/supplier relationships
3. Design/production relationships

Manufacturing cells

The ideal basis for operating a manufacturing cell (see section 12.11 of this chapter) is to use an FMS (flexible manufacturing system). This consists of a group of CNC machines and handling devices (e.g. robots or conveyors) which are controlled by a single computer. The way in which FMS has evolved is shown in Table 12.8.

This computer (which may be a mini or a PC) contains all the programs needed to manufacture a range of parts, i.e. programs for machining, parts handling, inspection and tool handling. The computer may be linked, via a network, to a larger computer that controls other cells and process units, to give a fully integrated system which ensures that parts and products are manufactured against schedules, in line with JIT kanban principles.

Table 12.8 The evolution of FMS

Stage	Machines	Computers	Handling/tooling
1. Manual	Operator control	None	Manual
2. Numerical control	Single NC machine	NC programs manually entered to machine	Mechanized
3. CNC control	Single CNC machine	CNC programs in machine computer	CNC-controlled handling of product and tools
4. DNC control	Number of CNC machines	DNC programs loaded from off-line computer	DNC-controlled work pool and tooling
5. Early FMS	Cell group of CNC machines	Mainframe controls DNC and transport	Transport moves parts between processes
6. Current FMS	Cell group of CNC machines	Mini or PC controls DNC and workflow	Integrated workflow and tooling transport sytems

The sequence of events in such a system could be as follows.

1. Main computer informs cell computer that a specified quantity of a particular part is to be manufactured.
2. Cell computer finds the appropriate programs and loads them to the various units in the cell; this is achieved by using DNC (direct numerical control) software. These programs could include CNC machining and inspection programs, tooling requirements programs and handling device programs.
3. When all programs have been loaded and tested by the computer, a signal is given to load the first part or unit of material. This may be a manual operation or could involve computer-controlled withdrawal from a pallet or rack.
4. The sequence of machining and handling operations proceeds, each unit working in conjunction with its preceding and succeeding units, until the requisite quantity has been made and inspected.
5. The cell closes down until the next instruction to manufacture is received.

It should be noted that most CNC machines include a tool magazine and tool changer, which can be programmed to install and use the correct tool for each operation on a particular component; this also applies to robot grippers and inspection machine probes.

The cell computer should also be programmed to deal with rejects at any stage of the cycle, and to 'halt' the system if a problem occurs. It can deal with such problems as tool breakages, and in some cases can make machine adjustments if measurements show that tolerance limits are likely to be exceeded.

FMS principles can also be used to control the flow of work between cells, e.g. by programming AGVs (automated guidance vehicles), and to control batches of products being moved into the finished goods area (automated warehouse equipment and automated sorting equipment for assembling despatch loads).

Customer/supplier relationships

The development of good customer/supplier relationships is one of the cornerstones of a successful JIT system, and this can be aided by the implementation of well-planned, computer-based communication systems. This is particularly true when suppliers are required to meet frequent delivery schedules, in quantities that may not be known until just before the delivery vehicle is planned to leave the supplier's premises. In such cases, an EDI (electronic data interchange) system can be used by the customer to make the 'call-off' or to enquire about the relevant stock position.

Computers can also be used by the purchasing function to ensure that the correct drawings or specifications are sent to the supplier with an order; this is achieved by transmitting the data electronically from the purchaser's CAD system to the supplier's, thus ensuring that the latest version is used.

The importance, in JIT, of internal customer/supplier relationships has already been covered in this book, and in this area the computer can also make a useful contribution. For example, changes in production plans and schedules can be quickly initiated if production planning (the internal supplier) is informed immediately of production problems via direct computer links from production management (the internal customer). Similarly, such links can be used by sales (internal supplier) to inform producton planning (internal customer) that changes are required due to a problem with the external customer.

Design/production relationships

In a JIT situation, it is essential that designs be related not only to customer requirements, but also to production capabilities. This means that new designs must be tested against these capabilities before production starts, including verification on tooling and production software. This used to mean that production could be disrupted during these tests, a loss that could be described as waste in a JIT company. However, such waste is now unnecessary, since all the test work can be simulated on computer at negligible cost, and designs, tooling, etc., modified (if necessary) before production starts.

Computer systems are also invaluable for cutting 'time-to-market' using CAD to prepare drawings, CAM to prepare machine and other programs, and CAPP to design the production processes. Computers can be used to simulate how a product will behave in service, e.g. by wind tunnel or crash simulation; this also reduces time-to-market.

PROBLEMS

1. The company for which you work is considering the possibility of using the JIT approach. However, the finance director is concerned about the change from economic batch quantities in production, which do not fit in with the JIT philosophy. Prepare a report explaining how, by reducing batch sizes, the company will gain 'overall' (see chapter 9, section 9.9 for EBQ methods).
2. The same company is planning to use a 'two-card' kanban system and wishes to explain this process to the production workforce. The proposed system will use actual cards rather than tote pans, and you are asked to prepare a flow chart showing the movement of the cards between two work cells. Your chart should indicate the signals received and the actions thus initiated.
3. In section 12.7 of this chapter a list of seven *measures of performance* is given. A supplier has asked for these to be defined and how the information will be obtained; you are asked to respond.
4. The Deming principles for the achievement of total quality are listed in chapter 13 (Table 13.7). Examine these and identify those that are relevant to the concept of JIT, together with brief reasons for your choices.

5. In this chapter (section 12.14) a flow chart is given as an example of a JIT organization (Table 12.7). You are asked to prepare a similar chart for the design function, starting at the point where a design specification is received and ending with the issue of drawings, NC machine programs and tooling lists. You should also suggest the composition of the cell that will carry out these functions.

FURTHER READING

Black, J. T., *The Design of the Factory with a Future*, McGraw-Hill, New York, 1991.

Harrison, A., *JIT in Perspective*, Prentice-Hall, London, 1992.

Hutchins, D., *Just in Time*, Gower, London, 1988.

Kirton, J. and Brooks, E., *Cells in Industry: Managing Teams for Profit*, McGraw-Hill, London, 1994.

Schneiderjans, M. J., *Topics in Just in Time Management*, Allyn and Bacon, Boston, MA, 1993.

Shingo, S., *A Study of the Toyota Production System*, Productivity Press, Cambridge, MA, 1989.

Williams, D. J. and Rogers, P., *Manufacturing Cells: Control, Programming and Integration*, Butterworth Heinemann, Oxford, 1991.

Womack, J. P., Roos, D. and Jones, D. T., *The Machine that Changed the World: The MIT $5m, 5 Year Report on the Future of the Automobile Industry*, Rawson Associates, Cambridge, MA, 1990.

CHAPTER

THIRTEEN

THE MANAGEMENT OF QUALITY

THE LINK

In the preceding chapters an attempt has been made to establish the basic principles of sound management and to look at the various functions that were traditionally considered to be a part of production management. This traditional approach did not consider quality as one of the production functions; in fact, in those days, the relationship between those responsible for production and those responsible for quality as often confrontational and antagonistic. As some saw it, the task of production was to produce the maximum quantity of product, and if some was of poor quality then it was the task of the inspectors to find it (if they could) and reject it.

This approach was the road to ruin, and is now accepted as such. Quality is no longer seen as a drag on production, but as an essential to survival, and in this chapter it is recognized that production cannot operate successfully without the application of quality control techniques. In other words, it is recognized that quality is now a part of the production function.

In fact, quality should be a company-wide 'way of life' which encourages every employee to recognize that he or she can contribute to a culture that aims for perfection in design, manufacture and service to the customer. This is the TQM (total quality management) approach which rejects the idea that quality is up to 'them', and accepts that it is up to 'us'.

13.1 INTRODUCTION

In recent years the topic of quality has been the subject of more books and papers, more seminars, more 'systems', and more discussions than all the other aspects of manufacturing management put together! This is not a criticism, but merely an observation. In this chapter an attempt is made to clear away some of the misconceptions, so that the real functions of quality management can be seen and the objectives set.

Until fairly recently, the costs of obtaining acceptable standards of quality in manufacturing were generally considered to be a part of the natural order, and therefore accepted as a necessary overhead that every company had to carry. However, this is no longer the case. It was mainly a recognition that the real costs were much higher than previously thought that drew the attention of management to this topic. In fact, in today's highly competitive environment, such costs would be totally unacceptable.

Therefore, having defined the objectives of quality management, this chapter examines the old ideas on quality in order to identify what these real costs were, and then follows through the various stages in the development of modern systems, to conclude by considering the spin-off benefits that have been derived from the implementation of such systems.

13.2 QUALITY MANAGEMENT—THE OBJECTIVES

The primary objective of any quality management system should be to define a series of performance standards, which are compatible with the products being manufactured and the customers' requirements; furthermore, this objective must be achieved at minimum cost. This can be accomplished only by a company which recognizes that total commitment and involvement are essential at all levels, and is prepared to apply the latest techniques. However, these performance standards will not be static because the demands of customers will change, and the level of competition will increase. Thus, as a secondary objective, it will be necessary to operate on the basis of continuous improvement in order to maintain a competitive lead.

This can perhaps be more easily defined by considering the term 'world-class'. Any company wishing to achieve real and lasting success should have as part of its overall objective the attainment of standards of quality that can stand comparison with the world's best. This means that the company should be working continuously to improve its standards at all levels and in all areas, for example:

- product design—design fit for function
- production engineering—process fit for design
- supplier relationships—supplies fit for production
- customer service—service fit for products.

These standards and their application are considered in more detail later in this chapter.

13.3 QUALITY MANAGEMENT—A HISTORICAL VIEW

Until comparatively recently the function of quality was considered to be merely a matter of acceptability. Thus, dimensions were checked to ensure that they were within limits and appearance was judged on a subjective basis. This approach did not look seriously at the causes of quality failures, but accepted that there would always be a certain level of wastage due to scrap, rectification and defects. In other words, management did not appear to recognize that good quality standards could be built into a product at the design stage, and maintained by ensuring that processes and bought-in materials were specified to maintain those standards. Quality was mainly in the hands of inspectors who worked on a controlled basis of sampling and measuring, causing hold-ups in production without any lasting improvement in quality performance.

This began to change when management realized that the true costs of quality management were much higher than previously thought (see section 13.4 below) and that the development of computers allowed the whole approach to be re-examined.

13.4 QUALITY MANAGEMENT—THE TRUE COSTS

Until fairly recently the true costs of quality were not generally recognized, probably because the wrong question was being asked. The question that should have been asked was 'What is the total cost to the company of quality failures?' and not 'What is the cost of an inspection

Table 13.1(a) The costs and savings of a total quality approach

Savings	Costs
Cost of producing scrap and defective items	Redesign of products
Cost of rectifying faulty items	Alteration of processes and tools
Cost of goods-inwards inspection and returns	Development of self-inspection procedures
Cost of customer complaints and claims	Development of supplier auditing systems
Cost of inspectors and lost time	Development of improved customer relations
	Continuous improvement programmes

Table 13.1(b) The costs and savings of a quality inspection system

Savings	Costs
Costs of defective items reduced	Inspectors' wages and lost time
Rectification work reduced	

system?'. Tables 13.1(a) and (b) show how this change in the question can radically affect management thinking.

The major difference of approach is obvious. In Table 13.1(a) the broad view is taken; all the factors that contribute to the costs of quality, and all the savings that can be achieved, have been taken into account. This is not the case with Table 13.1(b), which looks at the situation in a narrow and selective way.

13.5 COMPUTERS IN QUALITY MANAGEMENT

The development of computer technology in recent years has radically changed management thinking on the quality functions, since it is now possible to perform tasks that were at worst impossible, and at best time-consuming and difficult. This type of computer technology can be used in several ways, such as the following.

- To program and run inspection machines (for example a coordinate measuring machine (CMM) similar to that in Fig. 13.1) which can check the dimensional accuracy of a component in a fraction of the time it would take using manual methods. This allows such machines to be installed as a part of a production line or cell, and every component to be checked as it is produced. Such machines can then use the data collected to correct the production machine settings if the trend suggests that an 'out of control limits' situation is developing.
- To collect and analyse data from a series of measuring devices, so that information can be obtained and used to improve the process or the product. These improvements may be either short-term, to prevent the process going out of control, or the long-term, to identify potential product faults, that would need to be corrected by redesign.
- To carry out simulations to ensure that the product will perform in accordance with the specification. Simulation can be used to check the performance of a mechanism or electronic circuit under various conditions, to examine the effects of a component failure, or to minimize the effects of accidental damage (e.g. vehicle impact) by design improvements.
- To carry out complex engineering calculations using techniques now readily available under the generic title of CAE (computer aided engineering). These include FEA (finite element analysis), which is used to calculate stresses in a component under a variety of load condi-

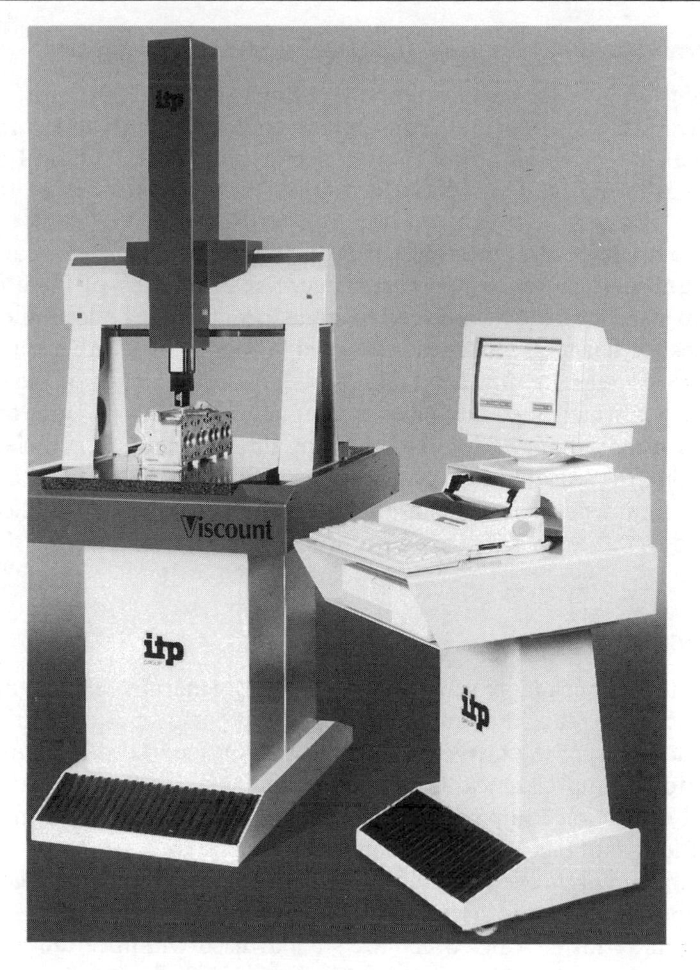

Figure 13.1 A Co-ordinate Measuring Machine (CMM). This is a moving bridge type CMM, the component being mounted on a static worktable. The bridge and probe head being moved over the component in a pre-programmed path so that the electronic probe can measure the co-ordinates at various points on the component. These can be compared with data on the linked computer in order to calculate any dimensional inaccuracies (photography by courtesy of ITP Ltd. Rugby).

tions, and CFD (computational fluid dynamics), which can provide information about the flow of liquids and gases, heat distribution and the effects of cooling.

It is therefore apparent that quality can, with the help of computer-based systems, be designed into a product and maintained throughout its production.

13.6 QUALITY MANAGEMENT—MEASUREMENT AND RECORDING

There are basically two types of measurement that have to be carried out as a part of any quality control system: quantitative and qualitative.

Quantitative measurement

These measurements all involve some form of dimensional accuracy, the most obvious entailing lengths, widths, heights, diameters and radii. Others could deal with concentricity and profile (for example, are all the cams on a cam shaft concentric with the shaft and do they have the correct profile?), hardness, the depth of surface finish (paint or plating) or the accuracy of a screw thread.

There is a wide range of equipment designed to cater for this type of measurement, including plug, thread, and dial gauges, calipers, and micrometers. Many of these now have digital read-outs; they can also be directly connected to some form of data collection or logging device. There are machines for hardness testing and for comparing profiles with a standard workpiece.

Vision or video systems are now available that allow a 2-D image of the component to be projected onto a screen which, if connected to a computer, can allow the user to determine—by comparison with a standard—whether there are any dimensional inaccuracies.

It is thus possible to measure virtually any dimensional characteristic of a component, and to record it on a computer for subsequent analysis, often at a speed that allows the measurements to be carried out at the work place, by a production operator, during the machine cycle.

Qualitative measurement

These measurements are much more subjective in that they generally relate to appearance rather than dimensional accuracy; thus they are unlikely to affect performance or 'fit'. For example, it may be necessary to measure the texture and form of a surface in order to ensure that it meets the specification (for example, optical lenses or bearing surfaces for lubrication). This can be done using special inspection equipment and computer software which can measure variations to an accuracy of less than 0.001 μm (micrometres).

Another example involves examination of painted surfaces for the 'orange peel' effect. It is now possible to obtain a portable laser scanning instrument (see Fig. 13.2) that calculates the values of the effect in terms of scales which are standardized within certain industries (e.g. the ACT rating).

However, some qualitative checks can only be carried out by the human eye, e.g. glaze faults in pottery, or dirt specks in a painted surface. In many cases it is important to identify the location of the fault as well as its nature, and computer software is available for this purpose (Fig. 13.3). In this figure a pre-printed form is shown, but current computerized systems will display (on screen) various views of the model concerned. The inspector can then use a 'light pen' to mark the fault locations, and to select codes for fault definition from a list on the screen.

It will be noted that all the tests and measurements described in this section are non-destructive, and this must always be the aim. With the systems and equipment now available, destructive testing is rarely necessary, except perhaps during the development phase of a new product.

13.7 QUALITY CONTROL PRINCIPLES

Like all management control systems, quality control is based on the closed-loop principle A typical application shown in Fig. 13.4 shows how the faults found by inspection should be recorded and analysed, and the results used to tackle apparent areas of weakness so that, whether the weakness is in design, production or supply, the company can move closer to the

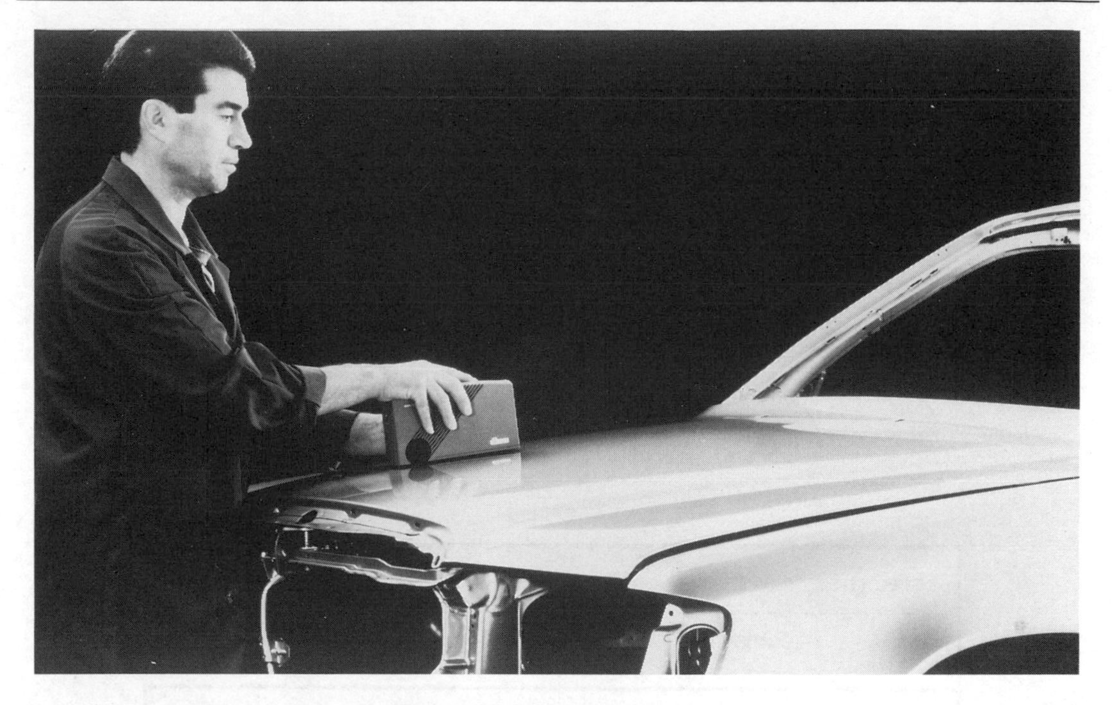

Figure 13.2 Scanning a painted car body shell for 'Orange Peel'. The instrument being used here is the Wave-Scan which detects the pattern of light and dark areas in the same way as the human eye. (photograph by courtesy of Sheen Instruments Ltd. Teddington).

ultimate (and unachievable!) objective of zero defects at zero cost. Generally such defects are now measured in parts per million (ppm).

The work of the quality control department can be broken down into six functional control areas, each of which is examined in detail in the remaining sections of this chapter:

- design
- suppliers
- processes
- traceability
- documentation
- customers.

13.8 QUALITY CONTROL AND DESIGN

A number of 'tools' or techniques can be used to check out a design from a quality point of view, and it is one of the functions of the quality control department to ensure that these techniques are used by the designers, and to assist where necessary. The tools concerned are:

1. failure modes and effects analysis (FMEA)
2. critical parameter management (CPM)
3. quality function deployment (QFD).

VISUAL INSPECTION REPORT

MODEL 104 SX - TWO DOOR
COLOUR IRISH GREEN
CHASSIS NO. 245784NP27
BATCH NO: C/487/231

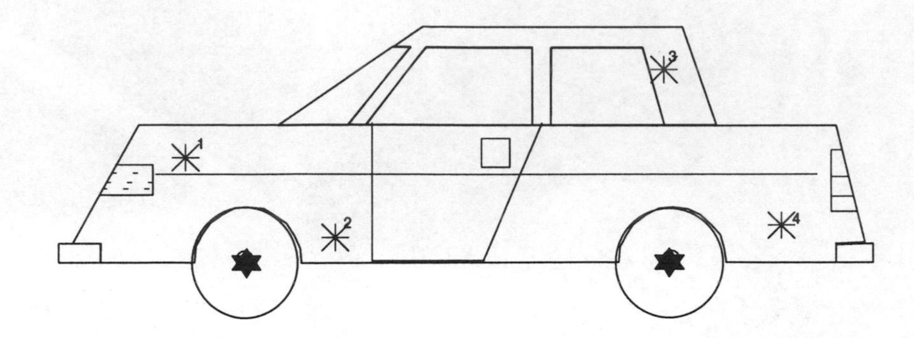

INSPECTED BY: J Roberts Date: dd/mm/yy Time: hh-mm

NO.	LOCATION	FAULT CODE	DESCRIPTION
1	UPPER FRONT WING	07	PAINT RIPPLE
2	LOWER FRONT WING	23	SURFACE SCRATCHING
3	REAR ROOF	24	PAINT BUBBLE
4	LOWER REAR WING	32	PAINT PINHOLES
5			

Figure 13.3 Recording the results of a visual inspection.

Failure modes and effects analysis

FMEA is designed to predict the failures that might occur in a sub-assembly or component, the effects that such failures might have in the operation of the full assembly, and what steps could be taken to prevent the failure and its consequent effects. FMEA is a formalized approach to the problem of evaluating a design from a quality of performance point of view. It answers the question 'is this product fit for the purpose for which it was designed?' and is based on a matrix which evaluates each part in terms of a risk priority number (RPN).

This RPN indicates the degree of correlation between a part and a failure mode, the scale and symbols commonly used being:

9 double circle strong correlation
3 single circle some correlation
1 triangle possible correlation.

Figure 13.5 shows how this rating scale has been applied to a simple product, in this case a wooden pencil, and how a weighting factor has been applied to the RPN values. Thus, the score

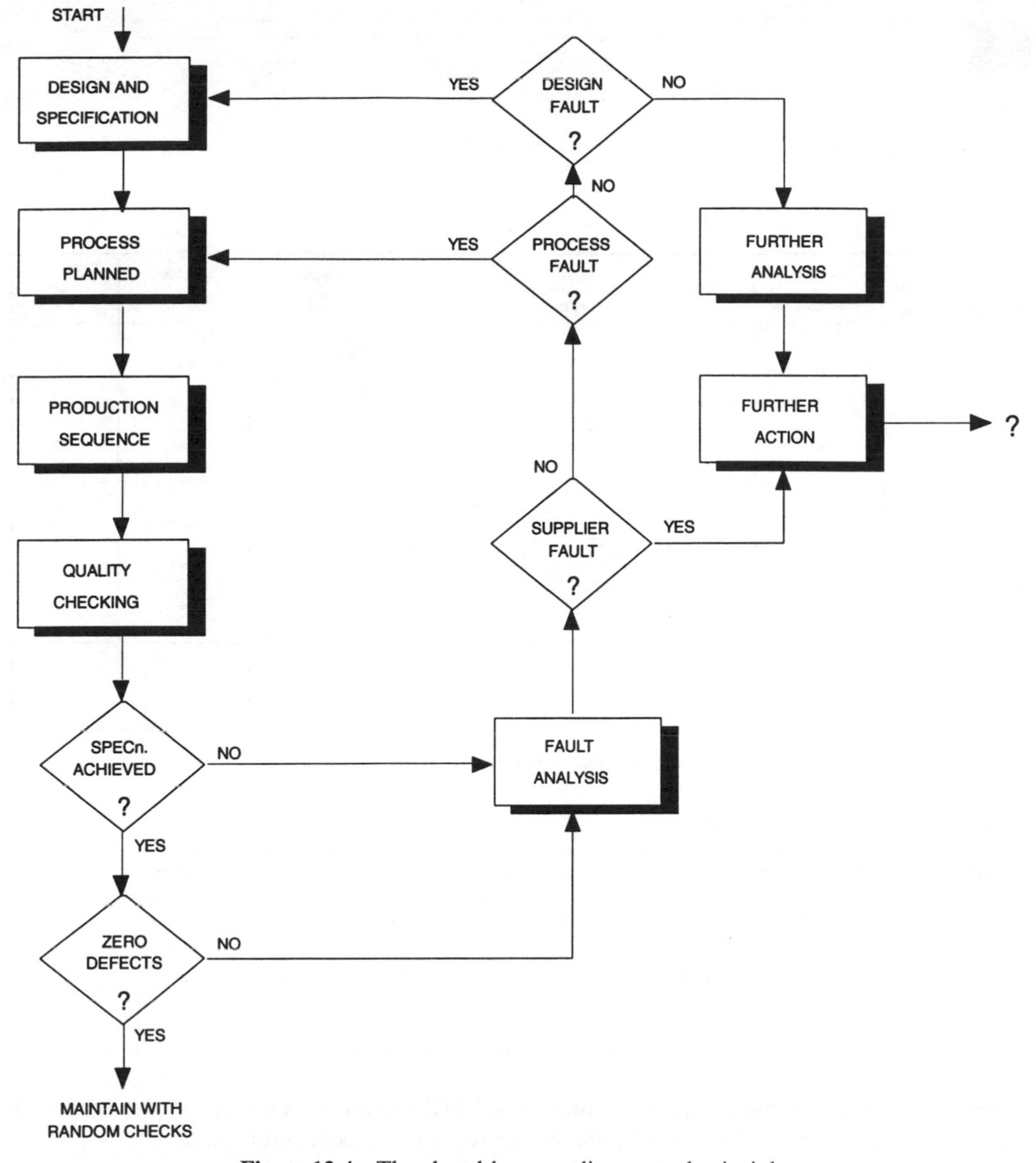

Figure 13.4 The closed loop quality control principle.

for each part/failure mode has been multiplied by the part weighting to give a value, and the total value for each part has then been calculated as a percentage of the total for all the parts. This shows that the graphite lead, at 42 per cent, has the highest priority, with the wood the second highest at 36 per cent; the eraser, which has the most correlations, has a priority of only 12 per cent because of its lower weighting. (This example is from Pugh (1991), which sets out to define the design process within the wider framework of engineering activities.)

An alternative approach to FMEA is shown in Fig. 13.6. In this example the rating scale is from 1 to 10, using three conditions for each failure mode (as given in Table 13.2). In this approach, the RPN is obtained by rating each potential failure mode against the three conditions, and then multiplying these ratings together, as shown in Fig. 13.6.

Parts \ Failure Modes	Weighting Factor	Breaks	Falls Apart	Smudges	Wears Out	Weighted Rating	%
1. Lead	16	★ 144		○ 48	★ 144	336	42
2. Wood	16	★ 144	★ 144			288	36
3. Eraser	4	○ 12	○ 12	★ 36	★ 36	96	12
4. Eraser Holder	9		★ 81			96	10
						801	100

Figure 13.5 Example of FMEA matrix (wooden pencil).

Table 13.2 An FMEA rating scale

Occurence	Severity	Detection	Rating
Almost never	Hardly noticeable	Absolutely obvious	1
Occasionally	Dissatisfaction	Visible but could go unnoticed	5 or 6
Often	Serious effects	Undetectable	10

Figure 13.6 also shows that, as a result of the FMEA exercise, a change in the product has been suggested and evaluated, which reduces the RPN to an acceptable value.

Critical parameter management

The critical parameters of a design can be defined as the measurable factors that must be kept within specific limits if the product is to perform in accordance with specification. For example, a ball-point pen, to perform satisfactorily, should be designed to perform the following functions:

1. meter ink on to the paper
2. provide a means of holding the pen while writing
3. contain an adequate supply of ink
4. prevent ink spillage

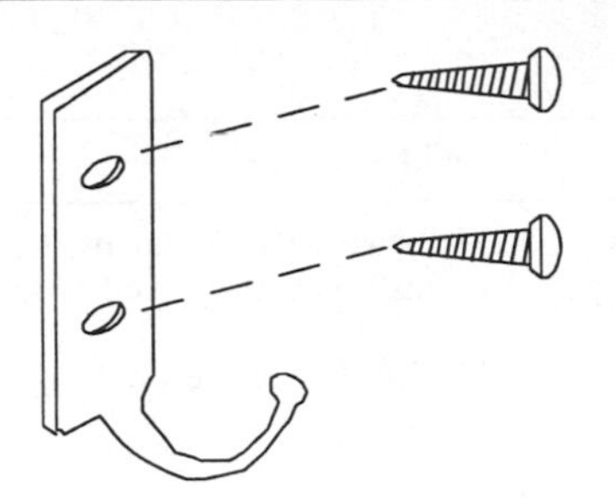

A COMPANY SUPPLIES COAT HOOKS WHICH ARE FITTED TO THE WALL WITH TWO SCREWS WHICH ARE NOT SUPPLIED.

THE POTENTIAL FAILURE MODE ANALYSIS COULD THEREFORE BE AS SHOWN BELOW:

POTENTIAL FAILURE MODE	POTENTIAL EFFECTS OF FAILURE	POTENTIAL CAUSES OF FAILURE	CURRENT CONTROLS	EXISTING CONDITIONS			
				OCC.	SEV.	DET.	RPN
HOOK COMES OFF WALL	COAT FALLS ONTO FLOOR	SCREWS TOO SHORT	ADVISE USER OF SCREWS TO USE	5	7	7	245
		ONE SCREW OMITTED	WORK CHECKED BY CUSTOMER	2	7	3	42

RECOMMENDED ACTIONS AND STATUS	RESPONSIBLE AREA	REVISED CONTROLS	RESULTING CONDITIONS			
			OCC.	SEV.	DET.	RPN
SUPPLY HOOK IN PACK WITH CORRECT SCREWS	J.GREEN PACKING DEPT.	NONE	1	7	7	49

Figure 13.6 Failure modes and effects analysis example.

5. allow the flow of ink
6. protect the point from damage.

From this list, further analysis suggests that the essential controllable parameters to achieve the required performance are the ball diameter/surface finish, the ball seat diameter, and the viscosity/surface tension of the ink. Each of these has a nominal measurable value and control limits can be set on either side of this value. There are three distinct stages in the management of the critical parameters, as given in Table 13.3.

This table also shows the work involved at each stage in identifying and defining the critical parameters (CPs), setting up the necessary test equipment, and using the data (the CPs, their nominal values and their tolerances) to establish and refine the design intent. In a complex product there may be a considerable number of sub-assemblies and each of these may have several critical parameters, making in total, perhaps, several hundred.

To manage such a product development effectively it is advisable to track each sub-assembly and record its level of stability or status, in terms of the number of its CPs that are defined,

Table 13.3 Critical parameter management stages

Development	Implementation	Audit
CP identification	Implement CPs on drawings	Measure CPs on hardware
Design experimental hardware	Compare drawings with design intent	Identify shortfalls and risks
Assess manufacturability	Identify shortfalls and risks	Modify hardware or design intent
Optimize CPs	Modify design intent to meet latitude requirement	
Test performance		
Redefine CPs		
Establish design intent		

pending, or undefined. This can be done on a bar chart (perhaps from a spreadsheet) as shown in Fig. 13.7. (More detailed explanations of CP management and QFD are given by Fox (1993). This takes a new look at what design really is and proposes new methods of working for engineering designers.)

Quality function deployment

QFD is a series of tools that can be used to examine a product, or a process, in order to determine where time and effort should be expended for the best return, or equally where they should not because the return would be likely to be minimal. Like many of these techniques QFD originated in Japan, and is based on charting and scoring in a manner similar to that used in the FMEA approach (see above).

The starting point in a QFD exercise is to prepare a matrix showing what is required against how these requirements will be met, as shown in Fig. 13.8. In this example, QFD is being used to study a projected new vehicle, but it could equally be used to examine a new process. In fact, there may be a series of such matrices matching the various internal 'customer/supplier' relationships in the chain, as given in Table 13.4. Each 'customer' prepares a list of what is required, and each 'supplier' prepares another list of how this requirement will be met; a matrix is then set up for the pairs of lists as shown in Fig. 13.9.

In each of these matrices the relationships are identified as strong, medium or weak and it is thus possible to follow a linkage from list 1 (customer requirements) to list 6 (parts specifications). However, it is also necessary for the QFD user to identify where items in a matrix are interrelated, positively or negatively, and the degree of these relationships, using another series

Table 13.4 Customer/supplier relationships in QFD

Customer	Supplier
1. Customer requirements	2. Supplier specifications
2. Supplier specifications	3. System specifications
3. System specifications	4. Subsystem specifications
4. Subsystem specifications	5. Manufacturing specifications
5. Manufacturing specifications	6. Part specifications

In each case, the 'customer' specifies *what* is required from its 'supplier', which then specifies *how* this requirement is to be met.

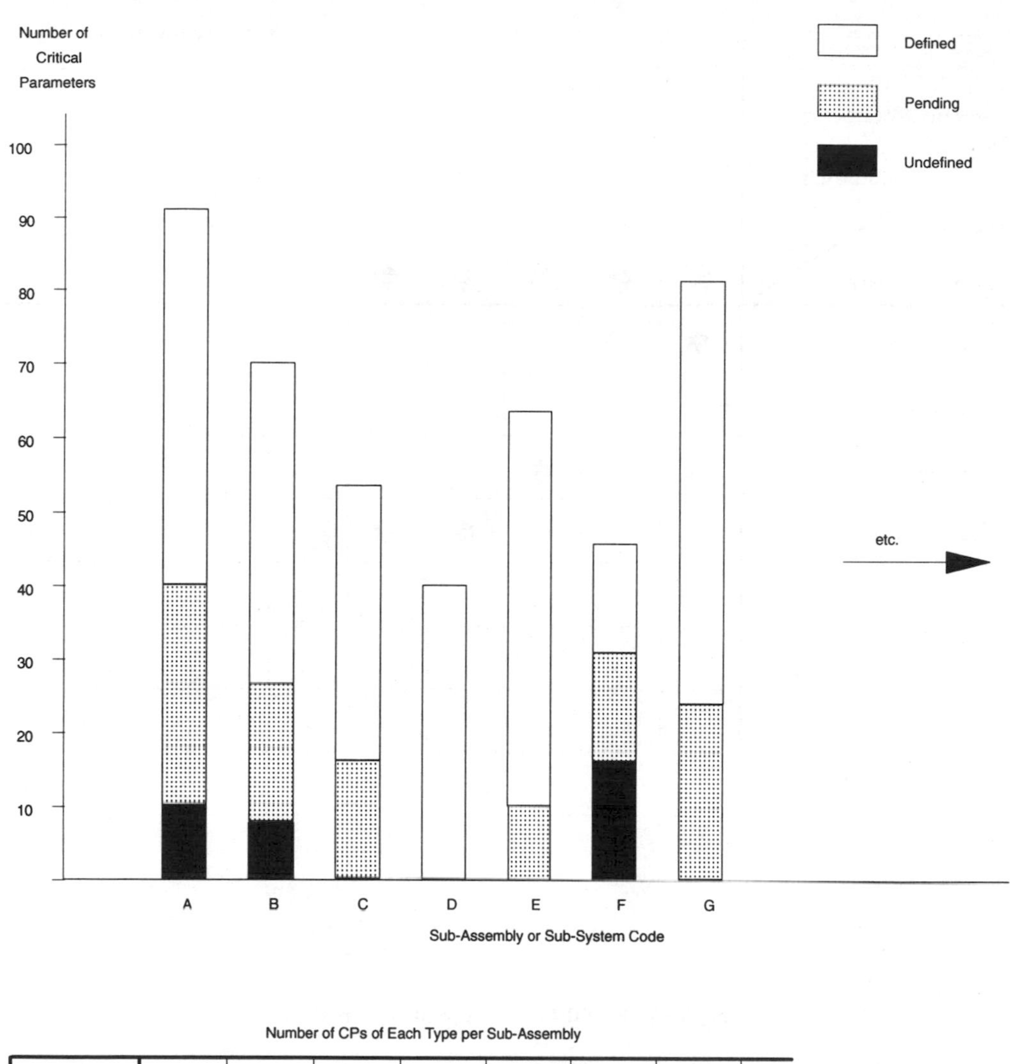

Number of CPs of Each Type per Sub-Assembly

Sub-Assy.	A	B	C	D	E	F	G	
Defined	51	44	37		etc.		etc.	
Pending	30	18	16		etc.		etc.	
Undefined	10	8	0		etc.		etc.	
Total CPs	91	70	53		etc.		etc.	

etc.

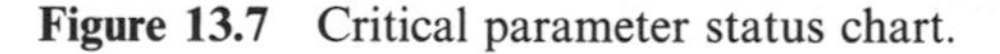

Figure 13.7 Critical parameter status chart.

of symbols as shown in Fig. 13.10. A positive relationship means that the two items help one another, whereas a negative relationship means that they are in conflict and oppose one another.

The full matrix, including not only the relationships but also the ratings and competitive analysis, can then be drawn. From this it is possible to set priorities, so that the items with the greatest potential for success in meeting the requirements can be developed and used. An example of this complete matrix is shown in Fig. 13.11.

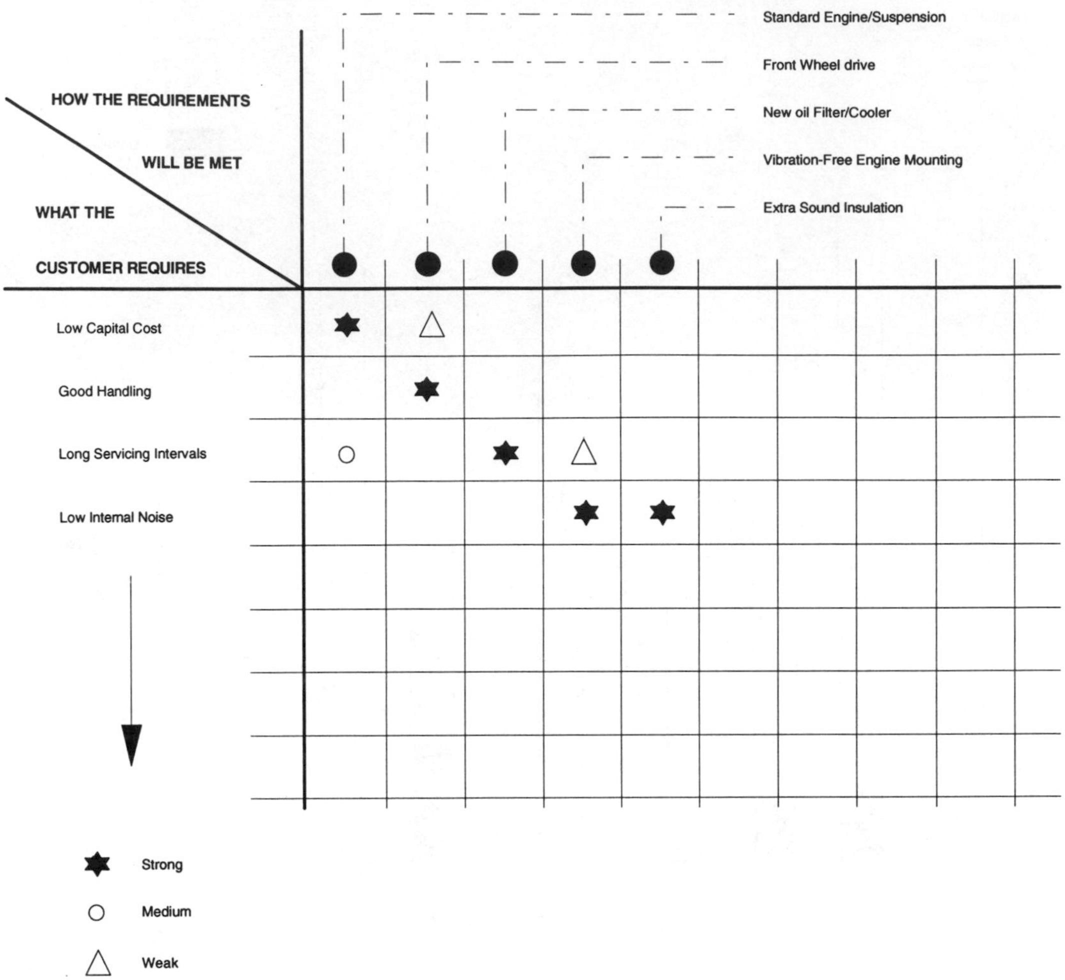

HOW THE REQUIREMENTS WILL BE MET

WHAT THE CUSTOMER REQUIRES

	Standard Engine/Suspension	Front Wheel drive	New oil Filter/Cooler	Vibration-Free Engine Mounting	Extra Sound Insulation
Low Capital Cost	✱	△			
Good Handling		✱			
Long Servicing Intervals	○		✱	△	
Low Internal Noise				✱	✱

✱ Strong

○ Medium

△ Weak

Figure 13.8 QFD what versus how matrix.

13.9 QUALITY CONTROL AND SUPPLIERS

There are three possible areas where quality control could be involved with suppliers, although it should not be necessary to apply all three since (2) and (3) probably make (1) unnecessary:

1. goods-inwards inspection
2. vendor rating
3. supplier audits.

Goods inwards inspection

If satisfactory supplier audits and vendor rating systems are in operation, it should not be necessary to carry out goods-inwards inspection on goods received from recognized suppliers. However, if the other systems have not been implemented it may be necessary to inspect key items qualitatively and some random quantity checks are also advisable, particularly on new suppliers or where only vendor rating is being used.

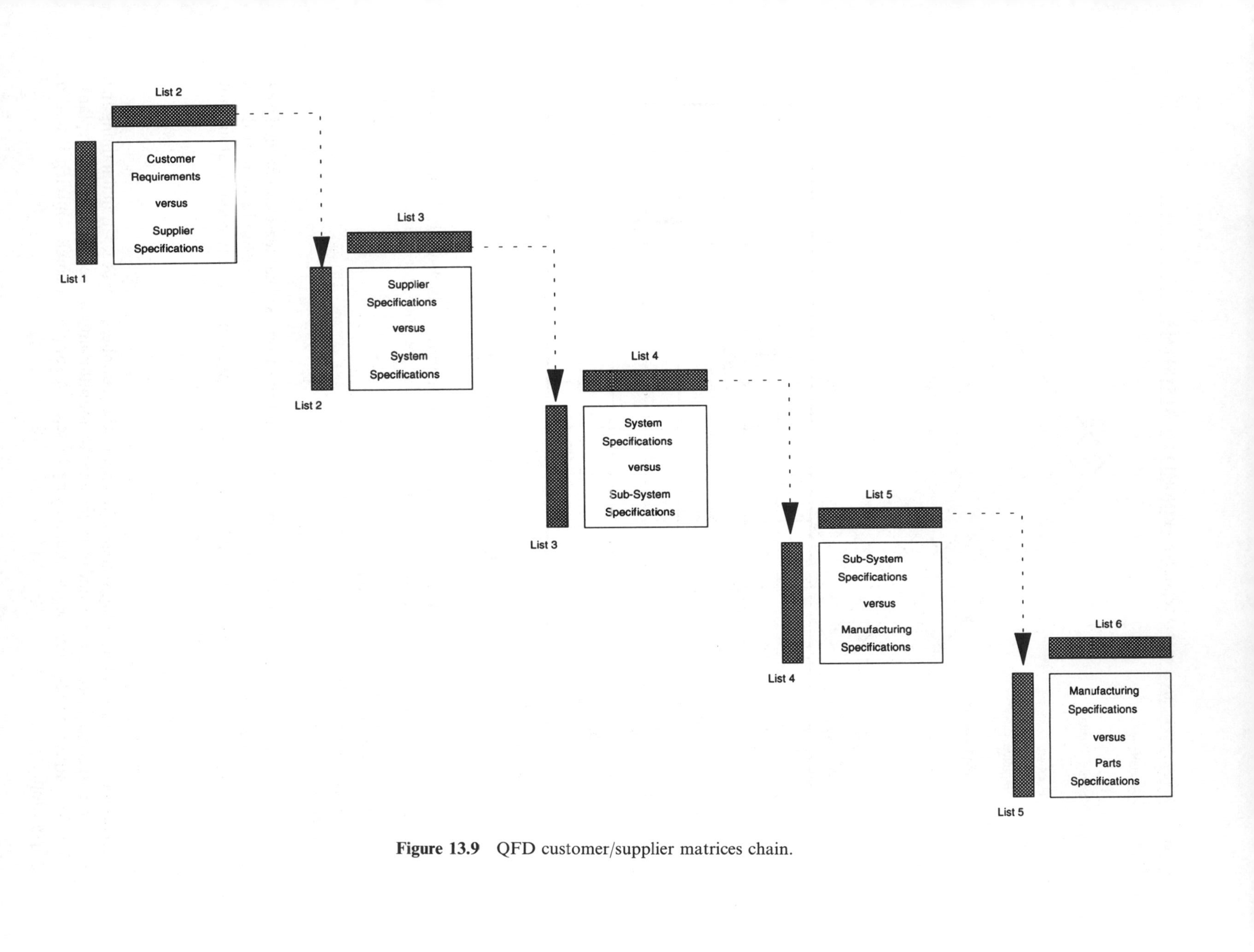

Figure 13.9 QFD customer/supplier matrices chain.

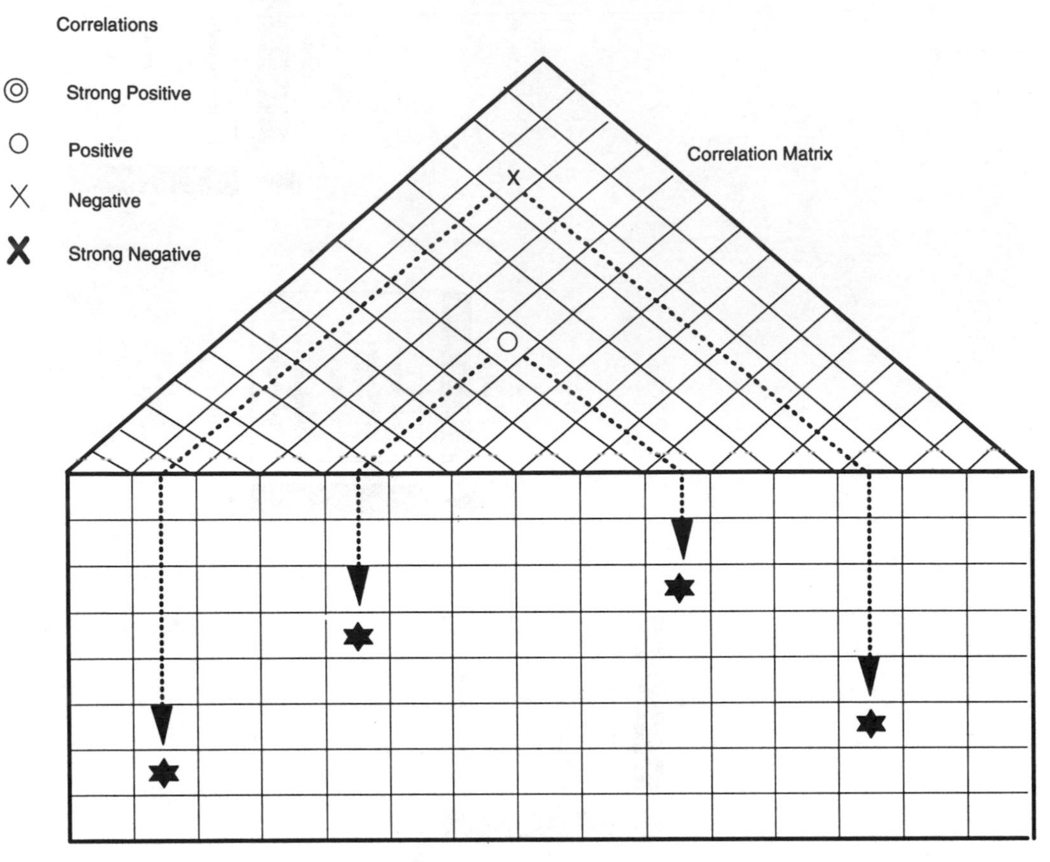

Figure 13.10 Inter-relationships in QFD.

If some form of ABC classification system is in operation (high-value items, class A; medium, class B; low, class C), an inspection policy can be specified, for example:

- *class A* 100 per cent inspection
- *class B* sample inspection
- *class C* no inspection.

If possible, computer records should be kept for all GI inspections, so that these can be analysed as a part of the overall quality management system and to allow any trends to be identified. An area of the goods-receiving bay should be set aside for inspections, and any goods that are not acceptable should be transferred to a secure quarantine area.

Vendor rating

The objective of a vendor rating system is to ensure that suppliers are, at least, maintaining the required standards of quality and service, and to monitor any agreed improvement plans. Vendor rating can also be used by Purchasing to select a supplier if single sourcing is not in operation.

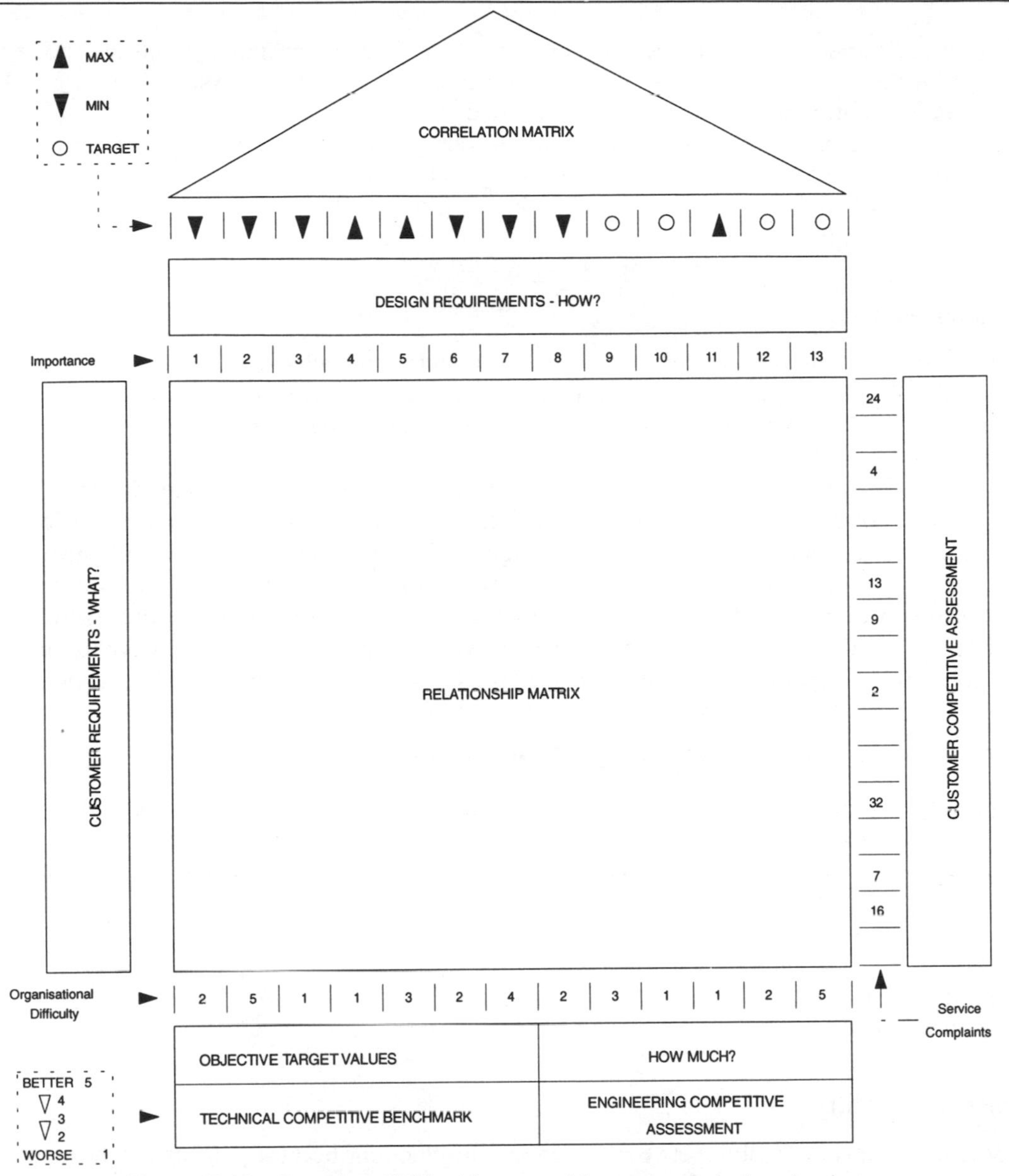

Figure 13.11 Complete QFD with competitive and technical evaluation.

If a computerized purchasing system is in use, suitable software for vendor rating may be included in the module, but if not the company may need to devise a system, based on a spreadsheet package that allows for entry of data in a number of categories, for example:

- *price stability* are prices held stable for reasonable periods of time?
- *delivery promises* are these reasonable for the type of supply?
- *delivery on time* are deliveries generally on time?
- *quality of goods* are quality standards maintained?
- *accuracy of documents* are advice notes and invoices accurate?
- *certification* are test certificates in order?

However, it is unlikely that all these categories will be used by one company, and the choice will depend on the type of business. An example of a typical vendor rating system is shown in Fig. 13.12, which shows how points might be allocated.

The assessments should be carried out regularly (say twice yearly) on the information collected during the six month period, and if a purchasing system is available, some of these data could be obtained by creating special reports, e.g. a delivery performance (delivery on time) report, as shown in Fig. 13.13.

Supplier audits

The objective of supplier auditing is to assist suppliers to achieve the customer's required standards of quality and service. This is usually done by a series of visits to the suppliers' premises by the customer's quality audit team during which performance in relevant areas is assessed and measured.

If the 'score' in a particular area is below standard, the audit team may recommend changes in systems, procedures and policies designed to improve the position. This is obviously a sensitive area and can only work if there is a high level of mutual trust, respect and partnership.

If a supplier meets the required standards, a certificate and rating may be awarded (e.g. Ford Motor Company's Q1) which is often used by a supplier as a selling feature. However, in order to retain this status (approved supplier) it will be necessary for the supplier to maintain and continuously improve performances. This topic is also covered in chapter 8 (section 8.13).

13.10 QUALITY CONTROL AND PROCESSES

To maintain quality control over a process it is necessary to set up a system of measurement, and to analyse the measurements, in order to detect where faults are likely to occur if an apparent adverse trend is not reversed. Four techniques must be considered in this context:

1. process capability
2. statistical process control (SPC)
3. sampling
4. calibration control.

Process capability

Process capability is the link between design and production, because it measures how capable the production process is of meeting the design tolerance. In other words, there are two constraints: the design tolerance (limits) that are set in accordance with the specification (by means of a critical parameter—see section 13.8 above), and the process capability of the machine or equipment that will be used to make the item concerned.

The capability of a production machine or process is usually defined in terms of a 'normal distribution' curve of the type considered in appendix B and illustrated in Fig. B2. The probabilities given by such a curve are related to standard deviations σ, and in some manufacturing control systems it is conventional to relate the output data measurements to a 3σ distribution situation.

The process capability index C_{pk} is therefore defined in these terms by giving it a value of 1 if the 3σ tolerance can be met, 1.33 if 4σ can be met, and so on. Thus, the higher the C_{pk} value, the higher is the capability of the machine or process to keep within the tolerances set by the

VENDOR RATING SYSTEM	SUPPLIER NAME					CODE
A. PRICE						
	HIGHER		SAME		LOWER	MARK
ARE PRICES REASONABLE AGAINST COMPETITION?	1	2	3	4	5	
	HIGHER		SAME		LOWER	MARK
PRICE INCREASES COMPARED WITH INFLATION RATE?	1	2	3	4	5	
	NO		FAIR		GOOD	MARK
GOOD DISCOUNTS FOR BULK ORDERS?	1	2	3	4	5	
B. DELIVERY PROMISES						
	LONGER		SAME		SHORTER	MARK
DELIVERY PROMISES REASONABLE/COMPETITIVE?	1	2	3	4	5	
	YES		SAME			MARK
EXTENDED FOR LARGER QUANTITIES?	1	2	3			
C. DELIVERY PERFORMANCES	MOSTLY LATE		SOMETIMES LATE		NEVER LATE	MARK
DELIVERED WITHIN DELIVERY PROMISE?	1	2	3	4	5	
	NEVER		SOMETIMES		OFTEN	MARK
DELIVERIES FLEXIBLE IF REQUESTED?	1	2	3	4	5	
	NEVER		SOMETIMES		OFTEN	MARK
DELIVERY SCHEDULES AMENDED IF REQUIRED?	1	2	3	4	5	
D. QUALITY						
	POOR		FAIR		GOOD	MARK
QUALITY UP TO REQUIRED STANDARD?	1	2	3	4	5	
	NO		YES			MARK
RETURNS CREDITED QUICKLY? NO FUSS?	1	2	3			
	NO		FAIR		GOOD	MARK
STANDARD QUALITY SYSTEM USED? WELL APPLIED?	1	2	3	4	5	
E. ACCURACY						
	NO		YES			MARK
ITEMS AND QUANTITIES AS ORDERED?	1	2	3			
	NO		YES			MARK
ADVICE NOTES ACCURATE?	1	2	3			
	NO		YES			MARK
INVOICES ACCURATE?	1	2	3			
	NO		YES			MARK
ORDER NUMBERS QUOTED ON DOCUMENTS?	1	2	3			

MAXM.POSS.	63	MINM.POSS.	15	% OF MAXM.		TOTAL	

Figure 13.12 Typical vendor rating scoring system.

JANUARY to JUNE 1994

SUPPLIER	TOTAL ORDERS	LATE (WEEKS)				ON TIME	EARLY (WEEKS)				PERCENTAGES	
		> 3	2-3	1-2				1-2	2-3	> 3	LATE	EARLY/ ON TIME
J.SMITH & CO	25	--	--	5	6	7	6	1	--	--	44.0	56.0
B.EVANS LTD	10	--	--	1	3	4	2	--	--	--	40.0	60.0
XYZ ENG.LTD	40	1	2	4	3	25	4	1	--	--	25.0	75.0
ABC SUPPLIES	18	--	--	1	3	11	2	--	1	--	22.2	77.8
FABCON LTD	5	--	--	--	1	3	1	--	--	--	20.0	80.0
JONES & CO	16	--	--	1	2	11	2	--	--	--	18.8	81.2
THE 100 GRP.	30	--	2	1	2	22	1	1	1	--	16.7	83.3
47 SUPPLRS.	420	7	16	38	44	259	38	11	5	2	25.0	75.0

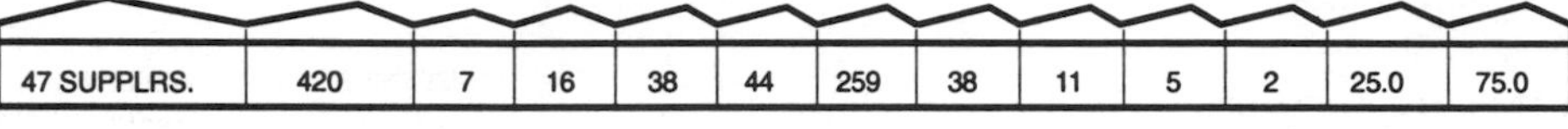

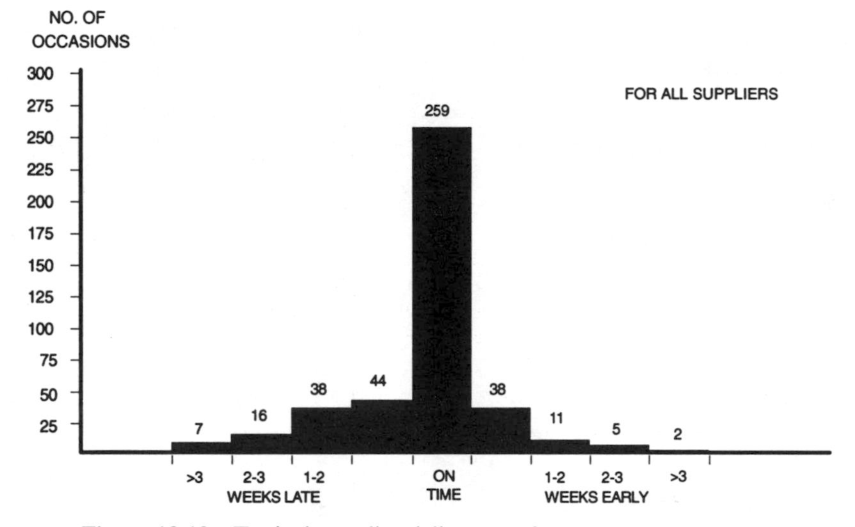

Figure 13.13 Typical supplier delivery performance report.

design specification. In fact, in some types of industry, the six sigma system is used (six standard deviations), sometimes known as the 'Motorola six sigma'. This demands a very high functional specification (i.e. very tight control limits) where the C_{pk} value is 2.0.

The capability index for a process can only be found by experimentation, i.e. by measuring a representative sample of the items produced by that process and plotting the results on a control chart. Then the variations found must be plotted as a bar chart, to check that the conditions fit the normal distribution pattern. These two steps are shown in Fig. 13.14.

As an example of this procedure (taken from Fox, 1993), the diameter of a hole in a component is design-specified as 20 mm with a design tolerance of ± 2 mm. Measurements of the diameter on a number of components gave a mean of 19.5 mm and a sigma value of 0.47.

The formulae used and the calculations performed to calculate the capability index for this or any other process are given in Table 13.5, where UCL is the upper control limit, LCL is the

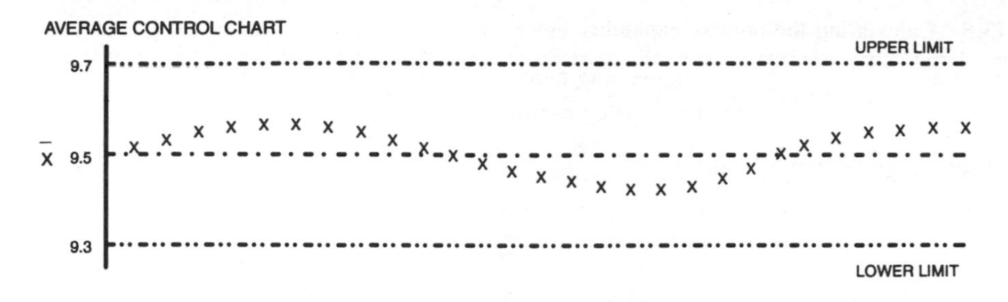

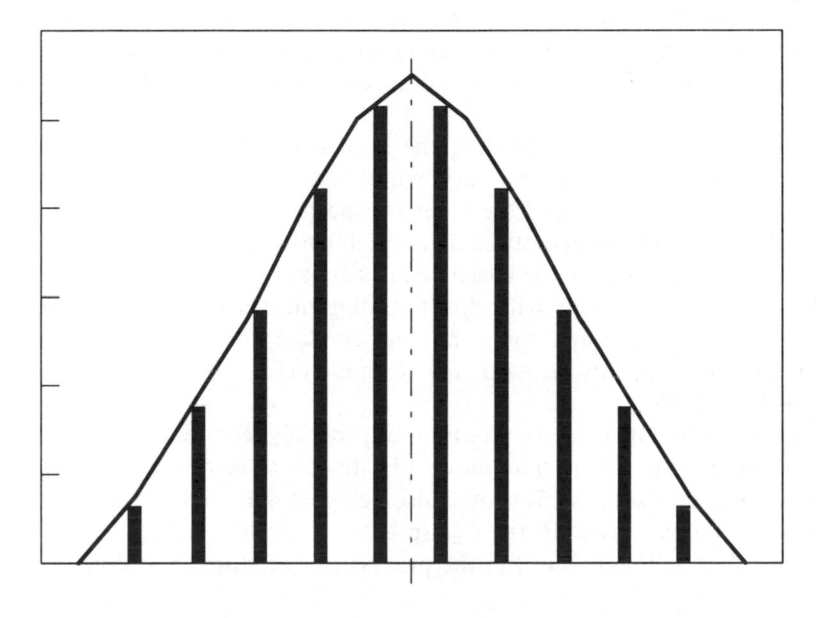

Figure 13.14 Determining process capability.

lower control limit and x is the mean value of the measurements. The higher the value of C_{pk}, the better is the capability and therefore the lower of the two values (1.06) is accepted, and can be used in the SPC process described in the next subsection. This process is within capability—but not by much! If it had been below 1.0 consideration would have had to be given to modifying the process, or changing the design.

Statistical process control

SPC can be defined as a procedure that uses statistical methods to analyse the measured results of a process, and uses that analysis to control and manage it. Because it is statistically based,

Table 13.5 Calculating the process capability index

$$C_{pk} = \frac{\text{engineering needs}}{\text{manufacturing capability}} = \frac{x - \text{LCL}}{3\sigma} \text{ or } \frac{\text{UCL} - x}{3\sigma}$$

Thus, in the example of the 20 mm hole:

$$C_{pk} = \frac{19.5 - 18}{3 \times 0.47} = \frac{1.5}{1.41} = 1.06 \text{ (for } LCL\text{) or}$$

$$C_{pk} = \frac{22 - 19.5}{3 \times 0.47} = \frac{2.5}{1.41} = 1.77 \text{ (for UCL)}$$

SPC can predict the probability of an event in the future, and thus allow for adjustments to be made in the process that will prevent that event from occurring if it is undesirable. SPC is yet another example of the basic management technique of 'feedback'.

Table 13.5 showed the way in which a process capability index can be calculated by comparing engineering (design) needs with manufacturing capability. It also showed how the upper and lower control limits, which were a part of the engineering needs, were used in these calculations.

SPC is therefore operated by plotting the results of test measurements of a sample of a production batch on to control charts, as shown in Fig. 13.15(a), but in this case two charts are used, one for the average value and the other for the range of readings, related to the control limits. Figure 13.15(b) shows some of the data which has been used to prepare these two charts.

A number of personal computer-based systems are available for running SPC, and many of these can be used in conjunction with direct reading measuring devices, which pass the data collected into the host computer via a number of data collection terminals. Alternatively, normal measuring devices can be used, the readings being keyed into a terminal similar to that shown in Fig. 13.16.

Such systems (with data collection terminals) generally feed the results of the analysis back to the process operator, in a form that allows adjustments to be made to the machine or process before control limits are exceeded. It should also be noted that a part of the computer analysis is to carry out continuous checks on the C_{pk} (process capability) values so that, if this value is showing a trend that will go below 1.0, the process can be stopped and investigated.

Sampling

Sampling is required for quality control, because it is highly uneconomic to carry out 100 per cent checking on all products and processes. The procedures for selecting sample sizes are clearly defined in various British and international standards such as BS 6001 and ISO 2859. These standards specify sampling plans based on the attributes of the products, indexed by the AQL (acceptable quality level). The AQL is the 'only just acceptable' quality level and is the poorest quality that can be accepted as the process average.

Inspection by attributes means that each unit of product inspected is classified as acceptable or defective; i.e. each unit is considered to have one of these attributes. This type of inspection system does not depend on the degree of acceptability or defectiveness, i.e. it is a 'pass or fail' system.

There is an alternative system of sampling, known as inspection by variables, which does take degrees into account, and operates by considering measurements made on the inspected

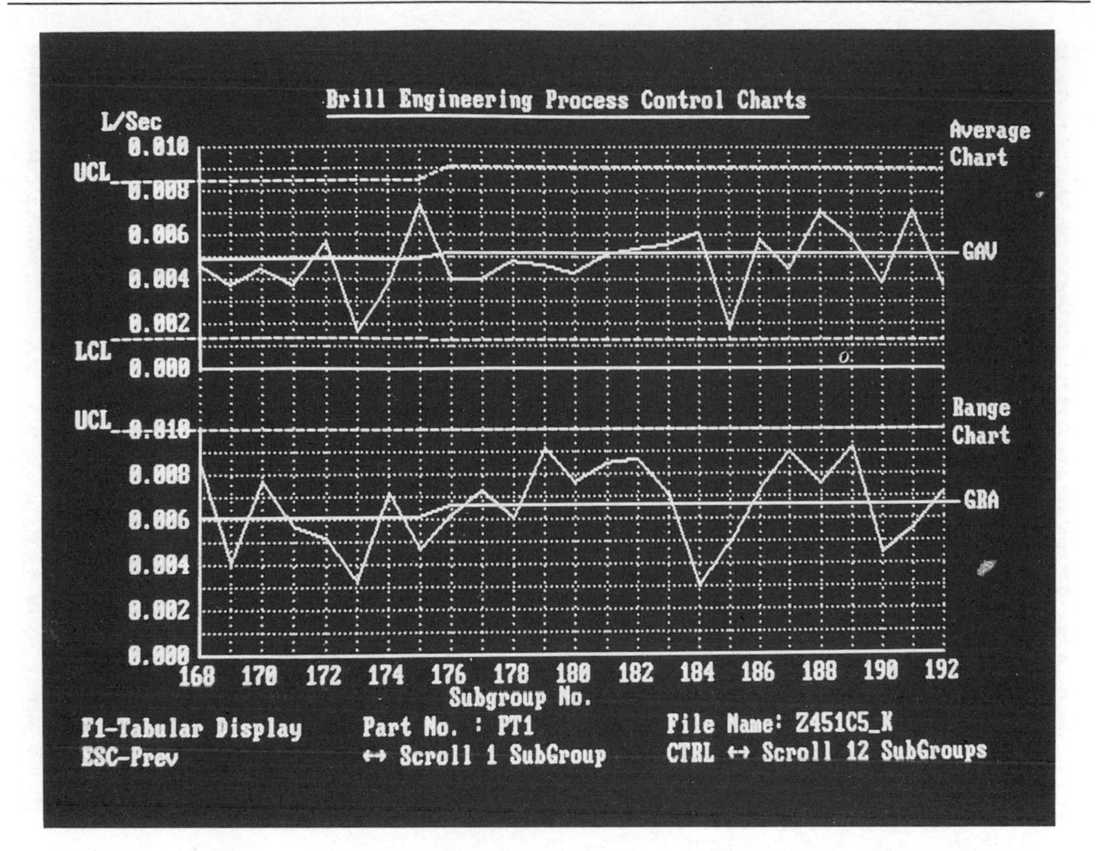

Figure 13.15a Typical SPC process control charts. The upper (average) chart shows the trend of the average values, that is, whether the process is stable or is moving towards an 'out of control limits' situation. The lower (range) chart shows the range of values in relation to the upper and lower control limits so that variations and non-conforming readings can be identified (photograph by courtesy of Brill Engineering, Preston).

items; this system is now widely used. Such a system is defined in MIL-STD-414 (*Sample procedures and tables for inspection by variables*).

Before preparing a sampling plan it is necessary to know:

1. the AQL, for example 1 per cent defectives
2. the inspection level, which defines the relationship between batch size and sample size
3. the type of inspection to be used (see note (a) below)
4. the type of sampling to be used (see note (b) below)
5. the batch size.

Note (a): Inspection types. There are three types of inspection — normal, tightened, and reduced — and these are defined in terms of the operating characteristic (OC) curves for each type, from which the sampling plans are derived; two typical OC curves are shown in Fig. 13.17. The first of these is for a plan designed to give a high probability of *rejection* if any batch is submitted having a quality *worse* than the AQL; the second is designed to give a high probability of *acceptance* if any batch is submitted having a quality *better* than the AQL.

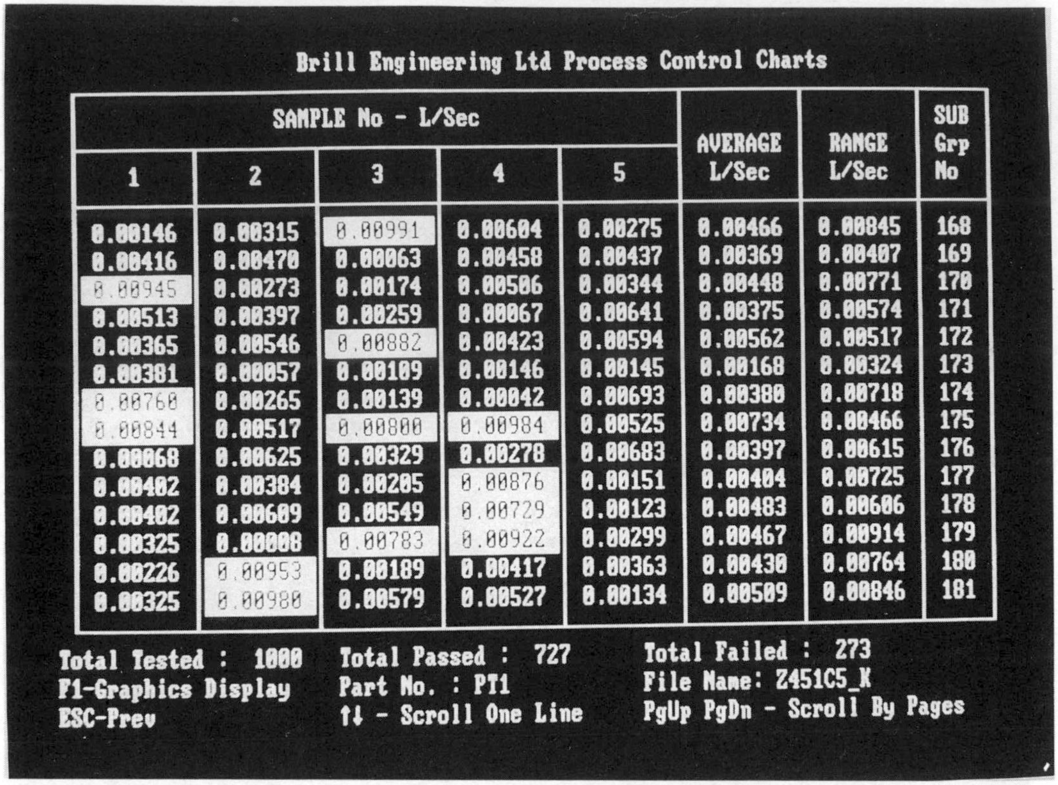

Brill Engineering Ltd Process Control Charts

SAMPLE No - L/Sec					AVERAGE L/Sec	RANGE L/Sec	SUB Grp No
1	2	3	4	5			
0.00146	0.00315	0.00991	0.00604	0.00275	0.00466	0.00845	168
0.00416	0.00470	0.00063	0.00458	0.00437	0.00369	0.00407	169
0.00945	0.00273	0.00174	0.00506	0.00344	0.00448	0.00771	170
0.00513	0.00397	0.00259	0.00067	0.00641	0.00375	0.00574	171
0.00365	0.00546	0.00882	0.00423	0.00594	0.00562	0.00517	172
0.00381	0.00057	0.00109	0.00146	0.00145	0.00168	0.00324	173
0.00760	0.00265	0.00139	0.00042	0.00693	0.00380	0.00718	174
0.00844	0.00517	0.00800	0.00984	0.00525	0.00734	0.00466	175
0.00068	0.00625	0.00329	0.00278	0.00683	0.00397	0.00615	176
0.00402	0.00384	0.00205	0.00876	0.00151	0.00404	0.00725	177
0.00402	0.00609	0.00549	0.00729	0.00123	0.00483	0.00606	178
0.00325	0.00008	0.00783	0.00922	0.00299	0.00467	0.00914	179
0.00226	0.00953	0.00189	0.00417	0.00363	0.00430	0.00764	180
0.00325	0.00980	0.00579	0.00527	0.00134	0.00509	0.00846	181

Total Tested : 1000 Total Passed : 727 Total Failed : 273
F1-Graphics Display Part No. : PT1 File Name: Z451C5_K
ESC-Prev ↑↓ - Scroll One Line PgUp PgDn - Scroll By Pages

Figure 13.15b Typical SPC process control chart data. This shows a typical display of the process control data, the failed tests bring in reverse video. (photograph by courtesy of Brill Engineering, Preston).

In general terms, tightened inspection means that the sample size is increased while keeping the same acceptance number as normal inspection, whereas for reduced inspection the sample size is reduced.

Note (b): Sampling types. There are four possible sampling types—single, double, multiple and sequential, and where sampling plans are available the choice must be based on such factors as simplicity, average sample size, variability of sample size, ease of drawing sample units, duration of test, and the incidence of multiple defects. These four types are defined in Table 13.6.

Given this information, it is then possible to use the tables in the standard publications to determine the sampling plan, which will specify sample size and the numbers of defectives allowable for acceptance or rejection.

For example, pins were inspected for high and low diameter and, based on the sampling plan, a sample size of 125 was specified on a batch of 1400 with an acceptance number of 5 and a rejection number of 6. A random sample of 125 pins was therefore taken from the batch of 1400 and gauged, the results being as follows:

within limits 122
above upper limit 2
below lower limit 1

Table 13.6 Definitions of sample types

Single	The conventional method whereby a single sample, of the size specified in the sampling plan, is drawn from the batch, inspected, and the results checked against the plan.
Double	In this case a smaller first sample is drawn than would be used in single sampling and if this is found to be sufficiently good or bad, the batch may be accepted or rejected at once. If the result is intermediate a second sample is taken before a decision is made.
Multiple	In this case the procedure is similar to that for double sampling except that more than two samples may be drawn, the plans usually specifying up to seven. Thus, if after a sample has been checked the result is still in doubt, a further sample should be taken. All samples are the same size and smaller than those used in double sampling.
Sequential	In this case there are no fixed sample sizes, but a continual watch is kept on the results as they occur, a decision being made to accept or reject on the basis of a scoring system where points are added for passes and deducted for failures. The batch is then accepted or rejected if the score reaches a target figure (accept) or falls to zero or below (reject).

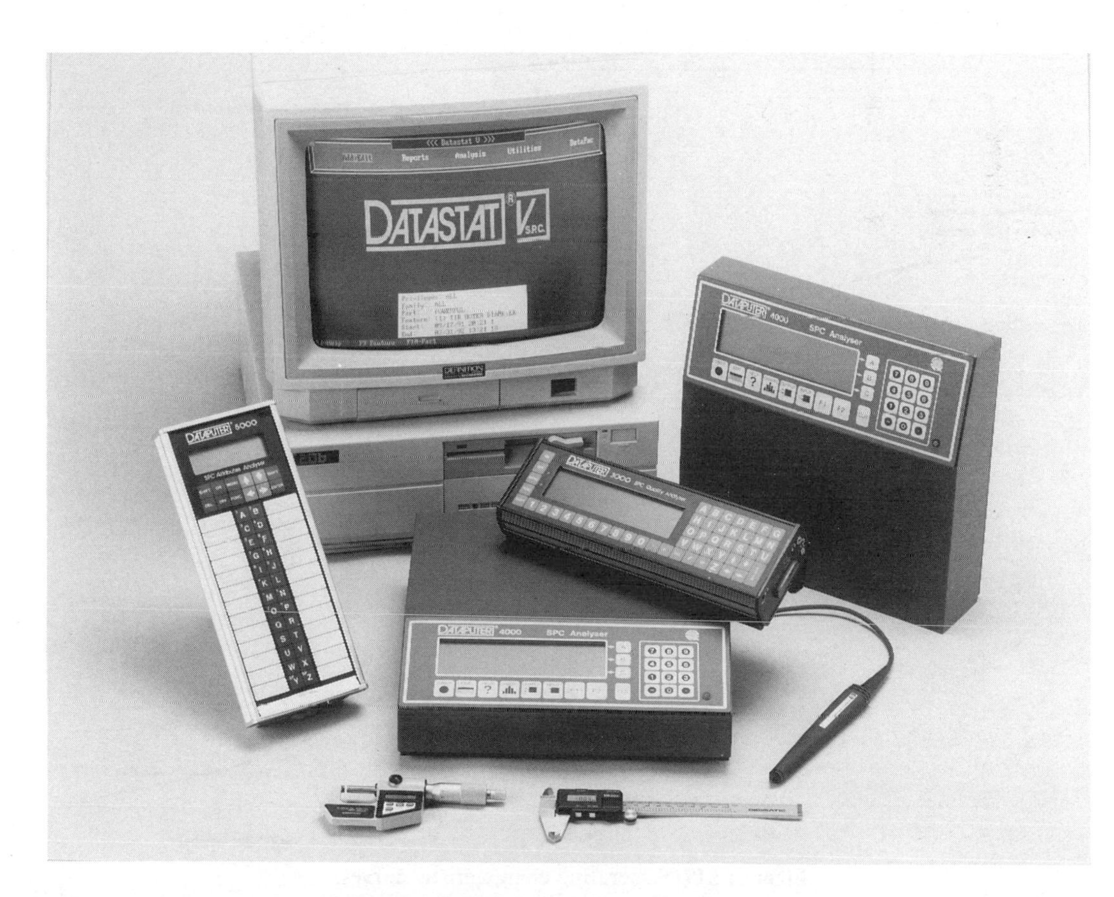

Figure 13.16 An SPC data collection system. This shows a range of Dataputer SPC data collection equipment from Elcometer. It not only allows an operator to enter data directly to the main SPC system (in this case Datastat V) for processing but also feeds data back to the operator so that adjustments can be made to the process if required (photograph by courtesy of Elcometer Instruments, Manchester).

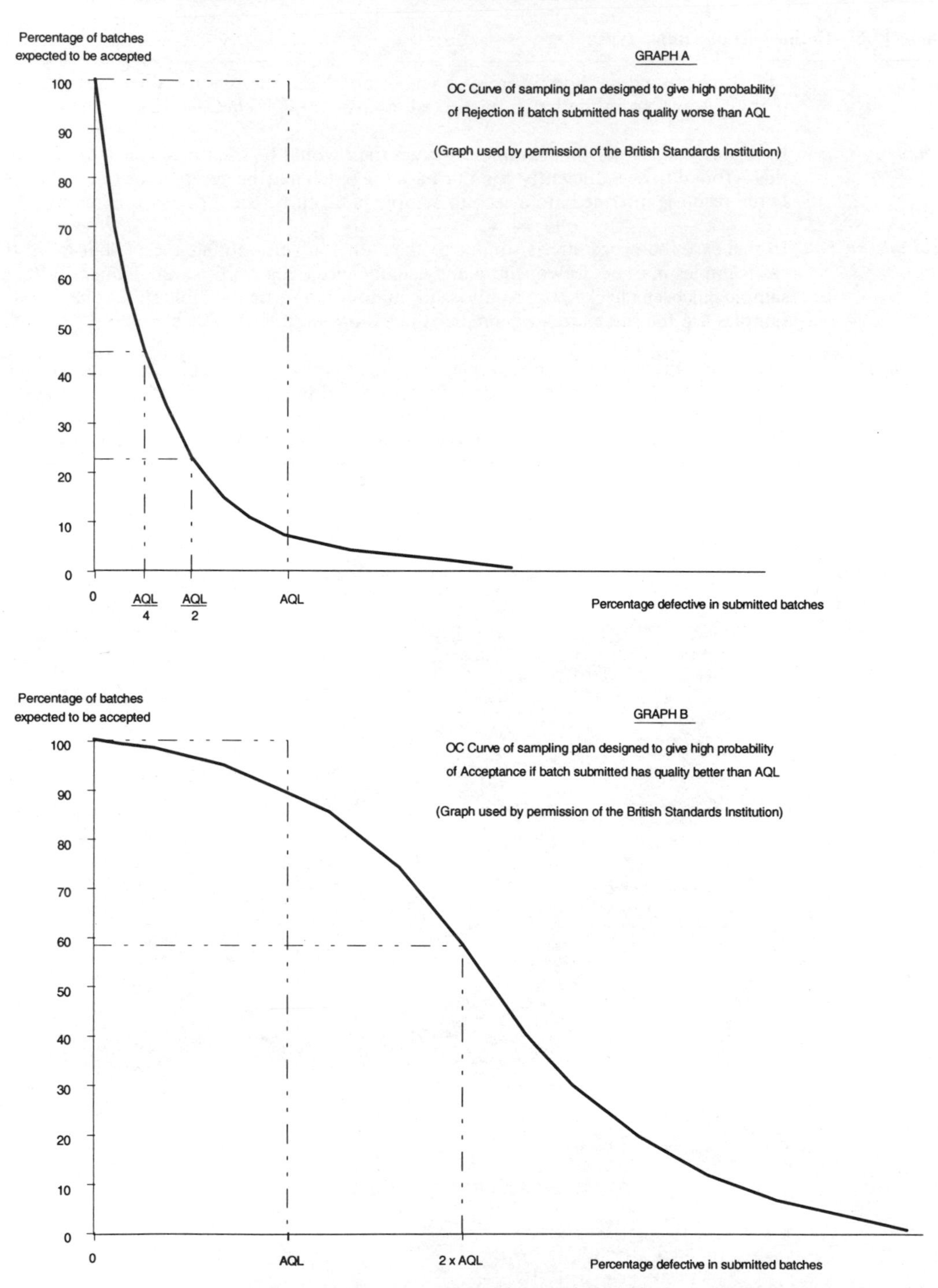

Figure 13.17 Operating characteristic curves.

There are thus three defectives in the sample and these were rejected, but because 3 was less than the acceptance number (which was 5), the remainder of the batch (1397) was accepted.

In order to minimize the costs of inspection without reducing quality levels, it is possible to operate a 'system' that allows the intensity of inspection to be reduced if results are good but increased if results are poor. This system of 'switching rules' is illustrated in Fig. 13.18.

In this system, inspection starts at the normal level but, if the results of a series of inspections meet specific targets, inspection can be reduced and remain at this reduced level until a lot fails to pass. Normal inspection is then resumed, but is monitored, and if the lot failure rate does not improve or drops further, the inspection level is tightened. Once again the situation is monitored, and if there is no improvement the process is stopped and the reasons for the quality failures investigated. However, if the position improves, inspection can return to the normal level. (The various sampling techniques and definitions described and illustrated in this subsection have been adapted from the relevant British Standards, and the author wishes to thank the BSI for allowing him to use this information.)

Calibration control

In order to ensure that all measuring instruments used in quality control systems are accurate, it is necessary to implement a calibration control system. This can best be done by applying the principles used for planned maintenance to all such instruments and machines. Calibration should be carried out only at a registered calibration centre, but a company can apply for such registration if the necessary facilities are available on site.

To conform to the procedures specified in BS5750 or ISO9000, all measuring instruments and machines must carry a code number, and full details of the item must be held on a register, which should also include information about the types of tests to be carried out; the test pieces to be used and the frequency of such testing should also be specified. Historical records should be kept for every registered item, showing that the correct checks have been carried out; the results of such checks, together with details of issues and receipts from the instrument stores, should be included in these records.

However, merely checking gauges etc. on a regular basis may not be enough, if they are not being used correctly or if they have a lack of stability (reliable accuracy). For this reason, some companies use a procedure known as 'R&R studies' (reliability and reproducibility). There are several variants in this type of study, but all are designed to determine how reliable the device is, as follows:

- *accuracy* to determine the difference between the observed average of measurements and the true average of the same feature on the same part
- *linearity* to determine the differences in accuracy over the whole range of the device
- *stability* to determine the difference in the average of several sets of measurements, using the same device on the same parts at different times
- *repeatability* to determine the variation in measurements, using the same device, when it is used by the same operator on several occasions to measure the same feature on the same part
- *reproducibility* to determine the averages of measurements taken by different operators, or in different locations, using the same device on the same feature of the same part.

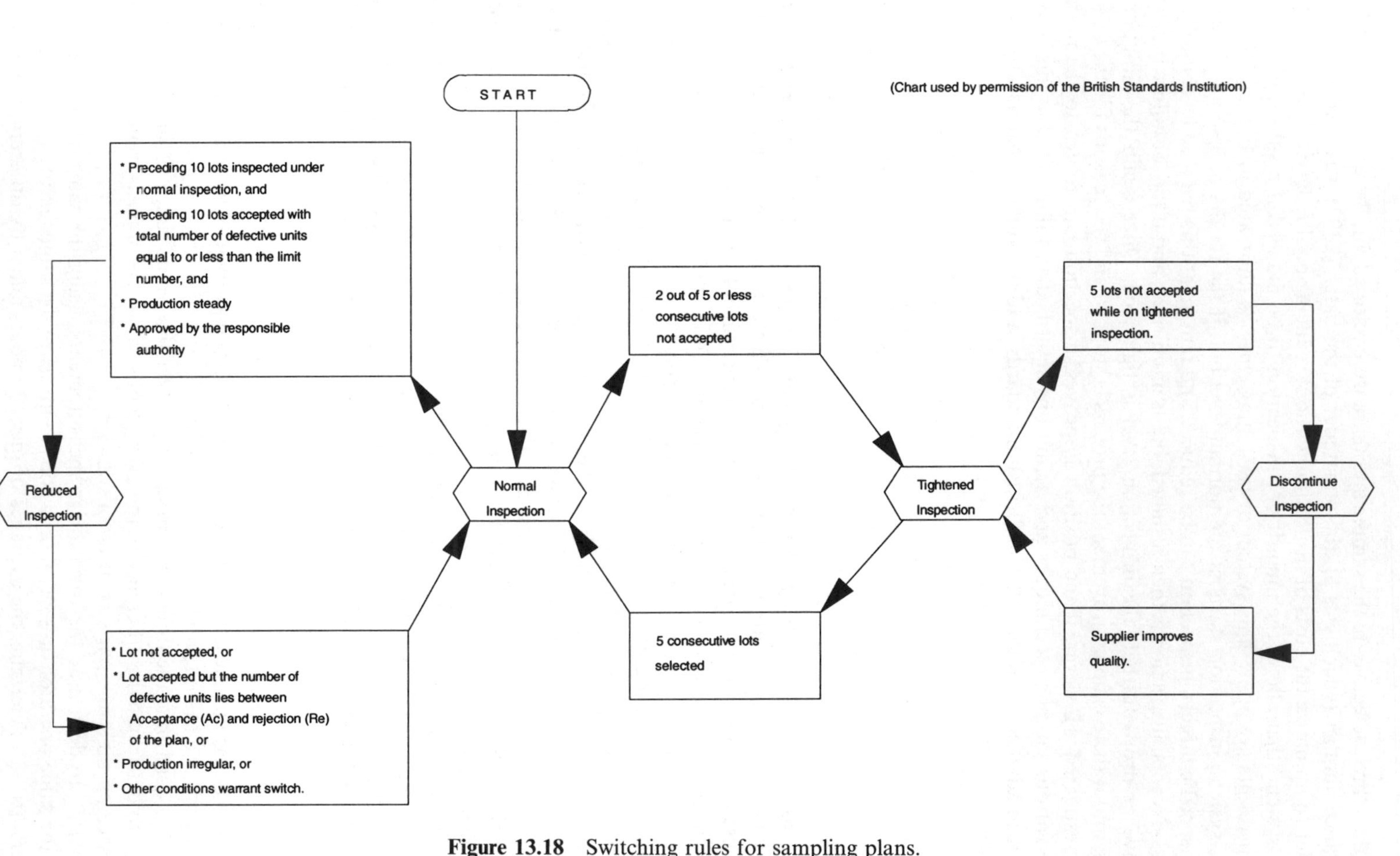

Figure 13.18 Switching rules for sampling plans.

13.11 QUALITY CONTROL AND TRACEABILITY

In many industries and product areas it is essential for quality management to be able to identify which materials and processes have been used on a batch of products; also, in some cases, it is necessary to know the customers who have bought those products, so that if a problem arises it can be investigated with all the available facts being known.

The required level of traceability depends to a considerable extent on how 'safety critical' the product is. For instance, all products in the aerospace or food industries would have very high traceability rating levels, whereas the products from a decorative tile maker might have basic ratings. Alternatively, a maker of nuts and bolts might need high traceability on some products and intermediate or even basic on others, depending on the application.

The required traceability level for a product can be classified into one of the following three categories.

- *Basic* In this case traceability is required only to tackle problems involving cost, for example if scrap levels are too high. In such cases the customer is unlikely to be affected.
- *Intermediate* In this case the problem could involve customer claims and rejections but there is no safety factor involved, for example a high number of complaints about product life requiring the supply of replacements.
- *High* In this case the problem will certainly involve customers and there could be a risk of consequential claims, for example an accident caused by faulty brake parts on a car, or an outbreak of food poisoning. In such a situation it may be necessary to recall some or all of the products for examination, and perhaps modification or rejection.

The techniques that may be used to obtain traceability information will cover the following areas.

Purchasing

All purchased items used to manufacture the product will be received into stock and held as totally discrete lots, each being recorded against a lot number; this will be cross-referenced to the supplier's delivery note and probably some form of documentation, for example a certificate of conformance, a test certificate or a chemical analysis. Such lots should never be mixed in a production batch where traceability is required, and should therefore be held in separate stock locations.

Production

Full records should be kept of all production batches such that, at some future date, it would be possible to know the operations or processes carried out, by whom, on which machines, with which tooling and on what day/shift. These records should also include full details of any inspections carried out (first-off, patrol, sampling) and any other information that may be relevant. Production batches should never be mixed if traceability is required.

Sales

Where full or intermediate traceability is required, each product should be coded with a serial or batch number and each of these numbers should be traceable to a customer. In addition, Sales should maintain complete records of all customer complaints and claims, so that this information can be used by quality management to identify potential problem areas. An example of a typical computerized lot traceability record is shown in Fig. 13.19.

13.12 QUALITY CONTROL AND DOCUMENTATION

Apart from the obvious records that should be kept by quality management (inspection records, calibration records, warranty claim investigations, etc.), the QC department should be responsible for maintaining the quality manuals and standard procedures. These should specify exactly

PRODUCT B.14680	BATCH NUMBER 24583	QUANTITY 100	DATE ISSUED 23/4/91

MATERIALS RECORD:

PART No.	DESCRIPTION	SUPPLIER	QTY.	UOM.	LOT No.
SC.20437	XXXXXXXXXXXXXXX	ABT.SUPPLIES	200	EA.	AB/23042
TP.23098	XXXXXXXXXXXXXXX	WALKER ENG'G.	500	EA.	WE/20962
K.98706	XXXXXXXXXXXXXXX	BROWN STEEL	0.50	TON	BS/23649
PE.48742	XXXXXXXXXXXXXXX	CONWAY LTD.	100	EA.	CO/23954
AR.90643	XXXXXXXXXXXXXXX	HARRIS-LOWE	500	ML	HL/25487

PROCESS RECORD:

OPN. No.	OPERATION DESCRIPTION	MACHINE	TOOLING	OPERATOR	DATE	
					START	END
10	TURN AND FACE	M.2564	STD.	G.SMITH(O14)	28/04/91	29/04/91
20	DRILL	M.3507	STD.	A.JONES(231)	01/05/91	02/05/91
30	MILL GROOVE	M.1904	T.0987	D.BROWN(094)	15/05/91	16/05/91
40	GRIND	M.9824	G.1487	P.HALL(132)	21/05/91	22/05/91

INSPECTION RECORD:

OPN. No.	OPERATION DESCRIPTION	MACHINE	INSPECTOR	QUANTITY		SCRAP CODE	REWORK & INSP.REPORT
				GOOD	SCRAP		
10	TURN AND FACE	M.2564	J.EAST(23)	99	1	07	23/1875
20	DRILL	M.3507	J.EAST(23)	99	0	03	24/1876
30	MILL GROOVE	M.1904	I.SMITH(15)	97	2	11	26/2456
40	GRIND	M.9824	P.ROWE(10)	96	1	04	21/4276

SALES RECORD:

SERIAL NUMBER	CUSTOMER NAME A/C. CODE	DATE OF DESPATCH	DELY.NOTE NUMBER	COMPLAINT/CLAIM RECORD NUMBER
M.27506	ABK SUPPLIES (AB29064)	DD/MM/YY	DN.45209	CMP/2589 (24/9/93)
M.27507	ABK SUPPLIES (AB29064)	DD/MM/YY	DN.45209	

Figure 13.19 Typical computerized lot traceability record.

how work should be done and what records should be kept: this information is needed as a part of the BS 5750 certification process. As an example, a typical standard procedure is shown in Fig. 13.20.

QC should also be responsible for issuing certificates of conformance (C of C) or test certificates which are required when certain types of products are despatched to a customer.

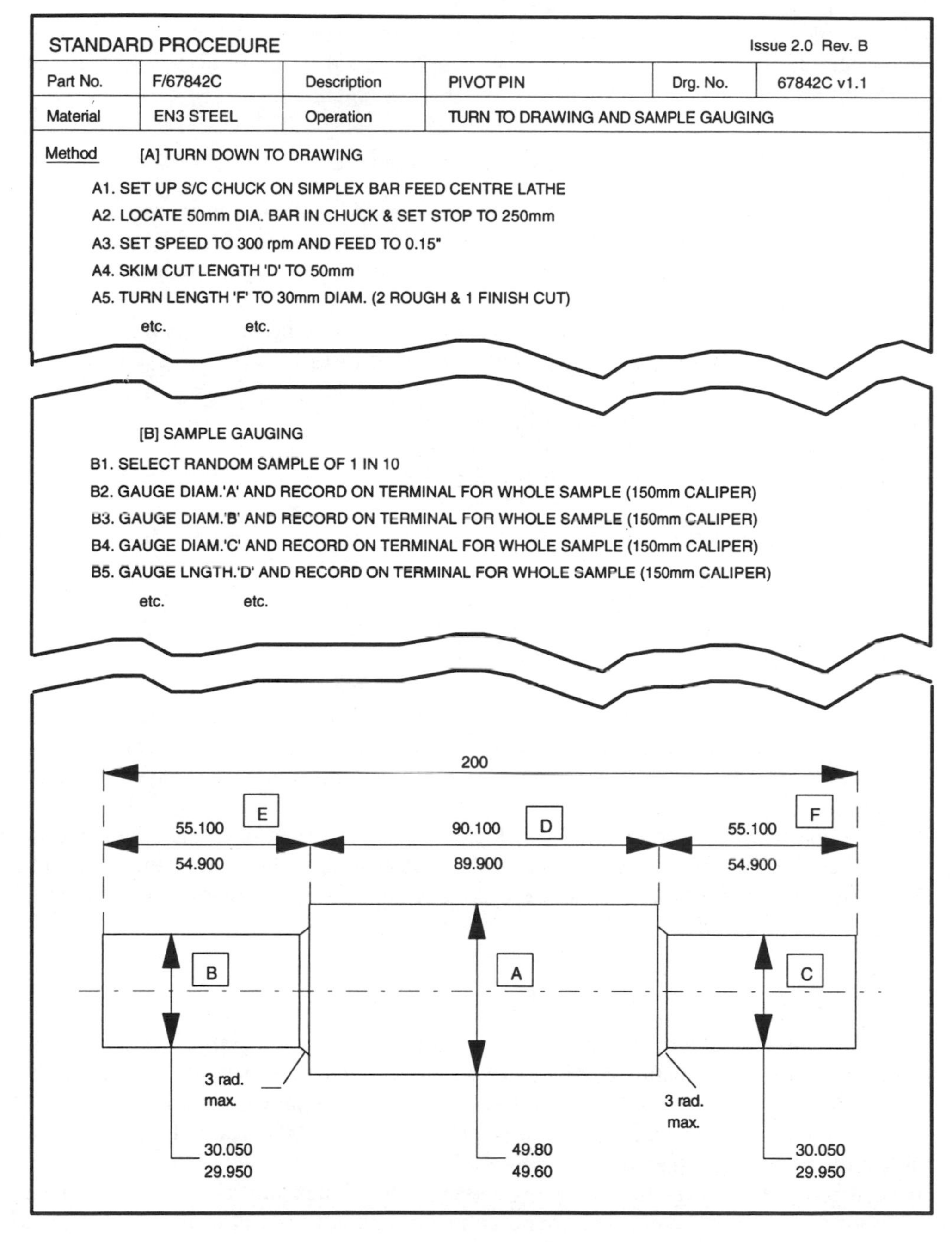

STANDARD PROCEDURE — Issue 2.0 Rev. B

Part No.	F/67842C	Description	PIVOT PIN	Drg. No.	67842C v1.1
Material	EN3 STEEL	Operation	TURN TO DRAWING AND SAMPLE GAUGING		

Method [A] TURN DOWN TO DRAWING

A1. SET UP S/C CHUCK ON SIMPLEX BAR FEED CENTRE LATHE
A2. LOCATE 50mm DIA. BAR IN CHUCK & SET STOP TO 250mm
A3. SET SPEED TO 300 rpm AND FEED TO 0.15"
A4. SKIM CUT LENGTH 'D' TO 50mm
A5. TURN LENGTH 'F' TO 30mm DIAM. (2 ROUGH & 1 FINISH CUT)
etc. etc.

[B] SAMPLE GAUGING

B1. SELECT RANDOM SAMPLE OF 1 IN 10
B2. GAUGE DIAM.'A' AND RECORD ON TERMINAL FOR WHOLE SAMPLE (150mm CALIPER)
B3. GAUGE DIAM.'B' AND RECORD ON TERMINAL FOR WHOLE SAMPLE (150mm CALIPER)
B4. GAUGE DIAM.'C' AND RECORD ON TERMINAL FOR WHOLE SAMPLE (150mm CALIPER)
B5. GAUGE LNGTH.'D' AND RECORD ON TERMINAL FOR WHOLE SAMPLE (150mm CALIPER)
etc. etc.

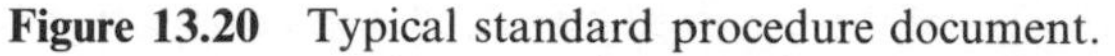

Figure 13.20 Typical standard procedure document.

These certificates are a form of guarantee, stating that the products listed conform to a standard that has been specified by the customer and agreed by the supplier.

QC may also be involved in the issue or receipt of concession documents, which are used to certify that a product or batch does not fully meet specification but will be accepted by the customer, often at a lower rate. Concessions can also be used internally, but in this case the production department is the 'supplier' and the QC department the 'customer'.

For example, it may be that a batch fails the sample check acceptance level by a small margin and should therefore be rejected, but because of special circumstances is accepted under the concession system.

13.13 QUALITY CONTROL AND THE CUSTOMER

One of the main functions of the QC department is to ensure that customers receive goods of a standard up to or above their expectations, at a competitive price. It is therefore important that QC know what these expectations are, and this can be achieved only by market surveys and investigations into complaints.

Quality, from a customer point of view, is difficult to define because it is partly subjective and therefore hard to measure, but is generally judged on three factors:

- *quantitative* does it perform its design functions effectively?
- *qualitative* does it look or feel good, without blemishes?
- *duration* does it have an acceptable life expectancy?

It is the function of QC to answer these questions, and to do this it must be able to set up the necessary testing, checking, and control procedures, on the basis of some of the principles put forward by Dr Genichi Taguchi. This Taguchi methodology is based on a concept known as the quality loss function, which states that every design parameter has a target value, with upper and lower control limits on either side of the target.

If, as ought to be the case, the target value is set to match the desires of the customer, then any deviation will suggest a reduced level of satisfaction, and the higher the deviation the higher will be the level of dissatisfaction. In other words, to keep the customers happy, the aim should be to keep as close to the target as possible. Full details of this method are given in chapter 5 (section 5.4).

The QC department should therefore work closely with the sales and marketing department to monitor these deviations, and to compare them with customer complaints and the results of customer satisfaction surveys so that, where necessary, designs or processes can be modified to reduce the deviation values.

13.14 THE BENEFITS OF TOTAL QUALITY MANAGEMENT

The tasks that make up the quality management function are now complex, due to the fairly recent realization that the function covers all aspects of an enterprise, from the specification of a new product through to after-sales services. Every area of a company is affected by this need to 'build in' quality in both products and services, and this need is complicated by the fact that all such endeavours must be integrated.

This can be illustrated in terms of the process capability index (see section 13.10), where the ratio between design requirements and manufacturing capability is a crucial factor in ensuring that the resultant products are of the required standard. Other departments will also be involved in this exercise, for instance the maintenance engineers, who need to ensure that process equipment is serviced to meet the demands of accuracy and reliability.

This process of involving and integrating all employees in a total quality strategy is known as TQM (total quality management), and requires full cooperation at all employee levels, from top to bottom. This means that effective and enthusiastic leadership must be provided at the top, with complete commitment from all, including suppliers. The benefits can be well worth the effort in terms of increased output (fewer rejects), increased sales (more customer loyalty), lower costs, and improved relations with customers, suppliers and employees.

13.15 APPLYING TOTAL QUALITY MANAGEMENT

TQM can be applied in any type of business—manufacturing, distribution, retailing or the provision of services—since the same principles apply in every case. These principles can be summarized in terms of the steps involved:

1. involve everybody from the start, and make it clear that their contributions are crucial to the success of the project
2. explain why such a project is necessary in terms of the need to meet world-class competition and the benefits that can be obtained
3. spread the responsibility, by allowing each department and work area to attack its own problems, within an overall plan (see note (a) below)
4. develop the theme that quality is not just a measure of product acceptability, but involves such aspects as waste, cleanliness, tidiness and good housekeeping (see note (b) below)
5. set targets that are achievable in terms of these factors and publicize the achievement of such targets
6. develop the concept of continuous improvement by encouraging each department to set new targets and to achieve these targets.

Note (a): The spreading of responsibility, in this context, usually involves the creation of 'quality circles'—a Japanese idea which, when modified to suit Western ideas, allows employees, at department or section level, to meet daily to discuss the problems of yesterday and the challenges of today. These meetings, if free from management interference, often generate the best improvement suggestions.
Note (b): The Japanese use the term '5S' when considering improvements in industrial housekeeping. This involves four basic principles (the fifth 'S' is for a Japanese word which means commitment to the other four):

- *seiri* the removal of unnecessary things
- *seiton* putting everything in order
- *seiketsu* maintenance of cleanliness
- *seiso* cleaning.

Companies that are now using the TQM approach have found that many of the best ideas and improvement suggestions have come from the least likely source. They have thus learnt that quality is not the preserve of the 'experts' since it is often the case that—given the proper motivation—the person closest to the job can see what needs to be done to improve it.

The TQM concept is also examined in chapter 15, section 15.12.

13.16 THE DEMING PRINCIPLES FOR TOTAL QUALITY

A total quality management system relates not just to the technical quality of the products, but more to the quality of performance of every function in the company. This recognizes that the quality and cost of a product, and the level of service provided, depend on its design, the lead times, the process reliability, the choice of materials, and the effectiveness of the supporting functions (suppliers, administrative staff, indirects, etc.).

The principles that govern the implementation of a total quality system, and which have proved effective over numerous applications, were originally laid down by Dr W. E. Deming and developed by others such as Dr J. M. Duran and Dr K. Ishikawa. These 13 principles are given in Table 13.7.

Table 13.7 The Deming principles for total quality

1. Innovate in all areas including training, and provide resources to assist. Maintain an innovative and vigorous training programme.
2. Learn the 'zero-defect' philosophy and the need for continued improvement.
3. Do not rely on mass inspection for quality. Put quality prevention on-line via SPC etc.
4. Reduce the number of suppliers and develop them for continuous improvement of service as well as cost.
5. Use statistical techniques to identify sources of waste and cure both system faults and local faults at source.
6. Ensure that organizational and management systems support innovation and continuous improvement.
7. Provide supervision with on-line techniques for problem identification and problem solution via their teams.
8. Create openness by encouraging questions and the reporting of problems.
9. Attack waste by the use of multidisciplinary teams.
10. Avoid exhortive slogans as a substitute for team approaches.
11. Beware of over-bureaucratic imposition of work standards.
12. Provide elementary statistical training to all employees.
13. Make maximum use of statistical data to focus on priority problems and direct the effort of all talent in the company.

What distinguishes a total quality company from a traditional one is the way in which its people think and act; the value that such people place on quality of performance in every activity, and what they do to improve the quality of their work. This attitude can be described as a culture, which defines how people at work share with each other, and how they do things on a daily basis.

The following comments (traditional and quality cultures) show how the different attitudes can lead to different results:

1. *Attitude* 'Quality is not my responsibility'
   Result People let mistakes go—they fail to check their own work.
2. *Attitude* 'No-one listens to what we say'
   Result Low involvement in continuous improvement programmes.
3. *Attitude* 'Management is there to help me to achieve my goals'
   Result Active support for total quality initiatives.

PROBLEMS

1. Prepare a preliminary FMEA for a new design of electric kettle and, using the principles and diagram from Fig. 13.6, enter your estimates of 'potential effects' and points values to calculate the RPNs for the following potential failure modes (for this purpose the kettle is of the jug type with a removable lid)
   (a) kettle is knocked over when empty
   (b) kettle is knocked over when full
   (c) kettle is switched on when empty
   (d) kettle fails to switch off when water has boiled.
2. Based on the results of your FMEA study in question 1, suggest what improvements could be made in the design and recalculate the RPNs.
3. Your company is initiating a supplier auditing system and wishes to tell its main suppliers what is involved. Prepare an outline for a document that is to be sent to each supplier explaining the procedure and the areas that are to be audited.
4. An SPC system has been implemented in your company, and as quality manager you have to prepare a C_{pk} (process capability index) for a new component. This is a pin of 10 mm diameter. The design tolerance is ± 0.5 mm, and a series of sample measurements gave the following readings. Determine the C_{pk} for these conditions (see Fig. 9.9 in chapter 9 for a method of calculating σ); see appendix A for the solution).

Reading no.	1	2	3	4	5	6	7	8	9	10
Value (mm)	10.1	9.8	10.0	9.8	10.1	10.0	9.9	9.7	9.8	9.8

5. A TQM system is to be implemented in your company and everyone is involved. Suggest how the following employees could contribute to the system in terms of improved levels of quality of product or service:
   (a) telephone operator/receptionist
   (b) delivery driver
   (c) wages clerk
   (d) fork-lift truck driver.

REFERENCES AND FURTHER READING

Bendell, A., Disney, J. and Pridmore, W. A. (eds), *Taguchi Methods*, IFS Publishing, Bedford, 1989.

Bergman, B. and Klefsjo, B., *Quality: from Customer Needs to Customer Satisfaction*, McGraw-Hill, London, 1994.

Besterfield, D. H., *Quality Control*, 4th edn, Prentice-Hall, Englewood Cliffs, NJ, 1994.

Dale, B. G. (ed.), *Managing Quality*, 2nd edn, Prentice-Hall, London, 1994.

Evans, J. R. and Lindsay, W. M., *Management and Control of Quality*, West, Boston, MA, 1993.

Feigenbaum, A. V., *Total Quality Control*, McGraw-Hill, New York, 1983.

Fox, J., *Quality through Design*, McGraw-Hill, London, 1993.

Logothetis, N., *Managing for Total Quality*, Prentice-Hall, London, 1992.

Oakland, J. S., *Total Quality Management*, 2nd edn, Butterworth Heinemann, Oxford, 1993.

Pugh, S., *Total Design*, Addison Wesley, Wokingham, 1991.

Sugiyama, T., *The Improvement Book*, Productivity Press, Cambridge, MA, 1989.

CHAPTER

FOURTEEN

THE MANAGEMENT OF COSTS

THE LINK

In previous chapters, the emphasis has been on the management of the functions that contribute effective production by controlling the means of production. However, it is equally important to control the financial aspects of these and other operations in order to ensure that the company achieves its targets in terms of turnover and profit.

There are two types of accounting—financial and management—and it is important for management to recognize the difference between these in terms of approach and objectives.

Financial accounting is designed to produce a series of accounts that show how the company is performing from a financial point of view. There is a legal requirement for these accounts to be prepared annually for tax purposes, but in most cases they are prepared monthly or quarterly so that the Directors can assess the success (or otherwise) of their company and declare dividends for shareholders. These accounts look at the overall situation, and are presented in the form of 'reports' such as profit and loss accounts, manufacturing accounts and balance sheets. The difference between financial accounting and auditing should noted: financial accounts are usually prepared internally by the company, but must be audited annually by independent accountants to meet the legal requirements.

On the other hand, management accounting is designed to look at the facts behind the figures, for example, the costs of the various operations and how these determine the costs of the products. These costs can then be compared with the selling prices to determine individual product profitabilities and, where costs are too high, the management accountant should be able to suggest how they could be be reduced, or what other actions need to be taken to meet profitability targets.

14.1 INTRODUCTION

In this chapter the need to identify costs and to allocate them appropriately is considered in terms of the needs of the company type. Various methods of costing are examined in terms of their applications, and this is followed by a detailed definition of thc different types of costs that can be found; these sections include suggested methods of dealing with each cost type. The various uses of costing are then considered including estimating, cost control, and budgeting and the chapter concludes with suggestions concerning the communication that is needed between departments to ensure that the costing and management accounting functions are fully effective.

14.2 THE BASIC PRINCIPLES

In order to manage a business effectively it is necessary to maintain full control over all types of cost. This means that a system must be set up that allows costing management to:

1. ensure that all activities which generate costs include—within the operating procedures—a cost-collection mechanism
2. classify each cost item so that it can be suitably analysed for costing and accounting purposes
3. analyse all costs so that they can be allocated to the correct activity and in the correct proportions (if shared by several activities).

Cost collection

Costs are incurred whenever goods are received, services are used, processes are carried out, labour is employed or buildings are occupied, and in some cases (but not all!) these costs are covered by suppliers' invoices. In fact, the first task of the management accountant in any new enterprise is to list all possible cost items and identify how best these costs can be recorded, classified and analysed.

Costs can be broken down into a number of categories. Figure 14.1 shows a typical list, by category, with the associated means of collection. This is not intended to be a complete list, but is merely an indication of the cost types and categories; a more detailed examination is given later in this chapter.

Cost classification

In the example in Fig. 14.1, a series of cost categories is shown, but this is only a part of the classification process. In fact, what is required is a cost structure and coding system that will indicate where a cost has come from, and how it has been allocated.

This is an essential part of the financial accounting system known as the nominal ledger, which is used to bring together all elements of income and expenditure from other ledgers (sales, purchases, journals, etc.) and from which a company's monthly and annual accounts will be developed. It is therefore necessary to ensure that the two systems (cost and financial accounting) are both operating on the same basis.

Cost analysis

In order to decide how costs should be analysed, it is necessary to know the basis on which the costing system will operate, and this depends on the type of manufacturing business. For instance, should the costs be analysed on the basis of products, jobs, batches, processes, or a perhaps a combination of some of these, bearing in mind the ultimate objectives of the costing process?

These objectives can be defined as follows:

1. to determine the profitability of a product or service
2. to form the basis for the company accounts
3. to assist with the preparation of estimates
4. to assist with management decisions (e.g. make or buy)
5. to evaluate and manage a project

IDENTIFICATION OF COSTS		
CATEGORY	DESCRIPTION	MEANS OF COLLECTION
DIRECT LABOUR	WAGES PAID TO EMPLOYEES DIRECTLY INVOLVED WITH PRODUCTION OF PRODUCTS.	FROM JOB BOOKINGS AND PAYROLL INFORMATION
DIRECT MATERIALS	VALUE OF MATERIALS ISSUED TO PRODUCTION AND USED DIRECTLY ON PRODUCTS.	FROM PURCHASE INVOICES OR STOCK VALUATIONS AND STOCK ISSUES.
INDIRECT LABOUR	WAGES PAID TO EMPLOYEES INDIRECTLY INVOLVED WITH PRODUCTION (eg. Stores, Supervision etc)	FROM PAYROLL
INDIRECT MATERIALS	VALUE OF MATERIALS ISSUED TO PRODUCTION AND USED INDIRECTLY (eg. Consumable Tools, Cutting Oil etc)	FROM STOCK VALUATIONS AND STOCK ISSUES.
WORKS OVERHEADS	ELECTRICITY, GAS, WATER, LOCAL RATES, CONTRACT CLEANING AND REPAIRS, TOOLROOM COSTS, etc.	FROM INVOICES
WORKS EXPENSES	MACHINE AND PLANT DEPRECIATION, CANTEEN SUBSIDIES, PRODUCTION ADMINISTRATION, INSURANCES, etc.	BY CALCULATION AND FROM ACCOUNTS
ADMINISTRATION	STAFF SALARIES, OFFICE HEAT & LIGHT, OFFICE MACHINE COSTS, etc.	FROM PAYROLL AND ACCOUNTS
DESPATCH	PACKING LABOUR & MATERIALS, TRANSPORT, DESPATCH ADMINISTRATION etc.	FROM PAYROLL AND ACCOUNTS

Figure 14.1 Identification and classification of costs.

6. to control and reduce costs
7. to compare actual performances with budgets.

To meet these objectives it is necessary for the management accountant to look at the processes used to make the products and then to define an analysis method for each. There are basically four such process types:

- *single-item batch–single product* where an operation is carried out at a work centre on a defined batch of one item, e.g. machining a quantity of a bearing housing that is used on only one product
- *single-item batch–multi-product* where an operation is carried out at a work centre on a defined batch of one item that is to be used on several products, e.g. assembly of a ball race
- *multi-item batch or process* where an operation is carried out on a number of different items at the same time that may be used on one or a number of different products, e.g. a heat-treatment furnace or a plating plant
- *continuous process* where a series of linked operations is being carried out either on a random mixture of product variants e.g. a car assembly line, or on a single product.

In all these cases it is possible to collect the costs of the operation or process, but in some it is also necessary to split these costs, so that they can be allocated to the different items or products concerned; this can be done only if suitable analysis methods have been set up.

14.3 TOTAL OR ABSORPTION COSTING

This is the name given to a costing method wherein every item of cost incurred in an organization is taken into account in arriving at the cost of an individual product or service. This is probably the method most commonly used in manufacturing industry, although many accountants feel that it is too heavily reliant on the pro-rata allocation of indirect cost and expenses, which can give rise to inaccurate product costs. For example, is it reasonable to allocate the costs of stock management on the basis of the number of issues from stores?

The essential terminology used in this method can be defined as follows.

1. *Cost centres* These can be a group of people or machines, a single person or machine, or a combination. They are usually associated with a particular location and may or may not be the same as work centres (as defined in chapter 6, section 6.4). The basis for setting up cost centres must be the realization that any work being processed through a cost centre will attract a predetermined cost per hour, and therefore the cost centre should not include diverse machines that are likely to have different cost rates. For example, a press shop may contain presses of differing sizes and capacities and if these are all put into one cost centre—at an average hourly rate—then the work done on the small presses will be too expensive and that done on the large presses too cheap. This is because a small press has a lower initial purchase price, occupies less space, and uses less power and probably less labour. The answer, in this case, must be to split the presses into several cost centres by size/capacity.
2. *Cost units* These must be set up and used to measure the direct costs of output or activity, and in many cases will be the same as the units of measure (UOMs) used for production or purchasing (each, tonnes, litres, etc.). Units will also be needed for indirect cost allocation (square or cubic metres, hours, etc.) so that, for example, the costs of a building can be allocated to a cost centre in terms of the area occupied.
3. *Cost apportionment* This is the term sometimes used for the process of allocating proportions of a cost to various cost centres, for instance the costs of heating in terms of floor area or space occupied.
4. *Overhead absorption* This is the term used to cover the technique whereby various overhead costs are combined into a rate per cost unit, for example a machine-hour rate as shown in Table 14.1.
5. *Prime cost* This is the term used for the total cost that can be directly related to a unit of production, i.e. the cost of direct labour, materials and expenses. This is the absolute minimum cost of making an item and assumes that all other costs, being indirect, could be carried by other products.

Figure 14.2 shows how costs are built up in a total or absorption costing system.

14.4 STANDARD COSTING

This is the name given to a costing method whereby a theoretical standard is set up for each cost element, so that comparisons can be made between these standard costs and the equivalent actuals. This method has a number of other useful applications, for example as a performance measure and as a means of controlling and reducing costs by analysing the reasons for the variances. However, a standard costing system may be more difficult to set up than a total

Table 14.1 Build-up of a machine-hour rate

Cost item	Annual cost (£)	Cost item	Hourly cost (£)
Depreciation	5000	M/C setter	7.50
Maintenance	1500	M/C operator	6.00
Insurance	800		
Power	3500		
Consumables	1750		
Tools	2850		
Totals	15 400		

Annual hours worked: 40 h/week × 48 weeks = 1920 hours
M/C setting: 12 h/week × 7.50 = £90/week
M/C operating: 20 h/week × 6.00 = £120/week (operates 2 machines)

Machine hour rate (£/h)

Annual cost £15 400 ÷ 1920 hours	= 8.02
Setting £90/week ÷ 40 hours	= 2.25
Operating £120/week ÷ 40 hours	= 3.00
Total	= £13.27 per hour

system because of the need to establish workable standards, to set up additional accounts (for variances) and to establish procedures for analysing the variances.

The advantages of using standard costing can be summarized as follows.

- After an initial settling-in period, the amount of cost investigation work is reduced, since the management-by-exception principle will allow all actuals, that are within an agreed tolerance of standard to be accepted automatically. (A problem with the management-by-exception approach to variance investigations is that if the comparisons are taken at too high a level they may show that actual and standard costs are within tolerance although in fact there are compensating variances at lower levels. This can be seen in Fig. 14.4, where at prime cost level there is no variance whereas at the materials and labour levels there are marked variances.)
- They can form a better basis for decision-making than actuals since the standards are consistent throughout, whereas actuals can be affected by unusual or freak conditions.
- They can be used to compare performances between products, work sections or departments.
- They can be used to reduce costs and eliminate waste by means of the variance analyses.

Figure 14.3 shows how actuals, standards and variances interact in a standard costing system. Figure 14.4 shows an example of variance analysis in standard costing.

14.5 ACTIVITY-BASED COSTING (ABC)

Activity-based costing (or ABC for short) can be defined as a method of applying overhead and indirect costs to a job or product on the basis of the work/costs generated by that activity. For example, the overheads generated by the purchasing department of a company could be rewritten in terms of the activities carried out rather than in the traditional form based on chargeable items. This can be seen from an examination of Table 14.2.

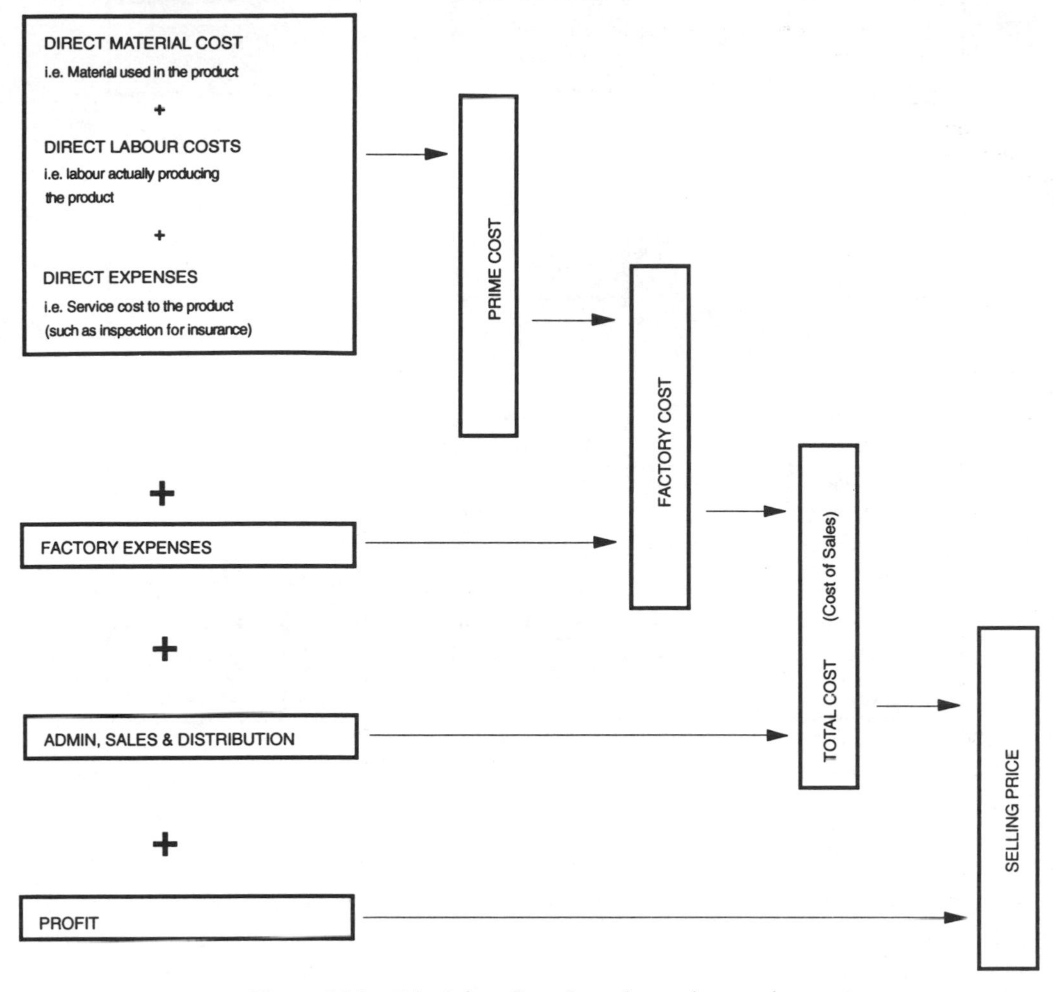

Figure 14.2 Principles of total or absorption costing.

Table 14.2 Activity based costing methods

Traditional method		Activity-based method	
Item	£m	Item	£m
Salaries	5.0	Audit suppliers	1.2
Office supplies	2.1	Negotiate with suppliers	1.1
Depreciation	1.6	Place orders	1.2
Telephone etc.	0.2	Expedite delivery	2.6
Electricity, rent	0.4	Resolve problems	3.2
Total	9.3		9.3

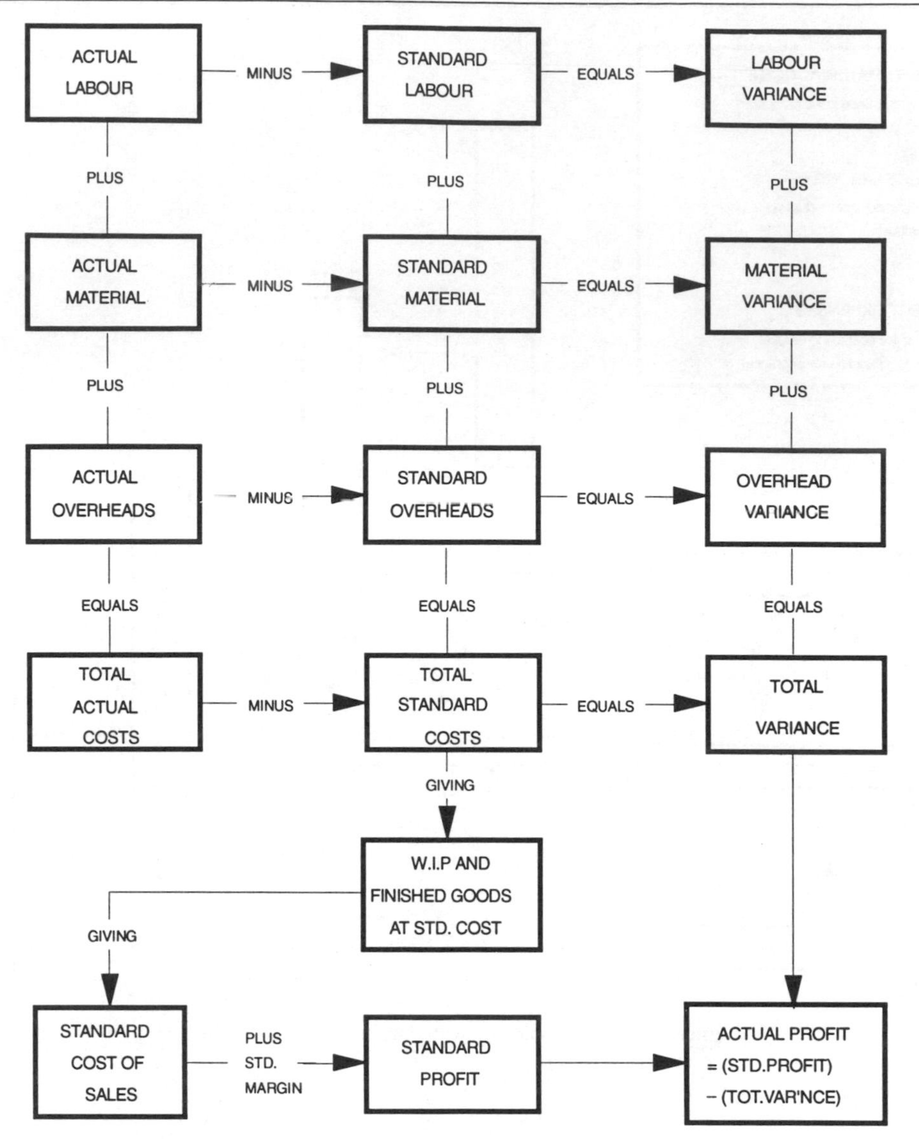

Figure 14.3 Principles of standard costing.

In the traditional method, product costs (including overheads) are derived from the hours spent in a cost centre and the hourly rate for that centre. In contrast, ABC uses activity-based cost pools and activity cost-based rates to apply overheads. The advantages and disadvantages of ABC are listed in Table 14.3.

Once the cost pools (e.g. for purchasing or stores) have been determined, the factors driving these costs (the cost drivers) can be identified, for example the number of purchase orders or stock issues. Then, by linking the drivers to individual jobs or product lines, more accurate and realistic costs can be obtained.

In ABC there are various levels at which cost-driven activities can be identified, for example:

Table 14.3 Activity-based v. traditional costing

Advantages of ABC
1. Gives new performance measures based on cost-driver rates.
2. Emphasizes activities across departments.
3. Enhances overhead visibility.
4. Allows control and reduction of overhead costs by eliminating or reorganizing activities.
5. Gives a better understanding of overhead costs.
6. Gives a fairer and more accurate measure of costs by job or product line.

Disadvantages of ABC
1. Requires judgement to select the correct cost drivers.
2. Can become extremely complex if badly applied; object should be 'keep it simple'.
3. Systems support is required; this can be costly.
4. ABC is an attention-directing system; for decision-making, future not past costs are required.
5. Draws attention to people and their activities and can therefore cause personnel problems.

EXAMPLE OF VARIANCE ANALYSIS:

THE STANDARD PRIME COST OF ONE UNIT OF A PRODUCT INCLUDES THE FOLLOWING ITEMS:

ITEM	QTY.	RATE	£
LABOUR + OVERHEAD	10 HOURS	AT 15.00	150.00
MATERIAL A	20 Kgs.	AT 4.50	90.00
MATERIAL B	30 Kgs.	AT 6.00	180.00
		TOTAL	420.00

IN A PARTICULAR WEEK 15 UNITS ARE PRODUCED AT THE FOLLOWING COST:

ITEM	QTY.	RATE	£
LABOUR + OVERHEAD	135 HOURS	AT 15.00	2025
MATERIAL A	330 Kgs.	AT 4.50	1485
MATERIAL B	465 Kgs.	AT 6.00	2790
		TOTAL	6300

THE STANDARD COST OF 15 UNITS IS 15 x 420 = £6300

SO THERE IS NO DIFFERENCE BETWEEN THE ACTUAL AND STANDARD PRIME COSTS. HOWEVER, THE VARIANCE ANALYSIS REVEALS THAT THERE ARE COMPENSATING ERRORS THUS:

LABOUR + OVERHEAD VARIANCE = –15 HRS. @ 15.00/HR = –225 (Favourable)

MATERIAL 'A' VARIANCE = +30 Kgs. @ 4.50/Kg = +135 (Adverse)

MATERIAL 'B' VARIANCE = +15 Kgs. @ 6.00/Kg = +90 (Adverse)

Figure 14.4 Variance analysis in standard costing.

1. *unit* at this, the lowest level, overhead costs are affected by the number of units being processed, perhaps measured in terms of the volume of materials handled.
2. *batch* at this level the overhead costs are affected by the number of batches processed, perhaps measured in terms of the number of machine set-ups
3. *product line* at this level the overhead costs are affected by the number of product lines being processed, perhaps measured in terms of bills of materials
4. *factory* at this, the top level, the existence of a production facility affects the overhead costs, for example the salaries of factory management.

A process chart showing how a typical ABC system might operate is shown in Fig. 14.5.

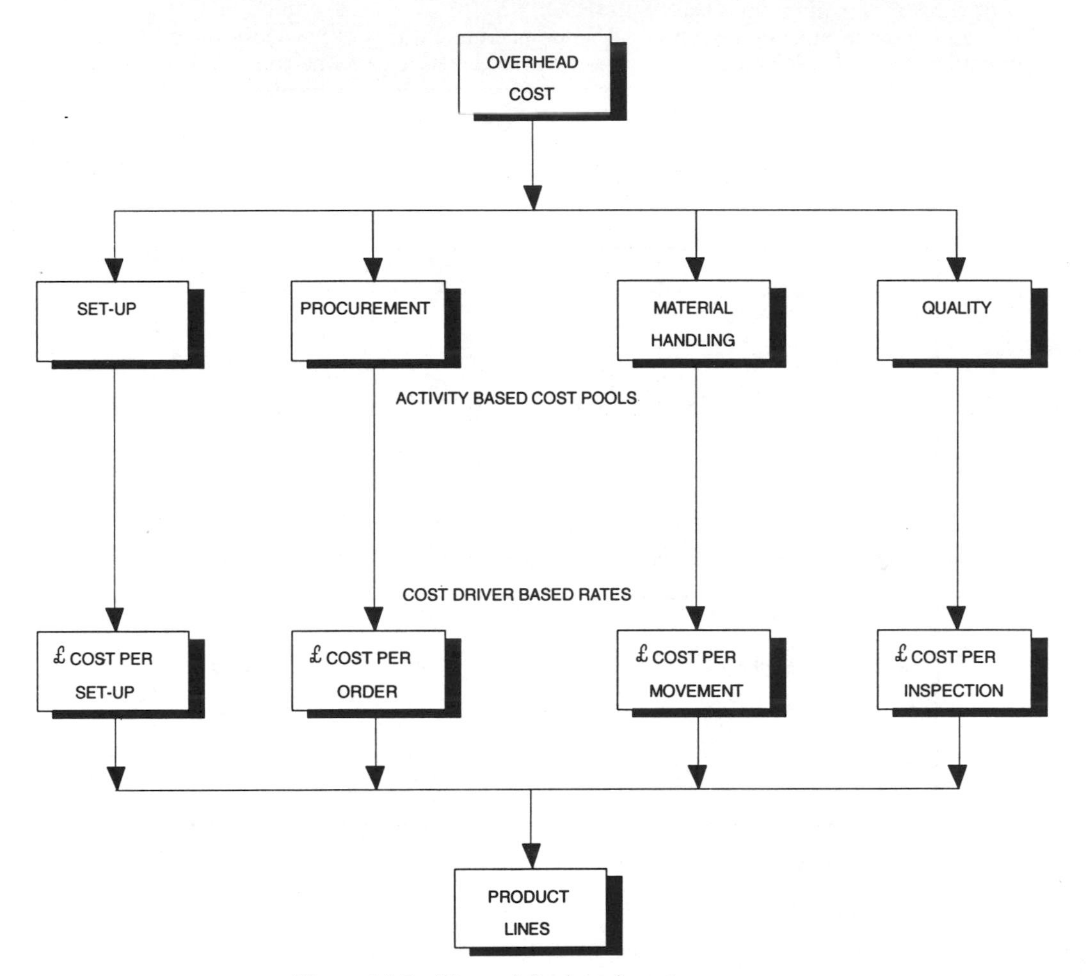

Figure 14.5 The activity-based costing process.

14.6 ESTABLISHING STANDARD COSTS

As can be seen from Fig. 14.3, standard costs need to be established for the three main cost elements — labour, materials and overheads — to give a total standard.

Standard labour costs

In this context, labour cost is taken to be the cost of direct labour only, and therefore excludes such items as shop-floor labour and supervision. The most effective means of establishing direct labour costs is to apply one of the work measurement techniques described in chapter 7 (section 7.9) which will allow standard times to be obtained. These can then be extended by the appropriate labour rates.

An important factor in establishing labour costs is that they should be realistic and therefore achievable. Thus, for example, it is necessary to take account of such factors as efficiency and utilization percentages when setting up the standards. If this is not done, the variances will always be adverse and the standards will fall into disrepute because they are unachievable.

An efficiency of 100 per cent assumes that the operator or machine, when working, is capable of a maintaining an overall performance of 100 on the 75/100 standard scale. If this is not the case then a lower efficiency value should be set. Similarly, a utilization of 100 per cent assumes continuous operation by the operator or machine with no unscheduled breaks or non-productive time. If this cannot be consistently achieved, a lower utilization value should be set.

There may be some disagreement about the level of the labour rates to be applied to the standard times, i.e. what should be included, and it is necessary to agree a standard approach such as that given in Table 14.4. It should be noted that these figures would not just be those for a typical week, but would be based on the average of a number of weeks, such that the rate per hour would be representative in terms of performance and overtime hours/payments.

Table 14.4 Setting a labour rate per hour

Wage element	Weekly value (£)	Comments
Basic wage	150.00	£4.00 × 37.5 hours
Overtime (A)	5.33	4 hours @ time and a third
Overtime (B)	6.00	4 hours @ time and a half
Bonus	54.60	45.5 hours @ £1.20 (95 performance)
Total	215.93	For 45.5 clock hours
Hourly rate	£4.75/clock hour	(at 95 performance)

Payments for sickness, holiday pay, employer's pension contributions, etc., should be recovered as overheads.

Standard material costs

As with standard labour, standard material cost means direct material costs only, and such items as consumable tools are excluded unless they can be directly related to a product or job. All materials used in a product are either manufactured or purchased at a standard cost, and in the case of purchased items the standards should be set realistically on the basis of the best price obtainable, for an agreed order size, from an approved supplier.

These purchasing standards should take account of the costs of rejects if this is a regular occurrence, but not the costs of purchase administration, which should be recovered as an overhead.

Standard overhead costs

All costs not included as standard labour or material should be included as overheads, and for each of these a standard should be calculated. The way in which these should be allocated depends to some extent on the mode of operation of the company concerned, but as a general rule allocation should be at the lowest possible level, the levels being as follows:

1. machine or employee (lowest)
2. cost centre
3. department
4. factory
5. company (highest).

A list of possible overheads for which standards would be required, together with suggested means of allocation, is shown in Fig. 14.6.

Standard overhead settings can be in various forms, but where possible should be easily applied as shown in the example in Table 14.5 (in this case applied at cost-centre level). Thus, the example shows that every job or product passing through one of these cost centres will collect a standard heating charge, based on the number of standard hours it spends in that cost centre, being processed. There is no charge if process time is not being booked.

Table 14.5 Setting overhead rates

The total standard cost of heating in a factory is £30 000 per year. This is to be allocated on the basis of floor area at cost-centre level.

Work centre	Floor area (square metres)	% of total area	Heating cost (£)	Standard hours per year	Heating cost/ standard hour (£)
A	1100	11.7	3510	3360	1.04
B	1450	15.4	4620	5840	0.79
C	950	10.1	3030	1850	1.64
D	2400	25.5	7650	7235	1.06
E	1850	19.7	5910	5980	0.99
F	1650	17.6	5280	4390	1.20
Total	9400	100.0	30 000	28 655	1.05

It is thus possible to build up a total standard overhead cost for each cost centre, and thence a total standard cost for the product or job.

14.7 BUILDING A TOTAL STANDARD COST

If a computerized system is in operation and the necessary database information has been set up, it is easy to find the standard cost of a product by using a process known as 'cost roll-up'. This operates by using three sets of data:

- the bill of materials
- the production routeing
- the standard labour/overhead cost rates and the standard material costs.

OVERHEAD EXPENSE ITEM	BASIS OF ALLOCATION (APPORTIONMENT)	UNITS
BY MACHINE/MAN HOUR		
DEPRECIATION	BY MACHINE	VALUE
HOLIDAY PAY	BY EMPLOYEE	COST
EMPLOYER'S CONTRIBUTIONS	BY EMPLOYEE	COST
SICK PAY	BY EMPLOYEE	COST
BY WORK CENTRE:		
RENT AND RATES	FLOOR AREA	SQ. METRES
LIGHTING	LIGHTING CAPACITY	WATTAGE
HEATING	FLOOR AREA	SQ. METRES
INDIRECT LABOUR	OBSERVATION/BOOKING	HOURS
REPAIRS	PREVIOUS RECORDS	COSTS
MAINTENANCE	JOB BOOKINGS	HOURS
LUBRICANTS & OILS	NUMBER OF MACHINES	UNITS
CLEANING MATERIALS	NUMBER OF MACHINES/PEOPLE	UNITS
INSURANCE	VALUATIONS OF EQUIPMENT	VALUES
CANTEEN	NUMBER OF PEOPLE	UNITS
CRANES & TRUCKS	OBSERVATION/BOOKING	HOURS
SUPERVISION	% TIME OCCUPIED	HOURS
MANAGER'S SALARY etc.	% TIME OCCUPIED	HOURS
SMALL TOOLS	TOOL STORES ISSUES	VALUES
TOOL ROOM - LABOUR	JOB BOOKINGS	HOURS
TOOL ROOM - MATERIALS	MATERIALS ISSUED	VALUES
STORES	MATERIALS ISSUED	VALUES
BY COMPANY:		
DESIGN/DEVELOPMENT	ALLOCATED AS A GENERAL OVERHEAD IN PERCENTAGE TERMS ON THE VALUE OF PRODUCTS SOLD	
ACCOUNTS		
SALES		
PURCHASING		
DISTRIBUTION		
PRODUCTION PLANNING		
DATA PROCESSING		

Figure 14.6 Overhead items and suggested methods of allocation.

The process operates by making a series of calculations, starting at the lowest level in the product BOM and rolling upwards to the top level:

1. the standard batch size is applied to the BOM to calculate the quantity of each material required
2. each of these quantities is multiplied by its standard material cost per unit, and these values are totalled to give the total standard material cost

3. the standard batch size is applied to the various operations on the production routeings for the manufactured items, in the BOM, to calculate the standard processing hours
4. each of these times is multiplied by the standard hourly rate for the operation concerned and these values are totalled to give the standard labour plus overhead cost
5. finally, these two costs are added together and divided by the quantity in the batch, to give a standard unit cost.

This process is illustrated in detail in Fig. 14.7.

14.8 COLLECTING ACTUAL COSTS

The process of collecting actual costs, either for comparison with the standards or on the basis of some other costing system, is much simpler if a computer with a suitable costing module is available, but it can be carried out manually if necessary. In either case, it is necessary to identify the three cost types—materials, labour and overheads—at whatever level is necessary, and this is usually done in the cost accounting process by setting up a series of accounts (ledgers) which accumulate the costs.

The level at which costing is carried out will determine the complexity or otherwise of this system. For example, at its lowest level all costs could be collected into one account for the factory and this could then be compared with the total revenue from sales to determine the profit (or loss!). However, most companies believe that this is inadequate, since it does not allow any form of investigation to take place: for example, which products or operations are profitable, and which are not. They will therefore aim higher so that costs (and thus profits or losses) can be determined at cost-centre or product level.

Actual material costs

The actual costs of materials can be obtained by analysing purchase invoices, but will probably vary. It is therefore essential to agree a method that will allow the value of materials issued to a works order to be determined. Four possible methods are available.

- *First in-first out (FIFO)* Batches of items are taken into stock at their actual cost price and these prices are recorded for use when the issues are made. This is done on the basis of the oldest stocks being issued first.
- *Last in-first out (LIFO)* Exactly the same method as FIFO, except that issues are made on the basis of newest stocks being issued first.
- *Average cost (AVCO)* Batches of items are taken into stock at their actual cost price and these prices are then used, in conjunction with the prices of batches already in stock, to calculate an average unit cost for all stock, which is then used to value all issues.
- *Replacement cost (REPCO)* Batches of items are taken into stock at their actual cost price and this is then compared with a predetermined replacement cost, the differences (favourable or adverse) being posted to a materials variance account. This replacement cost is used to value all issues.

Examples of the way in which these methods operate are shown in Fig. 14.8.

Any of these methods can be used for standard costing, by setting up variance accounts and posting the variances between standard and actual to them, as shown in Table 14.6. From the information in this set of accounts, it is then possible to determine which products have var-

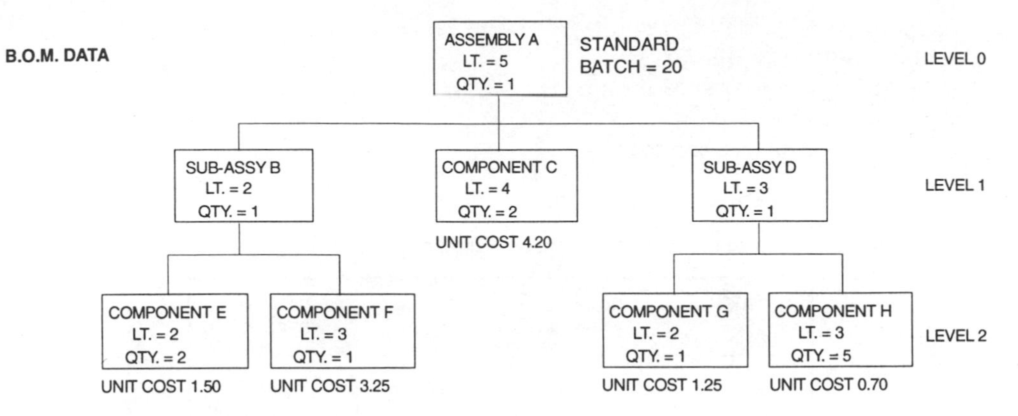

ROUTEINGS DATA

TO MAKE SUB-ASSY B (FROM E & F): QTY = 20

OPN. NO.	COST CR.	COST/ HR.	S.U. HRS.	OPN. HRS.	TOT. HRS.	OPN. COST
10	234	8.00	1.0	5.00	6.00	48.00
20	175	12.00	0.7	4.00	4.70	56.40
30	359	16.00	1.0	6.00	7.00	112.00

TOTAL LABOUR/OVERHEAD COST = 216.40
= 10.82 EACH.

TO MAKE SUB-ASSY D (FROM G & H): QTY = 20

OPN. NO.	COST CR.	COST/ HR.	S.U. HRS.	OPN. HRS.	TOT. HRS.	OPN. COST
10	540	8.50	0.5	1.00	1.55	13.18
20	145	10.00	0.7	2.00	2.70	27.00
30	300	16.00	0.4	3.00	3.40	54.40

TOTAL LABOUR/OVERHEAD COST = 94.58
= 4.73 EACH.

TO MAKE ASSY. A (FROM B, C, & D): QTY = 20

OPN. NO.	COST CR.	COST/ HR.	S.U. HRS.	OPN. HRS.	TOT. HRS.	OPN. COST
10	235	7.00	0.3	1.00	1.30	9.10
20	320	10.00	1.0	1.50	2.50	25.00

TOTAL LABOUR/OVERHEAD COST = 34.10
= 1.70 EACH.

COST ROLL-UP

LEVEL NO.	ITEM CODE	QTY. REQD.	COSTS				TOTAL COST	COSTS TO BE ROLLED-UP TO NEXT LEVEL
			UNIT OF MATERIAL	UNIT OF LABOUR	TOTAL MATERIAL	TOTAL LABOUR		
2	E	40	1.50	NONE	60.00	NONE	60.00	SUB-ASSY B COSTS 125.00 FOR 20 = 6.25 EA.
2	F	20	3.25	NONE	65.00	NONE	65.00	
2	G	20	1.25	NONE	25.00	NONE	25.00	SUB-ASSY D COSTS 95.00 FOR 20 = 4.75 EA.
2	H	100	0.70	NONE	70.00	NONE	70.00	
1	B	20	6.25	10.82	125.00	216.40	341.40	ASSY.A COSTS 699.00 FOR 20 = 34.95 EA.
1	C	40	4.20	NONE	168.00	NONE	168.00	
1	D	20	4.75	4.73	95.00	94.60	189.60	
0	A	20	34.95	1.70	699.00	34.00	733.00	

THUS ASSY.A HAS A UNIT STANDARD COST OF 733.00/20 = 36.65

Figure 14.7 The cost roll-up process.

1. POSSIBLE METHODS OF VALUATION OF STOCK ISSUES:

(a) FIFO (FIRST IN–FIRST OUT)

(b) LIFO (LAST IN–FIRST OUT)

(c) AVCO (AVERAGE COST)

(d) REPCO (REPLACEMENT COST)

2. EXAMPLES BASED ON THE FOUR METHODS:

THE GOODS INWARDS & STOCK RECORDS FOR A GIVEN ITEM ARE AS FOLLOWS:

LOCATION	QUANTITY AT LOC'N	ACTUAL COST	DATE OF RECEIPT
A1	100	1.00	15/12/92
A2	50	1.20	13/01/93
A3	150	1.15	12/02/93
A4	100	1.40	27/02/93

TOTAL STOCK = 400

3. A REQUISITION TO ISSUE 250 IS RAISED.

WHAT MATERIAL UNIT COST SHOULD BE USED ?

METHOD	CALCULATION	UNIT COST
FIFO	$\frac{(100 \times 1.00) + (50 \times 1.20) + (100 \times 1.15)}{250} = \frac{275}{250}$	1.10
LIFO	$\frac{(100 \times 1.40) + (150 \times 1.15)}{250} = \frac{312.5}{250}$	1.25
AVCO	$\frac{(100 \times 1.00) + (50 \times 1.20) + (150 \times 1.15) + (100 \times 1.40)}{400} = \frac{472.5}{400}$	1.18
REPCO	REPLACEMENT COST GIVEN AS 1.40	1.40

Figure 14.8 Methods of valuation of stock issues.

iances, and to investigate why these variances have occurred. However, it should be noted that there may also be variances due to purchasing, and this must also be taken into consideration.

For instance, the actual material cost content of a product may be too high because an above-standard price was paid to the material supplier or because an above-standard quantity was used, or perhaps a combination of both.

Actual labour costs

The actual costs incurred in processing should be calculated from the job bookings at each cost centre, in terms of hours booked and rate per hour. Such bookings can be in the form of daily work sheets, operation card entries or direct entry to the computer by means of shop-floor data capture (SFDC) terminals. However, records should be kept for all hours worked, even if some of these are spent on non-productive work, and these records should also show the grade of

Table 14.6 Typical standard costing variance account—materials

Date	Reference	Material	Quantity (units)	Standard unit cost	Actual unit cost	Variance (£)
2/2/95	WO.2479	AB.1234	150	1.20	1.25	+0.05
5/2/95	WO.2532	BK.2468	100	4.25	4.40	+0.15
11/2/95	WO.2578	PR.2622	25	0.55	0.50	−0.05
15/2/95	WO.2604	AB.1234	75	1.20	1.20	0.00
18/2/95	WO.3096	BK.2468	120	4.25	4.35	+0.10
	etc.		etc.		etc.	
Summary						
(Totals)		AB.1234	225	270.00	277.50	+7.50
		BK.2468	220	935.00	962.00	+27.00
		etc.		etc.		etc.
Totals				15 842	16 209	+367.00
Variance analysis (£)						
Material price		+117.25				
Material issued quantity		+148.75				
etc.		etc.	*Total*: £367.00			

labour and/or machine used, in order to identify which cost rate should be applied. This is particularly important if standard costing is being used, so that variances can be correctly identified. Table 14.7 illustrates how such a variance account might be kept.

As can be seen from this example, the objective is to identify the sources of variances, in order to eliminate excess costs and pinpoint areas where standards may be too high.

Actual overhead costs

Works overheads are recovered in product or job costs by adding a percentage to the labour costs for a cost centre on some agreed basis (e.g. area or tonnes produced). It is therefore possible to determine whether the costs of such overheads have been recovered, by comparing the actual costs of such items with the costs recovered.

If standard costing is being used there are three sets of values to be compared:

- the actual invoiced cost
- the actual cost recovered
- the standard cost.

However, only the first two will be applicable if another costing system is being used. As an example, using the values from Table 14.5, these comparisons are given in Table 14.8. These figures show that there is a considerable cost overrun against standard, but that this is to some extent compensated for by extra production hours.

14.9 COST OF SALES AND SELLING PRICES

The term 'cost of sales' is used to define all the costs that can be incurred by a company in its operations, either on a global basis or at lower levels, e.g. for a product group (family) or a

Table 14.7 Typical standard costing variance account—labour

Account for Cost Centre DR.124 Standard hours—setter, 35/week
—operator, 12/week
Standard rates —setter, £15.50/standard hour
—operator, £14.50/standard hour

Date	Reference	Hours booked	Booking code	Grade	Standard rate/hour	Actual rate/hour	Variance (£)
7/3/95	WO.2467	6.20	01	Operator	14.50	15.25	+0.75
7/3/95	WO.2478	2.50	01	Setter	15.50	14.50	−1.00
7/3/95	NP.9462	0.25	27	Operator	0.00	14.50	+14.50
	etc.		etc.		etc.		etc.
Summary							
		37.00		Operator	507.50	584.25	+76.75
		11.30		Setter	186.00	172.40	−13.60
	Totals	48.30			693.50	756.65	+63.15

Variance analysis (£)

Non-standard hours (operator)	+19.45
Non-standard hours (setter)	−3.55
Non-productive time (operator)	+15.25
Non-standard machine	+17.52
etc.	etc.

Total £63.15

Table 14.8 Typical standard costing variance account—overheads (heating cost variance analysis)

Work centre	Standard cost/hour	Standard hours/year	Standard cost/year	Actual hours/year	Recovered actual (£)	Variance
A	1.04	3360	3494	3720	3868.80	374.80
B	0.79	5840	4614	4589	3625.31	988.69–
C	1.64	1850	3034	2907	4767.48	1733.48
D	1.06	7235	7669	7721	8184.26	515.26
E	0.99	5980	5925	5864	5805.36	119.64–
F	1.20	4390	5268	5199	6238.80	970.80
Total	1.05 (average)	28 655	30 004	30 000	32 490.01	+2486.01

Actual invoice value (£) 33 560.50
Recovery (£) 32 490.01

Excess hours (30 000 − 28 655) = 1345 (above standard)

Cost variances (against standard) £3556.50
(against recovered) £1070.49

single product. If the cost of sales is deducted from the sales turnover—at any level—then the difference is the profit margin at that level (company, product group, market area, product, etc.).

Ideally, a company will aim to establish such margins at reasonable levels, across the board, but this is not an ideal world and therefore selling prices tend to be set by market forces, with higher profits in areas where the competition is low.

Thus, in normal circumstances, the profit margin is determined by two factors: *market forces*, which set the selling prices and *cost efficiency*, which sets the cost of sales. From this it would appear that reasonable levels of profit can be achieved only by effective cost control, and this is generally true. However, there are ways of increasing sales of specific products, or in specific markets, by selective price cutting leading to increased volume and thus lower unit costs.

The technique employed to achieve this objective is known as 'marginal costing' and although this is not strictly speaking a costing process, it is vital that those wishing to use the technique have a sound knowledge of costing and accurate information about costs at the appropriate level (e.g. product or market).

14.10 MARGINAL COSTING

The concept

In order to understand how marginal costing works, it is necessary to understand the difference between fixed and variable costs. These can be defined as follows:

- *fixed costs* these are costs that do not vary with production volume, e.g. the costs associated with the factory buildings
- *variable costs* these are costs that vary in direct proportion to production volume, e.g. direct labour and materials
- *semi-variable costs* these are costs that vary to some extent with production volume but are not directly proportional, e.g. indirect labour or the costs of tooling.

In marginal costing, the fixed costs borne by a product are known as its contribution, and the variable costs are known as the marginal costs.

The way in which these cost types operate to determine profit margins against volume is shown in the break-even chart in Fig. 14.9. In this example it has been assumed, for the sake of simplicity, that all costs are either fixed or variable and it can be seen that the break-even point (where there is neither profit nor loss) is reached when sales volume reaches about 46 000 units.

It will be apparent that, under normal circumstances, every product sale makes a contribution towards these costs, both fixed and variable. However, if one product were to be reduced in price by part of its contribution to the fixed costs, and as a result sales were to increase, then the contribution to these fixed costs could remain the same or possibly even increase. This concept is illustrated in Fig. 14.10, which shows how marginal costing can be used to answer a marketing policy question. In this case, if the sales quantity of product B is increased by one-third, the price can be cut by 10 per cent.

The purpose

Although marginal costing methods call for a clear distinction between the treatment of fixed and variable costs, and avoid the problems associated with the allocation of fixed costs to

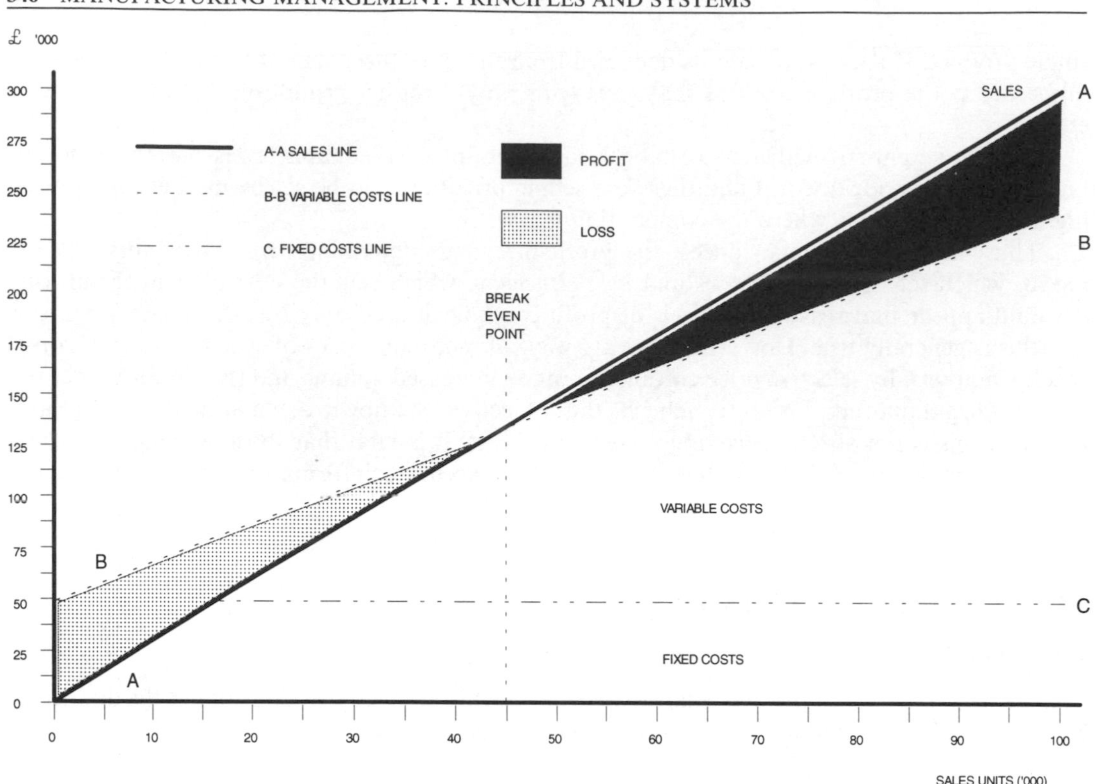

Figure 14.9 Costs and sales break-even chart.

departments or products, this is not their main purpose. Marginal costing is not designed to disclose total costs and is not therefore a costing method in the true sense.

Its real value is as a management tool which can be used to assist in decision-making, such as choosing between alternative courses of action, as the following examples show.

1. In the selection of the most profitable combination of sales prices and volumes: if a company can sell a particular volume of products at a certain price or, by reducing the price, increase this volume, a comparison of the respective total contributions (not the contributions per sales unit) will disclose the more profitable course.
2. If a company is considering an advertising campaign, the benefits to be gained from the increase in the contributions by the products concerned can be anticipated (due to extra sales), and these can be compared with the costs of the campaign.
3. If spare production capacity is temporarily available, the effect of accepting orders at less than normal prices can be determined. Provided that fixed-cost contributions are fully recovered from existing sales levels, and regular business is not adversly affected, it may be advantageous to accept additional business at a price in excess of the marginal (variable) cost but below the total cost, since such a course will increase net profit by the amount of the 'contribution'.
4. If a decision needs to be made between making a product internally or purchasing from outside, a comparison of the purchase price with the marginal cost of manufacture will, all other factors being equal, disclose the more profitable course of action.
5. If a company owns several production units, marginal costing can be used to decide where products should be made on the basis of the lowest marginal costs.

TABLE I - NORMAL CONTRIBUTIONS	PRODUCT A (Qty Sold = 5,000)		PRODUCT B (Qty Sold = 4,000)		PRODUCT C (Qty Sold = 16,000)	
	TOTAL	UNIT	TOTAL	UNIT	TOTAL	UNIT
TOTAL SALES VALUE BY PRODUCT	150,000	30.00	80,000	20.00	240,000	15.00
MARGINAL OR VARIABLE COST OF PRODUCT SALES	90,000	18.00	48,000	12.00	144,000	9.00
FIXED COST CONTRIBUTION FROM PRODUCT SALES	37,500	7.50	20,000	5.00	60,000	3.75
PROFIT BY PRODUCT	22,500	4.50	12,000	3.00	36,000	2.25
TOTAL SALES VALUE - ALL PRODUCTS			470,000			
MARGINAL OR VARIABLE COST OF ALL PRODUCT SALES			282,000			
FIXED COST CONTRIBUTION FROM ALL PRODUCT SALES			117,500			
TOTAL PROFIT			70,500			

QUESTION: MANAGEMENT BELIEVES THAT A PRICE REDUCTION FOR PRODUCT 'B' WOULD LEAD TO AN INCREASE IN SALES BY ATTRACTING NEW CUSTOMERS.

IF THE PRICE IS CUT BY 10% WHAT INCREASE IN SALES VOLUME OF 'B' WOULD BE REQUIRED TO MAINTAIN OVERALL PROFITS AT THE CURRENT LEVEL?

TABLE II - MARGINAL COSTING	PRODUCT A (Cty Sold = 5,000)		PRODUCT B (Qty Sold = 5,333)		PRODUCT C (Qty Sold = 16,000)	
	TOTAL	UNIT	TOTAL	UNIT	TOTAL	UNIT
TOTAL SALES VALUE BY PRODUCT	150,000	30.00	95,994	18.00	240,000	15.00
MARGINAL OR VARIABLE COST OF PRODUCT SALES	90,000	13.00	63,996	12.00	144,000	9.00
FIXED COST CONTRIBUTION FROM PRODUCT SALES	37,500	7.50	20,000	3.75	60,000	3.75
PROFIT BY PRODUCT	22,500	4.50	12,000	2.25	36,000	2.25
TOTAL SALES VALUE - ALL PRODUCTS			485,994			
MARGINAL OR VARIABLE COST OF ALL PRODUCT SALES			297.996			
FIXED COST CONTRIBUTION FROM ALL PRODUCT SALES			117,500			
TOTAL PROFIT			70,498			

ANSWER: SALES OF PRODUCT 'B' MUST INCREASE FROM 4000 TO 5333 (33.33%)

Figure 14.10 Marginal costing application.

6. If a factor of production is scarce or limited, it will be found that it will be more profitable to use that factor on the product lines that show the greatest contribution.

14.11 BUDGETARY CONTROL

In some companies costing is used to operate a system of budgetary control in order to reduce costs and increase efficiency, and if standard costing is being used this can form a suitable basis for budget setting. Thus, a budget can be set up annually for each department, or cost centre, showing the items and levels of expenditure that should be achieved (or even improved on), and these budgets should then be agreed by the managers concerned. Then, on a monthly basis, a budget statement can be issued to each budget manager, showing the budgeted and actual expenditure against each item, with variances. Managers may then be required to explain any excessive values.

However, this system does not operate very effectively if the factors that 'drive' expenditure vary to any great extent. For example, the value of materials used will be proportional to the output level, and if this level goes up then the budget for this item will be exceeded. It is therefore more realistic to operate a system of flexible budgets which relate the items that are variable or semi-variable to a factor by which the budget can be adjusted if circumstances change.

Budget statements are normally 'rolled up' from the lowest level to the highest. Thus, statements would be produced for all cost centres and these would then be grouped—with additional items—at the departmental level. Finally, a company statement would be produced covering all departmental figures, plus items applicable at company level only.

A typical departmental budget statement is shown in Fig. 14.11.

14.12 SHOP-FLOOR PERFORMANCE MEASUREMENT

The management accounting function is not merely one of recording and analysing cost data, since there is also a duty to identify areas of weakness and to recommend actions to overcome such problems. The function should therefore measure production performance in terms of a set of ratios, which can then be used to implement a series of continuous improvement policies. The basis for the selection of the most suitable data for use in this exercise will depend on the company type, but an example of a typical performance report is shown in Fig. 14.12. This type of report can easily be prepared on a spreadsheet.

14.13 EOQ—COSTS OF ORDERS AND STOCK-HOLDING

In chapter 9 (section 9.9), the difficulties involved in calculating economic order quantities were considered, and involved the methods that should be used to determine the values for the 'cost of placing an order' and the 'costs of holding stock'.

Placing an order

The most obvious way of calculating this cost would be to calculate the total annual costs of the department concerned (purchasing or works orders) and then to divide this by the number of orders placed in a year. However, this may not be accurate, since some of these costs may be unrelated (or only partially related) to the order-placing process. It is therefore better to look at

BUDGET STATEMENT

DEPARTMENT: Sales | BUDGET PERIOD: May 1994 (Period 2)

MANAGER: C Peters

ITEM	QTY.	BUDGET £	ACTUAL £	VARIANCE £	COMMENTS
SALARIES:					
Manager	1	2,500	2,500	NIL	
Grade A Staff	2	2,000	2,000	NIL	
Grade B Staff	5	3,750	3,000	(750)	1 LEAVER NOT YET REPLACED
STAFF OVERHEADS:					
Bonus Scheme		825	650	(175)	1 LEAVER GIVING EXTRA BONUS
DHS.		165	140	(25)	1 LEAVER NOT YET REPLACED
Pension		480	375	(105)	
SALES PROMOTION:					
Literature		200	250	50	EXCESS PRINT COST
Advertising		450	100	(350)	ADVERT CAMPAIGN DELAYED FOR EXTRA RESEARCH
Market Research		NIL	250	250	
TRAVELLING:					
Reps. Commission	3	3,200	3,800	600	HIGH VALUE OF SALES ITEMS
Car Mileage	21k	1,500	1,600	100	
Hotels		420	400	(20)	
SUMMARY:					
SALARIES:		8,250	7,500	(750)	
STAFF OVERHEADS:		1,470	1,165	(305)	
SALES PROMOTION:		650	600	(50)	
TRAVELLING:		5,120	5,800	680	
DEPT.TOTALS		15,490	15,065	(425)	2.7% FAVOURABLE

Figure 14.11 Typical budget statement.

each individual element of cost and to determine what effect it has (in percentage terms) on the cost of ordering, as shown in Table 14.9.

Holding stock

A similar technique should be used to determine the total costs of stock-holding, which will include the costs of operating the stores (labour, equipment, space, stationery, heating, stock-taking, etc.), the costs of wastage (obsolescence, shelf-life, etc.) and the costs of stock investment (the loss of interest that could otherwise be earned, or the cost of borrowing).

MONTHLY PERFORMANCE REPORT - JULY 1994							
FACTOR A		FACTOR B		PERFORMANCE			
					Actual	Target	%
Monthly Sales (£)	295,000	Total Employees	450	Monthly Sales/Employee (£)	656	750	87.5
Monthly Sales (£)	295,000	Direct Employees	250	Monthly Sales/Employee (£)	1,180	1,300	90.8
Monthly Sales (£)	295,000	Indirect Employees	150	Monthly Sales/Employee (£)	1,967	2,300	86.4
Monthly Sales (£)	295,000	Admin.Staff	50	Monthly Sales/Employee (£)	5,900	6,600	89.4
Monthly Wages (£)	117,750	Total Employees	450	Monthly Wages/Employee (£)	262	264	99.2
Monthly Wages (£)	71,750	Direct Employees	250	Monthly Wages/Employee (£)	267	255	104.7
Monthly Wages (£)	36,750	Indirect Employees	150	Monthly Wages/Employee (£)	243	250	97.2
Monthly Wages (£)	9,250	Admin.Staff	50	Monthly Wages/Employee (£)	185	180	102.8
Direct Mat'ls (£)	73,500	Monthly Sales (£)	295,000	Direct Materials/Sales	0.249	0.240	103.7
Indirect Mat'ls (£)	15,000	Monthly Sales (£)	295,000	Indirect Materials/Sales	0.051	0.045	113.3
Total Mat'ls (£)	88,500	Monthly Sales (£)	295,000	Materials/Sales	0.30		
Total Wages (£)	117,750	Monthly Sales (£)	295,000	Wages/Sales	0.40		
etc.		etc.	etc.	etc.	etc.		

Figure 14.12 Typical monthly performance report.

Table 14.9 The cost of placing an order

Cost item	Cost (£/year)	Efect ratio (%)	Effective cost (£/year)
Manager salary	25 000	50	12 500
Clerical salaries	60 000	80	48 000
Office space	12 000	75	9000
Heating	2000	75	1500
Stationery	4000	80	3200
Telephone, etc.	6000	70	4200
etc.	etc.	etc.	etc.
		Total	87 500

Number of orders/year = 3500, thus cost/order = £87 500/350 = £25

These total costs must then be set against the average value of the stocks held, to determine the percentage stock holding cost. For example, if the total costs (corrected by the effect ratios) were found to be £120 000 and the average stock level was at a value of £600 000, then the stock-holding cost would be (120 000 × 100)/600 000 = 20 per cent.

The determination of the average stock-holding figure can present problems if there is no computer system. However, if stocks are computerized it is fairly easy to value stocks at regular intervals (say weekly) and then to average these values over a longer period (say a year). This can also be done on a manual system, but it is a laborious process.

PROBLEMS

1. Prepare a draft advertisement for the recruitment of a management accountant for your company and a series of questions that could be used at an interview. Your company is involved in the manufacture and sale of a range of paints and wallpapers which are sold from stock. It uses a computer system which includes a standard costing module, but this has not yet been implemented.
2. You have been asked by the Board to examine the possibility of using activity-based costing as an alternative to the current absorption costing system. Prepare a brief report explaining how the ABC system might operate in your company, which manufactures a number of pre-packaged dairy food products such as cheeses, butter and yoghurt. Your report should explain the current method of overhead allocation and suggest how ABC might improve on this.
3. Prepare a break-even chart for a company on the basis of the information given below (see appendix A for solution).
   Annual sales: £300 000 at an output level of 100 000 standard hours.
   Fixed costs: £50 000 per annum.
   Variable costs: £1.50 per standard hour.
   Semi-variable costs: £12 500 fixed up to 50 000 standard hours, rising to £25 000 fixed for all standard hours above 50 000.
   (a) What is the break-even point in terms of sales value (£) and volume (standard hours)?
   (b) What would be the profit at 80 per cent of sales volume?
   (c) What would be the break-even values (value and volume) if fixed costs were cut by £5000?
4. You are a painter and decorator, working with a partner who is paid 30 per cent of all takings. You retain the other 70 per cent but from this must pay all expenses (travelling by van, consumables, paint/paper/paste, tools, etc.). Suggest a basis on which you could prepare estimates for new work.
5. Using the figures given in Fig. 14.10 (table 1) showing a marginal costing application, suggest answers to the following questions (see appendix A for solution).
   (a) If product B could be purchased from an outside supplier for £14, would this be more profitable than manufacturing it 'in house'?
   (b) If a TV advertising campaign costing £10 000 could increase the sales of product A by 25 per cent would this expenditure be justified?

FURTHER READING

Kaplan, R. S., *Advanced Management Accounting*, Prentice-Hall, London, 1982.
McNair, C. J., *Meeting the Technology Challenge: Cost Accounting in a JIT Environment*, USNAA, Montvale, NJ, 1988.
Mair, G., *Mastering Manufacturing*, Macmillan, London, 1993.
Sizer, J., *An Insight into Management Accounting*, Pelican, London, 1979.
Sneyd, P., *Principles of Accounting and Finance*, Routledge, London, 1994.

CHAPTER

FIFTEEN

THE MANAGEMENT OF CUSTOMER SERVICES

THE LINK

Without its customers a manufacturing company cannot exist. Unless it is in a monopoly situation it must try to ensure that it not only retains its existing customers, but also attracts customers from its competititors. This cannot be achieved merely by making a good product at a good price, although this is certainly important; it is also necessary for the manufacturer to build a reputation as a caring supplier that provides a good service, not only up to the point of delivery but beyond. In fact, there is a strong correlation between the level of service to the customer and the level of repeat orders.

Customers need to feel that they count — that they are not just numbers on a computer — and this can be achieved only by a manufacturer that genuinely recognizes this fact, and acts accordingly. It is certainly not good enough to rely on slogans and catchphrases such as 'in our business the customer is king' or 'your wish is our command'.

Thus, the caring manufacturer will encourage visits from its customers, will consult and involve them in new product development, and will deal with their complaints and suggestions quickly and efficiently. In other words, it will not only care, but be seen to care.

15.1 INTRODUCTION

In order to ensure that a business is successful, customer loyalty must be retained so that repeat orders do not pass to a competitor. To do this, it is necessary to provide a good product at the right price and to deliver on time. But this is probably not enough if a company wishes to attain world-class standards, since it is also necessary to service the customer after the goods have been delivered.

The need to provide this 'post-delivery' service depends on the type of business, therefore this chapter starts by examining the sort of services that should be provided by a range of different business types, with examples of each type. It continues by considering, in detail, the various means by which such services should be provided for the customer, including service agreements, warranty arrangements, customer complaints and customer information. It concludes by suggesting how customer loyalty can be measured and how a customer services department should be organized.

15.2 MATCHING CUSTOMER SERVICES TO THE BUSINESS TYPE

In order to make sure that the customer receives the required services, it is necessary for a supplier to examine some form of matrix showing types of service and business types. These can be defined as follows:

Customer services

- *Service agreement* An agreement to maintain an item of equipment, or a machine, for an agreed period at an agreed charge. The terms may vary considerably, for example the charge can cover labour only or labour and materials, and there may be extra charges for cover at weekends and over holiday periods.
- *Warranty arrangements* A warranty agreement will cover a fixed period and cover the costs of replacement parts and labour. However, it is unlikely that the costs of consequential losses (e.g. lost production) will be covered.
- *Commissioning* Where a complex plant item has been supplied, it is normal for the supplier to include commissioning, to ensure that it is operating to specification before hand-over to the customer.
- *Customer complaints* In some companies a customer complaints department deals with all problems notified by the customer, in order to minimize delays in response (which cause further customer dissatisfaction) and to try to apply an even-handed approach.
- *Customer information* Customer information can take a number of forms, e.g. the supply of operating manuals, spares lists, and training for users when the product is delivered. Information about new products, upgrades, price lists and special offers is produced subsequently.
- *Advertising* Where the end customers are difficult or impossible to contact it will be necessary to advertise new products and services, either by direct means (buying advertising space) or by means of an agency, which will issue press releases that may be used in the technical or general press.
- *Image creation* In some companies an image is created or a name established by means of sponsorship of events or by hospitality at sporting or other functions.
- *Exhibitions* Displays at exhibitions can be used to maintain contacts with existing customers and to attract new ones, but the main objective should be to show off new products or product developments.

Business types

In the context of customer services, business types can be classified into four distinct groups as shown, with examples, in Fig. 15.1. This method of classification is based on products and customers, and is therefore different from the methods considered earlier, which were based on sales types (e.g. make-to-order or make-to-stock).

Matrix of services v. business types

This matrix is shown in Fig. 15.2 and illustrates the relative importance of the various customer services to the various business types. From this, the importance of dealing with customer complaints can be seen, although it would be more positive to include in this

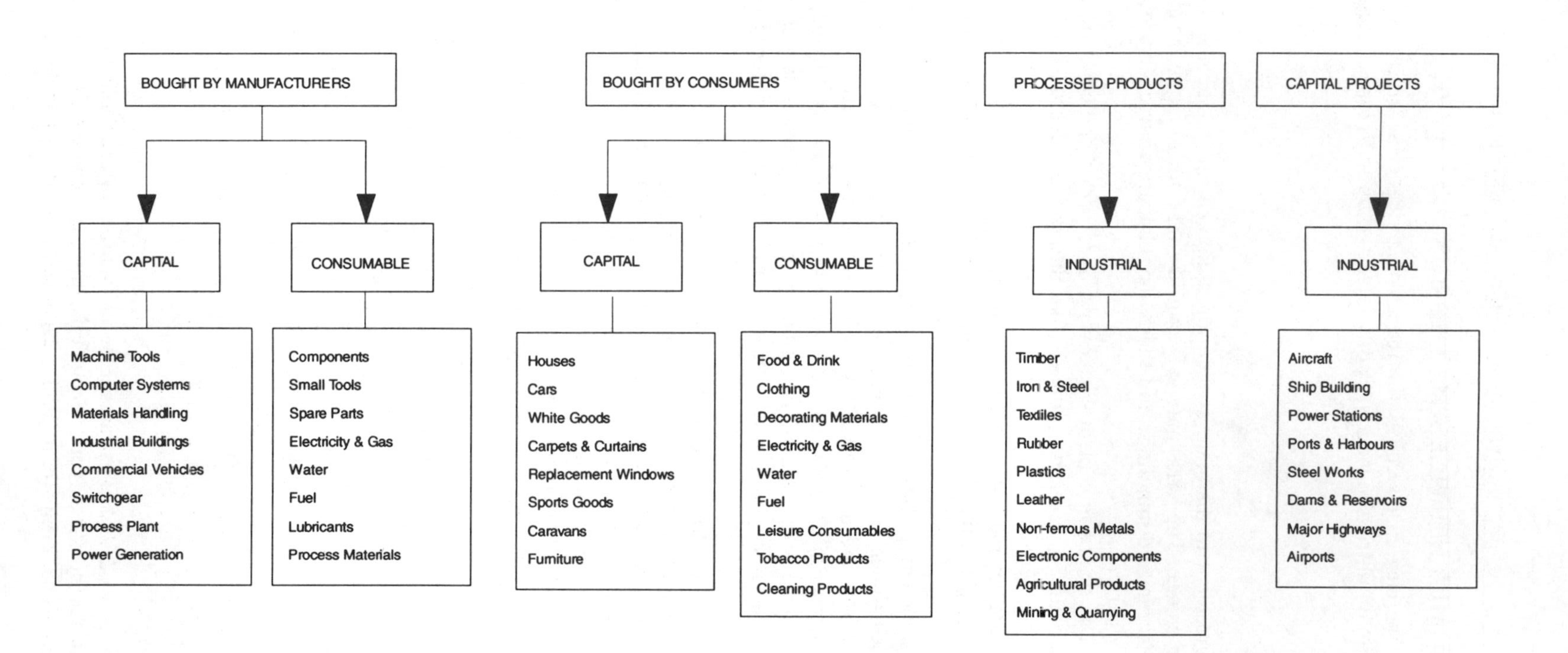

Figure 15.1 Business types for customer service matrix.

Customer Services	Business Types						
	Bought By M'facturers		Bought By Customers		Processed Products	Capital Projects	TOTALS
	Capital	Consumable	Capital	Consumable			
Service Agreements	2	0	1	0	0	2	5
Warranty Arrangements	2	0	2	1	1	2	8
Commissioning	1	0	0	0	0	2	3
Customer Complaints	2	2	2	2	2	2	12
Customer Information	2	2	2	0	1	1	8
Advertising	2	2	2	2	1	1	10
Image Creation	1	0	1	2	0	0	4
Exhibitions	2	0	2	2	1	1	8
TOTALS	14	6	12	9	6	11	---

Points System: No requirement = 0
Some Requirement = 1
Definite Requirement = 2

Figure 15.2 Matrix of customer services v. business types.

category customer comments and queries. This is not an area to be 'avoided' and many successful companies tend to solicit such comments by means of market surveys.

15.3 SERVICE AGREEMENTS

A service agreement is a contract drawn up between a supplier and a customer stating the service that will be provided, the limitations, and the charges for the service. Such a service can be on a regular basis (a form of planned maintenance) or 'as required' (a call-out system). In either case it is vital that the supplier maintains accurate records (preferably on computer), so that:

- agreements are renewed before they lapse
- service engineers are given full details of the item(s) covered
- scheduled calls are not missed
- customers are correctly invoiced
- history of work done is used for analysis purposes
- special factors are known (availability, safety, security, etc.)
- warranty claims are dealt with promptly.

Computer systems are available that have been specifically written for this purpose, but it is equally possible to operate on a standard database by setting up a series of files and using the built-in facilities to make enquiries and print work orders. In fact, there are strong similarities between a service agreement system and a planned maintenance system, and it is possible to use the same software for both functions (see chapter 7, section 7.6).

Figure 15.3 shows how such a system could operate, whereby the user (probably the service department) can create work orders either on a planned basis against contract schedules or on an unplanned 'call-out' basis. These orders can then be allocated to specific engineers, on the basis of availability, special skills or repeatability. This last is an important factor: customers

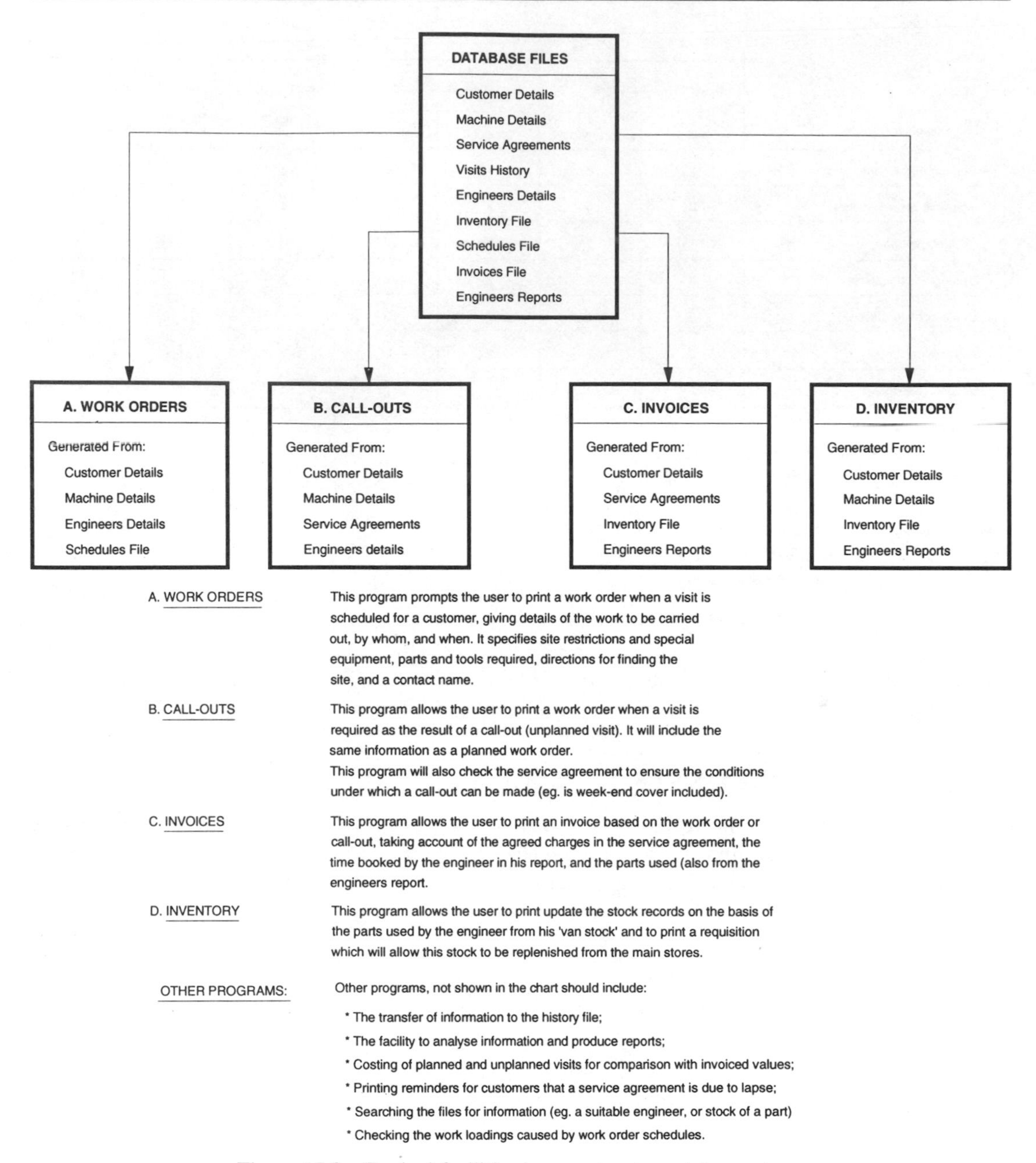

Figure 15.3 Typical facilities in a contract servicing system.

prefer to have a service engineer whom they know, and who knows them. This is a good example of the TQM (total quality management) approach, where the engineer builds up a good relationship with 'his' or 'her' customers.

Each engineer should carry a stock of standard spare parts and tools (van stock), which may be augmented for special jobs where necessary. The database files should contain all the information necessary to operate the system, and Table 15.1 gives some typical examples of the data required.

Table 15.1 Database files for customer service administration

File type	Data fields
Customer details	**Account code**, customer name, address, phone and fax numbers, contact name and position, access, restrictions, security, safety, travel directions
Machine details	**Account code**, machines installed details (model, when purchased, location, serial number, special features), service agreement number
Service agreement	**Account code**, service agreement number, conditions codes (i.e. conditions applicable), machine serial number, visits schedule (e.g. 12-weekly starting week 8), contract expiry date
Visits history	**Account code**, visits dates, work order number, engineer number, machine serial number, work-done codes, invoice number, engineer's comments
Engineers' details	**Engineer number**, name, specializations, pay rate code, vehicle number, home phone, holiday flag
Inventory file	**Part number**, description, stock levels, orders, locations, policies
Schedules file	**Account number**, service agreement number, call frequency, start date, work required, date of last call, standard time
Invoices file	**Account number**, invoice numbers, invoice values, payment flag
Engineers' reports	**Engineer number**, work order number, acccount number, machine serial number, parts used, hours booked, mileage, comments.

The advantages of such a system will be apparent to any company that needs to provide its customers with an after-sales service, based on service agreements, and if correctly installed and maintained will allow the service manager to obtain the answers to various questions instantly, such as the following.

- How many service engineers should be employed?
- Which is the best engineer to send on this job?
- When will a suitable engineer be available?
- Where is that engineer now? How can I make contact?
- What replacement parts are likely to be needed in the next period?
- What should be included in the annual budget?
- Which machines are causing problems, and why?
- Which customers should be contacted if a machine is modified or superseded?
- What is the cost of the service? Is it economic?

An efficient service agreement system can greatly improve customer relations and build loyalty. In fact, a team of well-trained and motivated service engineers, providing a first-class service, may often prove to be the best sales team a company can employ.

15.4 WARRANTY ARRANGEMENTS

Many customers expect some form of warranty when purchasing capital goods, and it is essential that the supplier provides a suitably drafted warranty document with each sale, stating clearly the terms under which the warranty applies and the conditions that could nullify the arrangement. An example of a typical warranty document is given in Table 15.2.

There may be some confusion between the terms 'warranty' and 'guarantee', but in fact the two are virtually interchangeable. They can be defined as 'an undertaking by a vendor that the item sold is the property of the vendor, is fit for use, and will be replaced or repaired if it fails to perform to an agreed standard, over a specified period, subject to specific conditions'.

The following points should be noted from the example in Table 15.2.

- The warranty is nullified if the purchaser cannot provide proof of purchase and prove that the warranty claim is within the specified period.
- It is nullified if specified conditions are not met, for example if the item has not been installed correctly.
- The purchaser has statutory rights under the consumer protection legislation, and these cannot be taken away by the terms of the warranty.
- Many suppliers now offer extended warranties at a specified cost. These are in fact a form of service agreement, whereby a customer can contract with the supplier for repair or replacement (subject to certain conditions) after the normal warranty has expired. Such a contract can be based on scheduled services or on an 'as-required' basis.

In order to administer a warranty system, the supplier should set up a recording system, preferably on computer, which can be similar to the system described in section 15.3 of this chapter for service agreements. These records should include details of:

- the goods originally sold, and to whom, including cross-references to any pre-delivery correspondence, quotations, orders, deliveries, etc.
- any service agreements made, with cross-references to records or reports of service visits.
- where relevant, the records of the manufacture of the product in terms of lot traceability and quality checks—these records are often linked to the serial number of the product
- any correspondence (verbal or written) between the customer and supplier after the goods have been supplied.

Table 15.2 Typical warranty document

Your guarantee
Your new appliance, as described on your receipt, is guaranteed against defect for 12 months following the date of purchase. During the guarantee period, repairs will be carried out FREE OF CHARGE provided that:

1. the appliance has not been subject to misuse or been wilfully or accidentally damaged by the purchaser
2. the appliance has not been tampered with or repaired by anyone other than [selling agent's name] staff, the manufacturer of the appliance, or its authorized repair agents.
3. this guarantee is presented, with proof and date of purchase, when service is required.

It should be noted that the terms of this guarantee in no way limit your legal and statutory rights as a consumer.

Extended warranty
In addition to this guarantee, you have the option of an insurance-based warranty which gives you an extra two or four years' protection in addition to your normal guarantee. If you wish to take up this option please contact [selling agent] for more details.

This information will be vital if a legal case develops over a warranty claim, or if claims are made after the warranty period has expired.

15.5 COMMISSIONING

If a supplier is required to commission a machine or a capital project, this should always be stipulated in the purchase order, and the details must be specified.

Thus, the order documents should define the performance standards to be achieved and, if necessary, how these are to be measured; by accepting the order, the supplier is legally committed (by contract) to meet these standards. For example, a power station should be capable, when commissioned, of producing an agreed output, sustainable for a specified period, at a stated efficiency level. Other factors may also be included, such as the ability to run at given manning levels or to operate within statutory levels of emissions.

The supplier should always ensure that the project plan allows time for such commissioning to take place, and should include the costs in the original quotation. This facet of customer service, although not relevant in many cases, is essential if a reputation as a reliable supplier is to be maintained.

15.6 CUSTOMER QUERIES AND COMPLAINTS

It is generally accepted that there will always be some customer complaints, although every effort should be made to reduce these to a minimum, by providing a quality product. It is equally accepted that customer satisfaction with a product will be enhanced if complaints are dealt with quickly, fairly and efficiently. The provision of an effective complaints procedure can positively increase customer satisfaction, and thus loyalty. Figure 15.4 illustrates the importance attached to customer satisfaction by a sample of chief executives.

No company should be satisfied with the level of customer complaints, however low, since this may conceal a much higher level. This is due to the now well-established fact that the majority of customers do not complain even if they have a problem, which may be partly due to the ineffectiveness of the company's complaints procedures.

Customer attitude surveys have shown that:

- *in the manufacturing sector* out of every 100 customers up to 70 have a complaint or problem, but of these between 50 and 60 (about 80 per cent) remain silent.
- *in the consumer sector* about 70 per cent of customers with a problem do not complain.

However, another survey showed a different situation, by suggesting that only one customer in 26 makes a complaint (4 per cent). Does this show that most customers are satisfied, or that they feel that their complaints are pointless?

If a supplier assumes that silence means satisfaction there is a serious risk that the business will fail, since an unhappy 'silent majority' will vote with their feet, and change to a competitor's product for their next purchase. No well-run business can accept such a risk, and it should therefore try to assess the true position by means of market or customer surveys. These should not only identify the real level of customer satisfaction, but also identify the 'hidden' problems, and their frequency. Figure 15.5 shows the results of such a survey carried out among passengers on flights in the USA. This illustrates how data from such a survey can be presented to highlight the problem areas.

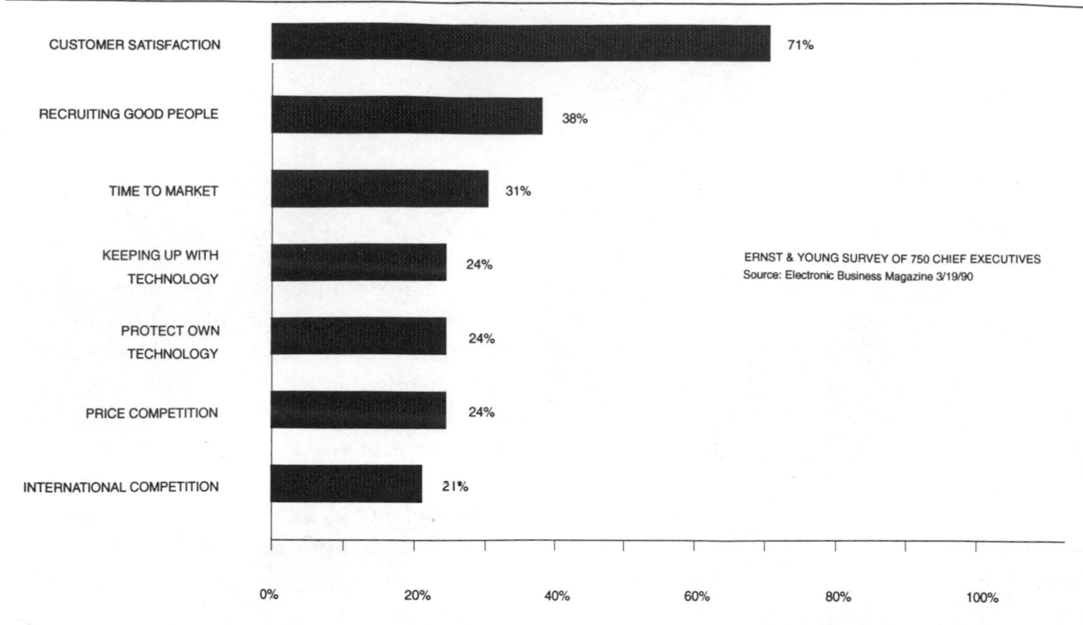

Figure 15.4 Customer satisfaction: the biggest challenge.

The importance of identifying customer complaints, and dealing with them quickly, cannot be overemphasized and this is reinforced by the informaton shown in Fig. 15.6, obtained in a survey of US consumers. From this information it can be seen that if complaints are resolved about 60 per cent of customers will buy again, and if these complaints are dealt with quickly the figure rises to over 80 per cent. However, if complaints are not resolved, or not identified, the 'buy again' figure falls to an average of about 30 per cent or less!

It might be thought that complaints can be eliminated by making products 'fault-free', at a competitive price, and by ensuring that the customer receives them when required, but this is not the case. Market surveys and analysis have shown that in some cases the complaint is not the fault of the manufacturer, but is due to some other factor, for example:

- the customer has bought a top quality-product, but it is not what was needed
- the customer has not assembled or installed the product correctly, or is using it incorrectly — this may be due to the customer, but could be because the written instructions from the supplier are inadequate
- the product has been damaged in transit, or in the distributor's warehouse; this may be due to bad packaging design or poor handling
- the customer has had a change of mind or of circumstances.

15.7 COMPLAINT ADMINISTRATION

Since it is generally accepted that complaints and problems cannot be totally eliminated, it is essential that they be settled quickly, by the provision of an effective series of administrative systems and procedures and — as in other areas of customer service — this can best be accomplished by means of a computer system. There are a number of such systems which can be run on a personal computer or a PC network. An example (Wang's CRIS) is shown in Fig. 15.7.

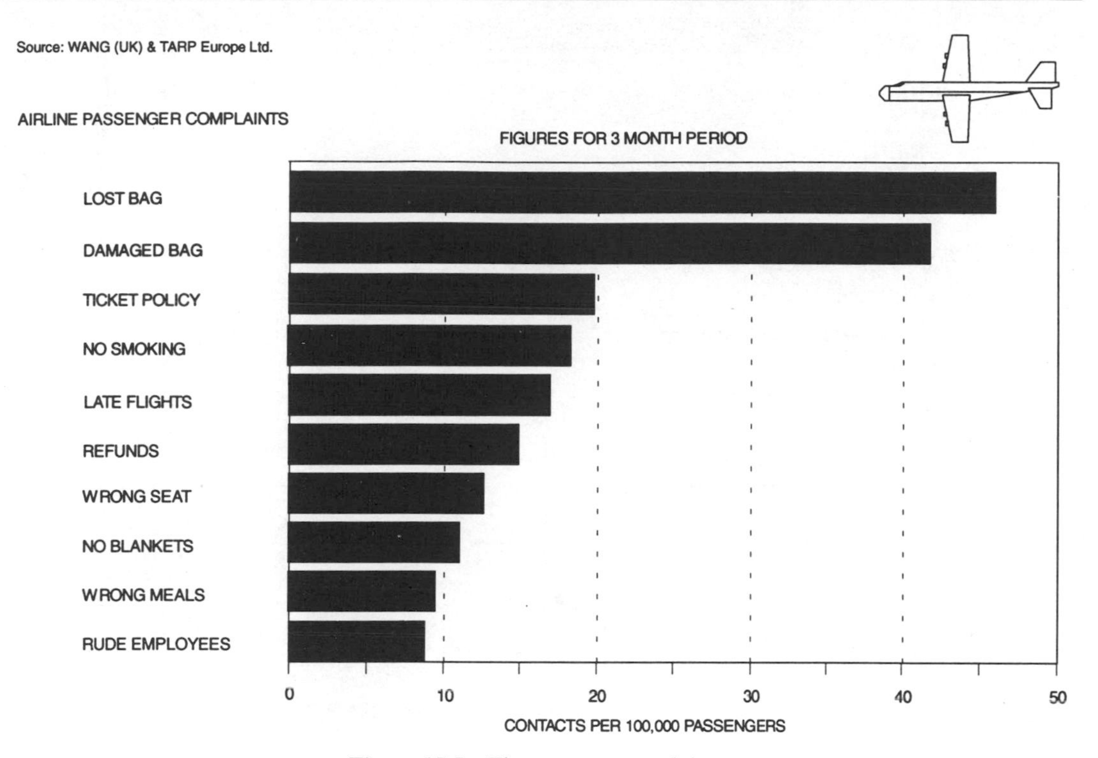

Figure 15.5 The top ten complaints.

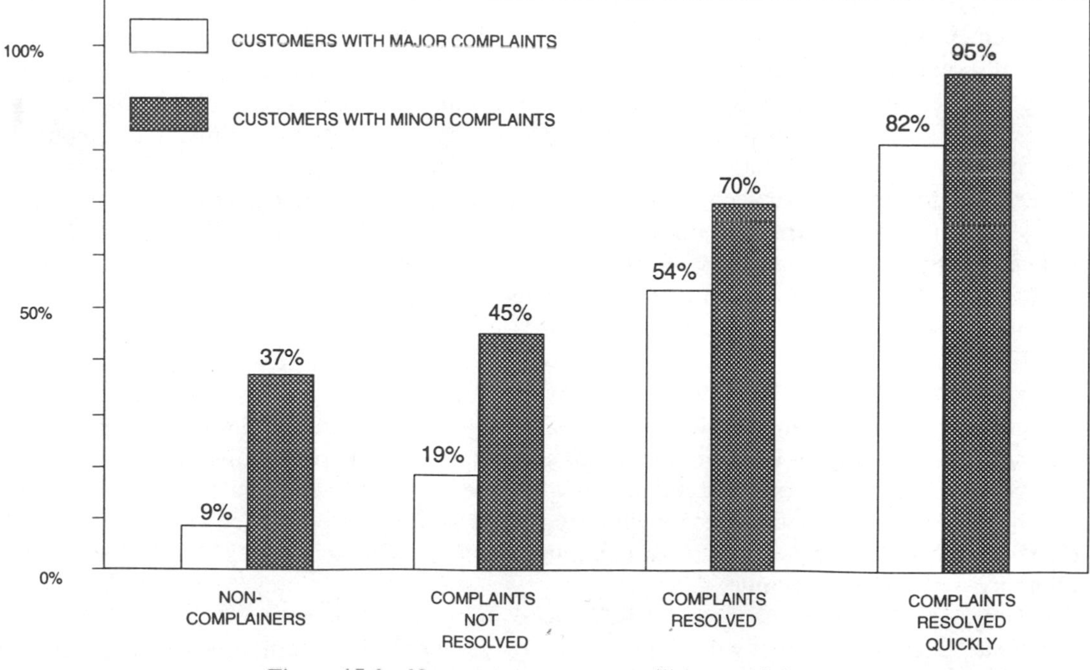

Figure 15.6 How many customers will buy again?

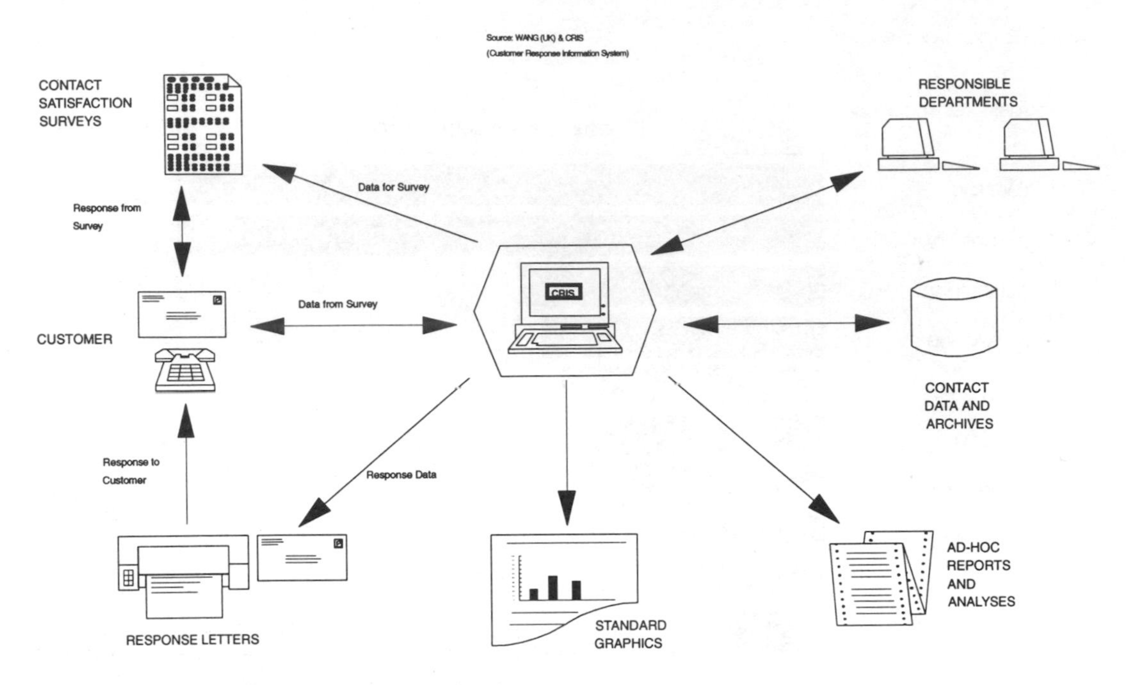

Figure 15.7 Example of customer complaint response system.

This shows a typical closed-loop system whereby data from customer complaints and queries (direct or via a survey) are recorded and analysed.

The information thus obtained is passed to the responsible department(s) (e.g. Design, Quality, Production, Legal) for action, and as a result the customers can be given suitable responses; further surveys can then be carried out to verify that the modifications and responses have achieved the required results. But such a system, which in a small company could be run manually, may not be adequate for an organization with a complex range of products and a large number of customers (e.g. an automotive company). Here, a complaint may involve numerous documents, and perhaps an extended period of negotiation; thus, a conventional system of correspondence files and reference manuals may become unmanageable, or at least difficult to control, due to such problems as missing files or lost documents. This can cause delays and frustration, which do not help when a customer (possibly not in the best of tempers) is kept waiting on the end of a telephone (probably at considerable expense) while a staff member searches for the appropriate files and reference manuals, and checks out the current position.

In order to deal with situations of this sort, it is necessary to provide a more advanced, computer-based, system and to develop new working practices. One such system developed by computer specialists Wang is illustrated in Fig. 15.8, which shows how all documents are scanned electronically and held in a series of computer files, one per complainant.

When a complaint is received initially, it is assigned to a member of staff (the 'owner'), who is supported by other staff members with specialized knowledge of particular problems. Each owner and work group member has a computer terminal, with access to a shared database of reference manuals and facilities.

Thus, when a complainant makes the initial contact by telephone or letter, a reference code is assigned such that when any subsequent contact is made the complainant is put in touch with

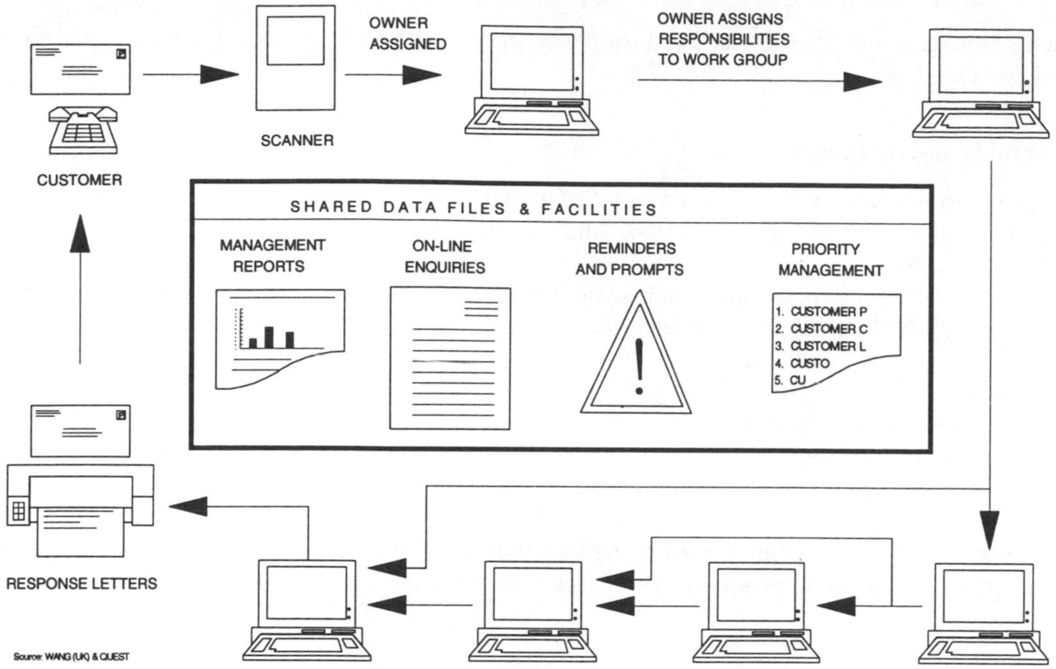

CUSTOMER
SCANNER
OWNER ASSIGNED
OWNER ASSIGNS RESPONSIBILITIES TO WORK GROUP
SHARED DATA FILES & FACILITIES
MANAGEMENT REPORTS
ON-LINE ENQUIRIES
REMINDERS AND PROMPTS
PRIORITY MANAGEMENT
1. CUSTOMER P
2. CUSTOMER C
3. CUSTOMER L
4. CUSTO
5. CU
RESPONSE LETTERS
Source: WANG (UK) & QUEST

Figure 15.8 Correspondence tracking (QUEST).

the owner, who can then progress the complaint with all the relevant data in view on his or her terminal. The owner can then update the file, and prepare a response letter to the customer, using the word processing facilities included in the system.

In this way the complainant is not kept waiting, and complaints can generally be resolved quickly and satisfactorily: where such a system has been installed, it has been found that the average time for settling a complaint can be reduced to less than three days. This is facilitated by the elements of the system that prompt an owner if no action has been taken within a pre-set period, and allocate priorities to each complaint.

By this means, a complainant's loyalty to the company is likely to be enhanced because problems are dealt with efficiently, on a personal basis, by an owner who is familiar with the customer's case. This overcomes one of the major difficulties that customers have when dealing with a large organization, by removing the feeling that they are just an impersonal record on a computer.

15.8 CUSTOMER INFORMATION

It is important to recognize that when customers receive a new 'capital-type' product they have high expectations, and provided that it has been delivered on time, at the right price and in apparently good condition, they are well disposed towards the supplier. However, this goodwill can soon be lost and frustration built up if the information provided with the product is inadequate.

Typical instances of this inadequacy can be found where installation or assembly instructions are badly written, or where user manuals are hard to understand. Most of us have personal experience of this (the author once bought a grass strimmer and found that the manual was in four languages — Danish, Swedish, Norwegian and Finnish!). It is therefore vital to ensure that such manuals are written in plain and simple language, without jargon, and are well illustrated and presented, as follows.

Assembly instructions

- All components should be clearly marked or labelled with a code and well drawn, three-dimensional sketches should include these codes and show the assembly methods and sequences.
- All small components should be placed in plastic bags and each bag labelled with its contents.
- Where low-cost items such as screws are provided (and they should be!) it is always better to have one too many than to be one short.
- A contents check-list should always be provided.
- Instructions should be broken down into a series of logical stages.
- It should not be assumed that the customer (if of the consumer type) has a well-fitted workshop with a full range of hand tools.

Author's note: It is sometimes hard to believe that such instructions have been tested by the supplier on a typical cross-section of the expected purchasers, but this should be standard practice.

User manuals

The task of preparing a user manual for a machine or a similar item should not be underrated. Far too many such manuals are left to gather dust on a remote shelf because they are not 'user-friendly', and many users prefer to learn by experimentation — sometimes with unfortunate results!

A user manual should be readable, and to achieve this objective, should be edited by a professional who can eliminate bad grammar and unfamiliar technical terms. Such a professional can also ensure that all the rules of good writing are observed, for example short sentences, good punctuation and a logical flow. The manual should be well illustrated and divided into sections, using a clear and sensible coding system, with an index that enables a user to find a reference without difficulty.

Other customer information

In addition to the information that should be provided with the product, it may be advisable to provide some other forms of after-sales service to ensure that the customer is satisfied with the product and is likely to remain loyal, such as the following.

- A 'help-desk' facility, whereby a customer with an immediate problem can discuss it on the telephone with the supplier. This facility is particularly relevant if the product is computer-based.
- An after-sales survey that allows the customer to comment on the product and suggest how improvements might be made. Such a service is also useful to the supplier, since it identifies new and continuing customers, highlights possible new development areas and identifies new applications.
- A regular newsletter which can be used to inform existing and potential customers of new product developments, new applications and success stories (preferably of human interest).

- The formation of user-groups, which can meet on a regular basis to discuss mutual problems and suggest new developments. Such group meetings also provide an excelllent opportunity for the supplier to announce new products, or product changes.

15.9 ADVERTISING AND EXHIBITIONS

Advertising

Advertising is usually costly, and unless it is properly targeted and professionally produced may be hard to justify. However, it is sometimes possible to obtain press coverage at low cost by means of press releases, provided that they are based on an interesting 'story'. Press releases that are a just a form of advertisement are unlikely to receive much coverage.

There are two types of advertisement— those designed to inform and those designed to create an image—and each has a role depending on the type of product being sold.

- *Information advertising* Such advertisements are generally for capital goods, either industrial or consumer, using the technical press for the former and newspapers or television for the latter.
- *Image creation* Such advertisements are generally for consumer consumables, where the objective is to establish a brand or supplier's name in the mind of the consumer.

The objective of information advertising is obviously to let potential customers know what is available, and in many cases to solicit enquiries for further information (for example by coupon return). Thus, such advertising can be effective for both consumer and capital items. The objective of image creation is to persuade a purchaser, in a retail outlet, to buy a particular product by supplier or brand name association. It is therefore more suitable for consumer goods.

Exhibitions

Exhibitions are also costly, and these costs cannot really be justified unless the return on the investment is reasonable. There are therefore two problems to be solved; firstly, how to ensure that participation is likely to be successful, and secondly, how to measure the level of this success.

To achieve success, it is necessary to answer a number of questions such as the following.

- What are the expectations?
- Which exhibitions should be used?
- What should be displayed and what size/type/location of stand?
- Who should be invited to visit the stand?
- What literature should be available?
- Who should 'man' the stand?

The answers to these questions, carefully thought out, should ensure that the expectations are met, and these expectations should be properly quantified in advance of the exhibition.

To measure the level of achievement it is necessary to compare the expectations with the achievements, and thus to obtain a 'score' for the exhibition as shown in Fig. 15.9; this can then be compared with the scores for other exhibitions to determine which provides the best return.

EXHIBITION REPORT AND SCORE

Exhibition:	Date:	Exhibition Location:	
Other Details:	Stand No./Location		Area & Cost
	Products Displayed:		
	Stand Manned By:		
	Literature Available:		

Report Item	Expectation	Actual	Score Value		Scores	
			Points	Per	Expected	Actual
No. of Enquiries			1	Enquiry		
No. of Quotations			5	Quotation		
Value of Quotations			5	1000 £		
No. of Orders			10	Order		
Value of Orders			10	1000 £		
Press Coverage			(see below)			
Press Coverage: None = 0; Some = 100; Fair = 200; Good = 400.			Totals		A	B

Exhibition Costs	Cost Details		Budget	Actual
Site Rental £				
Stand Preparation £				
Staff Salaries £				
Staff Expenses £				
Printing etc. £				
		Totals	C	D

Scores:

Expectation (C/A) = ________ = £ per Point (E)

Actual (D/B) = ________ = £ per Point (F)

Planning Efficiency Ratio (E x 100)/(F) = ________ x 100 = %

Exhibition Manager's Comments:

Figure 15.9 Exhibition evaluation sheet.

Table 15.3 Typical 'score' for an exhibition

Item	Expectation no.	Actual no.	Expectation score	Actual score
No. of enquiries	100	85	100	85
No. of quotations	20	22	100	110
Value of quotations	150 000	135 000	750	675
No. of orders	12	12	120	120
Value of orders	100 000	85 000	1000	850
Press cover	Fair	Fair	200	200
Total			2270	2040 (90%)
Costs				
Rental (5 days)	£1/sq. ft per day (1200 sq. ft)		6000	
	£1.20 ditto			7200
Stand preparation			1000	900
Staff salaries			1100	1150
Staff expenses			850	820
Printing			1500	1650
Total			10 450	11 730 (112%)
Scores				
Expectation	10 450/2270 = 4.60 (£ per point)			
Actual	11 720/2040 = 5.75 (£ per point)			
Efficiency ratio	4.60 × 100/5.75 = 80.9%			

Thus, based on the scoring system given in this figure, results for a particular exhibition might be as given in Table 15.3.

From this information it should be possible to prepare a 'league table' of all the exhibitions used, and thus to decide, given a limited budget, which should be used in the future. The same information could also be used to improve the expectation forecasts.

15.10 CUSTOMER LOYALTY MEASUREMENT

Numerous factors can affect customer loyalty, and it is necessary to take account of all of them to provide a practical measurement system. An example of such a system, which was designed for use by an automotive manufacturer selling through a network of agents, is shown in Fig. 15.10.

In this example, each agent, as a part of the agency agreement, is required to collect the necessary data so that the form can be completed, and the data from these forms is collated and analysed by computer in order to give overall figures. These are then used by the manufacturer to:

- assess overall loyalty to its products
- determine the level of customer service provided by the agent
- identify weak areas in the services provided
- develop improvement policies and advertising targets.

In a smaller company, a simpler system can be applied based on factors such as repeat sales, complaints, complaint response times, lost sales and new customers, in order to find answers to questions such as the following.

Company/Agent	Name		Phone		Contact	
	Address					

Sales Statistics

AVERAGE NUMBER OF PURCHASES MADE BY CUSTOMER IN A YEAR?

PERIOD OF CUSTOMER LOYALTY IN YEARS?

TOTAL NO. OF CUSTOMERS WHO PURCHASED YOUR PRODUCT IN A YEAR?

TOTAL PROFIT PER SALE?

Customer Survey Data

PROBLEMS AND COMPLAINTS:

% OF CUSTOMERS WHO EXPERIENCE PROBLEMS IN A YEAR

% OF CUSTOMERS WHO EXPERIENCE PROBLEMS & COMPLAIN?

% OF CUSTOMERS WHO DON'T EXPERIENCE PROBLEMS AND WHO RE-PURCHASE?

% OF CUSTOMERS WHO DO EXPERIENCE PROBLEMS DON'T COMPLAIN WHO WILL RE-PURCHASE?

CUSTOMER SATISFACTION:

% OF CUSTOMERS SATISFIED BY YOUR CUSTOMER RESPONSE PROCESS?

% OF CUSTOMERS MOLLIFIED BY YOUR CUSTOMER RESPONSE PROCESS?

% OF CUSTOMERS DISSATISFIED BY YOUR CUSTOMER RESPONSE PROCESS?

% OF SATISFIED COMPLAINERS WHO WILL RE-PURCHASE?

% OF MOLLIFIED COMPLAINERS WHO WILL RE-PURCHASE?

% OF DISSATISFIED COMPLAINERS WHO WILL REPURCHASE?

WORD OF MOUTH

NUMBER OF WORD OF MOUTH CONTACTS OF NON-COMPLAINERS?

NUMBER OF WORD OF MOUTH CONTACTS OF SATISFIED COMPLAINERS?

NUMBER OF WORD OF MOUTH CONTACTS OF MOLLIFIED COMPLAINERS?

NUMBER OF WORD OF MOUTH CONTACTS OF DISSATISFIED COMPLAINERS?

% OF CUSTOMERS LOST DUE TO NEGATIVE WORD OF MOUTH?

ADDITIONAL INFORMATION Please record additional information (eg. products/services, existing customer service process & customer/market attributes) on a seperate sheet.

Figure 15.10 Market damage baseline analysis.

- Why has the proportion of repeat orders fallen for the past two years?
- Why has the number of complaints about product X started to increase?
- How long does it take to deal with a complaint? Is this good enough?
- What are the most frequent reasons given for lost sales to existing customers?
- What are the most frequent reasons given by new customers for coming to us from a competitor?

These answers can then be used to develop the company's strengths in the customer services area, and to eliminate the weaknesses. This is an excellent area for applying the SWOT technique to maximize [S]trengths and minimize [W]eaknesses in order to take advantage of [O]pportunities and to counter [T]hreats.

15.11 ORGANIZING A CUSTOMER SERVICES DEPARTMENT

Customer services is usually a part of the sales department, and depending on product types and company size could be broken down into sections such as 'Warranty', 'Service Contracts', 'Complaints' and 'Marketing'. These sections should work closely together, and their responsibilities should include the following.

- *Warranty* Dealing with all warranty claims and the provision of service contracts. These activities would require links with the legal department (or company legal adviser).
- *Service contracts* Administering all service contract work including planned and unplanned calls, allocating engineers to jobs, and invoicing for work done. This would require links with Materials Control and Accounts, with information flow to Design and Quality Control.
- *Complaints* Responding to customer problems and complaints. This would require links with Quality Control, Design and, in some cases, Accounts and the legal adviser.
- *Marketing* In addition to its other activities such as sales promotion, forecasting and market research, this function should be responsible for customer surveys, advertising, exhibitions and complaints analysis.

15.12 CUSTOMER SERVICES AND TQM

The concept of total quality management requires not only quality of product, but also quality of service, and if this concept is applied in all areas of the company the benefits can be considerable, especially when it is recognized that customers can be both internal and external.

All manufacturing companies can be visualized as a number of chains, as shown in Fig. 15.11, and at each connecting point there is a 'supplier to customer' relationship. Whether these are internal or external, the principles of customer service apply. It will be noted that TQM is not a part of any 'chain' but is an overall concept, affecting every relationship. Thus, wherever a link exists, the principles should be applied. These principles can be defined as follows. Wherever a supplier/customer link (internal or external) exists:

- a strong partnership bond should be built up which will allow both partners to understand the other's problems and work together to overcome them
- the supplier should always put the needs of its customer at the top of its list of priorities
- the customer should always try to give his supplier the maximum possible notice of his needs and try to eliminate 'last minute' changes

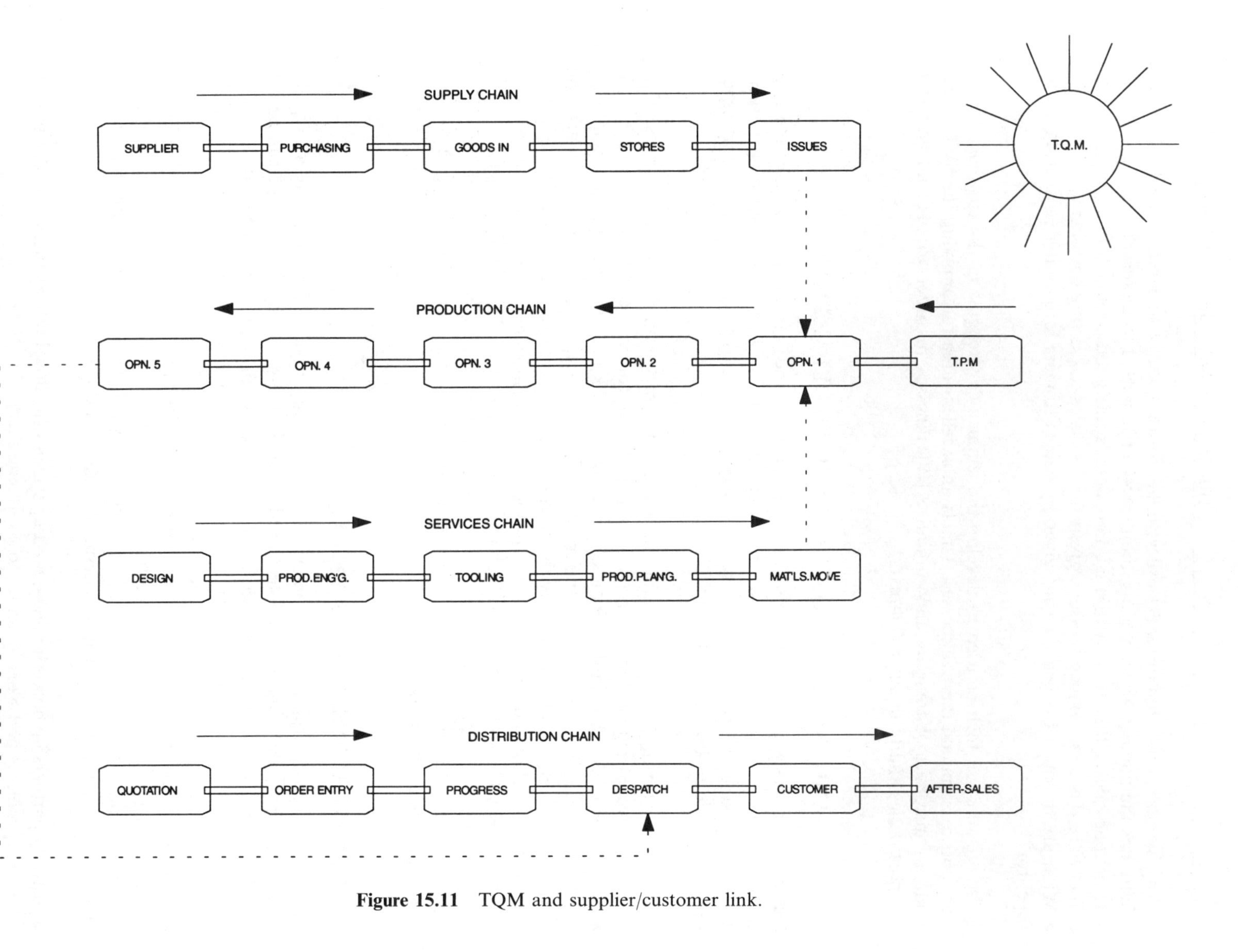

Figure 15.11 TQM and supplier/customer link.

- both customer and supplier should recognize that unity is strength and discord is weakness, and that unity can be achieved only by mutual trust and understanding.
- problems cannot be solved by asking 'who is to blame?': the questions that should be asked are 'how can we correct the situation?' and 'how can we stop it happening again?'.

A TQM culture cannot be imposed from above, but must be developed by motivation and encouragement at all levels, based on the realization that TQM is a means of moving a company to the top of the world-class league.

PROBLEMS

1. As the services manager for a new company, you are asked to prepare a draft service agreement for customers that wish to arrange for regular servicing of the machines you are supplying. This draft will then be passed to the legal adviser, so you do not need to write it in legal terms. Your company manufactures fork-lift trucks which can be supplied in a variety of models (e.g. gas or electric), and your agreement should include various options (full service, labour only, etc.).
2. Your company makes and supplies washing machines, and has received a complaint from a customer that a machine was damaged when unpacked at his home. He believes that this is transit damage but the sales agents deny responsibility, stating that the machine was delivered by them in good condition. Write a report suggesting what action should be taken and what reply should be given to the customer.
3. You have been given the task of writing an instruction manual for a new product about to be launched by your company. This is an electric fan heater for wall mounting in any location, including the bathroom. Prepare a draft for such a manual, covering the various aspects of installation and operation and stressing the 'dos' and 'don'ts' from a safety aspect.
4. A company that makes a range of portable electric tools for the DIY market has taken space at an exhibition designed to attract home improvers. Suggest how this space should be utilized and the stand organized, giving brief details of the displays, demonstrations and hand-outs you recommend for maximum impact.
5. Prepare an outline budget for particpation in this home improvers exhibition as described in your answer to question (4).

FURTHER READING

Murdick, R. G., *Service Operations Management*, Allyn and Bacon, Boston, MA, 1990.

Voss, C., Armistead, C., Johnson, B. and Morris, B., *Operations Management in Service Industries and the Public Sector: Text and Cases*, Wiley, Chichester, 1985.

APPENDIX A
SOLUTIONS TO PROBLEMS

Solution to Problem No. 3 - Chapter 2 - DCF METHOD

GIVEN THE FOLLOWING DATA: £K

... Capital Expenditure (Year 0) = 85 - 15 = 70

... Profit before Depreciation and Tax = 60 - 30 = 30

... Life of Investment = 5 Years

... Residual Value at end Year 3 = 5

YEAR	INVESTMENT	PROFIT	NET CASH FLOW	TEST DISCOUNTED @			PRESENT VALUE @		
				20% p.a.	40% p.a.	33% p.a.	20% p.a.	40% p.a.	33% p.a.
	£K	£K					£K	£K	£K
	A	B	C	D	E	F	G	H	I
0	−70	NIL	−70	1.00	1.00	1.00	−70	−70	−70
1		+30	+30	0.833	0.714	0.752	+24.99	+21.42	+22.56
2		+30	+30	0.694	0.510	0.565	+20.82	+15.30	+16.95
3		+30	+30	0.578	0.364	0.425	+17.34	+10.92	+12.75
4		+30	+30	0.482	0.260	0.320	+14.46	+ 7.80	+ 9.70
5	+5	+30	+35	0.402	0.186	0.240	+14.07	+ 6.51	+ 8.40
TOTAL	−65	+150	+80				+21.68	−8.05	+ 0.26

Thus the DCF Rate for this investment over a five year term = 33%.

Figure A.1 DCF method.

Solution to Problem No. 2 - Chapter 5 - FAST Diagram for Ball-Point Pen

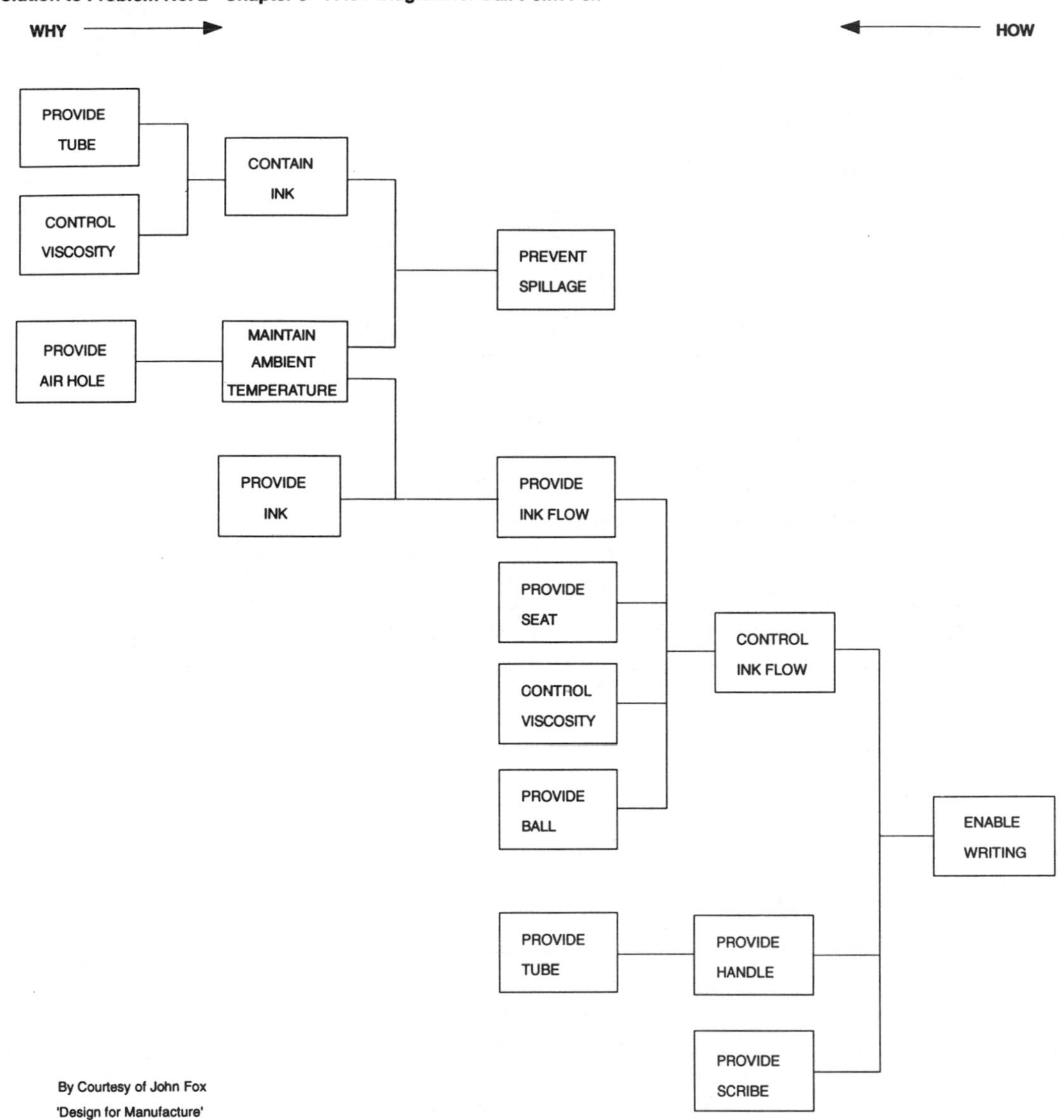

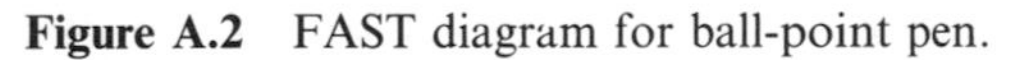

Figure A.2 FAST diagram for ball-point pen.

Solution to Problem No. 3 - Chapter 5 - Taguchi Loss Function & Specn. Limits

TAGUCHI LOSS FUNCTION:

Component has nominal length of 75mm and Tolerance of ±0.8mm.

If it falls outside these limits and reaches customer, cost of replacement is £20

The quadratic formula to be used is $L = k(y - m)^2$ where L = The Loss (£20)
k is the constant
y is the measured value (75.8mm or 74.2mm)
m is the target value (75mm)

thus, the constant (k) is found from $20 = k(75.8 - 75)^2$ or, k = 20/0.64 = 31.25

then, using this (k) value in the equation for various values of (L) $y = 75 \pm \sqrt{L/31.25}$

If L =	0	2	4	6	8	10	12	14	16	18	20
y =	75.000	75.253	75.358	75.438	75.506	75.566	75.620	75.669	75.716	75.759	75.800
or y =	75.000	74.747	74.642	74.562	74.494	74.434	74.380	74.331	74.284	74.241	74.200

These values of one can then be plotted as a Taguchi Loss Function graph.

TAGUCHI LOSS FUNCTION LIMITS:

The replacement cost of an item, if caught in production, is £3.50

The quadratic formula to be used is $L = k(y - m)^2$ where L = The Loss (£3.50)
k is the constant (31.25)
y is the Taguchi Limit value to be found
m is the target value (75mm)

Thus, $(y - 75) = \sqrt{3.50/31.25}$
or, $(y - 75) = \pm 0.335$

Therefore, the Upper Taguchi Limit is 75.335mm
and the Lower Taguchi Limit is 74.665mm

Figure A.3 Taguchi loss function and specification limits.

Solution to Problem No. 6 - Chapter 6 - PRIORITY CODING

WORKS ORDER STATUS

PRODUCT CODE: C.2218
DESCRIPTION: xxxxxxxxxxxxxxxxxxxxxxxxxx
QUANTITY REQD. 75
DUE DATE: WEEK 24/3

WORKS ORDER No. W.13142
ORDER STATUS: STARTED
DATE: WEEK 22/4

OPN. NO.	WK. CR.	OPERATION DESCRIPTION	SU. TIME	OPN. TIME	MOVE QUEUE	STRT DATE	FIN DATE	QUANTITY OK	QUANTITY SCR	ST	%age C'PLETE
10	RMST	xxxxxxxxxxxxxxxxxxxxx xxxxxxxxxxxxxxxxxxxxx	0.15	9.00	6.00	22/2	22/3	75	0	C	100
20	LCLA	xxxxxxxxxxxxxxxxxxxxx xxxxxxxxxxxxxxxxxxxxx	0.25	30.00	12.00	22/4	23/2	73	2	C	100
30	VMIL	xxxxxxxxxxxxxxxxxxxxx xxxxxxxxxxxxxxxxxxxxx	0.50	15.00	12.00	23/3	23/5	43	1	S	60
40	GRND	xxxxxxxxxxxxxxxxxxxxx xxxxxxxxxxxxxxxxxxxxx	0.45	18.75	10.00	24/1	24/3	0	0	N	NIL
50	INSP	xxxxxxxxxxxxxxxxxxxxx xxxxxxxxxxxxxxxxxxxxx	0.25	12.00	15.00	24/4	25/1	0	0	N	NIL

BASED ON 7.5 HOURS PER DAY:

LEAD TIME AT START: 141.35 HOURS OR 18.8 DAYS

CURRENT LEAD TIME: 93.05 HOURS OR 12.4 DAYS

WORKS ORDER PROGRESS:

WORKS ORDER IS 34.2% COMPLETED

PRIORITY VALUE (CR) = 73

Figure A.4 Priority coding.

Solution to Problem No. 2 - Chapter 9 - Vendor Service Levels

WEEK NO.	WEEKLY CALLS (X)	DEVIATION FROM AVGE. $(X - \bar{X})$	SQUARES OF DEVIATIONS $(X - \bar{X})^2$
1	125	-20	400
2	140	-5	25
3	164	+19	361
4	138	-7	49
5	121	-24	576
6	96	-49	2401
7	135	-10	100
8	157	+12	144
9	148	+3	9
10	168	+23	529
11	179	+34	1156
12	165	+20	400
Total	1736	Sum of Squares	6150
Avge $\bar{X}$	145		

SUM OF SQUARES DIVIDED BY (NO. OF WEEKS - 1) = $\frac{6150}{(12 - 1)}$ = 559 (variance)

SQUARE ROOT OF VARIANCE: $\sqrt{559}$ = 23.6 (Standard Deviation σ)

FROM THE k VALUES TABLE IN FIGURE 8.10 'k' FOR 95% = 1.65

THUS MINIMUM STOCK SHOULD BE SET AT AVERAGE + (1.65 σ)

= 145 + (1.65 * 23.6) = 145 + 39 = 184 units

COMMENT: This is probably excessive and a 90% VSL would be more realistic giving a minimum stock of 175.

Figure A.5 Vendor service levels.

Solution to Problem No. 4 - Chapter 13 - PROCESS CAPABILITY

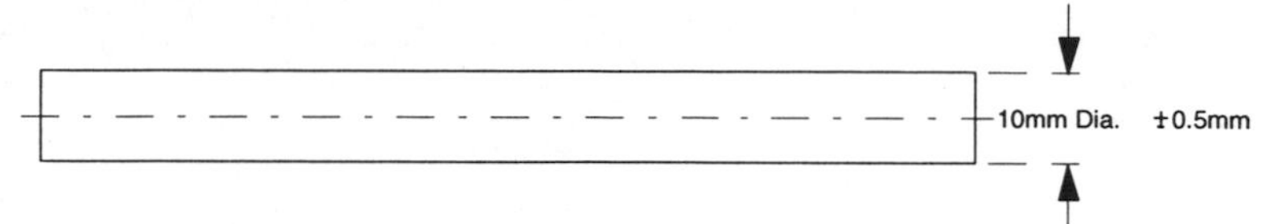

To calculate the C_{pk} for this item it is first necessary to calculate the value of σ (standard deviation) and then to calculate the value of 3σ

READING No.	READING VALUE (mm) (X)	DEVIATION FROM AVGE. $(X-\bar{X})$	SQUARES OF DEVIATIONS $(X-\bar{X})^2$
1	10.1	+0.2	0.04
2	9.8	−0.1	0.01
3	10.0	+0.1	0.01
4	9.8	−0.1	0.01
5	10.1	+0.2	0.04
6	10.0	+0.1	0.01
7	9.9	0	0
8	9.7	−0.2	0.04
9	9.8	−0.1	0.01
10	9.8	−0.1	0.01
Total	99.0	Sum of Squares	0.18
Avge $\bar{X}$	9.9		

1. DIVIDE SUM OF SQUARES BY (NO. OF READINGS - 1) $\frac{0.18}{(10-1)} = 0.02$ (Variance)

2. TAKE SQUARE ROOT OF VARIANCE $\sqrt{0.02} = 0.14$ (approx) THIS IS KNOWN AS THE STANDARD DEVIATION (σ) THUS,

THREE STD.DEVIATIONS (3σ) = 0.42

THEN THE C_{pk} VALUES CAN BE CALCULATED AS FOLLOWS:

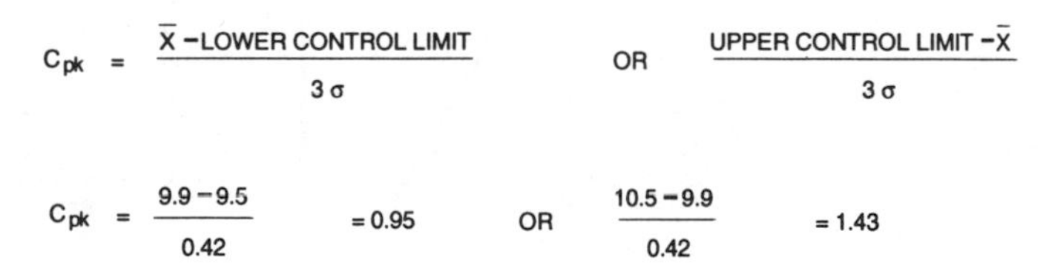

$$C_{pk} = \frac{\bar{X} - \text{LOWER CONTROL LIMIT}}{3\sigma} \quad \text{OR} \quad \frac{\text{UPPER CONTROL LIMIT} - \bar{X}}{3\sigma}$$

$$C_{pk} = \frac{9.9-9.5}{0.42} = 0.95 \quad \text{OR} \quad \frac{10.5-9.9}{0.42} = 1.43$$

Figure A.6 Process capability.

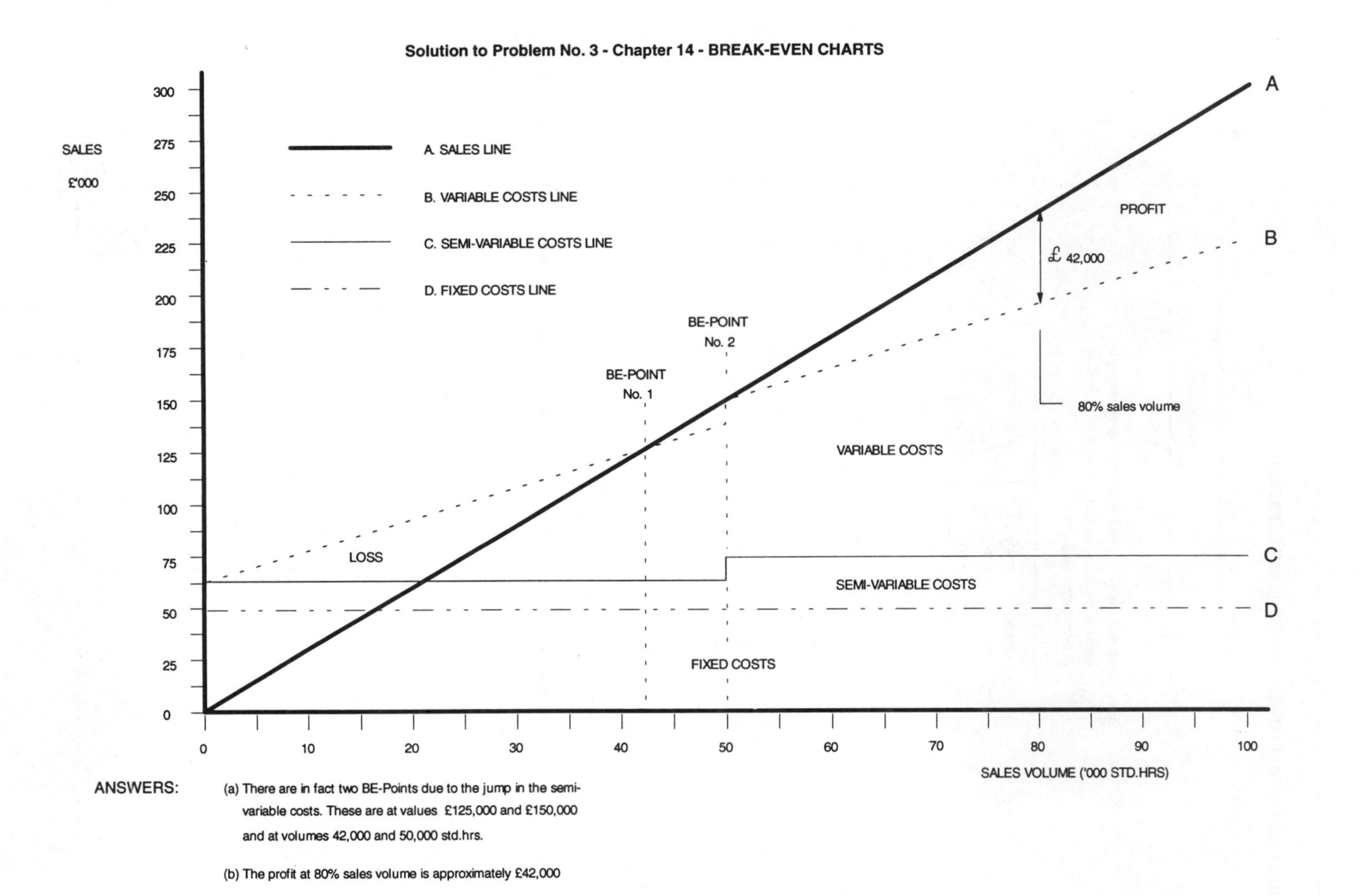

Figure A.7 Break-even charts.

Solution to Problem No. 5 - Chapter 14 - MARGINAL COSTING

Question 5(a)

TABLE I - MADE IN-HOUSE	PRODUCT A (Qty Sold = 5,000)		PRODUCT B (Qty Sold = 4,000)		PRODUCT C (Qty Sold = 16,000)	
	TOTAL	UNIT	TOTAL	UNIT	TOTAL	UNIT
TOTAL SALES VALUE BY PRODUCT	150,000	30.00	80,000	20.00	240,000	15.00
MARGINAL OR VARIABLE COST OF PRODUCT SALES	90,000	18.00	48,000	12.00	144,000	9.00
FIXED COST CONTRIBUTION FROM PRODUCT SALES	37,500	7.50	20,000	5.00	60,000	3.75
PROFIT BY PRODUCT	22,500	4.50	12,000	3.00	36,000	2.25
TOTAL SALES VALUE - ALL PRODUCTS			470,000			
MARGINAL OR VARIABLE COST OF ALL PRODUCT SALES			282,000			
FIXED COST CONTRIBUTION FROM ALL PRODUCT SALES			117,500			
TOTAL PROFIT			70,500			

TABLE I - PURCHASED OUTSIDE	PRODUCT A (Qty Sold = 5,000)		PRODUCT B (Qty Sold = 4,000)		PRODUCT C (Qty Sold = 16,000)	
	TOTAL	UNIT	TOTAL	UNIT	TOTAL	UNIT
TOTAL SALES VALUE BY PRODUCT	150,000	30.00	80,000	20.00	240,000	15.00
MARGINAL OR VARIABLE COST OF PRODUCT SALES	90,000	18.00	60,000	15.00	144,000	9.00
FIXED COST CONTRIBUTION FROM PRODUCT SALES	45,200 (*)	9.04	NIL	NIL	72,300 (*)	4.52
PROFIT BY PRODUCT	14,800	2.96	20,000	5.00	23,700	1.48
TOTAL SALES VALUE - ALL PRODUCTS			470,000			
MARGINAL OR VARIABLE COST OF ALL PRODUCT SALES			294,000			
FIXED COST CONTRIBUTION FROM ALL PRODUCT SALES			117,500			
TOTAL PROFIT			58,500			

ANSWER:

It would not be more profitable to buy product B outside since this would lead to a loss of some £8000 per year overall. This is due to the loss of contribution from product B of £20,000 which nust be loaded to the other products (*).

Figure A.8 Marginal costing.

Solution to Problem No. 5 - Chapter 14 - MARGINAL COSTING

Question 5(b)

TABLE I - BEFORE ADVERTISING CAMPAIGN	PRODUCT A (Qty Sold = 5,000)		PRODUCT B (Qty Sold = 4,000)		PRODUCT C (Qty Sold = 16,000)	
	TOTAL	UNIT	TOTAL	UNIT	TOTAL	UNIT
TOTAL SALES VALUE BY PRODUCT	150,000	30.00	80,000	20.00	240,000	15.00
MARGINAL OR VARIABLE COST OF PRODUCT SALES	90,000	18.00	48,000	12.00	144,000	9.00
FIXED COST CONTRIBUTION FROM PRODUCT SALES	37,500	7.50	20,000	5.00	60,000	3.75
PROFIT BY PRODUCT	22,500	4.50	12,000	3.00	36,000	2.25
TOTAL SALES VALUE - ALL PRODUCTS			470,000			
MARGINAL OR VARIABLE COST OF ALL PRODUCT SALES			282,000			
FIXED COST CONTRIBUTION FROM ALL PRODUCT SALES			117,500			
TOTAL PROFIT			70,500			

TABLE II - AFTER ADVERTISING CAMPAIGN	PRODUCT A (Qty Sold = 6,250)		PRODUCT B (Qty Sold = 4,000)		PRODUCT C (Qty Sold = 16,000)	
	TOTAL	UNIT	TOTAL	UNIT	TOTAL	UNIT
TOTAL SALES VALUE BY PRODUCT	187,500	30.00	80,000	20.00	240,000	15.00
MARGINAL OR VARIABLE COST OF PRODUCT SALES	112,500	18.00	48,000	12.00	144,000	9.00
FIXED COST CONTRIBUTION FROM PRODUCT SALES	37,500	6.00	20,000	5.00	60,000	3.75
PROFIT BY PRODUCT	37,500	6.00	12,000	3.00	36,000	2.25
TOTAL SALES VALUE - ALL PRODUCTS			507,500			
MARGINAL OR VARIABLE COST OF ALL PRODUCT SALES			304,500			
FIXED COST CONTRIBUTION FROM ALL PRODUCT SALES			117,500			
TOTAL PROFIT			85,500 (75,500 allowing for cost of campaign)			

ANSWER:

It would be more profitable to advertise product A for £10,000 since, if sales of A went up as a result by 25% there would be an increase in profit of £5,000.

Figure A.9 Marginal costing.

APPENDIX B

THE PRINCIPLES OF NORMAL DISTRIBUTION

If a series of data is plotted on a graph in terms of its distribution, this usually takes the form shown in Fig. B.1, which is known as a Poisson distribution. If the data are distributed unevenly the curve is said to be skewed; if they are evenly balanced the curve is said to be normal, i.e. a normal distribution.

In Fig. B.1 the heights of a group of people have been plotted in terms of their distribution and it can be seen that this distribution is normal, since the numbers of people on either side of the average (x-bar) line are nearly equal (2850:2808). If the distribution had been skewed this would not have been the case, as can be seen in the figure.

However, a number of deductions can be drawn from a normal distribution, that can be of considerable benefit to a user, and these concern the areas under the graph. If this total area is denoted as 100 per cent, then other areas can be determined mathematically in percentage terms and these give the probabilities that an 'event' will take place. For example, using the data shown in Fig. B.1, it is possible to calculate the probability of any person being between 64 and 70 inches in height, or of someone being above 72 inches.

This technique has many applications throughout industry, for example in the control of dimensional accuracy and in forecasting the likelihood of demand fluctuations. The objective of the calculations is to determine a value known as the standard deviation (denoted by σ the greek letter sigma), and once this has been found for any set of data, statistical tables can be used to determine the probabilities.

Figure B.2 shows some examples of this whereby, if σ has been found, a large number of event probabilities can be determined. In the population/height example $\sigma = 4$ (approximately) and therefore, using the values shown in Fig. B.2, various probabilities can be determined, as given in Table B.1.

Table B.1 Population/height probabilities

1. Probability that all people will be between 63 and 71 inches tall (67" ± 4")	68.2%
2. Probability that all people will be between 59 and 75 inches tall (67" ± 8")	95.4%
3. Probability that all people will be between 55 and 79 inches tall (67" ± 12")	97.0%
4. Probability that all people will be less than 71 inches tall (67" + 4")	84.1%
5. Probability that all people will be less than 75 inches tall (67" + 8")	97.7%
6. Probability that all people will be less than 79 inches tall (67" + 12")	98.5%

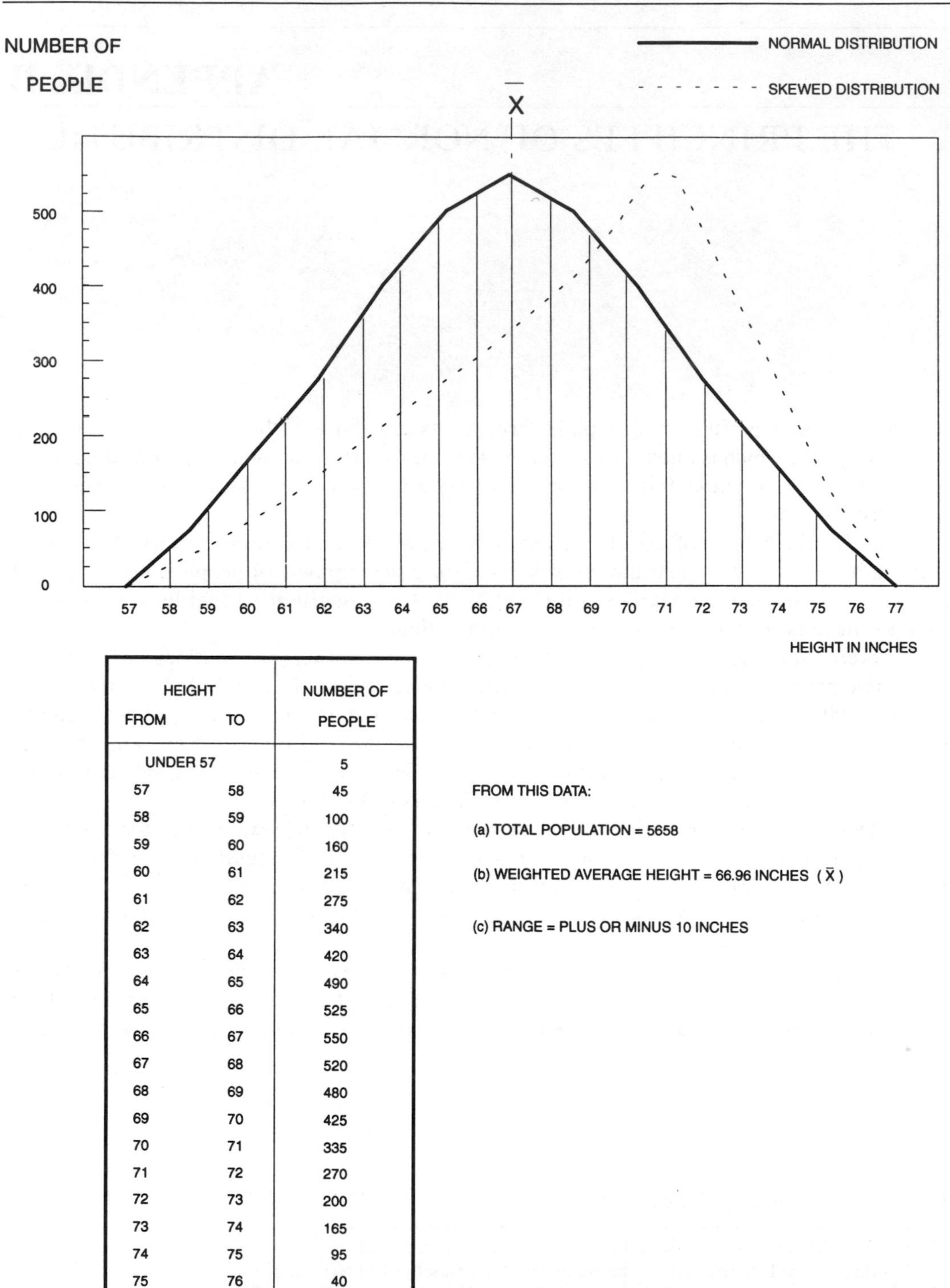

HEIGHT FROM	HEIGHT TO	NUMBER OF PEOPLE
UNDER 57		5
57	58	45
58	59	100
59	60	160
60	61	215
61	62	275
62	63	340
63	64	420
64	65	490
65	66	525
66	67	550
67	68	520
68	69	480
69	70	425
70	71	335
71	72	270
72	73	200
73	74	165
74	75	95
75	76	40
76	77	12
OVER 77		3

FROM THIS DATA:

(a) TOTAL POPULATION = 5658

(b) WEIGHTED AVERAGE HEIGHT = 66.96 INCHES ($\bar{X}$)

(c) RANGE = PLUS OR MINUS 10 INCHES

Figure B.1 Typical normal and skewed distribution.

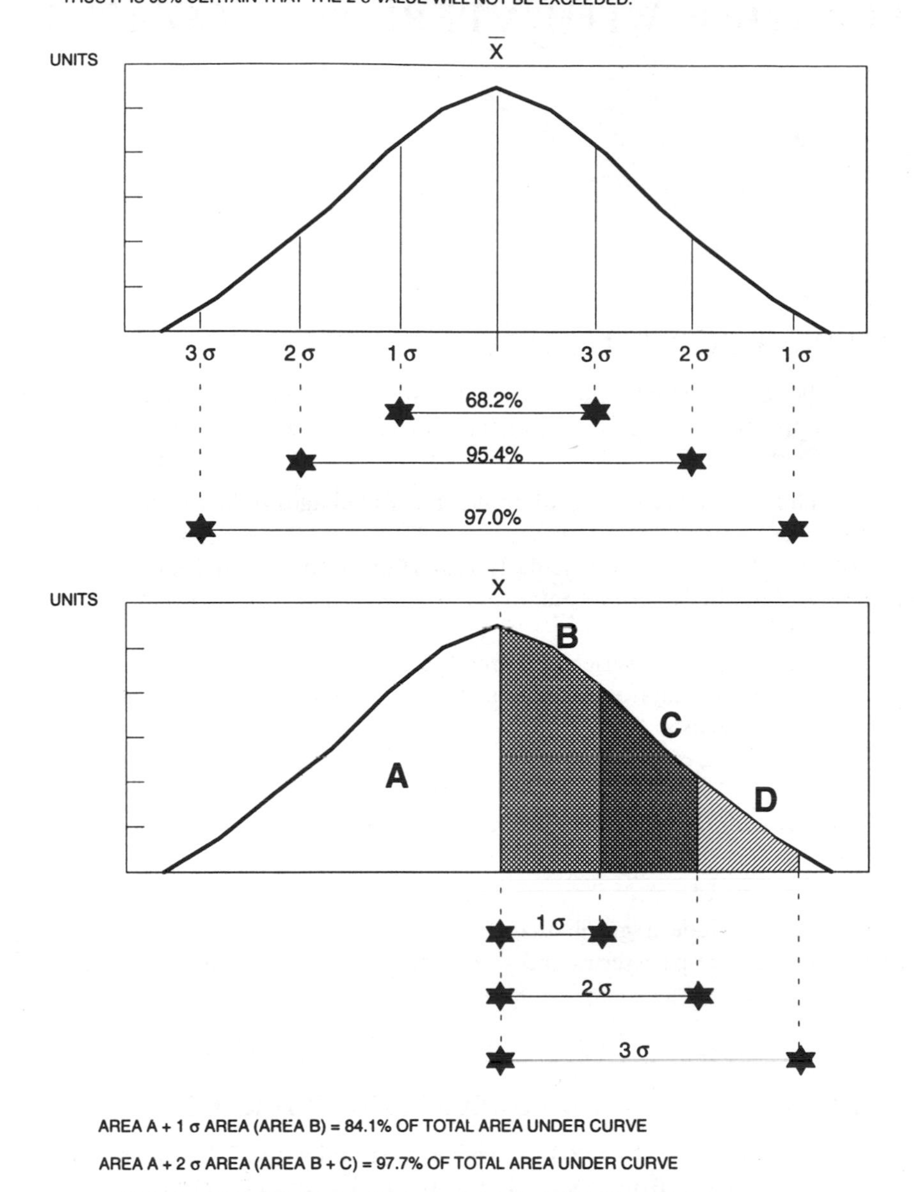

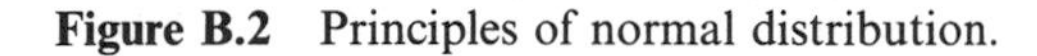

Figure B.2 Principles of normal distribution.

In the quality assurance field, this statistical approach has been developed into a sophisticated quality control technique, which can be used to indicate when a process is tending to exceed its control limits so that preventive action can be taken; this technique is known as SPC (statistical process control).

APPENDIX C

FORECASTING WITH A PERSONAL COMPUTER

C.1 CHOOSING A PC-BASED FORECASTING PACKAGE

There are a number of PC-based, time series, forecasting packages on the market. When making a selection it is important to ensure that certain key features are present, for example:

- the facility to choose the mathematical model: most packages offer this but they vary greatly in the method used
- the speed of operation: this is particularly important with seasonal models; it depends not only on the speed of the hardware but also on the way in which the software approaches the problem of modelling
- the ability to decompose the series into trend, seasonal and irregular components, and the forecasts into corresponding trend and seasonal components: some packages decompose only the series, not the forecasts
- the ability to detect and estimate outliers, with the option to change the threshold for detection
- the ability to present the data in graphical form, with clear identification of the various components.

The package should include a graph that shows the components on a common scale, and another that compares the past series and trend with the forecasts, and gives their margins of error.

C.2 EXAMPLE OF A PC-BASED FORECASTING PACKAGE

A menu-driven package called PROPHET, which runs under the DOS operating system, has been written and is available from ASRU Ltd at the University of Kent (in Canterbury). This package contains the following facilities:

- automatic model choice, using the latest method, including recommended transformation
- speed—it takes five seconds to fit a seasonal model to 120 observations and to detect outliers, using a PC with a 486 processor and a numeric coprocessor
- decomposition of both series and forecasts
- outlier detection, based on the irregular component of the fitted model

- menu-driven graphics containing the two overlay graphs (a graph that shows the components on a common scale, and another that gives a comparison of the past series and trend, with the forecasts and their margins of error).

This package has been used to prepare all the graphs in this appendix. These have been chosen as examples of the information that can be obtained by using statistical forecasting.

C.3 EXAMPLES OF OUTPUT FROM PROPHET (FORECASTING PACKAGE)

Figure C.1(a) and (b)

If a model is fitted with the last six data points excluded, forecasts can be compared with these six actuals. In these and the following figures, the bounds have been chosen to give a one-in-three chance of each point being out of bounds.

Figure C.1a Retail purchases of drug A.

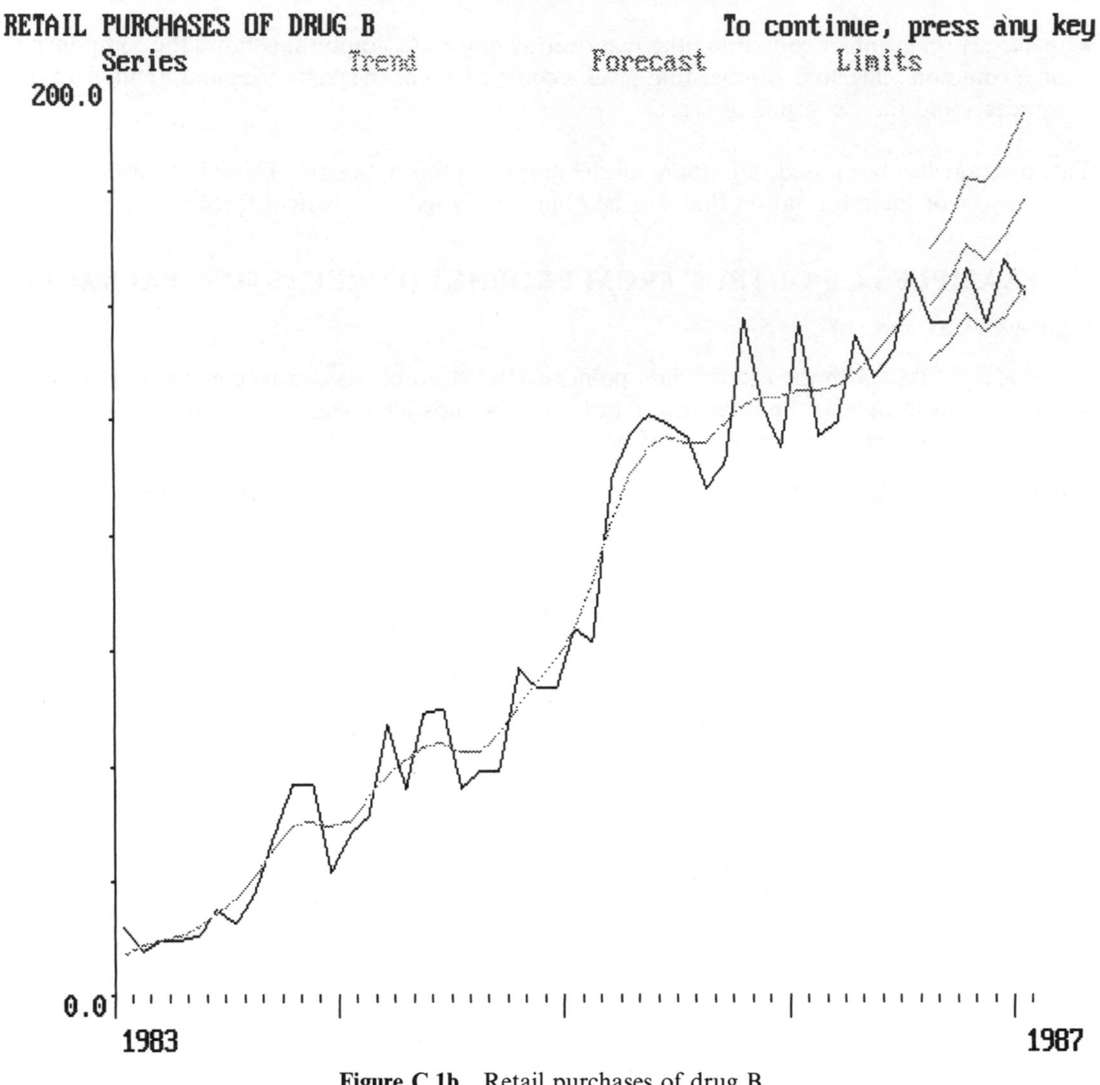

Figure C.1b Retail purchases of drug B.

Figure C.1(a) shows the monthly sales of a drug (A) declining because it was competing with a newer version. There is a hint of seasonality in the January–February dip in three out of the four years, but the series is really too short to estimate this, and to try to do so makes the forecast worse. Actual sales in the last six months oscillate unpredictably, but almost within the forecast bounds.

Figure C.1(b) shows the sales of another drug (B), which is mildly but significantly seasonal. The forecasts continue the upward trend, whereas the actual series seems to be flattening out, although it is still inside the error bounds until the last point. As further data are included in the model, the forecasts will be modified downwards.

Figure C.2(a) and (b)

Figure C.2(a) shows UK car production over a period of 20 years, with a trend forecast. The series has a generally downward trend, but is very erratic, partly due to numerous industrial disputes. It has significant seasonality, but this is obscured by the irregularity.

Figure C.2(b) shows how it is possible to 'zoom in', in this case on the most recent data from Fig. C.2(a). This shows more clearly the forecasts of the actual series, which are within the error bounds.

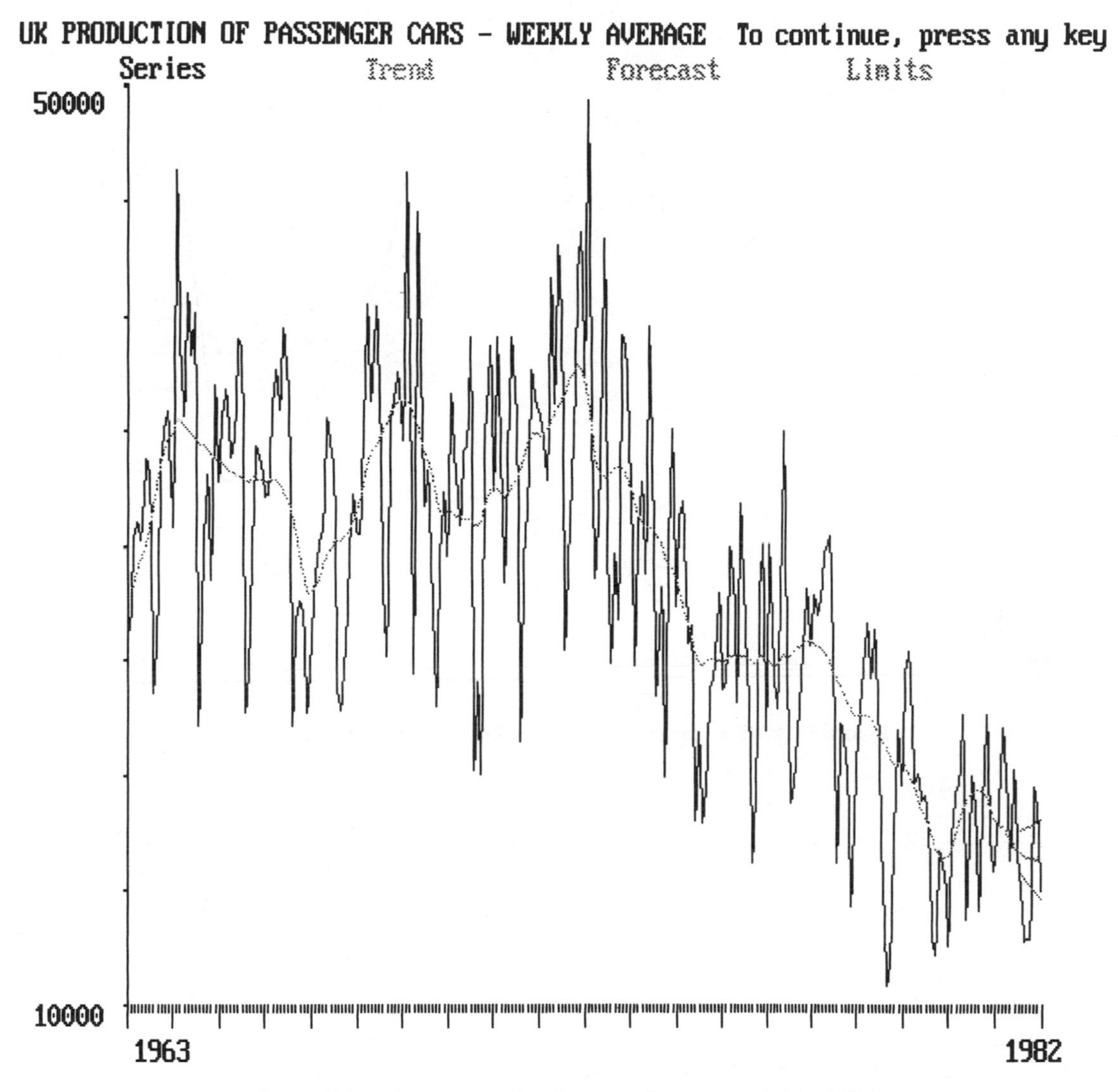

Figure C.2a UK car production weekly average (1963–1982).

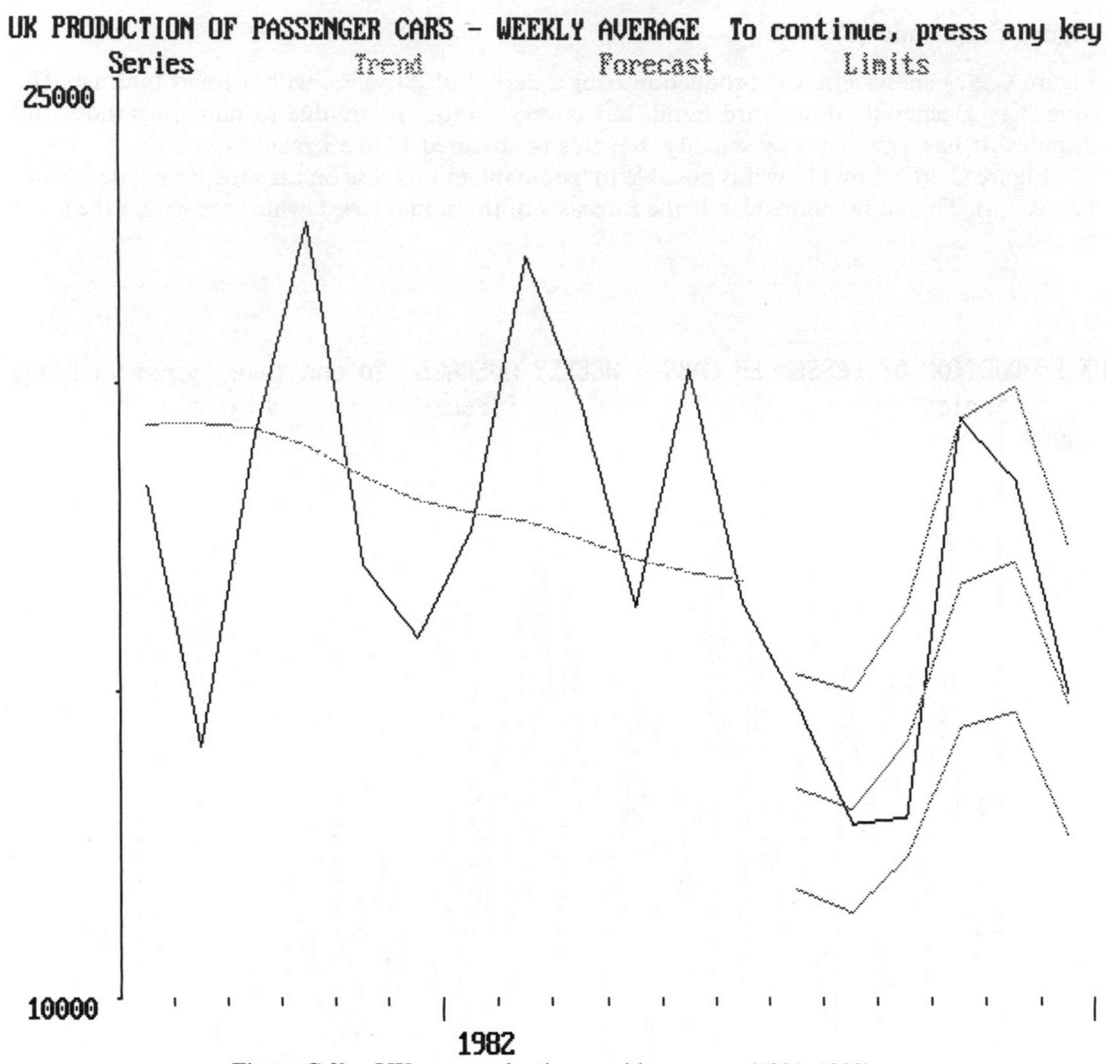

Figure C.2b UK car production weekly average (1981–1982).

Figure C.3(a) and (b)

Figure C.3(a) shows sulphuric acid production over a period of 12 years, which is also very erratic, with deep recessions in 1975 and 1981. The huge negative outlier in January 1979 is probably due to a strike, and would have been replaced by an interpolated value when the next month's data became available. In the forecast period the trend is still falling, but as the 'zoom in' in Fig. C.3(b) shows, the forecasts are quite accurate apart from the second month.

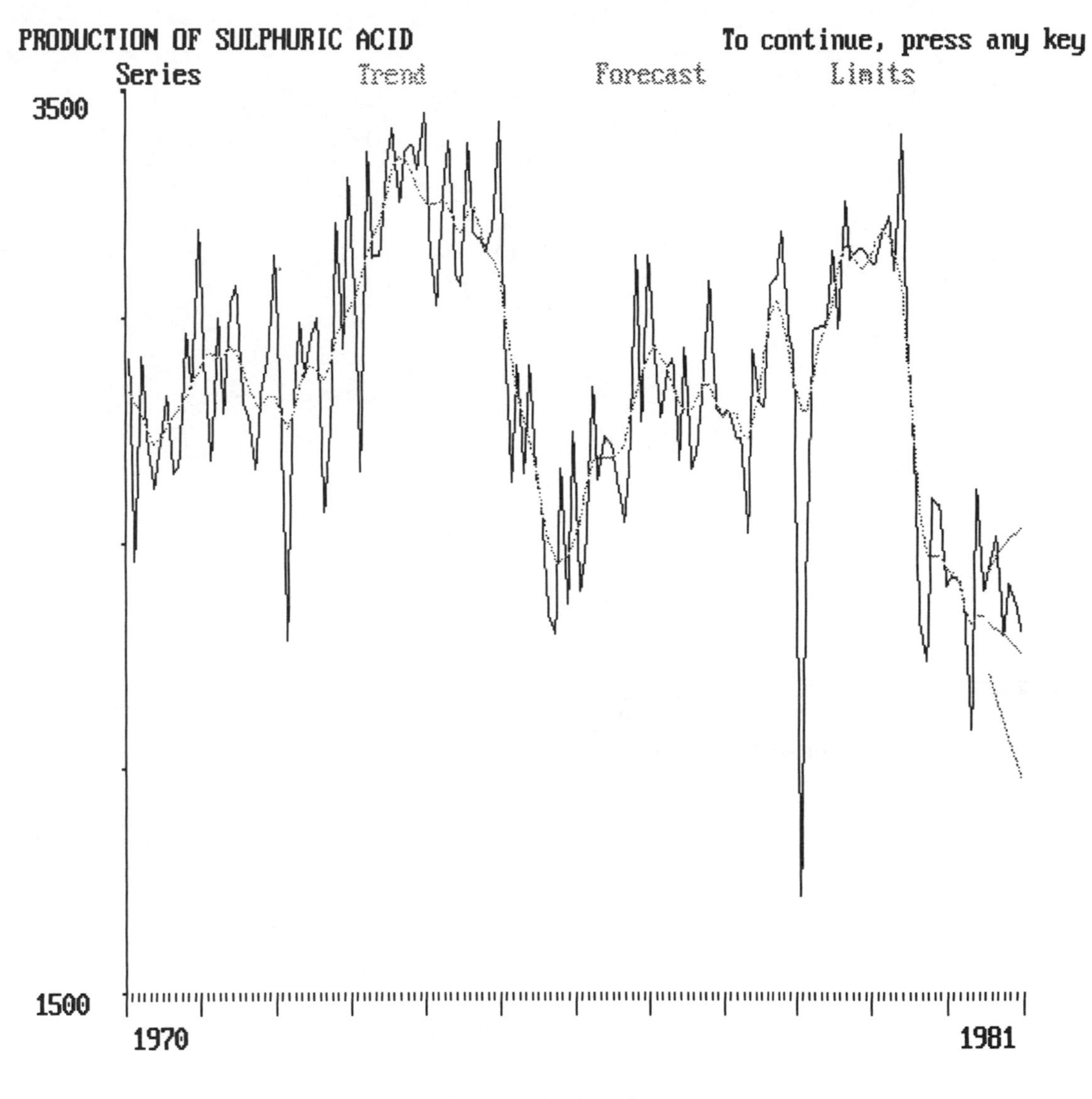

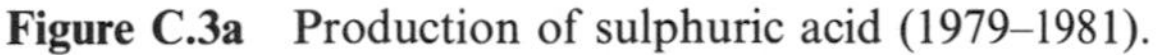

Figure C.3a Production of sulphuric acid (1979–1981).

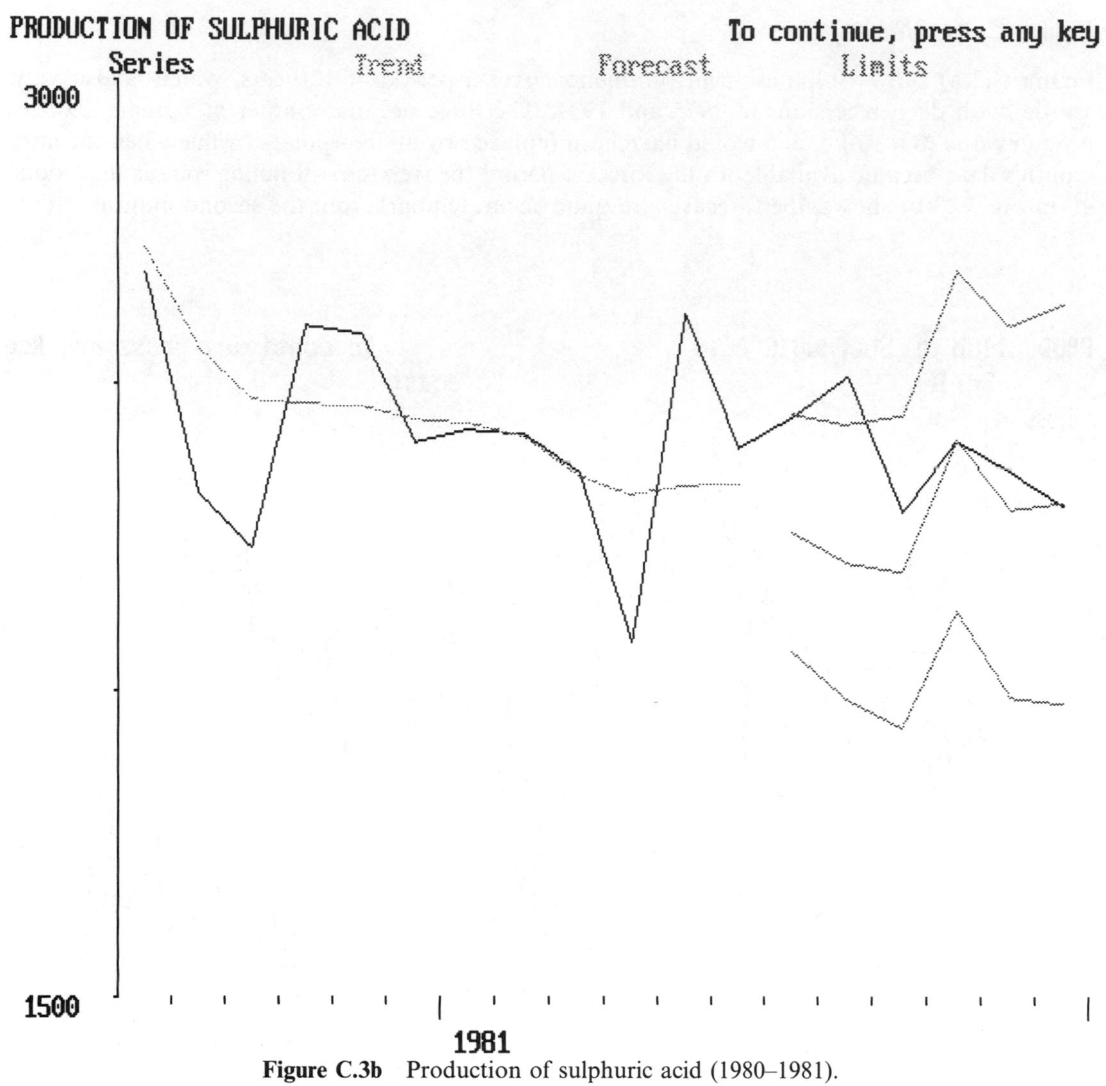

Figure C.3b Production of sulphuric acid (1980–1981).

Figure C.4(a) and (b)

Figure C.4(a) shows the effect of the seatbelt law in the UK on the number of car drivers killed or seriously injured. The initial estimate of the trend showed a rapid fall in January and February 1983, when the law came into force, so an extra variable was included in the model to measure the size of the step change; there is clear seasonality in the series (winter conditions). Figure C.4(b) shows that the forecasts are good, the apparent upward trend being entirely due to the onset of winter and the Christmas season.

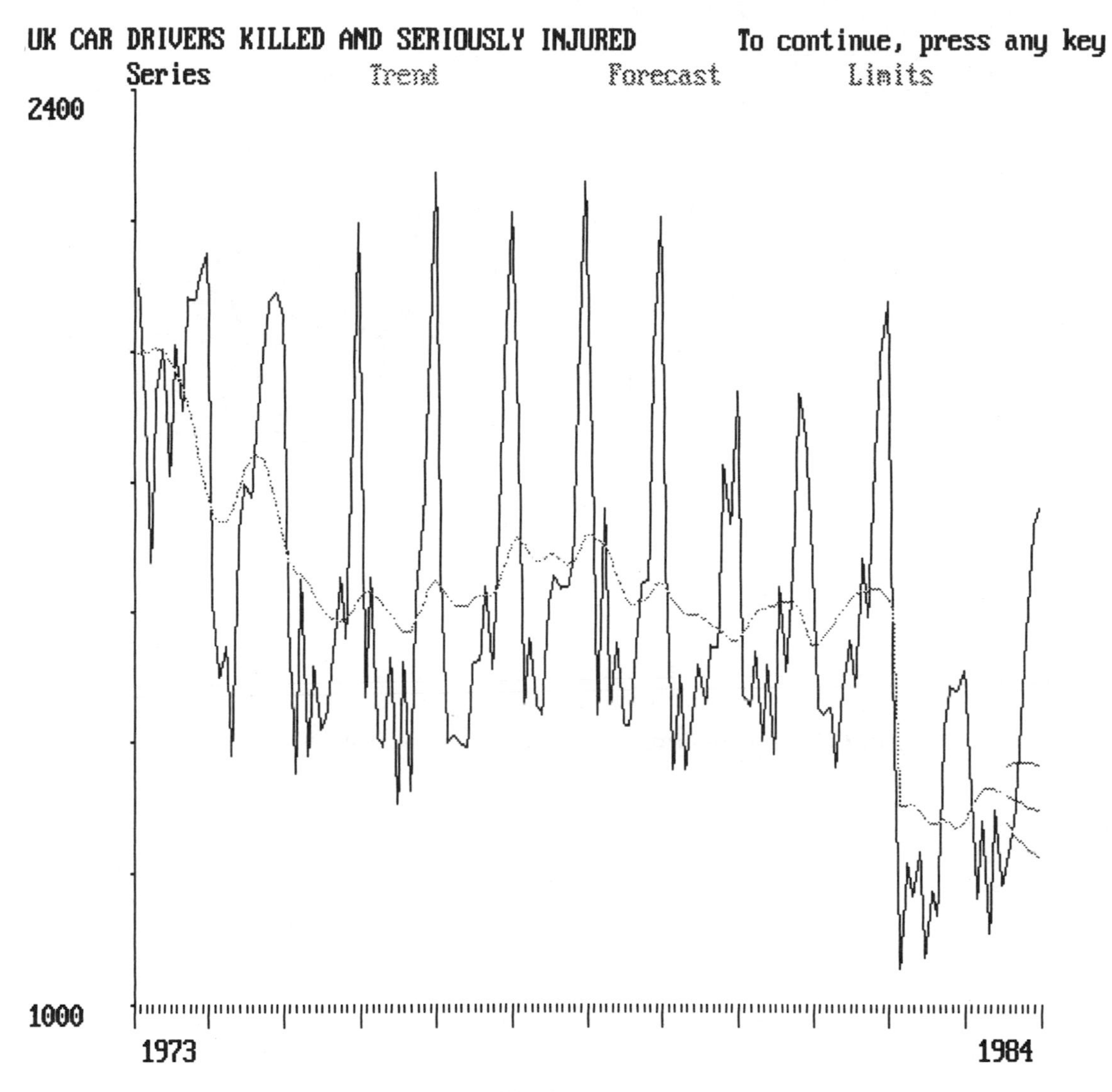

Figure C.4a UK drivers killed and seriously injured (1973–1984).

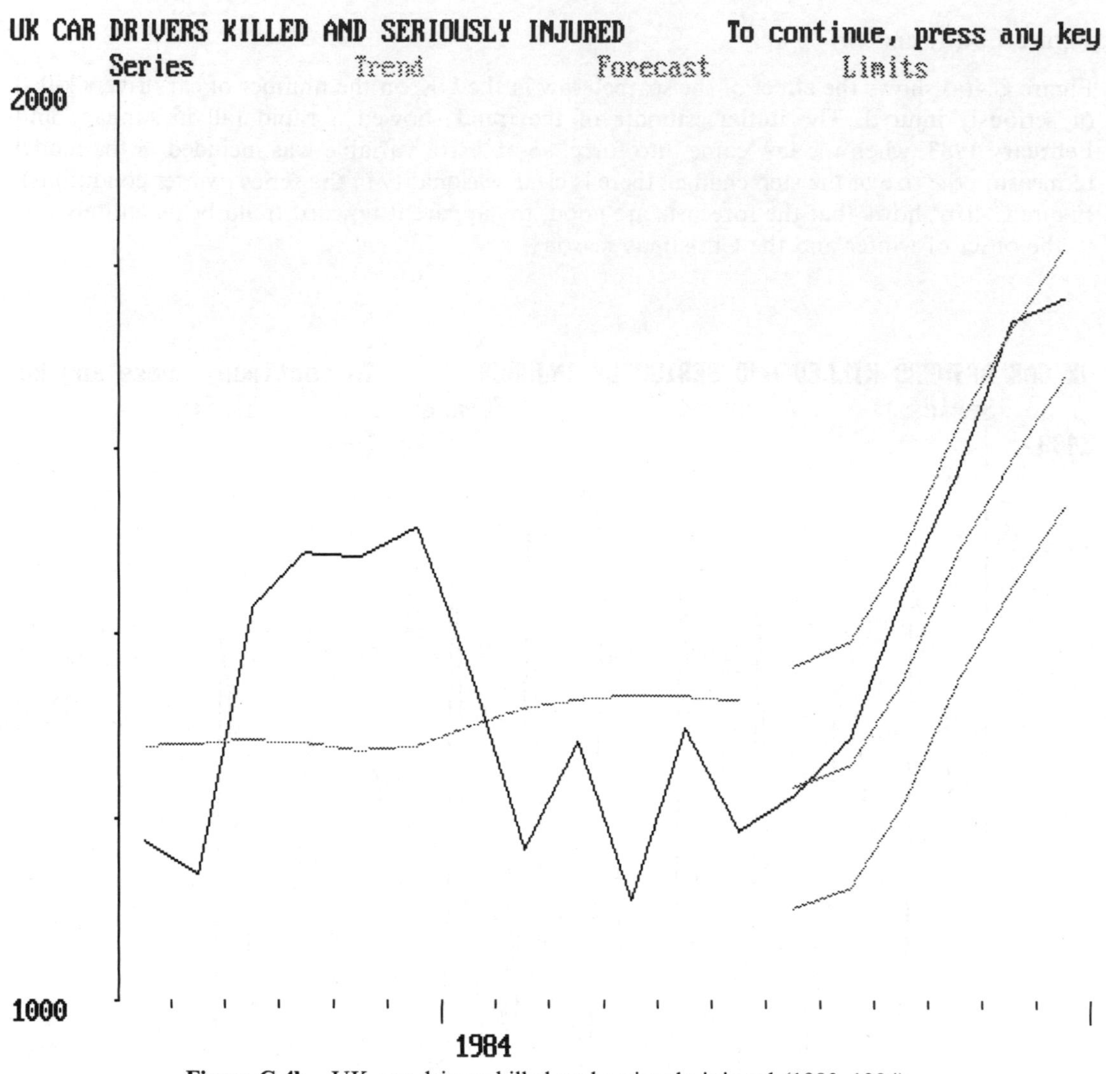

Figure C.4b UK car drivers killed and seriously injured (1983–1984).

APPENDIX D

TAGUCHI'S DESIGN OF EXPERIMENTS

One of the major advances in the improvement of quality was Dr Taguchi's work on the design of experiments. This broke away from the 'one factor at a time' approach which is a great time-waster, by using statistical methods that provide much more information at less cost.

The basis of any experiment is the parameters used, i.e. the range of conditions applied; for example the parameters used in an experiment to optimize engine performance might include a range of cylinder bores, strokes, fuel octanes and fuel–air ratios. Traditionally, it was assumed that the variation in experimental results was constant for different combinations of parameters, but Taguchi took a different view. In order to create what he called a 'robust' design, a combination of parameter values had to be found in order to obtain the desired product characteristics under a wide variety of conditions.

In Taguchi's opinion, an experiment that did not take account of every factor was inadequate, and he therefore stipulated that combinations of design parameter values should be chosen according to a fractional factorial design (known as an orthogonal array). An example is given in Table D.1. In addition, Taguchi identified another series of effects that could influence the experiment and cause variations; he called these disturbing or noise factors. These must be included in the experimental runs on a similar basis to the parameters. Table D.2 gives the orthogonal array for three noise factors.

Table D.1 Orthogonal array for a Taguchi experiment (parameters)

Run number	Parameter number			
	1	2	3	4
1	−1	−1	−1	−1
2	−1	0	0	0
3	−1	1	1	1
4	0	−1	0	1
5	0	0	1	−1
6	0	1	−1	0
7	1	−1	1	0
8	1	0	−1	1
9	1	1	0	−1

In this experiment there are four parameters and each of these has three possible values: high = 1, normal = 0, low = −1. Nine experimental runs are to be carried out, and each run uses a different combination of the parameter values. Thus, in run 1 all parameters are set out to their low value (−1); in run 2 parameter 1 remains at its low value while the other three are reset to their normal values (0).

Table D.2 Orthogonal array for a Taguchi experiment (noise factors)

Run number	Noise factor		
	1	2	3
1	−1	−1	−1
2	−1	1	1
3	1	−1	1
4	1	1	−1

In this experiment there are three noise factors and each of these has two possible values: high = 1, low = −1. Four experimental runs are to be carried out, each run using a different combination of noise factors. Thus, in run 1 all factors are set to the low value; in run 2 factor 1 remains low while the other two are reset to their high values.

These two arrays must be combined into one series of experiments such that, for each run against the four parameters, the noise factors must be varied. This means that there must be a total of 36 experimental runs:

(four noise factor variations) × (nine parameter variations)

This can be seen more clearly in Fig. D.1. As a result of these runs it is possible, from a study of the results, to select the parameter combination that gives the most robust design.

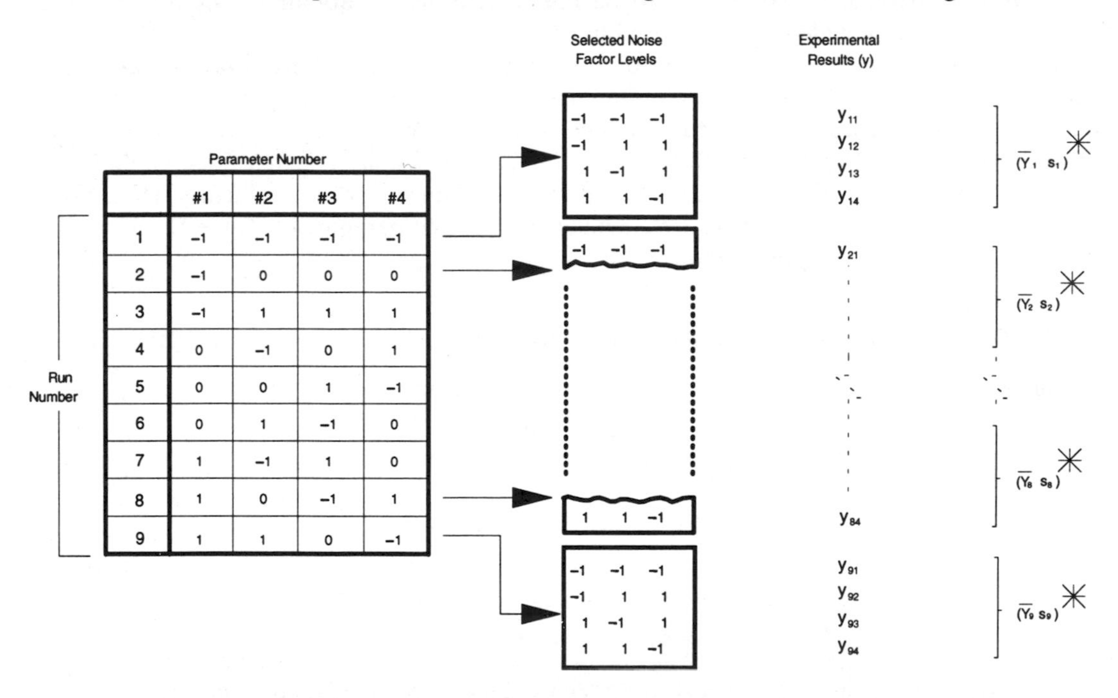

Figure D.1 Taguchi's design of experiments (parameter and noise values).

APPENDIX E

PRODUCTION PLANNING BY LINE OF BALANCE

E.1 THE PROBLEM

A manufacturing company receives an order for a particular type of pump, which requires it to deliver against a weekly schedule in varying quantities. Each pump is a 'three-level' assembly as illustrated in Fig. E.1; which shows how the assembly is built up, and the lead times involved.

In order to ensure that this delivery schedule is met, a detailed production plan must be prepared which will specify the volumes of materials, components, sub-assemblies and assemblies that should pass through the various 'production control points' at specified times. Adherence to this production plan can then be monitored by recording actual volumes passing through the control points and comparing these with what had been planned.

E.2 THE BASIC PLAN

The schedule specified that 7000 pumps were to be delivered over a 14 week period, at a rate of 400 per week for the first 7 weeks and 600 per week for the last 7 weeks. From the lead times for the various parts of the assembly (see Fig. E.1), a production plan was prepared for one pump using the 'back-scheduling' technique; this plan can be seen in Fig. E.2, showing that work must start (order first batch of component F), eight weeks before the date when the first delivery is due. This figure also shows the control points (numbered 1 to 11) required to monitor production.

E.3 THE DETAILED PLAN

Figure E.3 shows the next two steps in the process. Firstly, a cumulative delivery graph was drawn for the period during which deliveries should have been taking place. Secondly, in order to prepare a series of 'line-of-balance' charts for each of the 14 weeks, a master chart was prepared. The control point lead times on this chart were taken from the basic plan (Fig. E.2). The following methods were then used to prepare the 14 charts.

1. Work began by finding the cumulative totals for control point 11 for each of the 14 weeks; these were 400 for week 1, 800 for week 2, 1200 for week 3, and so on, up to 7000 for week 14 (see Fig. E.4).
2. The other control point cumulatives were then calculated, using the lead time values and cumulative delivery quantities, based on the following method. For a particular week number, the cumulative delivery quantity for control point 11 was found from the graph (Fig.

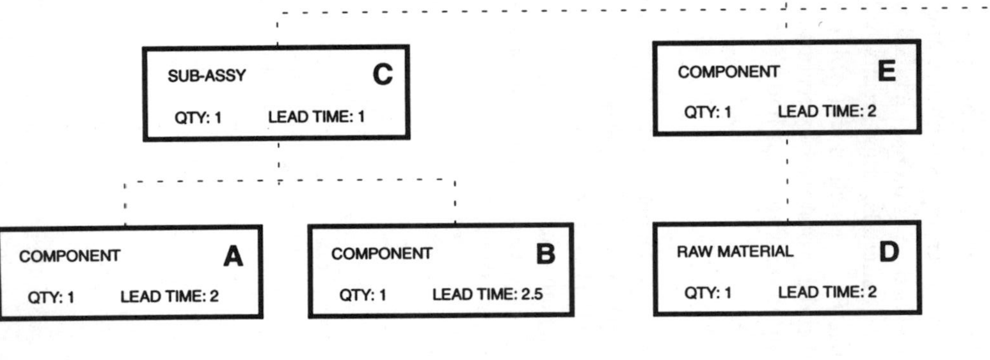

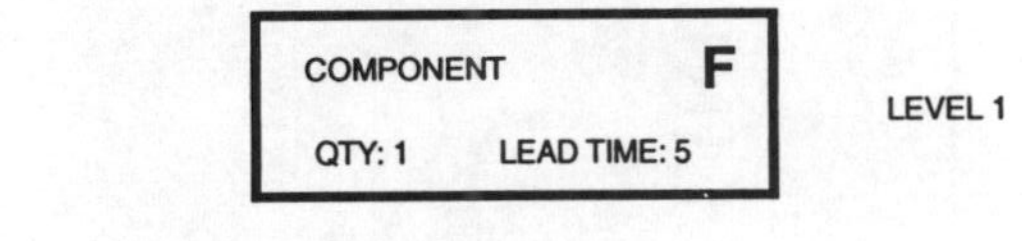

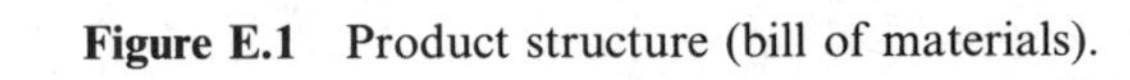
Figure E.1 Product structure (bill of materials).

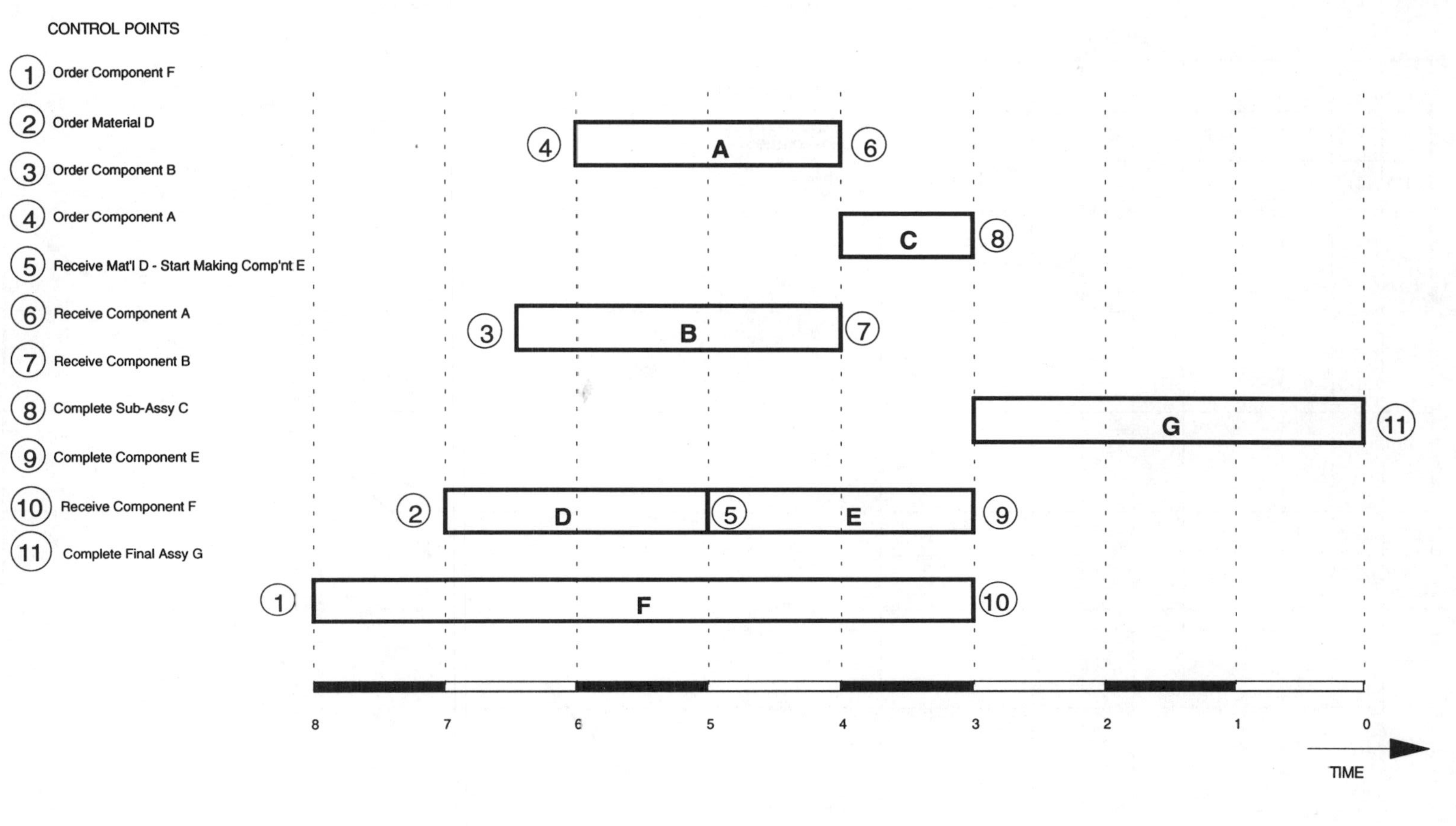

Figure E.2 Production plan (by back-scheduling).

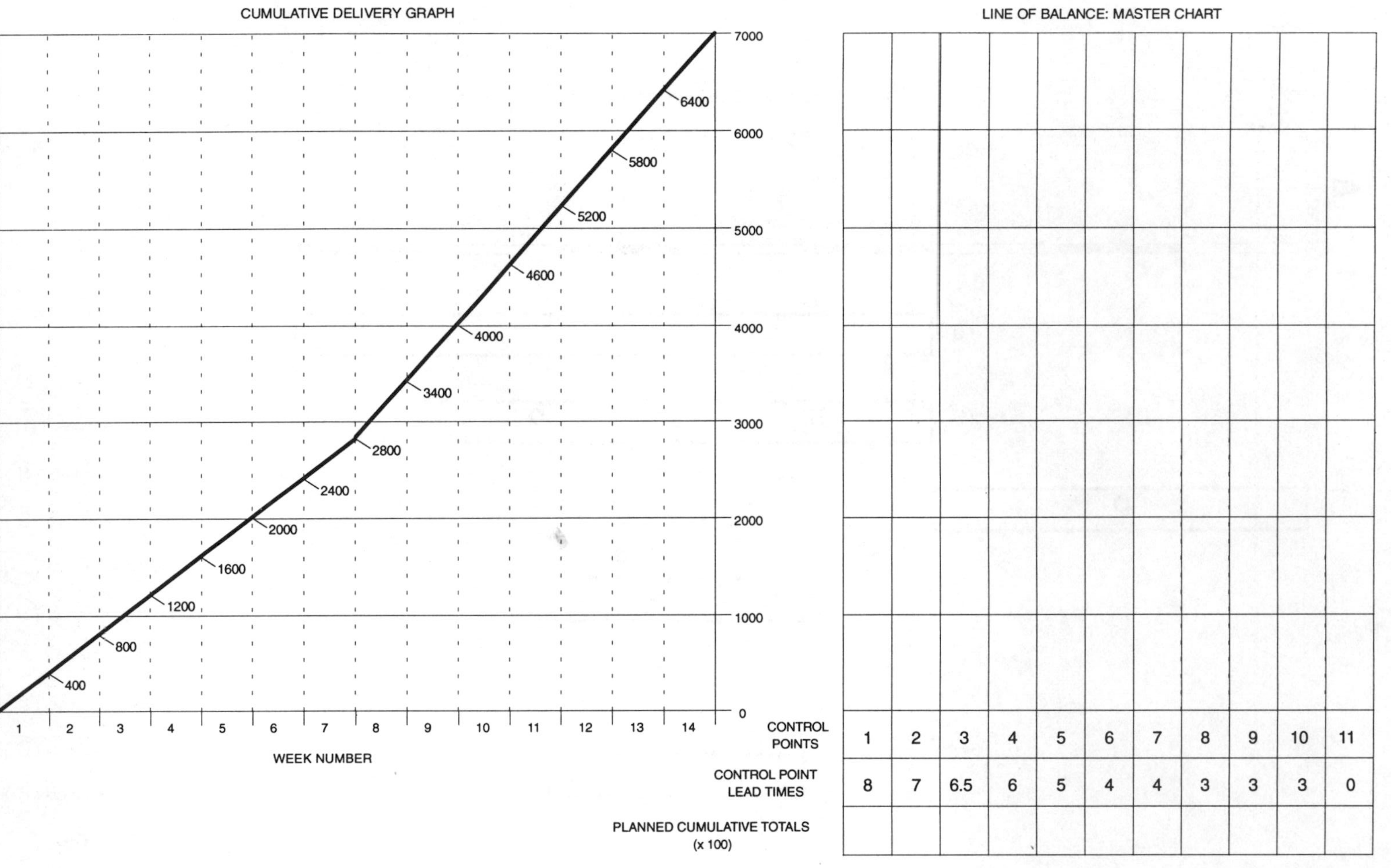

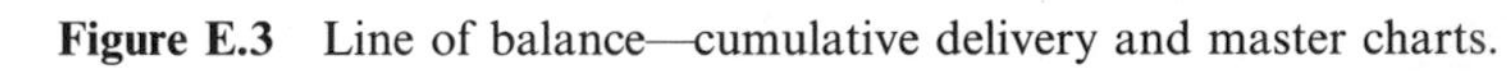

Figure E.3 Line of balance—cumulative delivery and master charts.

PLANNED CUMULATIVE TOTALS (x 100)

	WEEK NO.	1	2	3	4	5	6	7	8	9	10	11	CONTROL POINTS
DELIVERIES AT 400 UNITS PER WEEK	1	40	34	31	28	24	20	20	16	16	16	4	
	2	46	40	37	34	28	24	24	20	20	20	8	
	3	52	46	43	40	34	28	28	24	24	24	12	
	4	58	52	49	46	40	34	34	28	28	28	16	
	5	64	58	55	52	46	40	40	34	34	34	20	
	6	70	64	61	58	52	46	46	40	40	40	24	
	7	--	70	67	64	58	52	52	46	46	46	28	
DELIVERIES AT 600 UNITS PER WEEK	8	--	--	--	70	64	58	58	52	52	52	34	
	9	--	--	--	--	70	64	64	58	58	58	40	
	10	--	--	--	--	--	70	70	64	64	64	46	
	11	--	--	--	--	--	--	--	70	70	70	52	
	12	--	--	--	--	--	--	--	--	--	--	58	
	13	--	--	--	--	--	--	--	--	--	--	64	
	14	--	--	--	--	--	--	--	--	--	--	70	
		8	7	6.5	6	5	4	4	3	3	3	0	CONTROL POINT LEAD TIMES

Figure E.4 Calculating weekly cumulatives for line of balance.

E.3). Then, for any other control point, the cumulative quantity in that week was found from the same graph, by adding the number of lead time weeks for that control point to the week number being processed and reading off the value. For example, if week 6 is being processed, the graph shows that, for control point 11, the cumulative quantity is 2400. For control points 8 to 10, the lead time is 3 weeks which, when added to week 6, gives week 9; from the graph, the cumulative value for week 9 was 4000, and thus this was the value for control points 8 to 10. Similarly, for control points 6 and 7 the lead time was 4 weeks, and the cumulative value for 10 weeks (week 6 + 4 weeks' lead time) was 4600.

3. When the values had been calculated for each control point, for each of the 14 weeks, the master chart copies were completed as shown in Fig. E.5 (planned), by adding the 'bars' for the control points.
4. Finally, at the end of each week, the chart for that week was updated to include the actual cumulative values, as shown in Fig. E.5 (actuals).

From these planned/actual charts, any slippage against the planned schedule could be clearly seen and corrective action taken.

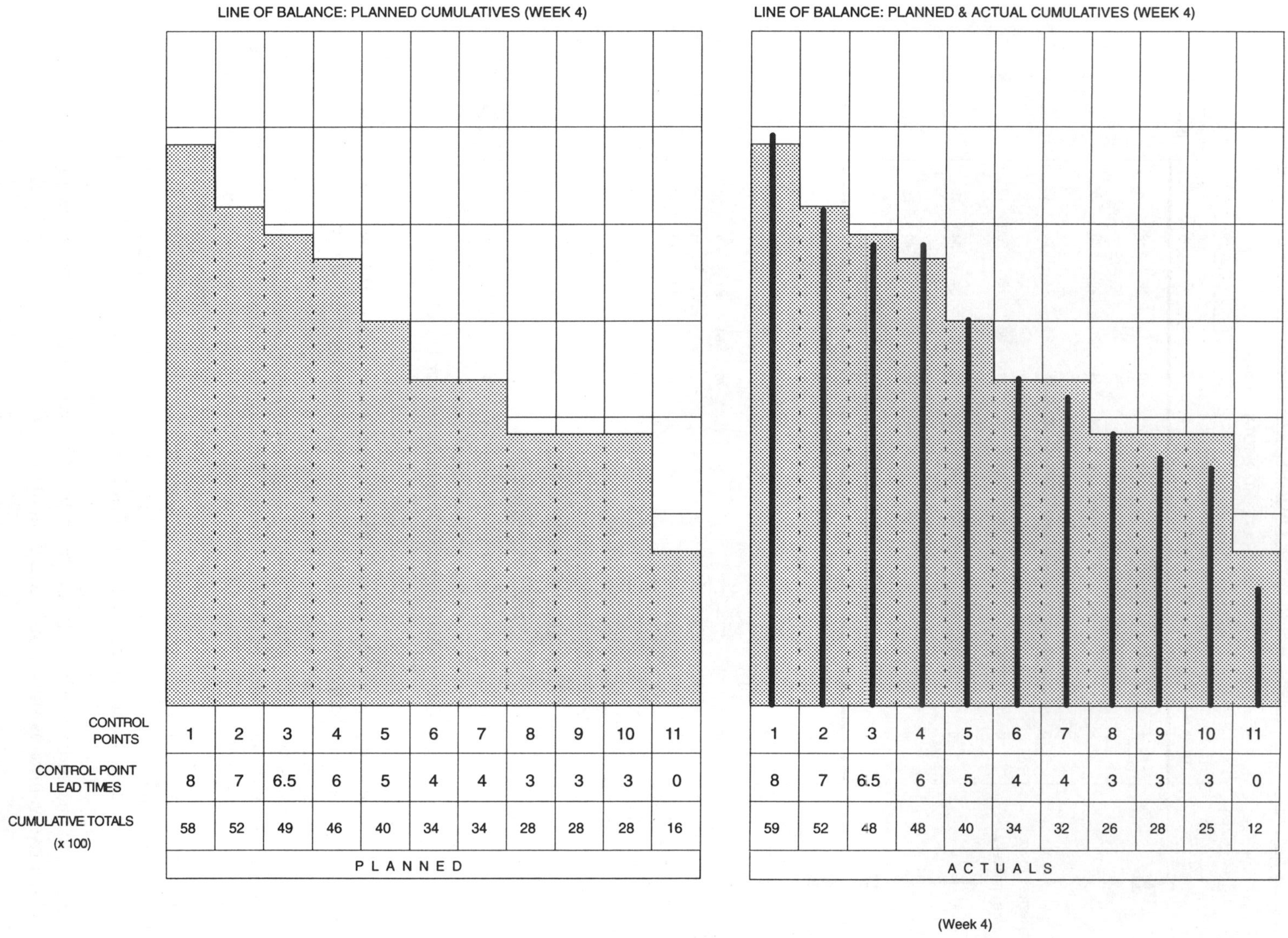

LINE OF BALANCE: PLANNED CUMULATIVES (WEEK 4)

CONTROL POINTS	1	2	3	4	5	6	7	8	9	10	11
CONTROL POINT LEAD TIMES	8	7	6.5	6	5	4	4	3	3	3	0
CUMULATIVE TOTALS (x 100)	58	52	49	46	40	34	34	28	28	28	16

PLANNED

LINE OF BALANCE: PLANNED & ACTUAL CUMULATIVES (WEEK 4)

CONTROL POINTS	1	2	3	4	5	6	7	8	9	10	11
CONTROL POINT LEAD TIMES	8	7	6.5	6	5	4	4	3	3	3	0
CUMULATIVE TOTALS (x 100)	59	52	48	48	40	34	32	26	28	25	12

ACTUALS

(Week 4)

Figure E.5 Line of balance—planned and actual production at control points (week 4).

Author's note: The work of preparing the line-of-balance charts is rather time-consuming if done manually, but it is not too difficult to set this task up on a spreadsheet; this will not only make the calculations automatically, but also draw both the planned and actual charts if a suitable graphics facility is available in the package.

APPENDIX F
GLOSSARY OF TERMS AND ABBREVIATIONS

2-D	two-dimensional—defines a computer screen view
3-D	three-dimensional—defines a computer screen view
ABC (1)	A system of classifying items (e.g. stock by usage value)
ABC (2)	activity-based costing
AGV	automated guidance vehicle
allocation	the process of reserving items in stock against an order
A/N	alpha/numeric (field containing mixture of letters and numbers in a computer record)
AQL	acceptable quality level—minimum acceptable quality in sampling
AR	auto-regression-used in statistical forecasting
ARIMA	auto-regression integrated moving average
ARMA	auto-regression moving average
ASME	American Society of Mechanical Engineers
AUV	annual usage value (used to classify stock items—unit value × annual usage)
AVCO	average cost—method of stock valuation
barcode	printed combination of dark and light lines, read as a code by an infra-red beam
batch	a quantity of items to be manufactured or purchased
BOM	bill of materials (breakdown of a product into a component structure)
BS5750	standard quality procedures system (see also ISO9000)
bulk issue	goods issued from stock in bulk rather than against an order
CAD	computer-aided design
CAE	computer-aided engineering
CAM	computer-aided manufacture
CAPP	computer-aided process planning
cell (1)	group of machines, located together, performing a sequence of processes
cell (2)	group of administrative staff, located together, performing a sequence of operations
CFD	computational fluid mechanics (computer-based design tool)
character	computer term for a single 'symbol' (letter, number, punctuation mark, etc.)
CIM	computer-integrated manufacture-integrated computerized manufacturing system

closed loop	a process or activity where feedback occurs (see feedback)
CMM	coordinate measuring machine — an inspection tool
CNC	computer numerical control (control of machines by computer)
C of C	certificate of conformance (certifying that product conforms to specification)
CoS	cost of sales — cost of product(s) — (selling price less profit)
cost centre	a group of people/machines with common costing characteristics
costs roll-up	method of costing an item from BOM and process routeing
CP	critical parameter — a key factor in a design specification
C_{pk}	process capability index — quality control measure
CPM (1)	critical path method of project planning
CPM (2)	critical parameter measurement (see CP)
CPU	central processing unit — the 'engine' in a computer
CR	critical ratio — a measure of dynamic priority in production
critical path	a sequence of tasks in a project plan that have no spare time (float)
CRP	capacity requirements planning (also see scheduling)
database	computer data storage and handling software system or package
DCF	discounted cash flow — method of financial planning and forecasting
DFA	design for assembly
DFF	design for fabrication
DFM	design for manufacture
DNC	direct numerical control (as CNC but by remote computer)
DoE	design of experiments (see Taguchi)
DOF	degrees of freedom (measure of robot flexibility)
DOS	a computer operating system (e.g. MS-DOS and PC-DOS)
EBQ	economic batch quantity — lowest cost production batch size
ECN	engineering change note
EDI	electronic data interchange — system for direct communication between computers
EOQ	economic order quantity — lowest cost purchasing batch size
FAST	function analysis system technology (a design technique)
FEA	finite element analysis (computer-based design tool)
feedback	process of iteration by feeding back data from an activity to control that activity
FG	finished goods (i.e. goods available for sale)
FIFO	first in–first out (method of stock management and valuation)
field	part of a computer record held in a file (e.g. part no.)
file	a group of computer records in a given format (e.g. supplier details)
fit	the degree to which a product or service meets a specification
flag	an information marker set on a computer record (e.g. order status)
float	term used in project planning for spare time in a non-critical element (task)
FMEA	failure modes and effects analysis (a quality improvement method)
FMS	flexible manufacturing system–computer-based control of a process
FoB	free on board (method of quoting an export price)
format	definition of way in which data is organized or displayed
free issue	see bulk issue
Gantt chart	a horizontal bar chart on a timescale used for planning

GI	goods inwards from a supplier
GL	general ledger—see NL
GRN (or GIN)	goods received (or inwards) note—record of goods received from a supplier
hardware	computer equipment (processors, printers, scanners, terminals, etc.)
I/O	input/output (a computer data-handling function)
ISO	International Standards Organization
IT	information technology—computer-produced information systems
ITT	invitation to tender-request to supplier for proposal/quotation
ISO9000	international version of BS5750 quality standard
JIT	just-in-time (an approach to more effective manufacturing management)
kit list	list of items to be issued by stores for production
kitting	method of issuing a batch of parts/materials to production
LCL	lower control limit—used in SPC to identify rejects (also UCL)
library	a file of standards on a computer system (e.g. drawing symbols)
LIFO	last in–first out—method of stock management and valuation
LoB	line of balance—production planning and monitoring technique
lot-for-lot	used in production planning to set batch size same as quantity required
MA	moving average (a non-statistical method of forecasting)
MAP	manufacturing automation protocol—an automation system 'interpreter'
matl reqn	material requisition (authority to draw items from stores)
matrix	a display of data showing relationships using vertical columns and horizontal rows
Mb	megabyte (one million bytes—computer measure of size of file or store)
modem	device that links a computer to a telephone network
module	a suite of computer programs or a group of techniques
MOST	Maynard optimized standard time—a work study technique
MP	maintenance prevention—a TPM (total productive maintenance) technique
MPS	master production schedule (a plan of products to be manufactured)
MRP I	materials requirements planning (computer-based planning tool)
MRP II	manufacturing resources planning (a development from MRP I)
MTM	methods–time–measurement—a work study technique
MTO	make-to-order—a manufacturing policy (i.e. no finished goods stock)
MTS	make-to-stock—a manufacturing policy (i.e. all sales from finished goods stock)
multiplexer	Device that links several computers via a telephone network
NC	numerical control (control of machines by program)
NCR	no carbon required—impregnated paper used in multiple document sets
nesting	method of cutting diverse shapes from a sheet for minimum wastage
network	a number of linked microcomputers with shared data and facilities
NL	nominal ledger—an accounting ledger for analysis
NPD	new product development
NVA	non-value added—a non-productive operation, e.g. rework
OC	operating characteristics—used in sampling plans
opn card	operation card (instruction to perform and book a production operation)

OPT	optimized production technology—a production planning/scheduling system
OSI	open systems interconnection—a standard for computer communication
outlier	a point in a time series that does not fit the pattern
parameter	rule used for calculation or defining a value
Pareto	the 80:20 rule defined by Count Pareto
PBR	payment by results—an incentive payment system
PC	personal computer (microcomputer with facilities to store and process data)
PCB	printed circuit board
PI	perpetual inventory (a stock-checking system)
picking	method of collecting goods together for despatch
pick list	list of items to be collected from stores for despatch
PLC	programmable logic controller—a process control computer
PMT	predetermined motion time—a work study technique
P/O	purchase order
PoD	proof of delivery (document signed by recipient)
POP	purchase order processing (computer module)
ppm	parts per million—a measure of defectives in a quality control system
progress	action to chase completion of a purchase or works order
pur reqn	purchase requisition (document requesting a purchase order to be raised)
QA	quality assurance
QC	quality control
QFD	quality function deployment—a design technique
quality circle	group of employees meeting daily to discuss their work
R&D	research and development
R&R	reliability and repeatability—studies used in instrument calibration
rating	work study technique for assessing rate of working on a standard scale
RCCP	rough cut capacity planning (an MRP II facility)
rectification	extra work done on an item by production or supplier
REPCO	replacement cost (a method of stock valuation)
returns	items returned by a customer or from production to stores
rework	(see rectification)
RLL	relay ladder language (for programming PLCs—low level)
RM	raw materials (e.g. steel sheet, coil and bar)
RoI	return on investment (a ratio used to evaluate a project)
ROL	reorder level (used for stock replenishment policy setting)
ROQ	reorder quantity (used for stock replenishment policy setting)
RPN	risk priority number (a factor used in the FMEA technique)
RRP	resource requirements planning (an MRP II technique)
rte card	route card (lists sequence of production operations for a works order)
routeing	master sequence of production operations for an item
SAE	Society of American Engineers
scheduling (1)	finite: a work plan based on limited production resources
scheduling (2)	infinite: a work plan based on unlimited production resources
scrap (1)	the wastage from an operation (e.g. swarf, offcuts)
scrap (2)	production items failing inspection that cannot be rectified
seasonality	measure of repeated pattern in a time series due to seasonal factors

SFDC shop-floor data capture (or collection) using computer terminals
six sigma a standard for process capability (six standard deviations)
software all computer programs, for operating and applications
SOP sales order processing (computer module)
SPC statistical process control — quality control technique
spreadsheet standard computer software for data analysis and graphics
STEP standard for the transfer of engineering product data
stillage wire or sheet metal cage on feet for storing and moving components
Taguchi (1) optimization methods devised by Dr Genichi Taguchi
Taguchi (2) design of experiments devised by Dr Genichi Taguchi
terminal video screen and keyboard linked to a remote computer
TPM total productive maintenance — system involving all employees to improve plant maintenance
TQM total quality management — system involving all employees to improve quality
traceability system to allow products to be traced back to original materials and processes
traveller see route card
UCL upper control limit — used in SPC to identify rejects (also LCL)
UNIX a computer operating system (also XENIX)
UOM unit of measure(ment) (kilograms, litres, metres, boxes, etc.)
vendor supplier (USA)
VFM value for money — means of evaluating projected expenditure
VSL vendor service level (used to determine stock requirements statistically)
window several displays of data on a single computer screen
Windows© A computer system (from Microsoft Corporation)
WIP work-in-progress (or process) — all items in production
work centre a group of people/machines with common capability
W/O works order
WP wordprocessor (or word processing)

INDEX